Beneficial Plant-microbial Interactions

Ecology and Applications

Beneficial Plant-microbial Interactions

Ecology and Applications

Editors

M. Belén Rodelas González

Department of Microbiology
Faculty of Pharmacy
University of Granada
Granada
Spain

Jesús González-López

Department of Microbiology
Faculty of Pharmacy
University of Granada
Granada
Spain

CRC Press
Taylor & Francis Group
Boca Raton London New York

CRC Press is an imprint of the
Taylor & Francis Group, an **informa** business
A SCIENCE PUBLISHERS BOOK

CRC Press
Taylor & Francis Group
6000 Broken Sound Parkway NW, Suite 300
Boca Raton, FL 33487-2742

Version Date: 20130624

International Standard Book Number-13: 978-1-4665-8718-2 (eBook - PDF)

Visit the Taylor & Francis Web site at
http://www.taylorandfrancis.com

and the CRC Press Web site at
http://www.crcpress.com

Preface

There are many books on plant-microbial interaction presently available, of which some are quite up-to-date, and several more comprehensive than ours. Why, then, have we thought it was worth coordinating the work of a new book on the subject?

In the past 15 or 20 years, plant-microbial interaction has become a fashionable area for research and now comprises a huge body of information. The very vast scope of this knowledge presents serious problems to PhD students, researchers and professionals in different scientific areas. In writing this book, we tried to keep these problems in mind and have therefore aimed at making this volume on plant-microbial interactions comprehensible rather than comprehensive. We have adopted several means toward this end by trying to emphasize the biological and practical significance of the subject.

It is in this context, that the present book aims to provide an insight into the mechanisms underlying the interactions of plants and microbes, the ecological relevance and roles of these symbioses, the adaptive mechanisms of plant-associated microorganisms to abiotic stress and their contribution to plant stress tolerance, and the potential of these interactions as tools in agrobiotechnology. A team of authors with wide experience in the area have contributed up-to-date reviews in 18 chapters devoted to different ecological and applied aspects of the rhizobia-legume symbiosis, ecto- and endo-mycorrhizas, and plant associations with diazotrophic or adiazotrophic plant-growth promoting rhizobacteria.

Our intention, then, has been to highlight those important principles, discuss their biological importance and integrate the subject through every aspect of the microbial-plant interaction. If we have succeeded in these aims, this book will be suitable not only for lecturers or researchers specialized in the field, but also for students of biological and agricultural sciences who need this information for study of subsidiary subjects. In order to maintain the balance of the book we have not concentrated exclusively on the needs of any one of these groups and we hope that the general approach we have adopted will be of benefit to all of them.

We would like to acknowledge the Spanish Society for Nitrogen Fixation (Sociedad Española de Fijacion de Nitrogeno, SEFIN) and the Spanish Thematic Network on Beneficial Plant-Microbial Interactions (Red Temática de Biotecnología de las Interacciones Beneficiosas entre Plantas y Microorgansimos) for their support. Finally, we would like to sincerely thank all the authors who contributed the chapters compiled in this book for contributing their outstanding expertise in the field, as well as for the many comments and suggestions they provided in the preparation of this volume. In particular we thank José Miguel Barea (Department of Soil Microbiology and Symbiotic Systems, Estación Experimental del Zaidín, CSIC, Granada) for his considerable help in coordinating the work of chapters dedicated to the mycorrhizal interactions. The enormous effort and devotion of all made the final version of the book possible.

January 2013

M. Belén Rodelas González
Jesús González López
Department of Microbiology
University of Granada
Spain

Contents

1

Nitrogen Fixing Endosymbiotic Bacteria: Old Chaps and New Findings

M.H. Ramírez-Bahena,[1] *A. Valverde,*[1] *M. Robledo,*[2,3] *L.P. Rivera,*[2,3] *E. Menéndez,*[2] *M. Medina-Sierra,*[1,4] *P.F. Mateos,*[2,3,*] *J.M. Igual*[1,3] and *R. Rivas*[2,3]

Introduction

"Heaven is under our feet as well as over our heads"
Henry Thoreau.

Nitrogen fixation is the second most important biological process on Earth, consisting in the reduction of N_2 to NH_4^+ by the nitrogenase enzyme complex. This process is one of the most important metabolic pathways for the maintenance of life in the biosphere. Interestingly, this crucial process can only be carried out by a few groups of life forms: bacteria. The nitrogen-fixing microorganisms are not a homogeneous taxonomic group; the only common feature is the presence of nitrogenases. These organisms can perform biological nitrogen fixation (BNF) either in free life or establishing

[1]Instituto de Recursos Naturales y Agrobiología, IRNASA-CSIC, Salamanca, Spain.
[2]Departamento de Microbiología y Genética, CIALE, Universidad de Salamanca, Spain.
[3]Unidad Asociada de I+D Universidad de Salamanca (USAL)-CSIC (IRNASA), Salamanca, Spain.
[4]Grupo Grica, Facultad de Ciencias Agrarias, Universidad de Antioquia, AA 1226 Medellín, Colombia.
*Corresponding author: pfmg@usal.es

symbiotic relationships with other organisms. Within this first group, we can find phototrophic organisms as in families *Rhodospirillaceae*, *Chlorobiaceae* and *Cianobacteriae*; chemoautotrophic organisms of the genus *Thiobacillus*, *Xanthobacter* and *Desulfovibrio*; and heterotrophic Gram negative bacteria such as *Azotobacter*, *Gluconacetobacter* (*Acetobacteraceae*), *Enterobacter*, *Klebsiella* (*Enterobacteraceae*) and Gram positive bacteria such as *Clostridium* and *Paenibacillus* (Sprent and Sprent 1990).

Within the symbiotic group heterotrophic bacteria that perform BNF by establishing symbiotic relationships with leguminous plants, collectively called rhizobia can be found. BNF involving host-specific symbiotic interactions between root-nodulating bacteria and legumes has received a lot of research attention due to the central role it plays in the maintenance of soil fertility. The Leguminosae (Fabaceae) is the third largest flowering plant family, which has the potential to contribute to soil N and increase subsequent or associated non-nodulating crops yields through symbiotic nitrogen fixation.

In 1542, the German botanist Leonhard Fuchsius published the first drawings of nodulated legumes (Fuchsius 1542). During the 17th century, Malpighi observed nodules on the bean roots (*Phaseolus vulgaris* and *Vicia faba*). In 1879, Frank found nodules on the roots of all healthy legumes and demonstrated that incinerating soil prevented the nodulation of *Pisum sativum*. In 1888 Hellriegel and Wilfarth showed that nodule formation results from a previous infection. However, it was Beijerinck, who presented the first proof that one bacteria called *Bacillus radicicola* causes nodules by preparing pure cultures of nodule organisms from *V. faba* and using them to infect fava beans growing in sterile soil (Beijerinck 1890). In 1889, Prazmowski inoculated *Pisum sativum* with pure cultures and showed that the bacteria penetrate legumes via infection threads in root hairs. In the same year, Frank renamed the first rhizobial species *Bacillus radicicola*, which nodulates *Vicia*, as *Rhizobium leguminosarum*. From this point the taxonomy of rhizobia has seen many changes, mainly due to the heterogeneity and complexity of this group of microorganisms. Also, the advances in microbiology and molecular biology have allowed us undoubtedly to uncover many unknowns about the complexity of rhizobia.

Gram-Negative Nitrogen Fixing Endosymbiotic Bacteria

Taxonomy of Rhizobia

Bergey's Manual of 1974 played a crucial role in rhizobial taxonomy. In this edition, all bacteria able to nodulate legumes were included in a single genus called *Rhizobium* (Frank 1889), within the family *Rhizobiaceae* proposed by Conn (1938). Different species were distinguished on the basis of a few phenotypic characteristics and the authors of this chapter pointed out the relevance of plant infection ability as essential for species classification.

Until [illegible] yr ago, no one knew much about the nodulating bacteria, all of which were classified within the genus *Rhizobium*. However, in the first edition of Bergey's Manual of Systematic Bacteriology of 1984, in order to separate slow-growing rhizobia species, *R. japonicum* was reclassified in a new genus called *Bradyrhizobium* (Jordan 1982). Fast-growing rhizobia previously designed *R. leguminosarum*, *R. trifolii*, and *R. phaseoli* were combined into a single species, *R. leguminosarum*, with three biovar designations based on their host range nodulation. From 1984 to 1991, two additional genera (*Azorhizobium* and *Sinorhizobium*) and five additional species (*R. galegae*, *S. fredii*, *S. xinjiangensis*, *R. tropici*, and *A. caulinodans*) of root- and stem-nodule bacteria were proposed. These reclassifications and the new descriptions were supported by phenotypic characteristics and 16S rRNA gene sequences. Further studies and additional isolations make taxonomy changes necessary. At this point it was considered that root- and stem-nodulating bacteria taxonomy was in transition. The classification of rhizobia based on plant infection was widely criticized, especially after it was shown that the genes required for nodulation and nitrogen fixation were mostly in transmissible plasmids. This led to the polyphasic taxonomy introduction as a stand tool for new rhizobia genera and species description, and the proposal of minimum standards: symbiotic performance with selected hosts, cultural and morphological characteristics, DNA:DNA relatedness, rRNA:DNA hybridization, 16S rRNA analysis, RFLP and multilocus enzyme electrophoresis (Graham et al. 1991).

The 16S rRNA gene sequence is a crucial tool for bacterial phylogeny study (Woese et al. 1984). It was possible to clarify rhizobial phylogeny and it is now clear that many diverse soil species or genera are able to nodulate the same legume if they harbor the adequate symbiotic genes.

Due to advances in molecular biology, the complexity of the rhizobia as heterotrophic bacteria has been revealed. That fact has made possible the identification of non-nodulating bacterial species within the classical rhizobia genera, all of them isolated from sources other than legume nodules. For example, *Bradyrhizobium betae* was isolated from a tumor-like formation in sugar beet (Rivas et al. 2004). It was not confirmed as the causal agent for malformations and no genes for nitrogen fixation or nodulation were found, so their status is unknown as symbiotic bacteria. There are many additional examples such as *Rhizobium selenitireducens*, a selenite-reducing bacteria isolated from bioreactors whose host for nodulation is unknown; *Rhizobium soli*, that neither harbors *nodD* and *nifH* genes nor nodulate the tested legumes; *Rhizobium borbori*, an aniline-degrading bacteria isolated from activated sludge; *Rhizobium rosettiformans*, isolated from hexachlorocyclohexane (HCH)-contaminated groundwater; *Rhizobium aggregatum* formerly *Blastobacter aggregatus*, in which no *nifH* genes have been found; and *Mooorhizobium thiogangeticum*, a sulfur oxidizing rhizobia with unknown nodulation.

Since 2000, the taxonomy of rhizobia has had significant changes: the genera *Agrobacterium* and *Allorhizobium* were reclassified into the genus *Rhizobium* (Young et al. 2001). However, we considered that the reclassification of *Agrobacterium* into *Rhizobium* could have been premature, except in the case of the former *Agrobacterium rhizogenes*, which clearly belongs to the genus *Rhizobium* (Farrand et al. 2003, Velázquez et al. 2005). Another change was the reclassification of genus *Sinorhizobium* into the genus *Ensifer*. According to the rules of bacteriological nomenclature code, the name *Ensifer* has priority over *Sinorhizobium* since it was described before by Casida in 1982 (Judicial Commission of the International Committee on Systematics of Prokaryotes 2008 and International Committee on Systematics of Prokaryotes 2011).

One of the most significant details in rhizobial taxonomy is that the principal phylogenetic tool (16S rRNA gene sequence) has limitations in differentiating closely related rhizobial species (Valverde et al. 2005, Ramírez-Bahena et al. 2008). These observations entail that the rhizobial rRNA gene phylogeny may not always accurately reflect prokaryotic phylogeny. However, housekeeping genes have been proposed as alternative phylogenetic markers. According to Stackebrandt et al. (2002), a good informative level of phylogenetic data would be obtained from the determination of five or more genes under stabilizing selection for encoded metabolic functions. Such genes should be at diverse chromosomal loci and widely distributed among taxa (multilocus sequence analysis, MLSA).

The complete list of valid species of rhizobia is constantly updated and recorded in the List of Prokaryotic Names with Standing in Nomenclature by Dr. Euzeby (http://www.bacterio.cict.fr). Considering the large number of legumes hitherto not studied and the few ecosystems analyzed up to date, we expect an exponential increase in the description of new rhizobial species.

Rhizobia or Non-rhizobia: Changing Rules?

When research focused on new environments and unexplored leguminous plants, the list of bacterial species able to nodulate legumes was expected to grow. However, the number of genera increased as well. As mentioned, the microorganisms capable of fixing nitrogen in symbiosis with leguminous plants are collectively known as rhizobia. The rhizobial species described so far were classified into the alpha-Proteobacteria. However, with the emergence of new nodulating-bacteria species classified within beta-Proteobacteria, the name of rhizobia needed to be more specific. In this way, the authors emphasize "classical-rhizobia" (belonging to the alpha-Proteobacteria) and non-rhizobia (included in the beta-Proteobacteria and alpha Proteobacteria).

Sy et al. (2001) reported that *Methylobacterium* formed nodules on *Crotalaria*, an African legume. *M. nodulans* harbors *nodABC* nodulation genes and the *nifH* gene encoding the structural nitrogenase enzyme. Nodulation by *Methylobacterium* spp. has been confirmed in several species of *Lotononis* (Yates et al. 2007). Moulin et al. (2001) reported evidence that two *Burkholderia* strains that were isolated from *Aspalathus carnosa* and *Machaerium lunatum* nodules, possessed nodulation genes. *Burkholderia* is also able to form ineffective nodules in *Macroptilium atropurpureum* roots and carries nodulation genes (*nodABC*) phylogenetically related to those found in legume symbionts belonging to alpha-Proteobacteria ("classic" rhizobia), supporting the hypothesis of lateral gene transfer in the rhizosphere, crossing the boundary between alpha and beta Proteobacteria.

Bontemps et al. (2010) collaborated on this idea with a comprehensive and thorough study of Mimosa-associated symbionts providing compelling evidence that *Mimosa* and *Burkholderia* species have an ancient history of symbiotic coexistence.

Chen et al. (2001) reported that *Ralstonia taiwanensis* could induce nodules in *Mimosa pudica* and *Mimosa diplotricha*. This species has been reclassified as *Cupriavidus taiwanensis*, a beta-Proteobacteria within the family *Burkholderiaceae*, Order *Burkholderiales*. *C. taiwanensis* carries nodulation genes (*nodBCIJHASUQ*) and one regulatory gene (*nodD*) on the pRalta mega-plasmid (Amadou et al. 2008) and conclusive evidence for nodule formation and effective nitrogen fixation came from the nodulation process analysis on its *Mimosa* hosts (Chen et al. 2003).

Additionally, several other legume-nodulating species of *Burkholderia* have been described: *B. mimosarum*, *B. phymatum*, *B. nodosa*, *B. tuberum*, *B. caribensis* and *B. sabiae* (Gyaneshwar et al. 2011). *Burkholderia* nodulate mainly mimosoid legumes all over the world. *B. phymatum* nodulates more than 40 *Mimosa* species. However, there is also evidence of nodulation on papilionoid legumes such as *Phaseolus vulgaris* (Tribe Phaseoleae) nodulated by *B. phymatum* and *B. tuberum* (Talbi et al. 2010, Gyaneshwar et al. 2011) or *Cyclopia*, *Podalyria* and *Virgilia* (Tribe Podalyrieae) nodulated by *B. tuberum* (Gyaneshwar et al. 2011). Recently, it was reported that some *Burkholderia* strains are more competitive than *Rhizobium tropici* for *Mimosa* nodulation (Elliot et al. 2009).

In 2001, we analyzed the 16S rRNA of two strains isolated from *Neptunia natans* in India, showing that they are classified within the genus *Devosia* from the alpha-Proteobacteria (Rivas et al. 2002). The 16S rRNA analysis revealed that they represent a new species, designated *D. neptuniae* (Rivas et al. 2003), which carries *nodD* and *nifH* genes closely related to *R. tropici*

CIAT899^T, an American species that nodulates *Leucaena*, suggesting that they were transferred from *R. tropici* to *D. neptuniae*. Later, this hypothesis was supported when *R. tropici* was reported to nodulate *Neptunia* in America (Zurdo-Piñeiro et al. 2004).

In the ensuing years, several "non-rhizobial" genera from alpha- and beta-Proteobacteria were reported to be legume endosymbionts. In 2002, nodulation of *Aeschynomene* by *Blastobacter denitrificans* was reported (van Berkum and Eardly 2002), although this species has been recently reclassified as *Bradyrhizobium denitrificans* (van Berkum et al. 2006).

In 2004, *Ochrobactrum* was found in *Acacia mangium* nodules but no information about their symbiotic genes was reported (Ngom et al. 2004). Then we found two novel species of this genus carrying symbiotic genes closely related to those of rhizobia, *O. lupini* able to nodulate *Lupinus* (Trujillo et al. 2005) and *O. cytisi*, that nodulates *Cytisus scoparius* in Spain (Zurdo-Piñeiro et al. 2007).

Phyllobacterium trifolii, was isolated in 2005 from *Trifolium pratense* nodules, was also found to be capable of nodulating *Lupinus albus*. Although this species harbors symbiotic plasmids containing *nodD* and *nifH* genes, it forms ineffective nodules in *Trifolium repens* and *Lupinus albus* roots (Valverde et al. 2005). One year later, in 2006, *Phyllobacterium leguminum* and *P. ifriqiyense* were isolated from root nodules of *Lathyrus numidicus*, *Astragalus algerianus* and *Argyrolobium uniflorum* (Mantelin et al. 2006). However, no nodulation or N-fixation genes were found and actually these strains have an unknown symbiotic status. In the same way, we reported *Phaseolus vulgaris* nodulation by two pathogenic *Agrobacterium rhizogenes* strains (currently *Rhizobium rhizogenes*). Interestingly *nodD* and *nifH* genes were found in their symbiotic plasmids and the simultaneous presence of tumorigenic or hairy root-inducing plasmids was shown (Velázquez et al. 2005). A few years ago, Lin et al. (2008) reported evidence that a new *Shinella* strain CCBAU 25048, described as *Shinella kummerowiae*, isolated from nodules on *Kummerowia stipulacea*, possessed *nodD* and *nifH* genes.

Recent published studies have shown beyond doubt that new examples are appearing. Ardley et al. (2012) reported the strains *Microvirga lupini*, *Microvirga lotononidis* and *Microvirga zambiensis*, which are Alphaproteobacterial root nodule bacteria that specifically nodulate and fix nitrogen with *Listia angolensis* and *Lupinus texensis*.

Currently, the last edition of the Bergey's Manual (Kuykendall 2005) includes the new order of *Rhizobiales*. Members of this order are morphologically and physiologically diverse and constitute the largest group within the α-2 Proteobacteria subgroup. Based on 16S rRNA gene sequence analysis, this order currently comprises 13 families with validly published names: *Aurantimonadaceae*, *Rhizobiaceae*, *Bartonellaceae*, *Brucellaceae*, *Phyllobacteriaceae*, *Methylocystaceae*, *Beijerinckiaceae*, *Cohaesibacteraceae*,

Bradyrhizobiaceae, Hyphomicrobiaceae, Methylobacteriaceae, Rhodobiaceae and *Xanthobacteraceae*.

The list of new taxa is continuously growing, and the reader is encouraged to consult websites maintained by B.S. Weir (Auckland, New Zealand) and the Center for Genomic Sciences (CCG, Cuernavaca, Mexico) for comprehensive and updated lists.

Gram-Positive Nitrogen Fixing Endosymbiotic Bacteria: *Frankia*

Actinorhizal Plants: a Brief Overview

The root nodule nitrogen-fixing symbiosis between the actinobacteria *Frankia* and about 220 plant species, collectively called actinorhizal plants, is a major source of fixed nitrogen for many terrestrial ecosystems. In contrast with legumes, which group together in one family (Fabaceae) inside the order Fabales, actinorhizal plants are phylogenetically much more diverse. They spread out through eight families from three rosid orders: Fagales, Cucurbitales and Rosales (Swensen 1996). Within the Fagales, actinorhizal genera are found in families Betulaceae (*Alnus*) and Casuarinaceae (*Casuarina, Allocasuarina, Ceuthostoma* and *Gymnostoma*); in the families Datiscaceae (*Datisca*) and Coriariaceae (*Coriaria*) within the Curcubitales; and the Rosales includes symbiotic genera in the families Elaeagnaceae (*Eleagnus, Hippophae* and *Shepherdia*), Rhamnaceae (*Ceanothus, Colletia, Discaria, Kentrothamnus, Retanilla, Talguenea* and *Trevoa*) and Rosaceae (*Dryas, Purshia, Cercocarpus, Chamaebatia* and *Cowania*). Excluding *Datisca glomerata*, which is herbaceous, actinorhizal plants are woody shrubs or trees broadly spread all over the Earth, except Antarctica.

In temperate regions, actinorhizal species play a key role as pioneer plants in the early stage of plant succession, providing combined nitrogen to the soil through shedding and decomposition of their roots and litter (Dawson 2008). In addition to its ecological role, some actinorhizal genera, such as *Alnus* and *Casuarina*, include species that have economical or practical value in forestry, reforestation and soil remediation (Benson and Silvester 1993, Gtari and Dawson 2011, Normand and Mullin 2008).

Systematics Physiology and Biochemistry of *Frankia*

Frankia strains have been cultured only since 1978 (Callaham et al. 1978). But even today, strains nodulating some genera of actinorhizal plants, particularly from the families Rosaceae and Rhamnaceae, have not been pure-cultured. According to the most recent taxonomic revision of the class *Actinobacteria*, *Frankia* is the sole genus of the family *Frankiaceae*, which jointly with *Nakamurellaceae, Cryptosporangiaceae, Sporichthyaceae, Geodermatophilaceae* and *Acidothermaceae* constitute the suborder *Frankineae* within the order *Actinomycetales*, one of the five orders included in the *Actinobacteria* class (Zhi et al. 2009). Before the development of molecular techniques, classification of *Frankia* strains relied on their specificity to nodulate the host-plant genera. Four host-infection groups were described (Baker 1987): *Frankia* strains able to nodulate (i) *Alnus* and *Myrica*, (ii) *Casuarina* and *Myrica*, (iii) *Elaeagnus* and *Myrica*, and (iv) those that nodulate only host genera from the family Elaeagnaceae (*Elaeagnus, Hippophaë* and *Sheperdia*). Once 16S rDNA sequences became available, new subdivisions within the genus *Frankia* were established, although they roughly match the groups primarily defined by host specificity. Analysis of almost complete 16S rRNA gene sequences from both pure culture and uncultured *Frankia* strains from root nodules displayed four clades. The first one includes typical nitrogen-fixing strains from *Alnus* and *Casuarina* host-infection groups, in which is included the genus type-strain *Frankia alni* ACN14a; a second group is formed by still uncultured endophytes of *Dryas, Coriaria* and *Datisca*; the third one includes those strains that belong to the *Elaeagnus* host-infection group; and finally, a group encompassing those atypical non-nitrogen-fixing strains (Normand et al. 1996). Nonetheless, the resolution of 16S rRNA gene sequences is not enough to distinguish species and thus, other molecular and phenotypic approaches have been used for systematic studies within the genus *Frankia* (Hahn 2008), such as LMW RNA profiling (Velázquez et al. 1998), cellulase isoenzymes profiling (Igual et al. 2001) or genomic fingerprinting (Igual et al. 2003). Polyphasic taxonomy approaches would hopefully lead to a better intragenus classification.

Frankia is a heterotrophic filamentous actinobacteria that grows *in vitro* under aerobic or microaerobic conditions on a variety of carbon sources such as sugars, short fatty acids and intermediates of tricarboxylic acid cycle (Valverde and Huss-Danell 2008). However, its growth rate *in vitro* is fairly slower when compared to other symbiotic bacteria as rhizobia. *Frankia* differentiates into three cell types: vegetative cells, which form a branched mycelium; sporangia, containing numerous ovoid spores; and thick-walled spherical structures, named vesicles, in which the nitrogen fixation occurs. Spores are the reproductive structure of *Frankia*. They exhibit low levels of endogenous respiration and are stimulated to germinate by host root extracts (Krumholz et al. 2003). Unlike most of rhizobia, *Frankia* can fix nitrogen aerobically because vesicles protect nitrogenase from oxygen inactivation. They are functionally equivalent to cyanobacteria heterocysts. Vesicles are surrounded by a multi-laminated structure made up of layers of different pentacyclic triterpenoids, designated generically as hopanoid lipids, which forms a physical barrier to oxygen diffusion (Berry et al. 1993). Moreover, *Frankia* synthesizes truncated hemoglobins that may regulate the oxygen concentration, thus contributing to protection of nitrogenase from the effects of oxygen excess (Coats et al. 2009, Tjepkema et al. 2002). *Frankia* also possesses an efficient uptake hydrogenase system that oxidizes the hydrogen produced during nitrogen fixation to protons and electrons, with the latter fed to the respiratory chain in energy production through oxidative phosphorylation, and the reductant generated in the process is used by the cells in various ways (Leul et al. 2005, Leul et al. 2009). All these features make *Frankia* an outstanding nitrogen-fixer both in free-living and symbiotic conditions.

Infection and Nodulation

Frankia can penetrate into the host plant roots through root hairs, the so-called intracellular infection pathway that occurs in actinorhizal Fagales; or along the middle lamella between adjacent epidermal cells, using the intercellular infection pathway, which is typically found in actinorhizal Rosales. The infection pathway of actinorhizal Cucurbitales is not yet clarified. The infection mode is host-plant dependent as it has been demonstrated: the same *Frankia* strain can infect different hosts by different infection pathways (Miller and Baker 1986). However, unlike rhizobia-legumes symbiosis, where signaling and recognition processes are quite well understood, detailed mechanisms in actinorhizal symbiosis are not elucidated yet.

There is evidence indicating chemically mediated interactions between *Frankia* and the host plant. In the actinorhizal plants infected in an intracellular way, the initial plant response to *Frankia* is the root hair curling, in response to a root hair deforming factor produced by the bacteria (Van Ghelue et al. 1997). As rhizobia and arbuscular mycorrhizal fungi, *Frankia* forms a root endosymbiosis that seem to share equal genetic basement (Markmann and Parniske 2009). Along with rhizobia, it has been recently demonstrated that lipochitooligosaccharides play a role as signal factors in arbuscular mycorrhizal fungi (Maillet et al. 2011). Therefore, it could be speculated that *Frankia* signal factors are also chitin-based (Pawlowski et al. 2011). Nonetheless, whatever these factors are, they remain uncharacterized. Several other compounds have been identified as involved in the molecular interaction between actinorhizal symbiotic partners, such as flavonoids, phenols, and hydroxycinnamic acid (Auguy et al. 2011, Benoit and Berry 1997, Hocher et al. 2006, Hughes et al. 1999, Popovici et al. 2011), as well as other undetermined compounds (Gabbarini and Wall 2011, Popovici et al. 2011, Zimpfer et al. 2002, Zimpfer et al. 2004).

Once the interaction between the host plant and *Frankia* has been established, nodule formation starts from the root pericycle. Although there is no homogeneity between the host plant families regarding structural and physiological characteristics, actinorhizal nodules are anatomically and ontogenetically modified lateral roots, in contrast with the stem-like structure of legume nodules. Another ultrastructural difference with respect to legume subfamilies Mimosaceae and Fabaceae nodules is that *Frankia* remains embedded within a cell wall-like matrix, while rhizobia are hosted within specialized membrane-bound compartments termed symbiosomes (Pawlowski and Sprent 2008).

New Insights from the "Omics"

The scientific knowledge on the actinorhizal symbiosis has grown slowly relative to rhizobia-legume symbiosis. There are several reasons starting with the fact that the actinorhizal microsymbiont *Frankia* is available for *in vitro* research only since 1978 (Callaham et al. 1978). Moreover, *Frankia* is a multicellular and filamentous microorganism that grows slowly under laboratory conditions and is also recalcitrant to genetic manipulations (Kucho et al. 2009). Similarly, macrosymbionts are woody perennial plants with long generation times and thus, difficult subjects for genetic studies. However, significant progress has been made recently. The main insights have arisen through new developed genetic transformation protocols of several actinorhizal plant species (Benabdoun et al. 2011, Perrine-Walker et al. 2011, Svistoonoff et al. 2010); and, focusing on the bacterial side, the availability of several *Frankia* strains genomes, and proteome and transcriptome analyses of free-living and symbiotic cells.

Genomics Studies

The publication of genome sequences and annotations of *Frankia* strains ACN14a, CcI3 and EAN1pec (Normand et al. 2007) led to a major breakthrough in actinorhizal symbiosis researching. Also, other strains are at different phases of the sequencing and annotation pipeline. Among them there is an unisolated *Frankia* strain living in *Datisca glomerata* root nodules (Persson et al. 2011).

The first relevant insight derived from such sequencing efforts deals with the genome sizes of three *Frankia* strains. Although the identity of 16S rRNA genes is high (97.8–98.9 percent), the size of their genomes is very disparate. Strain CcI3, which has the narrowest host range and geographical distribution, possesses the smallest genome (5.43 Mbp). This strain is representative of frankiae infecting Casuarinaceae plants, and which are only found in soils where suitable hosts are present. However, *Frankia* ACN14a is a medium host range strain, representative of *Frankia* infecting Betulaceae (*Alnus* spp.) and Myricaceae plants which are globally distributed in soils regardless of the presence of a suitable host plant, except in Australia. *Frankia* ACN14a genome size is 7.50 Mbp. Finally, the largest genome (9.04 Mbp) was found in *Frankia* EAN1pec, which represents the broadest host range of *Frankia* strains (infective in Elaeagnaceae, Myricaceae, Rhamnacea and Casuarinacea families) and are globally distributed, including Australia, in soils with or without host plants. So, the genome size of *Frankia* strains seems to be inversely correlated with their host range and the variety of environments where they naturally thrive. Thus, CcI3 genome has likely undergone a reduction process while saprotrophic growth capacity was reduced. Conversely, EAN1pec genome is expanded mainly by duplication and horizontal transfer of genes, gaining more than twice as many genes as either CcI3 and ACN14a (Normand et al. 2007). Supporting the hypothesis of *Frankia* genome size being related with their saprotrophic growth abilities is the even smaller genome size of the microsymbiont "*Candidatus Frankia* datiscae" Dg1 (5.32 Mbp), which is unable to grow *in vitro* (Persson et al. 2011).

The observed genome plasticity in *Frankia* is likely due to insertion sequences (IS) abundance, encoding proteins homologous to transposases (Bickhart et al. 2009, Normand et al. 2007). Transposases are more abundant in *Frankia* CcI3 and EAN1pec than in *Frankia* ACN14a, suggesting greater genome stability in the last one (Bickhart et al. 2009). Another striking feature is the absence of rhizobia *nod* genes homologs. Potential homologs of some common *nod* genes, like *nodB*, are present but they have low similarities with those found in rhizobia and are not arranged in operons. Moreover, *Frankia* lacks *nodA* homologs (Normand et al. 2007). The lack of canonical *nod* genes is even more remarkable considering that transcriptomic analyses indicate that the host plant-signaling pathway is conserved in legumes and actinorhizal plants (Hocher et al. 2011).

Proteomic Studies

Proteomic studies have firstly shown the heterogeneity of cell maturation in *Frankia* cultures since many intracellular proteins were found in supernatants, indicating older cells lysis (Alloisio et al. 2007, Mastronunzio et al. 2009). Despite this inherent pitfall with slow-growing filamentous bacteria, several noteworthy findings are derived from these studies. In *Frankia*, plant phenolic extracts elicit proteins expression that supposedly can attenuate plant defense responses (Bagnarol et al. 2007). In both free-living and symbiotic states, *Frankia* secretes very few proteins (Mastronunzio et al. 2008, Mastronunzio et al. 2009, Mastronunzio and Benson 2010) and, although cellulasic activity were previously observed in supernatants of *Frankia* cultures (Igual et al. 2001), no polysaccharide-degrading enzyme production has been detected in proteomic studies (Mastronunzio et al. 2009). It is known that *Frankia* infection triggers host defense-related gene expression (Ribeiro et al. 2011) but, on the basis of the aforementioned studies, it seems that *Frankia* strains present a relatively non-pathogenic visage to the plant (Benson et al. 2011).

Proteomic studies in nodules revealed that nitrogenase, tricarboxylic acid (TCA) cycle enzymes and electron transport components were among the most abundant proteins expressed by *Frankia*, reflecting the major process occurring in symbiosis (Mastronunzio and Benson 2010).

Transcriptomic Studies

Transcriptomic studies have been performed on *Frankia alni* ACN14a cultures grown on nitrogen-replete and nitrogen-limited media as well as on nodules formed by this strain in the actinorhizal plant species *Alnus glutinosa*, *A. nepalensis*, *Morella rubra* and *Myrica gale* (Alloisio et al. 2010). These studies have corroborated previous findings and also provide interesting new insights into the actinorhizal symbiosis. A first interesting insight is that, unlike rhizobia, there is not a symbiotic island in the genome of *F. alni* ACN14a since nodule-induced genes are separately clustered in several small chromosomal regions. Most of the upregulated genes are highly conserved in *Frankia* CcI3 and EAN1pec (Normand et al. 2007), representing the *Frankia* conserved core. *Frankia* nitrogen fixation-related genes as *nif* (nitrogenase), *hup2* (hydrogenase uptake), *suf* (sulfur-iron cluster), and *shc* (hopanoids synthesis) were found to be highly upregulated in symbiosis. Genes involved in energy production and conversion such as TCA cycle components, ATP synthase and electron-transport chain coding genes were also upregulated, likely to supply the high amount of energy demand of the nitrogen fixation process.

Regarding genes involved in ammonium assimilation and transport, the gene encoding for glutamine synthase I (*glnA*), a key enzyme in ammonium assimilation into glutamate, was found to be downregulated in both free-living and symbiotic conditions. However, other genes playing a role in this process, such as *glnII* and *gdhA* that codify for glutamine synthase II and glutamate dehydrogenase, respectively, shown to be differentially regulated in free-living and symbiotic conditions. This fact may indicate that ammonium assimilation regulation in *Frankia* is different if it is in a free-living stage or in symbiosis.

When compared to phylogenetically distant plant species, nodule transcriptomes of *F. alni* were quite similar. In comparison with transcriptomic data on *Sinorhizobium meliloti*, more genes were upregulated in *F. alni* nodules than in *S. meliloti* nodules, suggesting that *Frankia* is metabolically more active and autonomous than rhizobia, perhaps because it must protect by itself the nitrogenase from oxygen exposure (Alloisio et al. 2010).

Transcriptome analyses of free-living *Frankia* CcI3·have been carried out recently (Bickhart and Benson 2011). The results indicated that its transcriptomic profile varies depending on culture age and nitrogen source (ammonium added vs. nitrogen fixation), with aging being more influential overall. Even in ammonium-supplemented medium, *nif* transcripts were notably present, which may be due to microaerobic and nitrogen-depleted sites in mycelium. Also, higher levels of transposase expression were shown to be associated with nitrogen starvation and culture aging, suggesting that both factors provide conditions for on-going genome modification. As the authors point out, their results should call for caution in interpreting data from "-omics" studies, especially those obtained from root nodules, where cell heterogeneity would be expected to be quite high (Bickhart and Benson 2011).

Beyond the Actinorhizal Symbiosis: Broadening the Prospects of a Sustainable Agriculture

Root nodule symbiosis (RNS) rhizobia and actinorhizal symbiosis, turned out to be confined to a single large clade of the Rosid I lineage, termed the nitrogen-fixing clade. RNS arose in angiosperm history approximately 70 million years ago (Doyle 2011). As the molecular underpinnings of nodulation are explored deeper, it seems clear that RNS-related processes rely on conserved genetic programs (Markmann and Parniske 2009). Thus, the common symbiosis pathway controlling the early signaling stages of nodulation might have been recruited from the much older arbuscular mycorrhizal symbiosis. The recent characterization of mycorrhization factors (Myc), which, although structurally simpler than rhizobial Nod factors, are also lipo-chito-oligosaccharides (Maillet et al. 2011), reinforces such an assumption. Moreover, the evolution of nodule organogenesis might have involved the adaptation of pre-existing root developmental programs (Markmann and Parniske 2009).

The ability of plants to be able to form nitrogen-fixing root nodules of genetic programs that are universally conserved in angiosperms would facilitate engineering nitrogen fixation in important staple food plants like cereals. In extending symbiosis or associative nitrogen-fixing relationships to other plants, *Frankia* surpasses most diazotrophic bacteria characteristics (Normand and Mullin 2008), since (i) it has intrinsic oxygen-protection mechanisms, such as vesicles and hemoglobin production, and thus has the capacity to subsist and fix nitrogen under both symbiotic and nonsymbiotic conditions; (ii) it can adapt to much more host genetic backgrounds than rhizobia, as demonstrated by actinorhizal plants phylogenic diversity; (iii) given its diminished exoproteome and that it is not released into the cytoplasm of the plant cells, *Frankia* may be less likely to elicit defensive responses in non-host plants; and (iv) it can infect plants either through root hairs or intercellularly. Moreover, in stark contrast to legume nodules, actinorhizal nodules are shaped taking advantage of the universally extant genetic program of secondary roots formation. Therefore, gaining insights into the biology of the actinorhizal symbiosis will provide invaluable knowledge to generate artificially new nitrogen-fixing symbiotic associations (Markmann and Parniske 2009).

Acknowledgments

The authors gratefully acknowledge the financial support from Junta de Castilla y León grants (CSI287A11-2, SA183A11-2) and Spanish Government grant (AGL2011-29227). A. Valverde and L.P. Rivera received financial support from CSIC through, respectively, the programs JAE-doc and JAE-predoc, E. Menéndez acknowledges a FPI-MICINN PhD fellowship and M. Medina-Sierra was granted a fellowship from Colciencias and Universidad de Antioquia (Colombia).

References

Alloisio, N., S. Felix, J. Marechal, P. Pujic, Z. Rouy, D. Vallenet, C. Medigue and P. Normand. 2007. *Frankia alni* proteome under nitrogen-fixing and nitrogen-replete conditions. Physiol. Plant. 130: 440–453.

Alloisio, N., C. Queiroux, P. Fournier, P. Pujic, P. Normand, D. Vallenet, C. Medigue, M. Yamaura, K. Kakoi and K.I. Kucho. 2010. The *Frankia alni* Symbiotic Transcriptome. Mol. Plant-Microbe Interact. 23: 593–607.

Amadou, C., G. Pascal, S. Mangenot, M. Glew, C. Bontemps, D. Capela, S. Carrère, S. Cruveiller, C. Dossat, A. Lajus, M. Marchetti, V. Poinsot, Z. Rouy, B. Servin, M. Saad, C. Schenowitz, V. Barbe, J. Batut, C. Médigue and C. Masson-Boivin. 2008. Genome sequence of the beta-rhizobium *Cupriavidus taiwanensis* and comparative genomics of rhizobia. Genome Research 18: 1472–1483.

Ardley, J.K., M.A. Parker, S.E. De Meyer, R.D. Trengove, G.W. O'Hara, W.G. Reeve, R.J. Yates, M.J. Dilwort, A. Willems and J.G. Howieson. 2012. *Microvirga lupini* sp. nov., *Microvirga lotononidis* sp. nov., and *Microvirga zambiensis* sp. nov. are Alphaproteobacterial root nodule bacteria that specifically nodulate and fix nitrogen with geographically and taxonomically separate legume hosts. Int. J. Syst. Evol. Microbiol. (in press).

Auguy, F., K. Abdel-Lateif, P. Doumas, P. Badin, V. Guerin, D. Bogusz and V. Hocher. 2011. Activation of the isoflavonoid pathway in actinorhizal symbioses. Funct. Plant Biol. 38: 690–696.

Bagnarol, E., J. Popovici, N. Alloisio, J. Marechal, P. Pujic, P. Normand and M.P. Fernandez. 2007. Differential *Frankia* protein patterns induced by phenolic extracts from Myricaceae seeds. Physiol. Plant. 130: 380–390.

Baker, D.D. 1987. Relationships among pure cultured strains of *Frankia* based on host specificity. Physiol. Plant. 70: 245–248.

Beijerinck, M.W. 1890. Künstliche Infection von Vicia Faba mit Bacillus radicicola. Ernährungsbedingungen dieser Bacterie. Bot Zeitung. 52: 837–843.

Benabdoun, F.M., M. Nambiar-Veetil, L. Imanishi, S. Svistoonoff, N. Ykhlef, H. Gherbi and C. Franche. 2011. Composite Actinorhizal Plants with Transgenic Roots for the Study of Symbiotic Associations with *Frankia*. Journal of Botany 2011: doi:10.1155/2011/702947.

Benoit, L.F. and A.M. Berry. 1997. Flavonoid-like compounds from seeds of red alder (*Alnus rubra*) influence host nodulation by *Frankia* (Actinomycetales). Physiol. Plant. 99: 588–593.

Benson, D.R. and W.B. Silvester. 1993. Biology of *Frankia* strains, actinomycete symbionts of actinorhizal plants. Microbiol. Rev. 57: 293–319.

Benson, D.R., J.M. Brooks, Y. Huang, D.M. Bickhart and J.E. Mastronunzio. 2011. The Biology of *Frankia* sp. Strains in the Post-Genome Era. Mol. Plant-Microbe Interact. 24: 1310–1316.

Berry, A.M., O.T. Harriott, R.A. Moreau, S.F. Osman, D.R. Benson and A.D. Jones. 1993. Hopanoid lipids compose the *Frankia* vesicle envelope, presumptive barrier of oxygen diffusion to nitrogenase. Proc. Natl. Acad. Sci. USA. 90: 6091–6094.

Bickhart, D.M. and D.R. Benson. 2011. Transcriptomes of *Frankia* sp. strain CcI3 in growth transitions. BMC Microbiol. 11: 192.

Bickhart, D.M., J.P. Gogarten, P. Lapierre, L.S. Tisa, P. Normand and D.R. Benson. 2009. Insertion sequence content reflects genome plasticity in strains of the root nodule actinobacterium *Frankia*. BMC Genomics 10: 468.

Bontemps, C., G.N. Elliott, M.F. Simon, F.B. Dos Reis Júnior, E. Gross, R.C. Lawton, N.E. Neto, M. de Fátima Loureiro, S.M. De Faria, J.I. Sprent, E.K. James and J.P. Young. 2010. *Burkholderia* species are ancient symbionts of legumes. Mol. Ecol. 19: 44–52.

Callaham, D., P. DelTredici and J.G. Torrey. 1978. Isolation and cultivation in vitro of the actinomycete causing root nodulation in *Comptonia*. Science. 199: 899–902.

Chen, W.M., S. Laevens, T.M. Lee, T. Coenye, P. de Vos, M. Mergeay and P. Vandamme. 2001. *Ralstonia taiwanensis* sp. nov., isolated from root nodules of *Mimosa* species and sputum of a cystic fibrosis patient. Int. J. Syst. Evol. Microbiol. 51: 1729–1735.

Chen, W.M., E.K. James, A.R. Prescott, M. Kierans and J.I. Sprent. 2003. Nodulation of *Mimosa* spp. by the beta-proteobacterium *Ralstonia taiwanensis*. Mol. Plant Microbe Interact. 16: 1051–1061.

Coats, V., C.R. Schwintzer and J.D. Tjepkema. 2009. Truncated hemoglobins in *Frankia* CcI3: effects of nitrogen source, oxygen concentration, and nitric oxide. Can. J. Microbiol. 55: 867–873.

Conn, H.J. 1938. Taxonomic relationships of certain non-sporeforming rods in soil. J. Bacteriol. 36: 320–321.

Dawson, J.O. 2008. Ecology of actinorhizal plants. *In*: K. Pawlowski and W.E. Newton [eds.]. Nitrogen-fixing Actinorhizal Symbioses. Springer, Dordrecht, The Netherlands. pp. 199–234.

Doyle, J.J. 2011. Phylogenetic Perspectives on the Origins of Nodulation. Mol. Plant-Microbe In. 24: 1289–1295.

Elliott, G.N., J.H. Chou, W.M. Chen, G.V. Bloemberg, C. Bontemps, E. Martínez-Romero, E. Velázquez, J.P. Young, J.I. Sprent and E.K. James. 2009. *Burkholderia* spp. are the most competitive symbionts of *Mimosa*, particularly under N-limited conditions. Environ. Microbiol. 11: 762–778.

Farrand, S. K., P. van Berkum and P. Oger. 2003. *Agrobacterium* is a definable genus of the family *Rhizobiaceae*. Int. J. Syst. Evol. Microbiol. 53: 1681–1687.

Frank, B. 1889. Über die Pilzsymbiose der Leguminosen. Berichte der Deutschen Botanischen Gesellschaft 7: 332–346.

Fuchsius, L. 1542. De historia stirpium commentarii isignes. Michael Isingrin, Basel, Switzerland.

Gabbarini, L.A. and L.G. Wall. 2011. Diffusible factors from *Frankia* modify nodulation kinetics in *Discaria trinervis*, an intercellular root-infected actinorhizal symbiosis. Funct. Plant Biol. 38: 662–670.

Graham, P.H., M.J. Sadowsky, H.H. Keyser, Y.M. Barnet, R.S. Bradley, J.E. Cooper, D.J. de Ley, B.D.W. Jarvis, E.B. Roslycky, B.W. Strijdom and J.P.W. Young. 1991. Proposed Minimal

Standards for the Description of New Genera and Species of Root- and Stem-Nodulating Bacteria. Int. J. Syst. Bacteriol. 41: 582–587.

Gtari, M. and J.O. Dawson. 2011. An overview of actinorhizal plants in Africa. Funct. Plant Biol. 38: 653–661.

Gyaneshwar, P., A.M. Hirsch, L. Moulin, W.M. Chen, G. N. Elliott, C. Bontemps, P. Estrada-de los Santos, E. Gross, F. Bueno dos Reis, Jr., J.I. Sprent, J.P.W. Young and E.K. James. 2011. Legume-Nodulating Betaproteobacteria: Diversity, Host Range, and Future Prospects. Mol. Plant Microbe Interact. 24(11): 1276–1288.

Hahn, D. 2008. Polyphasic taxonomy of the genus *Frankia*. *In*: K. Pawlowski and W.E. Newton [eds.]. Nitrogen-fixing Actinorhizal Symbioses. Springer, Dordrecht, The Netherlands. pp. 25–48.

Hocher, V., F. Auguy, X. Argout, L. Laplaze, C. Franche and D. Bogusz. 2006. Expressed sequence-tag analysis in *Casuarina glauca* actinorhizal nodule and root. New Phytol. 169: 681–688.

Hocher, V., N. Alloisio, F. Auguy, P. Fournier, P. Doumas, P. Pujic, H. Gherbi, C. Queiroux, C. Da Silva, P. Wincker, P. Normand and D. Bogusz. 2011. Transcriptomics of Actinorhizal Symbioses Reveals Homologs of the Whole Common Symbiotic Signaling Cascade. Plant Physiol. 156: 700–711.

Hughes, M., C. Donnelly, A. Crozier and C.T. Wheeler. 1999. Effects of the exposure of roots of *Alnus glutinosa* to light on flavonoids and nodulation. Can. J. Bot. 77: 1311–1315.

Igual, J.M., E. Velázquez, P.F. Mateos, C. Rodríguez-Barrueco, E. Cervantes and E. Martínez-Molina. 2001. Cellulase isoenzyme profiles in *Frankia* strains belonging to different cross-inoculation groups. Plant Soil. 229: 35–39.

Igual, J.M., A. Valverde, R. Rivas, P.F. Mateos, C. Rodríguez-Barrueco, E. Martínez-Molina, E. Cervantes and E. Velázquez. 2003. Genomic fingerprinting of *Frankia* strains by PCR-based techniques. Assessment of a primer based on the sequence of 16S rRNA gene of *Escherichia coli*. Plant Soil. 254: 115–123.

Jordan, D.C. 1982. Transfer of *Rhizobium japonicum* Buchanan 1980 to *Bradyrhizobium* gen. nov., a genus of slow-growing, root nodule bacteria from leguminous plants. Int. J. Syst. Bacteriol. 32: 136–139.

Judicial Commission of the International Committee on Systematics of Prokaryotes. 2008. The genus name *Sinorhizobium* Chen et al. 1988 is a later synonym of *Ensifer* Casida 1982 and is not conserved over the latter genus name, and the species name '*Sinorhizobium adhaerens*' is not validly published. Opinion 84. Int. J. Syst. Evol. Microbiol. 58: 1973.

Judicial Commission of the International Committee on Systematics of Prokaryotes. 2011. Subcommittee on the taxonomy of *Agrobacterium* and *Rhizobium*. Int. J. Syst. Evol. Microbiol. 61: 3089–3093.

Krumholz, G.D., M.S. Chval, M.J. McBride and L.S. Tisa. 2003. Germination and physiological properties of *Frankia* spores. Plant Soil. 254: 57–67.

Kucho, K.I., K. Kakoi, M. Yamaura, S. Higashi, T. Uchiumi and M. Abe. 2009. Transient Transformation of *Frankia* by Fusion Marker Genes in Liquid Culture. Microbes Environ. 24: 231–240.

Kuykendall, L.D. 2005. Order VI. Rhizobiales ord. nov. *In*: D.J. Brenner, N.R. Krieg, J.T. Staley and G.M. Garrity [eds.]. Bergey's Manual of Systematic Bacteriology, 2nd edn., vol. 2 (The Proteobacteria), part C (The Alpha-, Beta-, Delta- and Epsilon-proteobacteria). Springer, New York. p. 324.

Leul, M., A. Mohapatra and A. Sellstedt. 2005. Biodiversity of hydrogenases in *Frankia*. Curr. Microbiol. 50: 17–23.

Leul, M., P. Normand and A. Sellstedt. 2009. The phylogeny of uptake hydrogenases in *Frankia*. Int. Microbiol. 12: 23–28.

Lin, D.X., E.T. Wang, H. Tang, T.X. Han, Y.R. He, S.H. Guan and W.X. Chen. 2008. *Shinella kummerowiae* sp. nov., a symbiotic bacterium isolated from root nodules of the herbal legume *Kummerowia stipulacea*. Int. J. Syst. Evol. Microbiol. 58: 1409–1413.

Maillet, F., V. Poinsot, O. Andre, V. Puech-Pages, A. Haouy, M. Gueunier, L. Cromer, D. Giraudet, D. Formey, A. Niebel, E.A. Martinez, H. Driguez, G. Becard and J. Denarie. 2011. Fungal lipo-chito-oligosaccharide symbiotic signals in arbuscular mycorrhiza. Nature. 469: 58–63.

Mantelin, S., M.F.L. Saux, F. Zakhia, G. Bena, S. Bonneau, H. Jeder, P. de Lajudie and J.C. Cleyet-Marel. 2006. Emended description of the genus *Phyllobacterium* and description of four novel species associated with plant roots: *Phyllobacterium bourgognense* sp. nov., *Phyllobacterium ifriqiyense* sp. nov., *Phyllobacterium leguminum* sp. nov. and *Phyllobacterium brassicacearum* sp. nov. Int. J. Syst. Evol. Microbiol. 56 (4): 827.

Markmann, K. and M. Parniske. 2009. Evolution of root endosymbiosis with bacteria: how novel are nodules? Trends Plant Sci. 14: 77–86.

Mastronunzio, J.E. and D.R. Benson. 2010. Wild nodules can be broken: proteomics of *Frankia* in field-collected root nodules. Symbiosis. 50: 13–26.

Mastronunzio, J.E., L.S. Tisa, P. Normand and D.R. Benson. 2008. Comparative secretome analysis suggests low plant cell wall degrading capacity in *Frankia* symbionts. BMC Genomics 9: 47.

Mastronunzio, J.E., Y. Huang and D.R. Benson. 2009. Diminished Exoproteome of *Frankia* spp. in Culture and Symbiosis. Appl. Environ. Microbiol. 75: 6721–6728.

Miller, I.M. and D.D. Baker. 1986. Nodulation of actinorhizal plants by *Frankia* strains capable of both root hair infection and intercellular penetration. Protoplasma. 131: 82–91.

Moulin, L., A. Munive, B. Dreyfus and C. Boivin-Masson. 2001. Nodulation of legumes by members of the beta-subclass of Proteobacteria. Nature. 411(6840): 948–950.

Ngom, A., Y. Nakagawa, H. Sawada, J.Tsukahara, S. Wakabayashi, T. Uchiumi, A. Nuntagij, S. Kotepong, A. Suzuki, S. Higashi and M. Abe. 2004. A novel symbiotic nitrogen-fixing member of the *Ochrobactrum* clade isolated from root nodules of *Acacia mangium*. J. Gen. Appl. Microbiol. 50: 17–27.

Normand, P. and B.C. Mullin. 2008. Prospects for the study of a ubiquitous actinomycete, *Frankia*, and its host plants. *In*: K. Pawlowski and W.E. Newton [eds.]. Nitrogen-fixing Actinorhizal Symbioses. Springer, Dordrecht, The Netherlands. pp. 289–303.

Normand, P., S. Orso, B. Cournoyer, P. Jeannin, C. Chapelon, J. Dawson, L. Evtushenko and A.K. Misra. 1996. Molecular phylogeny of the genus *Frankia* and related genera and emendation of the family *Frankiaceae*. Int. J. Syst. Bacteriol. 46: 1–9.

Normand, P., P. Lapierre, L.S. Tisa, J.P. Gogarten, N. Alloisio, E. Bagnarol, C.A. Bassi, A.M. Berry, D.M. Bickhart, N. Choisne, A. Couloux, B. Cournoyer, S. Cruveiller, V. Daubin, N. Demange, M.P. Francino, E. Goltsman, Y. Huang, O.R. Kopp, L. Labarre, A. Lapidus, C. Lavire, J. Marechal, M. Martinez, J.E. Mastronunzio, B.C. Mullin, J. Niemann, P. Pujic, T. Rawnsley, Z. Rouy, C. Schenowitz, A. Sellstedt, F. Tavares, J.P. Tomkins, D. Vallenet, C. Valverde, L.G. Wall, Y. Wang, C. Medigue and D.R. Benson. 2007. Genome characteristics of facultatively symbiotic *Frankia* sp. strains reflect host range and host plant biogeography. Genome Res. 17: 7–15.

Pawlowski, K. and J.I. Sprent. 2008. Comparison between Actinorhizal and Legume Symbiosis. *In*: K. Pawlowski and W.E. Newton [eds.]. Nitrogen-fixing Actinorhizal Symbioses. Springer, Dordrecht, The Netherlands.

Pawlowski, K., D. Bogusz, A. Ribeiro and A.M. Berry. 2011. Progress on research on actinorhizal plants. Funct. Plant Biol. 38: 633–638.

Perrine-Walker, F., H. Gherbi, L. Imanishi, V. Hocher, F. Ghodhbane-Gtari, J. Lavenus, F.M. Benabdoun, M. Nambiar-Veetil, S. Svistoonoff and L. Laplaze. 2011. Symbiotic Signaling in Actinorhizal Symbioses. Curr. Protein Pept. Sci. 12: 156–164.

Persson, T., D.R. Benson, P. Normand, B. Vanden Heuvel, P. Pujic, O. Chertkov, H. Teshima, D.C. Bruce, C. Detter, R. Tapia, S. Han, J. Han, T. Woyke, S. Pitluck, L. Pennacchio, M. Nolan, N. Ivanova, A. Pati, M.L. Land, K. Pawlowski and A.M. Berry. 2011. Genome Sequence of "*Candidatus* Frankia datiscae" Dg1, the Uncultured Microsymbiont from Nitrogen-Fixing Root Nodules of the Dicot *Datisca glomerata*. J. Bacteriol. 193: 7017–7018.

Popovici, J., V. Walker, C. Bertrand, F. Bellvert, M.P. Fernández and G. Comte. 2011. Strain specificity in the Myricaceae-*Frankia* symbiosis is correlated to plant root phenolics. Funct. Plant Biol. 38: 682–689.

Ramírez-Bahena, M.H., P. García-Fraile, A. Peix, A. Valverde, R. Rivas, J.M. Igual, P.F. Mateos, E. Martínez-Molina and E. Velázquez. 2008. Revision of the taxonomic status of the species *Rhizobium leguminosarum* (Frank 1879) Frank 1889AL, *Rhizobium phaseoli* Dangeard 1926AL and *Rhizobium trifolii* Dangeard 1926AL. *R. trifolii* is a later synonym of *R. leguminosarum*. Reclassification of the strain *R. leguminosarum* DSM 30132 (=NCIMB 11478) as *Rhizobium pisi* sp. nov. Int. J. Syst. Evol. Microbiol. 58: 2484–2490.

Ribeiro, A., I. Graca, K. Pawlowski and P. Santos. 2011. Actinorhizal plant defence-related genes in response to symbiotic *Frankia*. Funct. Plant Biol. 38: 639–644.

Rivas, R., E. Velázquez, A. Willems, N. Vizcaíno, N.S. Subba-Rao, P.F. Mateos, M. Gillis, F.B. Dazzo and E. Martínez-Molina. 2002. A new species of *Devosia* that forms a unique nitrogen-fixing root-nodule symbiosis with the aquatic legume *Neptunia natans* (L.f.) Druce. Appl. Environ. Microbiol. 68: 5217–5222.

Rivas, R., A. Willems, N.S. Subba-Rao, P.F. Mateos, F.B. Dazzo, R.M. Kroppenstedt, E. Martínez-Molina, M. Gillis and E. Velázquez. 2003. Description of *Devosia neptuniae* sp. nov. that nodulates and fixes nitrogen in symbiosis with *Neptunia natans*, an aquatic legume from India. Syst. Appl. Microbiol. 26: 47–53.

Rivas, R., A. Willems, J.L. Palomo, P. García-Benavides, P.F. Mateos, E. Martínez-Molina, M. Gillis and E. Velázquez. 2004. *Bradyrhizobium betae* sp. nov., isolated from roots of *Beta vulgaris* affected by tumour-like deformations. Int. J. Syst. Evol. Microbiol. 54: 1271–1275.

Sprent J.I. and P. Sprent. 1990. Nitrogen fixing organisms. Pure and applied aspects. Chapman and Hall. London.

Stackebrandt, E., W. Frederiksen, G.M. Garrity, P.A. Grimont, P. Kämpfer, M.C. Maiden, X. Nesme, R. Rosselló-Mora, J. Swings, H.G. Trüper, L. Vauterin, A.C. Ward and W.B. Whitman. 2002. Report of the ad hoc committee for the re-evaluation of the species definition in bacteriology. Int. J. Syst. Evol. Microbiol. 52: 1043–1047.

Svistoonoff, S., H. Gherbi, M. Nambiar-Veetil, C. Zhong, Z. Michalak, L. Laplaze, V. Vaissayre, F. Auguy, V. Hocher, P. Doumas, J. Bonneau, D. Bogusz and C. Franche. 2010. Contribution of transgenic Casuarinaceae to our knowledge of the actinorhizal symbioses. Symbiosis 50: 3–11.

Swensen, S.M. 1996. The evolution of actinorhizal symbioses: Evidence for multiple origins of the symbiotic association. Am. J. Bot. 83: 1503–1512.

Sy, A., E. Giraud, P. Jourand, N. García, A. Willems, P. de Lajudie, Y. Prin, M. Neyra, M. Gillis, C. Boivin-Masson and B. Dreyfus. 2001. Methylotrophic *Methylobacterium* bacteria nodulate and fix nitrogen in symbiosis with legumes. J. Bacteriol. 183: 214–220.

Talbi, C., M. J. Delgado, L. Girard, A. Ramírez-Trujillo, J. Caballero-Mellado and E.J. Bedmar. 2010. *Burkholderia phymatum* Strains Capable of Nodulating *Phaseolus vulgaris* are present in Moroccan Soils. Appl. Environ. Microbiol. 76(13): 4587.

Tjepkema, J.D., R.E. Cashon, J. Beckwith and C.R. Schwintzer. 2002. Hemoglobin in *Frankia*, a nitrogen-fixing actinomycete. Appl. Environ. Microbiol. 68: 2629–2631.

Trujillo, M.E., A. Willems, A. Abril, A.M. Planchuelo, R. Rivas, D. Ludeña, P.F. Mateos, E. Martínez-Molina and E. Velázquez. 2005. Nodulation of *Lupinus albus* by strains of *Ochrobactrum lupini* sp. nov. Appl. Environ. Microbiol. 71: 1318–327.

Valverde, C. and K. Huss-Danell. 2008. Carbon and nitrogen metabolism in actinorhizal nodules. *In*: K. Pawlowski and W.E. Newton [eds.]. Nitrogen-fixing Actinorhizal Symbioses. Springer, Dordrecht, The Netherlands. pp. 167–198.

Valverde, A., E. Velázquez, F. Fernández-Santos, N. Vizcaíno, R. Rivas, P.F. Mateos, E. Martínez-Molina, J.M. Igual and A. Willems. 2005. *Phyllobacterium trifolii* sp. nov., nodulating *Trifolium* and *Lupinus* in Spanish soils. Int. J. Syst. Evol. Microbiol. 55: 1985–1989.

van Berkum, P. and B.D. Eardly. 2002. The Aquatic Budding Bacterium Blastobacter denitrificans is a Nitrogen-Fixing Symbiont of *Aeschynomene indica*. Appl. Environ Microbiol. 68(3): 1132–1136.

van Berkum, P., J.M. Leibold and B.D. Eardly. 2006. Proposal for combining *Bradyrhizobium* spp. (*Aeschynomene indica*) with *Blastobacter denitrificans* and to transfer *Blastobacter denitrificans* (Hirsch and Muller 1985) to the genus *Bradyrhizobium* as *Bradyrhizobium denitrificans* (comb. nov.). Syst. Appl. Microbiol. 29(3): 207–215.

Van Ghelue, M., E. Lovaas, E. Ringo and B. Solheim. 1997. Early interactions between *Alnus glutinosa* and *Frankia* strain ArI3. Production and specificity of root hair deformation factor(s). Physiol. Plant. 99: 579–587.

Velázquez, E., E. Cervantes, J.M. Igual, A. Peix, P.F. Mateos, S. Benamar, A. Moiroud, C.T. Wheeler, J.O. Dawson, D. Labeda, C. Rodríguez-Barrueco and E. Martínez-Molina. 1998. Analysis of LMW RNA profiles of *Frankia* strains by staircase electrophoresis. Syst. Appl. Microbiol. 21: 539–545.

Velázquez, E., A. Peix, J.L. Zurdo-Piñeiro, J.L. Palomo, P.F. Mateos, R. Rivas, E. Muñoz-Adelantado, N. Toro, P. García-Benavides and E. Martínez-Molina. 2005. The coexistence of symbiosis and pathogenicity-determining genes in *Rhizobium rhizogenes* strains enables them to induce nodules and tumors or hairy roots in plants. Mol. Plant Microbe Interact.18: 1325–1332.

Woese, C.R., E. Stackebrandt, W.G. Weisburg, B.J. Paster, M.T. Madigan, V.J. Fowler, C.M. Hahn, P. Blanz, R. Gupta, K.H. Nealson and G.E. Fox. 1984. The phylogeny of purple bacteria: The alpha subdivision. Syst. Appl. Microbiol. 5: 315–326.

Yates, R. J., J.G. Howieson, W.G. Reeve, K.G. Nandasena, I.J. Law, L. Brau, L., J.K. Ardley, H.M. Nistelberger, D. Real and G.W. O'Hara. 2007. *Lotononis angolensis* forms nitrogen fixing, lupinoid nodules with phylogenetically unique, fast-growing, pink-pigmented bacteria, which do not nodulate *L. bainesii* or *L. listii*. Soil Biol. Biochem. 39: 1680–1688.

Young, J.M., L.D. Kuykendall, E. Martínez-Romero, A. Kerr and H. Sawada. 2001. A revision of *Rhizobium* Frank 1889, with an emended description of the genus, and the inclusion of all species of *Agrobacterium* Conn 1942 and *Allorhizobium undicola* de Lajudie et al. 1998 as new combinations: *Rhizobium radiobacter*, *R. rhizogenes*, *R. rubi*, *R. undicola* and *R. vitis*. Int. J. Syst. Evol. Microbiol. 51: 89–103.

Young, J.P.W. and K. Haukka. 1996. Diversity and phylogeny of rhizobia. New Phytol. 133: 87–94.

Zhi, X.-Y., W.-J. Li and E. Stackebrandt. 2009. An update of the structure and 16S rRNA gene sequence-based definition of higher ranks of the class *Actinobacteria*, with the proposal of two new suborders and four new families and emended descriptions of the existing higher taxa. Int. J. Syst. Evol. Microbiol. 59: 589–608.

Zimpfer, J.F., B. McCarty, C.M. Kaelke, L. Mulongwe, J.M. Igual, C.A. Smyth and J.O. Dawson. 2002. *Casuarina cunninghamiana* cladode extracts increase the *Frankia* infectious capacity of a tropical soil. Symbiosis. 33: 73–90.

Zimpfer, J.F., J.M. Igual, B. McCarty, C. Smyth and J.O. Dawson. 2004. *Casuarina cunninghamiana* tissue extracts stimulate the growth of *Frankia* and differentially alter the growth of other soil microorganisms. J. Chem. Ecol. 30: 439–452.

Zurdo-Piñeiro, J.L., E. Velázquez, M.J. Lorite, G. Brelles-Mariño, E.C. Schröder, E.J. Bedmar, P.F. Mateos and E. Martínez-Molina. 2004. Identification of fast-growing rhizobia nodulating tropical legumes from Puerto Rico as *Rhizobium gallicum* and *Rhizobium tropici*. Syst. Appl. Microbiol. 27: 469–477.

Zurdo-Piñeiro, J.L., R. Rivas, M.E. Trujillo, N. Vizcaíno, J.A. Carrasco, M. Chamber, A. Palomares, P.F. Mateos, E. Martínez-Molina and E. Velázquez. 2007. *Ochrobactrum cytisi* sp. nov., isolated from nodules of *Cytisus scoparius* in Spain. Int. J. Syst. Evol. Microbiol. 57: 784–788.

2

Biodiversity of Slow-Growing Rhizobia: The Genus *Bradyrhizobium*

D. Durán,[1] *L. Rey,*[1] *C. Sánchez-Cañizares,*[1] *B. Jorrín,*[1] *J. Imperial*[1,2] and *T. Ruiz-Argüeso*[1,*]

Introduction

Legume root symbiotic bacteria, now collectively known as rhizobia, were first described by Beijerinck in 1888, and later isolated from the root nodules of many legumes. With the introduction of soybean in the U.S.A., it was demonstrated that the culture only succeeded if their root nodule symbiotic bacteria were also introduced. Although then known as *Rhizobium japonicum*, these bacteria were quite different from other rhizobia: they were larger, grew slowly and did not acidify sugar-containing media. Similar bacteria were later isolated from other legumes, such as lupines, cowpeas, and many tropical trees and shrubs, and were eventually grouped into a separate genus, *Bradyrhizobium*. Despite its large diversity (see below), most

[1]Centre for Plant Biotechnology and Genomics (CBGP), Universidad Politécnica de Madrid and INIA, Campus de Montegancedo, Ctra. M-40, km 38, 28223 Pozuelo de Alarcón, Madrid (Spain).
[2]C.S.I.C
*Corresponding author: t.ruizargueso@upm.es

studies have been performed on *B. japonicum*, undoubtedly because of the importance of soybean as a crop.

Taxonomy

The genus *Bradyrhizobium* was created to assemble a group of bacteria able to fix N_2 in symbioses with several leguminous species and characterized by a slow growth rate and by an alkaline reaction in culture media containing mannitol as the carbon source (Jordan 1982). Based on their 16S rDNA gene sequences, the genus *Bradyrhizobium* of legume-nodulating belongs to a monophyletic group within the α-proteobacteria that also accommodates oligotrophic soil and aquatic bacteria (Sawada et al. 2003). Presently, the *Bradyrhizobium* genus includes α-proteobacteria capable of living in symbiotic and endophytic associations with legumes and nonlegumes, and posseses a physiological and symbiotic versatility and a broad geographical distribution. It includes strains isolated from nodules of herbaceous and woody legume species of tropical and temperate origin and aquatic legumes such as *Aeschynomene* species (van Berkum and Eardly 2002) and the nonlegume *Parasponia andersonii* (Trinick and Hadobas 1989). In addition, *Bradyrhizobium* also includes endophytic bacteria related to *B. elkanii* from wild and modern rice (*Oryza sativa* L.) (Tan et al. 2001).

Much attention has been focused on the genus *Bradyrhizobium* in the last few years. Using phenotypic characteristics and a broad range of genotyping methods and multiple genetic markers, a high genetic diversity has been demonstrated among strains (Vinuesa et al. 2005a, Menna et al. 2009). Studies on *Bradyrhizobium* bacteria are, nonetheless, still limited. Despite this noteworthy diversity and worldwide distribution, few species have been described, and only 12 *Bradyrhizobium* species have been defined and validated until now (Table 2.1).

Out of these, four are soybean symbionts (*B. japonicum, B. elkanii, B. liaoningense* and *B. huanghuaihainense*), one nodulates *Lespedeza cuneata* (*B. yuanmingense*) and two nodulate shrubs of the tribes *Genisteae* and *Loteae* (*B. canariense, B. cytisi*). *B. betae* is nonsymbiotic (Rivas et al. 2004), while two species that nodulate *Pachyrhizus erosus (B. pachyrhizi* and *B. jicamae,* Ramirez-Bahena et al. 2009*)* and another one that nodulates *Lablab purpureus* (Chang et al. 2011) have been described recently. The *B. denitrificans* novel species proposed by van Berkum et al. (2006) as a result of the reclassification of *Blastobacter denitrificans*, has already been validated. Apparently, species definition in this genus is limited by the low diversity of their 16S rRNA gene sequences (Willems et al. 2001).

Although the current definition of a new species is based both on phenotypic and genetic characteristics, a DNA-DNA hybridization comparison test is generally required in order to define a novel species;

Table 2.1 Validated species of the genus *Bradyrhizobium*.

Species	Hosts	References
B. betae	*Beta vulgaris*	Rivas et al. 2004
B. canariense	*Chamaecytisus, Lupinus*	Vinuesa et al. 2005b
B. cytisi	*Cytisus villosus*	Chahboune et al. 2011
B. denitrificans	*Aeschynomene indica*	van Berkum et al. 2006
B. elkanii	*Glycine max*	Kuykendall et al. 1992
B. huanghuaihaiense	*Glycine max*	Zhang et al. 2011
B. iriomotense	*Entada koshunensis, Macroptilium atropurpureum*	Islam et al. 2008
B. japonicum	*Glycine max*	Jordan 1982
B. jicamae	*Pachyrhizus erosus, Lespedeza* spp.	Ramirez-Bahena et al. 2009
B. lablabi	*Lablab purpureus, Arachis hypogaea*	Chang et al. 2011
B. liaoningense	*Glycine max*	Xu et al. 1995
B. pachyrhizi	*Pachyrhizus erosus*	Ramirez-Bahena et al. 2009
B. yuanmingense	*Lespedeza* spp.	Yao et al. 2002

however, an increasing number of arguments have been raised against its use (Coenye et al. 2005, Auch et al. 2010). Besides the sequence analysis of 16S rRNA genes, other genetic markers have frequently been used for taxonomy of legume endosymbionts. Among them, a multilocus sequence analysis, based on the sequences of several conserved housekeeping genes (more frequently, 16S rRNA, *glnII, atpD, recA, dnaK*) dispersed over the genome, has been shown to be a powerful method to improve phylogenetic analysis, and an effective tool for identification and taxonomic classification of *Bradyrhizobium* strains (Cooper and Fell 2004, Moulin et al. 2004, Vinuesa et al. 2005a, Menna et al. 2009, Stepkowski et al. 2011). However, the number and nature of housekeeping genes examined limit the possibilities of multilocus sequence analysis for species delineation and, at present, it cannot replace DNA-DNA hybridization. Instead, a procedure based on the analysis of partial genome sequences (Richter and Roselló-Mora, 2009) has been suggested. Recently other genomic approaches have been developed to define and understand the involvement of whole bacterial genomes in the symbiotic process (Coenye et al. 2005, MacLean et al. 2007).

Based on different simple or concatenated genetic markers (16S rRNA, ITS, or MLST), two large subgroups of *Bradyrhyzobium* strains have been repeatedly identified among large collections of strains (e.g., Kwon et al. 2005, Menna et al. 2009). In particular, for 16S rRNA gene sequences, all the type or reference strains of the recognized *Bradyrhyzobium* species are separated, with bootstrap support of 100 percent, in two distinct groups (Fig. 2.1). Group I includes *B. japonicum, B. betae, B. liaoningense, B. canariense, B. cytisi and B. yuanmingense*, while Group II strains are related to *Bradyrhizobium elkanii* USDA76^T, *B. pachyrhizi* PAC48^T and *B. jicamae*

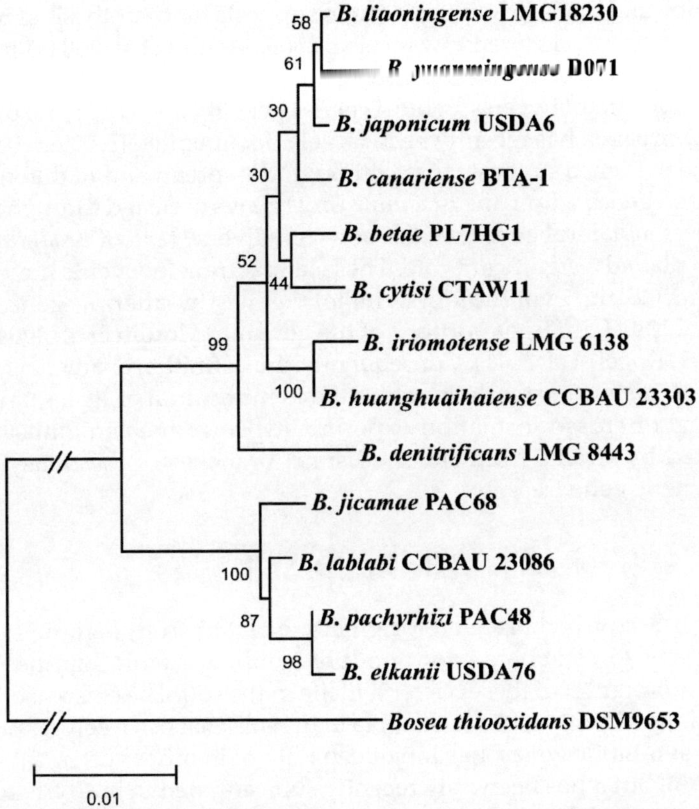

Figure 2.1 Neighbor-Joining (NJ) phylogenetic tree of 16S rRNA genes from type strains of validated *Bradyrhizobium* species showing the existence of two distinct groups. Bootstrap confidence levels (percent) are indicated at the internodes.

PAC68^T. Except for *B. elkanii* USDA76, all species within Group II have been described only recently, and it is likely that isolates from legumes, mainly shrubs, that are presently under study, such as *Lupinus mariae-josephae* and *Retama* spp. (Sánchez-Cañizares et al. 2011) may be ensembled into new species within this second group of *Bradyrhizobium*.

Symbiotic genes (nitrogen fixation and nodulation genes) have also been used to aid in *Bradyrhizobium* strain classification. Nodulation genes, particularly, constitute a unique component in the genome of rhizobia and there are no homologues present—or only distant counterparts—in other organisms. By using the *nodA* and *nodC* genes, distinct clades or groups could be differentiated among *Bradyrhizobium* strains (e.g., Moulin et al. 2004, Stepkowski et al. 2007). Surprisingly, phylogenies based on symbiotic genes display a remarkably lower diversity for *Bradyrhizobium* strains than phylogenies based on nonsymbiotic genes, and this dissimilarity between nodulation and housekeeping gene trees supports the hypothesis of multiple lateral gene transfers (Stepkowski et al. 2003, Moulin et al. 2004, Vinuesa et al. 2005a, Vinuesa et al. 2005b, Parker et al. 2006, Menna and Hungria 2011). The relationship between symbiosis gene diversity and core gene diversity in *Bradyrhizobium* has recently been largely documented (Parker 2012). The available evidence suggests that, although the spread and maintenance of nodulation genes within the *Bradyrhizobium* genus occurred through vertical transmission, lateral gene transfer between diverse taxa of *Bradyrhizobium* has also played a significant role. These lateral transfer events may as well have affected the evolution of symbiotic genes by altering genealogical relationships of different portions of the genome (Moulin et al. 2004, Sachs et al. 2011, Rogel et al. 2011). Consequently, the definition of any new species of *Bradyrhizobium* currently involves the identification of its legume hosts by means of cross-inoculation experiments, since the same host can be nodulated by strains from clearly distinct branches of the nonsymbiotic phylogenetic gene trees.

Life in the Soil

Although bradyrhizobia were originally isolated from legume nodules, their ability to establish a root-nodule symbiosis with legumes is just another adaptation of these bacteria to life in the soil. The existence of large populations of nonsymbiotic rhizobia in the soil, that can freely acquire and transfer symbiotic genes, has long been known in other systems (Sullivan et al. 1995), but it has been only recently substantiated in bradyrhizobia by the isolation and genomic characterization of the nonsymbiotic ancestral *Bradyrhizobium* sp. S23321 (Okubo et al. 2012). The scientific literature is rich in publications documenting the ability of bradyrhizobia to cope with stresses in the soil (metals, extreme pH, antibiotics), and to degrade recalcitrant organic compounds (e.g., Qu and Spain 2011). The abilities of *B. japonicum* as a denitrifier have been thoroughly characterized (Bedmar et al. 2005, Torres et al. 2011), as well as its ability to grow chemolithoautotrophically on CO_2 and H_2 (Hanus et al. 1979) or on CO_2 and thiosulphate (Masuda et al. 2010). Photosynthetic bradyrhizobia have been isolated from the stem nodules of different *Aeschynomene* species (Evans et al. 1990, Molouba et al. 1999) and the role of their photosynthetic apparatus in symbiosis, demonstrated (Giraud et al. 2000). Bradyrhizobia are motile, and contain not one, but two distinct sets of flagella (Kanbe et al. 2007), both of them relevant for competitive success in the rhizosphere (Althabeogoiti et al. 2011). Finally, quorum sensing-dependent regulation has been described in the bradyrhizobia. It is mediated by novel and unique molecules (Loh et al. 2002, Ahlgren et al. 2011, Lindemann et al. 2011) that also play a role in the establishment of symbiosis (Loh et al. 2002, Jitacksorn and Sadowsky 2008).

All of the above suggests that bradyrhizobia are endowed with a large number of metabolic capabilities that probably allow them to thrive in the changing, oligotrophic soil environments. But the true extent of these adaptations has only become apparent after the first complete genome sequences have been determined and analyzed (see under Genomics and Genetics). Bradyrhizobia contain very large genomes, and the genome of *B. japonicum* USDA110 is among the largest bacterial genomes sequenced, together with members of the Myxococcales and the Actinomycetales, all of them typical soil bacteria. Thus, it would appear that harboring large genomes—and an encompassing large encoding capacity—is a common strategy among soil bacteria. Apart from a large number of already expected clusters (see under Genomics and Genetics), several recognizable genomic islands have been identified (Kaneko et al. 2011). A large fraction of the chromosomal encoding capacity appears to be used for catabolic operons. Although the specific nature of substrates for these catabolic pathways remains to be determined, the KEGG database (http: / /www.genome.jp/ kegg/) lists a large number of methyl-accepting chemotaxis proteins (from 18 for *B. japonicum* USDA110 to 54 for *Bradyrhizobium* sp. BTAi1) and a huge number of ABC transport systems (up to 226 ABC transport permease genes both in *B. japonicum* USDA110 and *Bradyrhizobium* sp. BTAi1). This suggests that bradyrhizobia can actively swim towards, transport and degrade a large variety of exogenous compounds. Bradyrhizobia are respiratory bacteria, and they ensure that they can catabolize those compounds at any external oxygen tension by means of six to eight different and complementary terminal oxidase systems (Okubo et al. 2012), or with nitrate as terminal electron acceptor (Bedmar et al. 2005).

Symbiosis

As pointed above, bradyrhizobia were first recognized as a nitrogen-fixing root-nodule legume symbionts, and the genetic and developmental determinants for the establishment and maintenance of symbiosis have been thoroughly characterized, especially in the model *B. japonicum*—soybean symbiosis. However, bradyrhizobia establish symbioses with a very wide variety of legume plants, many of which have been poorly studied. This is, in turn, reflected by the large variety of Nod factors they form and of nodule types they induce.

Nod Factors

Bradyrhizobia-legume symbioses are initiated by molecular recognition events involving both partners. Briefly, legume roots exudate flavonoid molecules able to induce, in most cases, expression of nodulation genes (*nod*) in specific rhizobia. These genes lead to production of specific lipochito-oligosaccharides named Nod factors. Nod factors are in turn recognized by plant kinase receptors, triggering a nodule formation developmental program. A correct Nod factor chemical structure is required for induction of a particular plant response. Nod factors consist of an oligomeric backbone of β-1,4-linked *N*-acetyl-D-glucosaminyl (3–5) residues, *N*-acylated at the nonreducing terminal residue by a fatty acid (Fig. 2.2, Kamst et al. 1998). Quantitative and qualitative aspects of Nod factor population are strain specific. Differences can be found in the number of GlcNAc residues in the oligomeric backbone, in the nature of the fatty acyl group, and in the residues substituents (Table 2.2). Synthesis of Nod factors is encoded by *nod, nol* and *noe* genes. For an extensive revision on rhizobial Nod Factor structure, synthesis and roles, see Chapter 4.

The diversity of Nod factor populations produced by some bradyrhizobial strains is described by D'Haeze and Holsters (2002); for instance *B. elkanii* USDA61 produce some Nod factors where the C_1 at the reducing-terminal GlcNAc is substituted with a glyceryl (Stokkermans et al. 1996). An extensive study performed on soybean with natural and synthetic *Bradyrhizobium* Nod factors showed that a methyl fucosyl group at the reducing-terminal residue is required for nodulation, but the location of the double bond in the fatty acid is not critical (Stokkermans et al. 1995).

Recently, two different lineages of bradyrhizobia nodulating lupines from, respectively, acid and basic soils of the Iberian Peninsula have been described (Sánchez-Cañizares et al. 2011). Cross-inoculation tests showed that isolates from basic soil lupines are not compatible with acid lupines. Preliminary sequence data of a representative strain from the two lineages

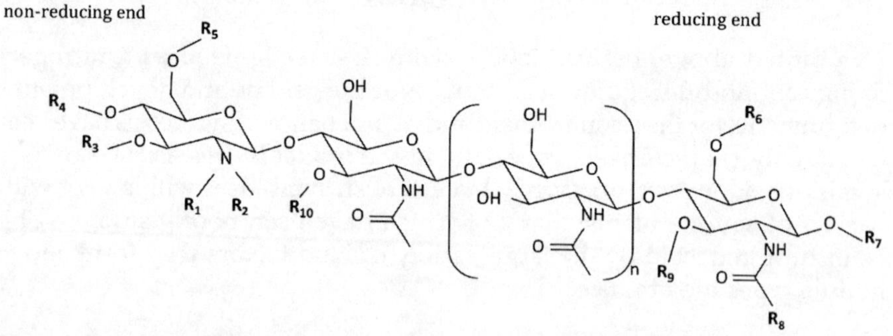

Figure 2.2 General structure of Nod factor produced by rhizobia. For details on the identity of the substitutions (R1–R10) and the oligomerization degree (n), see Table 2.2.

Table 2.2 Overview of Nod factor composition from bradyrhizobia and their specific substitutions (see Fig. 2.2).

Bradyrhizobium	**R1**	**R2**	**R3**	**R4**	**R5**	**R6**	**R7**	**R8**	**R9**	**R10**	**n**
Bradyrhizobium sp. (Acacia)	C18:1 C18:2 C16:1 C16:2	Me H	Cb H	H	Cb H	3O-S2O-MeFuc	H	H	H	H	2
Bradyrhizobium sp. (Lotus)* NZP 2309	C16:1 C16:1	Me	Cb	Cb	Cb	3O-AcFuc	H	Me	H	H	2
B. elkanii USDA61	C18:1 C16:0	Me H	Cb Ac	Cb Ac H	Cb Ac H	2O-MeFuc Fuc	Gro H	Me	H	H	1 2
B. japonicum USDA110	C18:1	H	H	H	H	2O-MeFuc	H	Me	H	H	1
B. japonicum USDA135	C18:1 C16:0 C16:1	H	H	H	Ac H	2O-MeFuc	H	Me	H	H	1

Data from D'Haeze and Holsters 2002 and (*) Bek et al. 2010; Abbreviations: Ac, acetyl; Cb, carbamoyl; Fuc, fucosyl; Gro, glycerol; Me, methyl; *n*, oligomerization degree.

indicate that the *nolL* gene, putatively encoding a Nod factor fucosyl acetyl transferase, is only present in the basic strain (our unpublished results). Production of different Nod factors could explain, at least partially, the differential symbiotic phenotype observed in cross-inoculation tests (Sánchez-Cañizares et al. 2011 and our unpublished data).

Most bradyrhizobia contain *nod* genes; however in 2007, several strains of photosyntetic bradyrhizobia isolated from stem nodules of two *Aeschynomene* species, *A. sensitiva* and *A. indica* in Africa and North America respectively, were described as lacking *nod* genes (Giraud et al. 2007). This characteristic is shared by all strains of cross-inoculation group 3 of bradyrhizobia that nodulate tropical legumes of the genus *Aeschynomene* and include nonphotosyntetic strains (Miche et al. 2010). Nod-independent symbioses were observed in all sampling areas and appeared to be linked to *Aeschynomene* host species. Nod factor–independent infection is intercellular and proceeds by crack entry, unlike infection threads generated by rhizobia Nod factor producers (Madsen et al. 2010).

Recent phylogenetic analyses based on the *recA* gene and AFLP fingerprints from 110 photosynthetic rhizobial strains showed that they form a separate cluster within the bradyrhizobia, regardless of their geographical origin. On the other hand, nonphotosynthetic, Nod-lacking strains isolated from *Aeschynomene* stem nodules form a differentiated cluster placed in an intermediate phylogenetic position between the *Aeschynomene* root-nodulating *Bradyrhizobium* spp. and the photosynthetic ones (Miche et al. 2010).

Nod Factor Receptors

Nod factors are recognized by Nod Factor Receptors (NFR), a specific class of plant receptor kinases that contain LysM (lysine motif) extracellular domains. NFRs play a major role in the legume–rhizobia recognition process which is also involved in determining host range and strain specificity (Radutoiu et al. 2007). Studies with *Lotus japonicus* and *L. pedunculatus* and their compatible symbionts, *Mesorhizobium loti* and *Bradyrhizobium* sp. (*Lotus*) respectively, showed that Nod factors from *Bradyrhizobium* sp. (*Lotus*) differ from *M. loti* Nod factors in a carbamoyl residue in position 6 of the nonreducing moiety, so differences in Nod factor structure could explain diverse symbiotic phenotypes. Further characterization showed that extracellular NFR regions from *L. japonicus* and *L. pedunculatus* were interchangeable, but complete compatibility could not be obtained in plants transformed with the heterologous receptor and inoculated with the heterologous endosymbiont. These experiments highlight that in addition to Nod factor–NFR recognition, one or more additional factors control the persistence of the symbiotic process (Bek et al. 2010).

Nodule Types and Infection Processes

Root nodules are mainly of two types, depending on whether they present a persistent meristematic activity (indeterminate nodules) or not (determinate nodules). The type of nodule formed is controlled by the host plant. Rhizobia must cross the legume root epidermis and cortex to access to the developing nodule primordia, where the bacteria are then released into the nodule cells in which they fix N_2. Rhizobial infection can occur through root hairs, cracks in the epidermis or interstitial infections between epidermal cells (Gage 2004). The great majority of nodulating legumes have indeterminate nodules and the most common mode of infection is through the formation of infection threads (Sprent 2008). It is shown in all nodulating mimosoid and caesalpinioid legumes examined so far and in over 50 percent of papilionoid legumes. Exceptions, all from papilionoids, are: determinate "desmodioid" nodules with root hair infection in phaseoloids (e.g., soybean) and *Lotus* spp.; "aeschynomenoid" nodules with crack infection in core dalbergioids (e.g., peanut); indeterminate nodules of various types (including collar nodules) with epidermal infection in the genistoids (e.g., lupine) and possibly the related genistoid tribe Crotalarieae (Doyle 2011).

Bradyrhizobia nodulate plants from all the main clades in legumes subfamilies and, interestingly, all those that deviate from the common model of indeterminate nodules produced after root hair infection. A well-known example is the *Bradyrhizobium*-soybean symbiosis, that results in determinate nodules after root hair infection. *Arachis hypogaea* (peanut) is usually nodulated by slow-growing rhizobia of the genus *Bradyrhizobium* through crack entry and without infection threads (Muñoz et al. 2011). The genus *Lupinus* (lupine), nodulated mainly by bradyrhizobia, forms a special type of indeterminate nodule known as lupinoid nodule, frequently encircling the subtending root by a collar-like structure. Entry of the bacteria into the plant root occurs at the junction between the root hair base and the adjacent epidermic cell and only sporadic "infection threads" have been observed (Fernández-Pascual et al. 2007). A case involving bradyrhizobia has been reported in which crack entry leads to the development of an indeterminate nodule, that of the Canary Islands indigenous plant *Chamaecytisus* (tagasaste). Tagasaste belongs to the genistoid group and invasion of its nodule cells by bradyrhizobia is similar to that in *Arachis* nodules (Vega-Hernández et al. 2001). Other plants in the genistoid group, such as *Genista*, also present bradyrhizobial indeterminate root nodules. Genistoid microsymbionts grouped together in a monophyletic cluster in phylogenetic trees based on nodulation loci. *Genista tinctoria* (dyer's weed) bradyrhizobia are able to establish symbioses not only with the native host, but also with other plants of the *Genisteae* tribe, such as *Lupinus luteus*, *Sarothamnus scoparius*, and *Chamaecytisus ratisbonensis*, and they were classified as the genistoid cross-inoculation group (Kalita et al. 2006). Finally, in subtropical and tropical zones of China, *Bradyrhizobium* strains have been described to nodulate *Crotalaria pallida*, a member of the Crotalarieae tribe within the "genistoid" group (Liu et al. 2007).

Genomics and Genetics

Improved understanding of the symbiotic interaction between rhizobia and their respective host legumes has implications for sustainable agricultural practices and global nitrogen cycles considered as integral components of ecosystem function. Whole genome sequencing can potentially provide new insights into the genetic features contributing to symbiotically relevant processes such as how bacteria adapt to the rhizosphere or how they enter the nodule and fix nitrogen within the host. The number of genome sequences of bacteria of the genus *Bradyrhizobium* available in public databases is constantly increasing. To date, the genome structures of six symbiotic bradyrhizobia (two of them are still in high-quality draft status) and of a nonsymbiotic isolate, have been determined. The main features of the five complete available genomes sequenced are listed in Table 2.1.

At the genomic level, two features of bradyrhizobia stand out: they contain a very large number of genes, and are encoded in a single, very large chromosome, although recognizable chromosomal islands and even plasmids can be present as well. As already mentioned, the large number of genes appears to be a common feature of soil bacteria, apparently related to their need to respond to a very complex and variable environment. It can be hypothesized that the large bradyrhizobial chromosome imposes a burden on the replication machinery, and that it is responsible for the observed longer generation times in these bacteria. This genome organization contrasts with that of other rhizobia, such as members of the genus *Rhizobium*, that contain a multipartite genome with large plasmids. Keeping the chromosome small and providing additional origins of replication should allow faster replication and reduced generation times, as it is indeed observed in the "fast" rhizobia. At the same time, by stabilizing a new set of advantageous genes in plasmids, multipartite genomes may enhance the bacterium's metabolic or symbiotic capabilities (MacLean et al. 2007). On the other hand, keeping all essential genes in one replicon provides a measure of genomic stability that may be advantageous in complex environments. The presence and role of accessory plasmids in *Bradyrhizobium* has been poorly documented; Cytryn et al. (2008) carried out a screening by PFGE-CHEF of 46 *Bradyrhizobium* strains from different geographic locations. Only 11 strains, five belonging to *B. japonicum* and six to *B. elkanii*, harbored plasmids of varying sizes.

Bradyrhizobium japonicum is agriculturally the most important rhizobial species, in view of its ability to form root nodules on soybeans (*Glycine max*). *B. japonicum* USDA110 was the first *Bradyrhizobium* to be completely sequenced (Kaneko et al. 2002). It is one of the largest bacterial genomes, and contains 8,317 potential protein-coding genes in a single chromosome (Table 2.3).

Two photosynthetic bradyrhizobia, strains ORS278 and BTAi1, have also been completely sequenced. Photosynthetic bradyrhizobia can be isolated from nodules that develop in the stems of different *Aeschynomene* species. Although large predicted coding sequences (6,752 and 7,729 respectively), the genomes of strains ORS278 and BTAi1 are smaller than *B. japonicum* USDA110 genome. Strain BTAi1 contains a single, circular plasmid of 228,826 bp, with a lower G+C content (60.7 percent) than the chromosome (64.9 percent) (Giraud et al. 2007, Cytryn et al. 2008). This is the only plasmid found so far in photosynthetic bradyrhizobia, as a search for plasmids by PFGE and *repABC* analyses in seven additional strains was negative (Mornico et al. 2012).

The genome of *B. japonicum* strain USDA6^T has also been sequenced. As this is the type strain of the species, and *B. japonicum* is the type species of the genus *Bradyrhizobium*, analysis of this genome is relevant for genus definition. Previous comparative genomic hybridization (CGH) analysis along with phylogenetic analysis based on the internal transcribed spacer region of ribosomal DNA sequence (ITS) had already revealed genomic variations between USDA6^T and USDA110. Consistent with this, and although genome sizes, G+C contents and numbers of protein-encoding genes are similar (Table 2.3), detailed comparative analyses between both strains indicate that USDA110 and USDA6^T are more divergent than previously thought (Kaneko et al. 2011).

Finally, the most recently published *Bradyrhizobium* genome is that of strain S23321, a nonsymbiotic, oligotrophic bacterium isolated from a paddy field soil in Japan. It has a smaller single circular chromosome of 7,231,841 bp (Okubo et al. 2012). Although phylogenetically close to *B. japonicum* USDA110, comparative analyses suggest that this bacterium contains an ancestral-type genome previous to the acquisition of a symbiosis island by horizontal gene transfer.

The rapid generalization of affordable DNA sequencing technologies has led to a widespread appearance of additional genomic efforts. This is the case of *B. japonicum* strain CPAC15, a highly competitive commercial strain applied to soybean crops established on several million ha in developing countries (Godoy et al. 2008). A total of 4,328 shotgun reads, resulting in 2,046,740 bp, covered approximately 13 percent of the genome with 1,371 coding DNA sequences identified (CDSs). Also, *Bradyrhizobium* sp. WSM471 (7,784,016 bp; NZ_AHLW00000000.1) and WSM1417 (8,030,181

Table 2.3 General features of *Bradyrhizobium* complete genomes.

	B. japonicum	***B. japonicum***	***Bradyrhizobium* sp.**	***Bradyrhizobium* sp.**	***Bradyrhizobium* sp.**
	USDA110	USDA6^T	ORS278	BTAi1	S23321
Genome size (bp)	9,105,828	9,207,384	7,456,587	8,493,515	7,231,841
G+C content (%)	64.1	63.7	65.5	64.9	64.3
Ribosomal RNA operons	1	2	2	2	1
Transfer RNAs	50	51	50	52	45
Total protein-coding genes	8,317	8,829	6,752	7,729	6,898
Plasmid no. (size in bp)	0	0	0	1 (228,826)	0
Accession No.	BA000040	AP012206	CU234118	CP000494 CP000495	AP012279
Reference	Kaneko et al. (2002)	Kaneko et al. (2011)	Giraud et al. (2007)	Giraud et al. (2007)	Okubo et al. (2012)

bp; PRJNA61989) are part of the GEBA-RNB whole genome sequencing program carried out by the DOE Joint Genome Institute, which nowadays comprises 40 *Bradyrhizobium* genome projects (http://genome.jgi-psf.org/genome-projects/pages/projects.jsf), some of them already classified as high-quality draft. Draft genome sequences of seven new *Aeschynomene* symbionts representing phylogenetically diverse bradyrhizobial strains have also been obtained in a recent work (Mornico et al. 2012). Comparative genomics allowed inferring the core, pan- and variable-genomes of these bradyrhizobia. The core genome was composed of 3,663 genes, of which 67 percent had an annotated function; 9,395 genes were found in the variable genome, reflecting a high rate of horizontal gene transfer and a great metabolic versatility among bradyrhizobia. Finally, pangenome gene pool estimates increased dramatically with each new addition of a bacterial genome. It currently stands at 20,162 for the *Bradyrhizobiaceae* family, and has not yet reached a plateau (Mornico et al. 2012), reflecting the large genomic diversity within this clade.

A large variation in genome sizes of *Bradyrhizobium* strains, a low level of synteny and the presence of numerous mobile genetic elements indicate that bradyrhizobial genomes are highly plastic. The ORS278 and BTAi1 genomes contain 21 and 29 putative horizontally acquired genomic islands, respectively, which confer functional advantages in the adaptation of these bacteria to their symbiotic or free-living ecological niches (Giraud et al. 2007). In *B. japonicum* USDA110, 60 percent of all transposase genes (100 of 167) are located within the symbiotic island, which represents 7.5 percent of the entire genome (Kaneko et al. 2002). When the proteomes of the five complete sequenced strains are compared (see Fig. 2.3), only 63.5 percent of the gene families present in the proteome of both *B. japonicum* strains shared an identity above 50 percent, while this value dropped to 59.2 percent when the photosynthetic strains were compared. These are the highest values found in these comparisons, given that they decrease to 35–38 percent when the *B. japonicum* proteomes are compared to any of the photosynthetic ones. Nonsymbiotic *Bradyrhizobium* strain S23321 is closer to USDA110 and USDA6^T (55–56 percent) than to ORS278 and BTAi1 strains (40–44 percent). Such large and heterogeneous genomes, characteristic of rhizobia, may, as already pointed out above, reflect the necessity for a large inventory of genes to maximize growth and survival in the complex environment of the soil, together with the additional genetic requirement imposed by the plant host (MacLean et al. 2007). This is the case of genes involved in: solute uptake, importing and exporting molecules, the production of cell-surface polysaccharides or nitrogen metabolism (Downie and Young 2001).

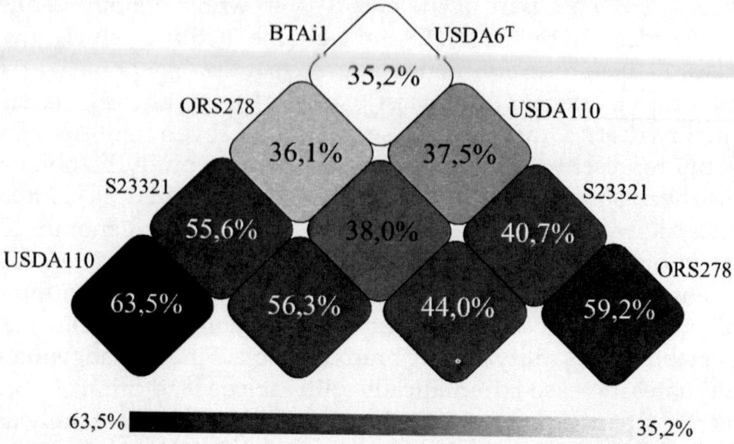

Figure 2.3 Pairwise homology between proteomes of complete sequenced *Bradyrhizobium* strains. The figure represents the relative percentage of gene families preserved between any two *Bradyrhizobium* strains, calculated according to Binnewies et al. (2005) and Ussery et al. (2009). Gray intensities are proportional to the percentage of gene families conserved. Strains used are those from Table 2.3.

Transport and Secretion Systems

Genes encoding transport systems constitute the largest (9 percent) class of genes in the *B. japonicum* genome, and most of them are ABC transporters. Regarding secretion systems, which in symbiotic bacteria may be involved in fine-tuning of host specific nodulation (see Chapter 9 for a review), gene homologues of the *ctpACDFGHI, pilA* type IV secretion system were found in *B. japonicum* USDA110 outside of symbiosis islands, along with genes for the *rhcC1C2JNQRSTUV* type III secretion system within the symbiosis island (Kaneko et al. 2002). The *rhc* system is also positioned at a single locus within the symbiosis island in strain USDA6^T (Kaneko et al. 2011). Homologues of *trbBCDEJLFGI, traG* and *traF* genes, essential for conjugative transfer of Ti-plasmid in *A. tumefaciens*, and which may play a role in the mobilization of the symbiosis island in *B. japonicum*, are also present in USDA110, but not in USDA6^T (Kaneko et al. 2002, 2011). If the *trb* genes are functionally related to T4SS, their deletion in USDA6^T might cause differences in host range between USDA6^T and USDA110 (Kaneko et al. 2011). ORFs encoding VirB1 to VirB11 components of the TraG/VirD4 family type IV secretion system have also been detected in the accessory plasmid of BTAi1 (Cytryn et al. 2008).

Regulatory Proteins

The need for versatile microorganisms to be able to colonize soil and water, as well as plant and animal tissues, and to adapt to a wide-range of environmental conditions determine their need for a wide array of transcriptional regulatory systems. Regulatory genes can represent up to 8 percent of the genome, decreasing to about 3 percent or as low as 1 percent in specialized pathogens (Stover et al. 2000). In *B. japonicum* they make up a substantial fraction of the genome, up to 6.8 percent of the total protein-encoding genes in USDA110 (Kaneko et al. 2002) and up to 5.9 percent in CPAC15 (Godoy et al. 2008).

Bacterial Adhesion and Surface Structural Elements

Major extracellular polysaccharides present in *B. japonicum* cells are cyclic β-glucan, lipopolysaccharide (LPS), and exopolysaccharide (EPS). Genes for synthesis of cyclic β-glucan and LPS are conserved in both *B. japonicum* genomes and also in *Bradyrhizobium* sp. BTAi1, and are similarly arranged in each gene cluster. However, most of the genes involved in EPS synthesis, which are conserved in USDA6^T and USDA110, are not found in *Bradyrhizobium* sp. BTAi1 (Kaneko et al. 2002, Kaneko et al. 2011). Regarding genes involved in degradation of plant cell walls during the infection process, *B. japonicum* USDA110 presents genes such as a putative polygalacturonase (*blr1993*), pectinesterase (*blr1994*), ligninase (*ligE; bll7427*), cellulase (*blr3367*), and endo-1,4-β-xylase (*bll2241*) (Kaneko et al. 2002).

A significant number of genes for flagella formation and motility, including those for chemotaxis, are present in most bradyrhizobia. Generally, two sets of flagellar genes compactly clustered in two separate regions are found, one for the lateral flagella and a second cluster for the polar flagellum. This is the case of USDA110 and S23321 (Okubo et al. 2012). However, photosynthetic strains ORS278 and BTAi1 only contain the complete cluster for polar flagella.

Metabolic Pathways

All the genes for carbon dioxide fixation enzymes are present in *B. japonicum* USDA110, including those related to the Calvin-Benson-Bassham (CBB) cycle (*cbbRcbbFPTALSX: bll2580–blr2587*), a PEP carboxylase gene (*blr2955*) and a carbonic anhydrase gene (*bll2065*). These genes might be required for chemoautotrophic growth of *B. japonicum* USDA110 in the presence of hydrogen gas as an energy source (Kaneko et al. 2002).

Strains BTAi1 and ORS278 display an extremely versatile carbon and energy metabolism, as they can grow as heterotrophs, autotrophs or phototrophs. Among their 1,343 common genes, photosynthetic genes are clustered in a 50-kb region designated as the photosynthesis gene cluster (PGC). The organization of this cluster is identical in both strains, and is highly conserved with the PGC of their closest photosynthetic neighbor, *Rhodopseudomonas palustris* (Giraud et al. 2007). *Bradyrhizobium* sp. S23321 also harbors a PGC nearly identical to that of *R. palustris*, although its gene arrangement differs from ORS278 and BTAi1. Strains S23321 would synthesize the simplest type of photosynthetic apparatus, the light-harvesting 1 (LH1) complex (Okubo et al. 2012). Strains ORS278 and BTAi1 also contain three putative bacteriophytochromes and two *kaiBC* circadian clock operons (Giraud et al. 2007). In addition, the BTAi1 genome presents a genomic island containing all the genes required for CO_2 fixation and a separate hydrogenase gene cluster. Complete sets of hydrogenase genes are present in *B. japonicum* USDA110 and CPAC15 (Giraud et al. 2007, Kaneko et al. 2002, Godoy et al. 2008). These uptake hydrogenase systems scavenge hydrogen produced during nitrogen fixation (Imperial et al. 2006). Surprisingly, an extra set of hydrogen uptake genes was found in USDA110 outside the symbiosis island (Kaneko et al. 2002). However, some *B. japonicum* strains, including USDA6^T do not harbor genes for an uptake hydrogenase (Kaneko et al. 2011) and the plasmid-borne hydrogenase described in strain BTAi1 does not seem to be significantly involved in symbiotic growth (Cytryn et al. 2008).

Nodulation and Nitrogen Fixation

Most rhizobial genomes characterized to date have nodulation and nitrogen fixation genes clustered either on plasmids or in large chromosomal symbiotic islands, whose genes are required for the establishment and maintenance of diazotrophic symbiosis with the host plant. This is the case of *B. japonicum* USDA110, which harbors a 681-kb island containing 655 putative protein-coding genes that include those related to symbiotic nitrogen fixation and DNA transmission. As shown in Fig. 2.4, its lower G+C content and its different codon usage, as compared with the rest of the genome, suggests an alien origin (Kaneko et al. 2002). A similar situation was found in the USDA6^T genome, with a conserved 695-kb DNA region with lower G+C content that includes most of the *nod, nif,* and *fix* genes (Fig. 2.4, Kaneko et al. 2011). In contrast, no equivalent islands were found in the chromosomes of strains ORS278 and BTAi1 (Giraud et al. 2007), which would explain why this lower G+C region is not conserved when compared to both *B. japonicum* strains (Fig. 2.4). Notably, *nodABC* genes, encoding Nod factors, which are present in all rhizobia characterized thus

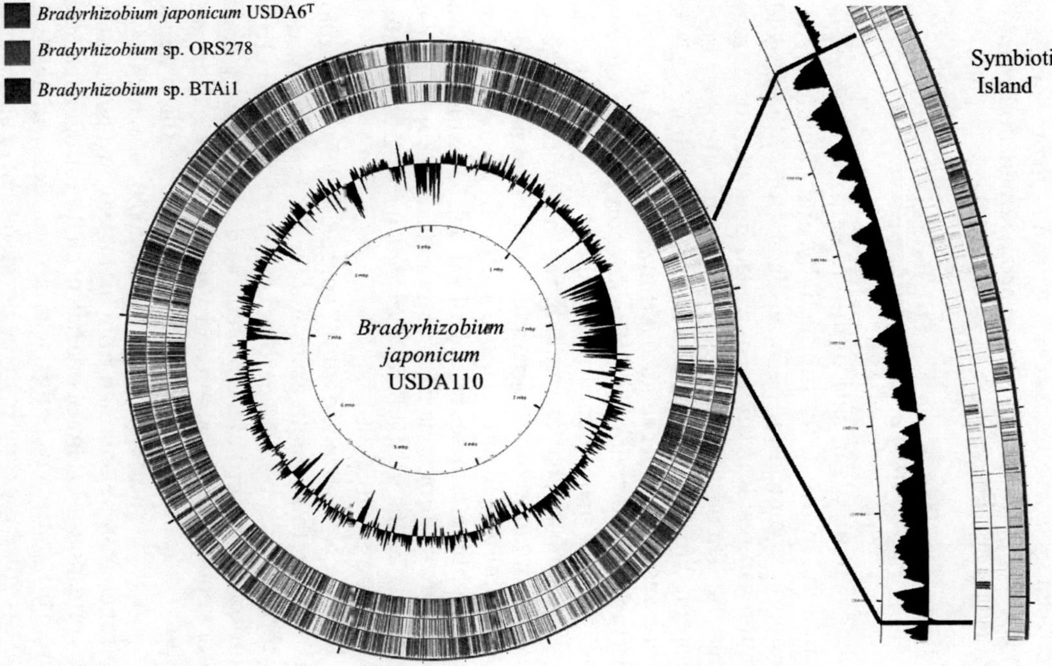

Figure 2.4 Comparative Genome Atlas of *Bradyrhizobium japonicum* USDA110. The comparative genome atlas was constructed using CGview software (Grant and Stothard 2008) to map pairwise blastN alignments between *Bradyrhizobium* strains. The chromosome of *B. japonicum* USDA110 (genome positions in Mb represented in the innermost circle) was compared to *B. japonicum* USDA6^T (red), *Bradyrhizobium* sp. ORS278 (green) and *Bradyrhizobium* sp. BTAi1 (blue). The average G+C content is shown in the second circle. The region harboring the symbiotic island was zoomed 6x.

Color image of this figure appears in the color plate section at the end of the book.

far, are missing in the genomes of these photosynthetic *Bradyrhizobium* strains. This indicates that these rhizobia use an alternative pathway to initiate symbiosis, and the involvement of a purine precursor, acting as a signal molecule during infection of *Aeschynomene* stems, has been reported (Giraud et al. 2007). As expected, no symbiotic island was detected in the ancestral genome of *Bradyrhizobium* sp. S23321. However, strong colinearity was observed between the adjacent sequences surrounding the symbiotic island in USDA110, suggesting a genome rearrangement following its insertion into the USDA110 genome (Okubo et al. 2012).

Agricultural Relevance

Nitrogen plays an essential role in agriculture. Together with water, it is one of the most limiting nutrients for productivity, and farming systems require the addition of chemical nitrogen fertilizers for optimal yields. External fertilization can result in an uncontrolled liberation of nitrogen into the environment. Therefore, effectively managing nitrogen in the environment is essential for any agricultural sustainability strategy. In the current scenario of excessive use of chemical fertilizers, with high economic and environmental costs, the development of ecofriendly and cost-effective agricultural practices is a major concern. For these reasons, there is a renewed interest in organisms that carry out nitrogen fixation and have a major role in the accumulation of nitrogen in microbial biomass and soil organic matter. Biologically fixed nitrogen is used directly by the plant, and therefore it is less susceptible to losses by volatilization, denitrification and leaching, especially when nitrogen is fixed symbiotically, as it is the case with bacteria belonging to the genus *Bradyrhizobium* (as well as the other rhizobia) and their legume symbionts.

Worldwide, legume crops are grown on approximately 250 million ha. Their importance, in terms of both grain and forage, is highlighted by the fact that this corresponds to 12–15 percent of the arable land on Earth and that they contribute one-third of the protein needs of humanity under minimal subsistence conditions (Graham and Vance 2003). It is estimated that, globally, legume crops require an input of about 100 million tons of nitrogenous nutrients (total nutrient; FAOstat data 2009), of which 44–66 million tons of nitrogen may be obtained from biological nitrogen fixation (BNF). This figure corresponds to almost one half of all the nitrogen fertilizer used in agriculture. Thus, nitrogen-fixing bacteria associated with legumes have the potential to contribute greatly to the development of agricultural practices that respect the environment and are economically more viable, especially in developing countries.

In most cases, rhizobium-legume symbioses fix more nitrogen than the plant can use, and this excess nitrogen is left in the soil and made available for other plants or later crops. Due to this fact, intercropping with a legume, an ancient agricultural practice, has the potential to decrease the need for applied fertilizer. It has also been observed that incorporation of plant growth promoting rhizobacteria (PGPR; see Chapter 13) to *Bradyrhizobium* inoculants (Nazmun et al. 2009, Argaw 2012) results in increased plant nodulation and nitrogen fixation under normal growth conditions (Li and Alexander 1988).

Three grain legume crops nodulated by bradyrhizobia are especially relevant, because of their global or local importance: soybeans, peanuts and lupines.

Soybeans

Soybean, *Glycine max* (L.) Merr., a subtropical member of the *Leguminosae* family, is one of the most important oil seed crops in the world. Its seeds contain 18 to 22 percent oil, highly desirable in diets, and 40 to 42 percent protein; hence it is one of the best sources for protein and oil (Ali 2010). The main producing countries (FAOStat data 2010) are the USA (90 million tons), Brazil (70 million tons) and Argentina (50 million tons). Soybean depends on its symbionts to satisfy its nitrogen requirements, and up to 450 kg N fixed/ha have been estimated (Argaw 2012). Three different species of *Bradyrhizobium* nodulate soybean. These include *B. japonicum* (Jordan 1982), *B. elkanii* (Kuykendall et al. 1992), and the extra slow growing bacteria *B. liaoningense* (Xu et al. 1995). All these species of *Bradyrhizobium* nodulating soybean belong to the biovar *glycinearum* (Rogel et al. 2011).

In the USA, Argentina and Brazil, soybean is an exotic plant species, and symbiotic bradyrhizobia had to be introduced together with the soybean crops through inoculation. The case of Brazil exemplifies an agricultural success story linked to the development of bradyrhizobial inoculant technology. Soybean and its products constitute Brazil's foremost agricultural export, grown on over 22 million hectares, about 45 percent of the country's cropped land (Hungria et al. 2006), and the country produces 20 percent of the world's soybeans (32.7 × 10^6 Mg), placing Brazil as the second soybean producer after the USA (75.4 × 10^6 Mg). Currently, farmers are able to obtain soybean yields in excess of 4,000 kg/ha, and this amount of grain contains between 250 and 280 kg N/ha (Hungria et al. 2006). Although contributions of BNF to the plant nitrogen economy have been estimated as amounting to between 30 and 94 percent (Hungria and Campo 2004), most modern Brazilian soybean varieties derive almost all of their protein nitrogen from BNF, which represents an economic benefit in terms of N-fertilizer savings of over US$ 2.5 billion per year. The BNF system is so efficient that attempts to increase grain yields by addition of nitrogen fertilizer are hardly ever successful if the plants have been effectively inoculated with the recommended *Bradyrhizobium* strains.

Soybean was introduced in Brazil 125 yr ago. Brazilian soils were originally devoid of rhizobia able to effectively nodulate soybean, but massive inoculation with a few strains used in commercial inoculants during the last decades resulted in the establishment of resident populations of *B. japonicum.* Strain-selection programs for soybean were established in Brazil with the first commercial plantings in the early 1960s. *B. japonicum* SEMIA566, was one of the first strains employed. It belongs to the same serogroup as USDA123, which is recognized as the most competitive in the USA (Weber et al. 1989), and has greatly contributed to the successful establishment of the crop in southern Brazil (Hungria et al. 2006). Currently, inoculants in Brazil contain a mixture with either *B. elkanii* SEMIA5019 or SEMIA587. In Argentina, soils are extensively inoculated with strain E109, a derivative of USDA138 (Melchiorre et al. 2011). In the USA, a recent study showed that soybeans nodulated with the highly effective strain USDA122 fixed 29 percent more N_2 and produced 31 percent more grain than plants nodulated with naturalized soil bradyrhizobia (Herridge and Roseb 2000).

Peanuts

Peanut (*Arachis hypogaea* L.) is an agriculturally valuable plant original from South America. In the global context (24 million ha, 37.6 x 10^6 Mg), China leads the production worldwide (about 37.5 percent of the overall world production), followed by India (19 percent) and Nigeria (11 percent). In the Americas, USA, Argentina and Brazil are the major producers (FAOStat data 2009).

Peanuts are able to establish symbiotic relationships with *Bradyrhizobium* sp. (Urtz and Elkan 1996), reducing their need for nitrogen fertilizers. Cell surface components in general, and the characteristic membrane lipid phosphatidylcholine in particular, play a fundamental role in the symbiotic interaction (Medeot et al. 2010), and this is related to the particular "crack entry" infection process (see above Nodule types and infection processes). *Bradyrhizobium* sp. SEMIA6144 is a slow-growing strain recommended for inoculating peanut roots (Gomes-Germano et al. 2006).

Lupines

Lupines, especially *Lupinus albus* L., are favored as a crop for use in sustainable and environment friendly agriculture because of their high potential for both biological nitrogen fixation, which may contribute up to 150–200 kg N/ha to the soil for use by future crops (Van Santen et al. 1993), and provision of nutritious food and feed (32–38 percent protein). The high nitrogen-fixation potential of lupine makes it ideally suited for low-input agriculture because mineral nitrogen fertilizer may not be needed at all for crop production. Symbiotic partners of lupines usually belong to the genus *Bradyrhizobium* (Barrera et al. 1997).

Acknowledgments

Supported by grants from Fundación BBVA, Comunidad de Madrid (Programa Microambiente S2009/AMB-1551) and AECID (A1/038234/11), all to TRA, MICINN (CGL2011-26932) to JI, and UPM (AL12-0210-01) to LR.

References

Ahlgren, N.A., C.S. Harwood, A.L. Schaefer, E. Giraud and E.P. Greenberg. 2011. Aryl-homoserine lactone quorum sensing in stem-nodulating photosynthetic bradyrhizobia. Proc. Nat. Acad. Sci. U.S.A. 108: 7183–7188.

Ali, N. 2010. Soybean processing and utilization. *In:* G. Singh [ed.]. The soybean botany, production and uses. CABI International, Wallingford, UK. pp. 345–374.

Althabegoiti, M.J., J.M. Covelli, J. Perez-Gimenez, J.I. Quelas, E.J. Mongiardini, M.F. Lopez, S.L. Lopez-Garcia and A.R. Lodeiro. 2011. Analysis of the role of the two flagella of *Bradyrhizobium japonicum* in competition for nodulation of soybean. FEMS Microbiol. Lett. 319: 133–139.

Argaw, A. 2012. Evaluation of co-inoculation of *Bradyrhizobium japonicum* and phosphate solubilizing *Pseudomonas* spp. Effect on soybean (*Glycine max* L. (Merr.)) in Assossa Area. J. Agric. Sci. Technol. 14: 213–224.

Auch, A.F., M. von Jan, H.P. Klenk and M. Goker. 2010. Digital DNA-DNA hybridization for microbial species delineation by means of genome-to-genome sequence comparison. Stand. Genomic. Sci. 2: 117–134.

Barrera, L.L., M.E. Trujillo, M. Goodfellow, F.J. Garcia, I. Hernandez-Lucas, G. Davila, P. Van Berkum and E. Martinez-Romero. 1997. Biodiversity of bradyrhizobia nodulating *Lupinus* spp. Int. J. Syst. Bacteriol. 47: 1086–1091.

Bedmar, E.J., E.F. Robles and M.J. Delgado. 2005. The complete denitrification pathway of the symbiotic, nitrogen-fixing bacterium *Bradyrhizobium japonicum*. Biochem. Soc. Trans. 33: 141–144.

Bek, A.S., J. Sauer, M.B. Thygesen, J.O. Duus, B.O. Petersen, S. Thirup, E. James, K.J. Jensen, J. Stougaard and S. Radutoiu. 2010. Improved characterization of *nod* factors and genetically based variation in LysM receptor domains identify amino acids expendable for nod factor recognition in *Lotus* spp. Mol. Plant Microbe Interact. 23: 58–66.

Binnewies, T.T., P.H. Hallin, H.H. Staerfeldt and D.W. Ussery. 2005. Genome update: proteome comparisons. Microbiology-SGM. 151: 1–4.

Chahboune, R., L. Carro, A. Peix, S. Barrijal, E. Velazquez and E. J. Bedmar. 2011. *Bradyrhizobium cytisi* sp. *nov.*, isolated from effective nodules of *Cytisus villosus*. Int. J. Syst. Evol. Microbiol. 61: 2922–2927.

Chang, Y.L., J.Y. Wang, E.T. Wang, H.C. Liu, X.H. Sui and W.X. Chen. 2011. *Bradyrhizobium lablabi* sp. *nov.*, isolated from effective nodules of *Lablab purpureus* and *Arachis hypogaea*. Int. J. Syst. Evol. Microbiol. 61: 2496–2502.

Coenye, T., D. Gevers, Y. Van de Peer, P. Vandamme and J. Swings. 2005. Towards a prokaryotic genomic taxonomy. FEMS Microbiol. Rev. 29: 147–167.

Cooper, J.E. and E.J. Feil. 2004. Multilocus sequence typing—what is resolved? Trends Microbiol. 12: 373–377.

Cytryn, E.J., S. Jitacksorn, E. Giraud and M.J. Sadowsky. 2008. Insights learned from pBTAi1, a 229-kb accessory plasmid from *Bradyrhizobium* sp. strain BTAi1 and prevalence of accessory plasmids in other *Bradyrhizobium* sp. strains. ISME J. 2: 158–170.

D'Haeze, W. and M. Holsters. 2002. Nod factor structures, responses, and perception during initiation of nodule development. Glycobiology. 12: 79R–105R.

Downie, J.A. and J.P. Young. 2001. Genome sequencing. The ABC of symbiosis. Nature. 412: 597–598.

Doyle, J.J. 2011. Phylogenetic perspectives on the origins of nodulation. Mol. Plant Microbe Interact. 24: 1289–1295.

Evans, W.R., D.E. Fleischman, H.E. Calvert, P.V. Pyati, G.M. Alter and N.S. Rao. 1990. Bacteriochlorophyll and photosynthetic reaction centers in *Rhizobium* strain BTAi 1. Appl. Environ. Microbiol. 56: 3445–3449.

Fernández-Pascual, M., J.J. Pueyo, M.-R. de Felipe, M.P. Golvano and M.M. Lucas. 2007. Singular features of the *Bradyrhizobium-Lupinus* symbiosis. *In*: J.A.Teixeira da Silva [ed.]. Dynamic soil, dynamic plant, vol 1. Global Sciences Books, Japan. pp 1–16.

Gage, D.J. 2004. Infection and invasion of roots by symbiotic, nitrogen-fixing rhizobia during nodulation of temperate legumes. Microbiol. Mol. Biol. Rev. 68: 280–300.

Giraud, E., L. Hannibal, L. Fardoux, A. Vermeglio and B. Dreyfus. 2000. Effect of *Bradyrhizobium* photosynthesis on stem nodulation of *Aeschynomene sensitiva*. Proc. Nat. Acad. Sci. U.S.A. 97: 14795–14800.

Giraud, E., L. Moulin, D. Vallenet, V. Barbe, E. Cytryn, J.C. Avarre, M. Jaubert, D. Simon, F. Cartieaux, Y. Prin et al. 2007. Legumes symbioses: Absence of Nod genes in photosynthetic bradyrhizobia. Science. 316: 1307–1312.

Godoy, L.P., A.T.R. Vasconcelos, L.M.O. Chueire, R.C. Souza, M.F. Nicolas, F.G. Barcellos and M. Hungria. 2008. Genomic panorama of *Bradyrhizobium japonicum* CPAC 15, a commercial inoculant strain largely established in Brazilian soils and belonging to the same serogroup as USDA 123. Soil Biol. Biochem. 40: 2743–2753.

Gomes-Germano, M., P. Menna, F.L. Mostasso and M. Hungria. 2006. RFLP analysis of the rRNA operon of a Brazilian collection of bradyrhizobial strains from 33 legume species. Int. J. Syst. Evol. Microbiol. 56: 217–29.

Graham, P.H. and C.P. Vance. 2003. Legumes: importance and constraints to greater use. Plant Physiol. 131: 872–7.

Grant, J.R. and P. Stothard. 2008. The CGView Server: A comparative genomic tool for circular genomes. Nucleic Acids Res. 36: W181–W184.

Gulati, S.L., S.K. Mishra, N. Gulati and M.C. Tyagi. 2001. Effect of inoculation of plant growth promoting rhizobacteria on cowpea. Ind. J. Microbiol. 41: 223–224.

Gyaneshwar, P., G. Naresh Kumar, L.J. Parekh and P.S. Poole. 2002. Role of soil microorganisms in improving P nutrition of plants. Plant Soil 245: 83–93.

Hanus, F.J., R.J. Maier and H.J. Evans. 1979. Autotrophic growth of H_2-uptake-positive strains of *Rhizobium japonicum* in an atmosphere supplied with hydrogen gas. Proc. Nat. Acad. Sci. U.S.A. 76: 1788–1792.

Herridge, D. and I.R. Roseb. 2000. Breeding for enhanced nitrogen fixation in crop legumes. Field Crops Res. 65: 229–248.

Hungria, M. and R.J. Campo. 2004. Economical and environmental benefits of inoculation and biological nitrogen fixation with the soybean: situation in South America. *In*: F. Moscardi, C.B. Hoffmann-Campo, O.F. Saraiva, P.R. Galerani, F.C. Krzyzanowski and M.C. Carrão-Panizzi [eds.]. Proceedings of the VII World Soybean Research Conference. Embrapa, Londrina, Brasil. pp. 488–498.

Hungria, M., R.J. Campo, I.C. Mendes and P.H. Graham. 2006. Contribution of biological nitrogen fixation to the N nutrition of grain crops in the tropics: the success of soybean

(*Glycine max* L. Merr.) in South America. *In*: R.P. Singh, N. Shankar and P. K. Jaiwal [eds.]. Nitrogen nutrition and sustainable plant productivity. Studium Press, Houston, Texas. pp. 43–93.

Imperial, J., J.M. Palacios, B. Brito, L. Rey, E. Cabrera, A.C. Ureta and T. Ruiz-Argüeso. 2006. El sistema hidrogenasa de las bacterias que nodulan leguminosas. *In*: E.J. Bedmar, J. González, C. Lluch and B. Rodelas [eds.]. Fijación de Nitrógeno: Fundamentos y Aplicaciones. SEFIN-Universidad de Granada, Granada, Spain. pp. 35–44.

Islam, M.S., H. Kawasaki, Y. Muramatsu, Y. Nakagawa and T. Seki. 2008. *Bradyrhizobium iriomotense* sp. *nov.*, isolated from a tumor-like root of the legume *Entada koshunensis* from Iriomote Island in Japan. Biosci. Biotechnol. Biochem. 72: 1416–1429.

Jitacksorn, S. and M.J. Sadowsky. 2008. Nodulation gene regulation and quorum sensing control density-dependent suppression and restriction of nodulation in the *Bradyrhizobium japonicum*-soybean symbiosis. Appl. Environ. Microbiol. 74: 3749–3756.

Jordan, D.C. 1982. Transfer of *Rhizobium japonicum* Buchanan 1980 to *Bradyrhizobium gen. nov.*, a genus of slow-growing, root nodule bacteria from leguminous plants. Int. J. Syst. Bacteriol. 32: 136–139.

Kalita, M., T. Stepkowski, B. Lotocka and W. Malek. 2006. Phylogeny of nodulation genes and symbiotic properties of *Genista tinctoria* bradyrhizobia. Arch. Microbiol. 186: 87–97.

Kamst, E., H.P. Spaink and D. Kafetzopoulos. 1998. Biosynthesis and secretion of rhizobial lipochitin-oligosaccharide signal molecules. *In*: B.B. Biswas and H.K. Das [eds.]. Plant-microbe interactions-subcellular biochemistry. Plenum Press, New York. pp. 29–71.

Kanbe, M., J. Yagasaki, S. Zehner, M. Goettfert and S.I. Aizawa. 2007. Characterization of two sets of subpolar flagella in *Bradyrhizobium japonicum*. J. Bacteriol. 189: 1083–1089.

Kaneko, T., Y. Nakamura, S. Sato, K. Minamisawa, T. Uchiumi, S. Sasamoto, A. Watanabe, K. Idesawa, M. Iriguchi, K. Kawashima et al. 2002. Complete genomic sequence of nitrogen-fixing symbiotic bacterium *Bradyrhizobium japonicum* USDA110. DNA Res. 9: 189–197.

Kaneko, T., H. Maita, H. Hirakawa, N. Uchiike, K. Minamisawa, A. Watanabe and S. Sato. 2011. Complete genome sequence of the soybean symbiont *Bradyrhizobium japonicum* strain USDA6^T. Genes. 2: 763–787.

Kuykendall, L.D., B. Saxena, T.E. Devine and S.E. Udell. 1992. Genetic diversity in *Bradyrhizobium japonicum* Jordan 1982 and a proposal for *Bradyrhizobium elkanii* sp. *nov.* Can. J. Microbiol. 38: 501–505.

Kwon, S. W., J. Y. Park, J. S. Kim, J. W. Kang, Y. H. Cho, C. K. Lim, M. A. Parker and G. B. Lee. 2005. Phylogenetic analysis of the genera *Bradyrhizobium*, *Mesorhizobium*, *Rhizobium* and *Sinorhizobium* on the basis of 16S rRNA gene and internally transcribed spacer region sequences. Int. J. Syst. Evol. Microbiol. 55: 263–270.

Li, D.M. and A. Alexander. 1988. Co-inoculation with antibiotic-producing bacteria to increase colonization and nodulation by rhizobia. Plant Soil. 108: 211–219.

Lindemann, A., G. Pessi, A.L. Schaefer, M.E. Mattmann, Q.H. Christensen, A. Kessler, H. Hennecke, H.E. Blackwell, E.P. Greenberg and C.S. Harwood. 2011. Isovaleryl-homoserine lactone, an unusual branched-chain quorum-sensing signal from the soybean symbiont *Bradyrhizobium japonicum*. Proc. Nat. Acad. Sci. U.S.A. 108: 16765–16770.

Liu, X.Y., E.T. Wang, Y. Li and W.X. Chen. 2007. Diverse bacteria isolated from root nodules of *Trifolium*, *Crotalaria* and *Mimosa* grown in the subtropical regions of China. Arch Microbiol. 188: 1–14.

Loh, J., R.W. Carlson, W.S. York and G. Stacey. 2002. Bradyoxetin, a unique chemical signal involved in symbiotic gene regulation. Proc. Nat. Acad. Sci. U.S.A. 99: 14446–14451.

MacLean, A.M., T.M. Finan and M.J. Sadowsky. 2007. Genomes of the symbiotic nitrogen-fixing bacteria of legumes. Plant Physiol. 144: 615–622.

Madsen, L.H., L. Tirichine, A. Jurkiewicz, J.T. Sullivan, A.B. Heckmann, A.S. Bek, C.W. Ronson, E.K. James and J. Stougaard. 2010. The molecular network governing nodule organogenesis and infection in the model legume *Lotus japonicus*. Nat. Commun. 1: 10.

Masuda, S., S. Eda, S. Ikeda, H. Mitsui and K. Minamisawa. 2010. Thiosulfate-dependent chemolithoautotrophic growth of *Bradyrhizobium japonicum*. Appl. Environ. Microbiol. 76: 2402–2409.

Medeot, D.B., C. Sohlenkamp, M.S. Dardanelli, O. Geiger, M.G. de Lema and I.M. Lopez-Lara. 2010. Phosphatidylcholine levels of peanut-nodulating *Bradyrhizobium* sp. SEMIA6144 affect cell size and motility. FEMS Microbiol. Lett. 303: 123–131.

Melchiorre, M., M.J. de Luca, G. Gonzalez Anta, P. Suarez, C. Lopez, R. Lascano and R.W. Racca. 2011. Evaluation of bradyrhizobia strains isolated from field-grown soybean plants in Argentina as improved inoculants. Biol. Fert. Soils. 47: 81–89.

Menna, P. and M. Hungria. 2011. Phylogeny of nodulation and nitrogen-fixation genes in *Bradyrhizobium*: supporting evidence for the theory of monophyletic origin, and spread and maintenance by both horizontal and vertical transfer. Int. J. Syst. Evol. Microbiol. 61: 3052–3067.

Menna, P., F.G. Barcellos and M. Hungria. 2009. Phylogeny and taxonomy of a diverse collection of *Bradyrhizobium* strains based on multilocus sequence analysis of the 16S rRNA gene, ITS region and *glnII, recA, atpD* and *dnaK* genes. Int. J. Syst. Evol. Microbiol. 59: 2934–2950.

Miche, L., L. Moulin, C. Chaintreuil, J.L. Contreras-Jimenez, J.A. Munive-Hernandez, M.D. Villegas-Hernandez, F. Crozier and G. Bena. 2010. Diversity analyses of *Aeschynomene* symbionts in Tropical Africa and Central America reveal that nod-independent stem nodulation is not restricted to photosynthetic bradyrhizobia. Env. Microbiol. 12: 2152–2164.

Molouba, F., J. Lorquin, A. Willems, B. Hoste, E. Giraud, B. Dreyfus, M. Gillis, P. de Lajudie and C. Masson-Boivin. 1999. Photosynthetic bradyrhizobia from *Aeschynomene* spp. are specific to stem-nodulated species and form a separate 16S ribosomal DNA restriction fragment length polymorphism group. Appl. Environ. Microbiol. 65: 3084–3094.

Mornico, D., L. Miché, G. Béna, N. Nouwen, A. Verméglio, D. Vallenet, A. A.T. Smith, E. Giraud, C. Médigue and L. Moulin. 2012. Comparative genomics of *Aeschynomene* symbionts: Insights into the ecological lifestyle of Nod-independent photosynthetic bradyrhizobia. Genes 3: 35–61.

Moulin, L., G. Bena, C. Boivin-Masson and T. Stepkowski. 2004. Phylogenetic analyses of symbiotic nodulation genes support vertical and lateral gene co-transfer within the *Bradyrhizobium* genus. Mol. Phyl. Evol. 30: 720–732.

Muñoz, V., F. Ibañez, M.L. Tonelli, L. Valetti, M.S. Anzuay and A. Fabra. 2011. Phenotypic and phylogenetic characterization of native peanut *Bradyrhizobium* isolates obtained from Cordoba, Argentina. Syst. Appl. Microbiol. 34: 446–452.

Nazmun, A., M. Rokonuzzaman and M.N. Hasan. 2009. Effect of *Bradyrhizobium* and *Azotobacter* on growth and yield of mungbean varieties. J. Bangladesh Agric. Univ. 7: 7–13.

Okubo, T., T. Tsukui, H. Maita, S. Okamoto, K. Oshima, T. Fujisawa, A. Saito, H. Futamata, R. Hattori, Y. Shimomura et al. 2012. Complete genome sequence of *Bradyrhizobium* sp. S23321: Insights into symbiosis evolution in soil oligotrophs. *Microbes Environ.* [Epub ahead of print], doi: 10.1264/jsme2.ME11321.

Parker, M.A. 2012. Legumes select symbiosis island sequence variants in *Bradyrhizobium*. Mol. Ecol. 21: 1769–1778.

Parker, M.A. and D.A. Kennedy. 2006. Diversity and relationships of bradyrhizobia from legumes native to eastern North America. Can. J. Microbiol. 52: 1148–1157.

Qu, Y. and J.C. Spain. 2011. Molecular and Biochemical Characterization of the 5-nitroanthranilic acid degradation pathway in *Bradyrhizobium* sp. strain JS329. J. Bacteriol. 193: 3057–3063.

Radutoiu, S., L.H. Madsen, E.B. Madsen, A. Jurkiewicz, E. Fukai, E.M. Quistgaard, A.S. Albrektsen, E.K. James, S. Thirup and J. Stougaard. 2007. LysM domains mediate lipo-chitin-oligosaccharide recognition and Nfr genes extend the symbiotic host range. EMBO J. 26: 3923–3935.

Ramirez-Bahena, M.H., A. Peix, R. Rivas, M. Camacho, D.N. Rodriguez-Navarro, P.F. Mateos, E. Martinez-Molina, A. Willems and E. Velazquez. 2009. *Bradyrhizobium pachyrhizi* sp. *nov.*

and *Bradyrhizobium jicamae* sp. *nov.*, isolated from effective nodules of *Pachyrhizus erosus*. Int. J. Syst. Evol. Microbiol. 59: 1929–1934.

Richter, M. and R. Rossello-Mora. 2009. Shifting the genomic gold standard for the prokaryotic species definition. Proc. Natl. Acad. Sci. U.S.A. 106: 19126–19131.

Rivas, R., A. Willems, J.L. Palomo, P. Garcia-Benavides, P.F. Mateos, E. Martinez-Molina, M. Gillis and E. Velazquez. 2004. *Bradyrhizobium betae* sp. *nov.*, isolated from roots of *Beta vulgaris* affected by tumour-like deformations. Int. J. Syst. Evol. Microbiol. 54: 1271–1275.

Rivas, R., M. Martens, P. de Lajudie and A. Willems. 2009. Multilocus sequence analysis of the genus *Bradyrhizobium*. Syst. Appl. Microbiol. 32: 101–110.

Rogel, M. A., E. Ormeno-Orrillo and E. Martinez Romero. 2011. Symbiovars in rhizobia reflect bacterial adaptation to legumes. Syst. Appl. Microbiol. 34: 96–104.

Sachs, J.L., J.E. Russell and A.C. Hollowell. 2011. Evolutionary instability of symbiotic function in *Bradyrhizobium japonicum*. Plos One 6(11): e26370. doi: 10.1371/journal.pone.0026370.

Sánchez-Cañizares, C., L. Rey, D. Duran, F. Temprano, P. Sánchez-Jiménez, A. Navarro, M. Polajnar, J. Imperial and T. Ruiz-Argüeso. 2011. Endosymbiotic bacteria nodulating a new endemic lupine *Lupinus mariae-josephi* from alkaline soils in Eastern Spain represent a new lineage within the *Bradyrhizobium* genus. Syst. Appl. Microbiol. 34: 207–215.

Sawada, H., L.D. Kuykendall and J.M. Young. 2003. Changing concepts in the systematics of bacterial nitrogen-fixing legume symbionts. J. Gen. Appl. Microbiol. 49: 155–179.

Sprent, J.I. 2008. 60Ma of legume nodulation. What's new? What's changing? J. Exp. Bot. 59: 1081–1084.

Stepkowski, T., A. Swiderska, K. Miedzinska, M. Czaplinska, M. Swiderski, J. Biesiadka and A.B. Legocki. 2003. Low sequence similarity and gene content of symbiotic clusters of *Bradyrhizobium* sp. WM9 (*Lupinus*) indicate early divergence of "lupin" lineage in the genus *Bradyrhizobium*. Ant. van Leeuwenhoek. 84: 115–124.

Stepkowski, T., C.E. Hughes, I.J. Law, L. Markiewicz, D. Gurda, A. Chlebicka and L. Moulin. 2007. Diversification of lupine *Bradyrhizobium* strains: Evidence from nodulation gene trees. Appl. Environ. Microbiol. 73: 3254–3264.

Stepkowski, T., M. Zak, L. Moulin, J. Kroliczak, B. Golinska, D. Narozna, V.I. Safronova and C. J. Madrzak. 2011. *Bradyrhizobium canariense* and *Bradyrhizobium japonicum* are the two dominant rhizobium species in root nodules of lupin and serradella plants growing in Europe. Syst. Appl. Microbiol. 34: 368–375.

Stokkermans, T.J., S. Ikeshita, J. Cohn, R.W. Carlson, G. Stacey, T. Ogawa and N.K. Peters. 1995. Structural requirements of synthetic and natural product lipo-chitin-oligosaccharides for induction of nodule primordia on *Glycine soja*. Plant Physiol. 108: 1587–1595.

Stokkermans, T.J.W., R. Orlando, V.S.K. Kolli, R.W. Carlson and N.K. Peters. 1996. Biological activities and structures of *Bradyrhizobium elkanii* low abundance lipo chitin-oligosaccharides. Mol. Plant-Microbe Interact. 9: 298–304.

Stover C.K., X.Q. Pham, A.L. Erwin, S.D. Mizoguchi, P. Warrener, M.J. Hickey, F.S. Brinkman, W.O. Hufnagle, D.J. Kowalik, M. Lagrou et al. 2000. Complete genome sequence of *Pseudomonas aeruginosa* PAO1, an opportunistic pathogen. Nature. 31: 959–64.

Sullivan, J.T., H.N. Patrick, W.L. Lowther, D.B. Scott and C.W. Ronson. 1995. Nodulating strains of *Rhizobium loti* arise through chromosomal symbiotic gene transfer in the environment. Proc. Natl. Acad. Sci. U.S.A. 92: 8985–8989.

Tan, Z., T. Hurek, P. Vinuesa, P. Muller, J.K. Ladha and B. Reinhold-Hurek. 2001. Specific detection of *Bradyrhizobium* and *Rhizobium* strains colonizing rice (*Oryza sativa*) roots by 16S–23S ribosomal DNA intergenic spacer-targeted PCR. Appl. Environ. Microbiol. 67: 3655–3664.

Torres, M.J., E. Bueno, S. Mesa, E.J. Bedmar and M.J. Delgado. 2011. Emerging complexity in the denitrification regulatory network of *Bradyrhizobium japonicum*. Biochem. Soc. Trans. 39: 284–288.

Trinick, M.J. and P.A. Hadobas. 1989. Competition by *Bradyrhizobium* strains for nodulation of the nonlegume *Parasponia andersonii*. Appl. Environ. Microbiol. 55: 1242–1248.

Urtz, B.E. and G.H. Elkan. 1996. Genetic diversity among *Bradyrhizobium* isolates that effectively nodulate peanut (*Arachis hypogaea*). Can. J. Microbiol. 42: 1121–1130.

Ussery, D.W., S. Borini and T.M. Wassenaar. 2009. Computing for comparative genomics: bioinformatics for microbiologists. Computational Biology, vol. 8. Springer, New York.

van Berkum, P. and B.D. Eardly. 2002. The aquatic budding bacterium *Blastobacter denitrificans* is a nitrogen-fixing symbiont of *Aeschynomene indica*. Appl. Environ. Microbiol. 68: 1132–1136.

van Berkum, P., J.M. Lelbold and B.D. Eardly. 2006. Proposal for combining *Bradyrhizobium* spp. *(Aeschynomene indica)* with *Blastobacter denitrificans* and to transfer *Blastobacter denitrificans* (Hirsch and Muller, 1985) to the genus *Bradyrhizobium* as *Bradyrhizobium denitrificans (comb. nov.)*. Syst. Appl. Microbiol. 29: 207–215.

Van Santen, E., D.W. Reeves and G.L. Mullins. 1993. White lupin, a potential new crop for Alabama. Alabama Agric. Expt. Sta. Highlights Agric. Res. 40(4): 15.

Vega-Hernandez, M.C., R. Perez-Galdona, F.B. Dazzo, A. Jarabo-Lorenzo, M.C. Alfayate and M. Leon-Barrios. 2001. Novel infection process in the indeterminate root nodule symbiosis between *Chamaecytisus proliferus* (tagasaste) and *Bradyrhizobium* sp. New Phytol. 150: 707–721.

Vinuesa, P., C. Silva, D. Werner and E. Martinez-Romero. 2005a. Population genetics and phylogenetic inference in bacterial molecular systematics: the roles of migration and recombination in *Bradyrhizobium* species cohesion and delineation. Mol. Phyl. Evol. 34: 29–54.

Vinuesa, P., M. Leon-Barrios, C. Silva, A. Willems, A. Jarabo-Lorenzo, R. Perez-Galdona, D. Werner and E. Martinez-Romero. 2005b. *Bradyrhizobium canariense* sp. *nov.*, an acid-tolerant endosymbiont that nodulates endemic genistoid legumes (Papilionoideae: Genisteae) from the Canary Islands, along with *Bradyrhizobium japonicum bv. genistearum*, *Bradyrhizobium* genospecies alpha and *Bradyrhizobium* genospecies beta. Int. J. Syst. Evol. Microbiol. 55: 569–575.

Weber, D.F., H.H. Keyser and S.L. Uratsu. 1989. Serological distribution of *Bradyrhizobium japonicum* from U.S. soybean production areas. Agron. J. 81: 786–789.

Willems, A., F. Doignon-Bourcier, J. Goris, R. Coopman, P. de Lajudie, P. De Vos and M. Gillis. 2001. DNA-DNA hybridization study of *Bradyrhizobium* strains. Int. J. Syst. Evol. Microbiol. 51: 1315–1322.

Xu, L.M., C. Ge, Z. Cui, J. Li and H. Fan. 1995. *Bradyrhizobium liaoningense* sp. *nov.*, isolated from the root-nodules of soybeans. Int. J. Syst. Bacteriol. 45: 706–711.

Yao, Z.Y., F.L. Kan, E.T. Wang, G.H. Wei and W.X. Chen. 2002. Characterization of rhizobia that nodulate legume species of the genus *Lespedeza* and description of *Bradyrhizobium yuanmingense* sp. *nov*. Int. J. Syst. Evol. Microbiol. 52: 2219–2230.

Zhang, Y., Y. Li, W.F. Chen, E. Wang, X. Sui, Q. Li, Y. Zhou and W.X. Chen. 2011. *Bradyrhizobium huanghuaihaiense sp. nov.*, an effective symbiotic bacterium isolated from soybean (*Glycine max* L.) nodules in Northern (Huang-Huai-Hai) Plain of China. Int. J. Syst. Evol. Microbiol. Doi: ijs.0.034546-0; published ahead of print, October 14, 2011.

3

Importance of Motile and Biofilm Lifestyles of Rhizobia for the Establishment of Symbiosis with Legumes

J. Nogales, D. Pérez-Mendoza, M.T. Gallegos and *M.J. Soto**

Introduction

Most bacteria can live either as free-living individual cells (planktonic) or as part of a sessile community, commonly referred to as biofilm. These two different lifestyles confer distinct advantages: motility allows bacteria to rapidly reach and colonize new niches, whereas microorganisms in biofilms demonstrate intrinsic resistance to unfavourable environmental conditions. What makes bacteria decide between sessile and motile lifestyles is an area of intense research, particularly in clinically relevant pathogenic bacteria that cause infections often problematic due to antibiotic-resistant biofilms. However, the choice between motility and biofilm formation may be a general modus operandi among microorganisms that live in varying environments like plant mutualistic rhizobia.

Dpto. Microbiología del Suelo y Sistemas Simbióticos. Estación Experimental del Zaidín, Consejo Superior de Investigaciones Científicas (CSIC), Profesor Albareda 1, 18008 Granada, Spain.
*Corresponding author: mariajose.soto@eez.csic.es

Rhizobia are motile bacteria that can exist either as free-living organisms in the soil or in symbiotic association with their host legume plants. It is likely that both lifestyles have remained in these mutualistic bacteria to provide a complex and efficient survival mechanism which can be fine-tuned according to environmental and/or endogenous signals. Under favourable conditions for the establishment of the *Rhizobium*-legume symbiosis, a switch from the planktonic to the plant root surface-adherent community lifestyles should take place in these soil bacteria. It is known that rhizobial motility is dispensable for nodule formation and nitrogen fixation, and the same might be true for the ability to form biofilms according to the scarce data available from rhizobia in this field. However, accumulating evidence suggests that the switch between the two different lifestyles is tightly regulated in rhizobia which may confer ecological advantages (e.g., bacterial survival among competitors, avoidance of plant defence responses). Considering that pathogenic and mutualistic bacteria often use similar strategies to establish associations with their hosts (Soto et al. 2009), the available wealth of data about regulatory pathways controlling biofilm formation in pathogenic bacteria will facilitate the discovery of molecular mechanisms and signals that govern the transition from the free-living to the plant-associated lifestyle in rhizobia. In this chapter, we review the existing knowledge on motility and biofilm formation in rhizobia, as well as the contribution of these bacterial behaviours to the *Rhizobium*-legume symbiosis.

Motility *versus* Biofilm Formation: Lessons from Host-Pathogen Interactions

Taxis is the movement of a microorganism in response to an external stimulus, which is an important survival mechanism and a distinct advantage for host-adapted species. Benefits of motility include increased efficiency of nutrient acquisition and avoidance of toxic substances, thus enabling successful competition with other microorganisms, the ability to translocate to preferred hosts and access optimal colonization sites within them, and the dispersal in the environment during the course of transmission. The costs of motility are also significant since they include the metabolic burden of synthesizing flagellar components and the energetic expense of fuelling flagellar motors. More than 50 genes, regulated and expressed in an ordered way, are required for flagellar formation and function. Therefore, their synthesis and functioning require energy consumption by the cell, about 2 percent of the biosynthetic energy expenditure in *Escherichia coli* (Macnab 2003). Another cost of motility is the presentation of polymeric and highly antigenic targets to the immune system. One of the major structural components of the flagellar filament is flagellin, which is the ligand for transmembrane receptors (Toll-like receptors in mammals, FLS2 in plants) of host cells. Like lipopolysaccharide (LPS), flagellin is considered one of the pathogen-associated molecular patterns (PAMP), which can trigger immune responses (Felix et al. 1999, Hayashi et al. 2001). It is therefore not surprising that synthesis of the motility apparatus is usually subject to strict control. Moreover, motility is regulated at several levels, including transcriptional control of flagellar genes and signal-transduction cascades that govern chemotaxis.

Flagella are regarded as an important virulence factor in bacterial pathogenesis, mainly because of their role in motility and chemotaxis, which is critical for colonization, penetration and invasion of tissues. However, recent evidence has denoted that the bacterial flagella participate in numerous additional processes, by acting as bacterial adhesins, promoting bacterial biofilm formation, translocating virulence proteins into host cells via special type III secretion systems, and triggering host immune responses through the Toll-like receptor signalling pathway (reviewed by Duan et al. 2012). Therefore, flagella might be involved in pathogenicity not just by enabling bacterial motility but also through other complex mechanisms.

Biofilms are assemblages of microorganisms enclosed in a matrix that function as a cooperative consortium. In the mature biofilm, cells are enclosed in an extracellular matrix composed of proteins, exopolysaccharides and extracellular DNA (eDNA) (Flemming and Wingender 2010). In addition to immobilizing the bacteria, the matrix is a scaffold that traps nutrients and various biologically active molecules, such as cell-cell communication signals. The matrix may resemble an external digestion system, as it also accumulates enzymes that can degrade various matrix components as well as any nutrients or other substrates; once degraded, the products are then in close proximity to the cells, facilitating uptake. Moreover, the matrix acts as a shield against toxins, antimicrobials and predators.

Microscopic studies reveal that biofilm formation occurs in a sequential process including transport of bacteria to a surface, initial attachment, formation of microcolonies and biofilm maturation (Tolker-Nielsen et al. 2000, Sauer et al. 2002). It has been shown that flagella contribute not only to early attachment as surface adhesins, but also to biofilm development and maturation in *Pseudomonas aeruginosa*, *Yersinia enterocolitica*, *Listeria monocytogenes* and *E. coli*. However, the contribution of flagella to biofilm formation also remains controversial, since other studies demonstrated that, although flagellum-mediated motility played an important role, it was not necessary for biofilm formation in *E. coli*, *P. aeruginosa* or *P. fluorescens* (reviewed by Duan et al. 2012).

It is now well accepted that the biofilm lifestyle is a feature common in most microorganisms in natural, medical and engineered systems.

Recent advances show that biofilms are structurally complex, dynamic systems sharing features with both primordial multicellular organisms and multifaceted ecosystems. Biofilm formation gives protection from a wide range of environmental challenges, such as UV exposure, metal toxicity, acid exposure, dehydration and salinity, phagocytosis and several antibiotics and antimicrobial agents (reviewed by Hall-Stoodley et al. 2004).

Studies of a variety of bacterial-host interactions demonstrate different roles for motility and biofilm formation at different points along the bacterial life cycle. This is particularly well studied in the model bacterium *P. aeruginosa* (reviewed by Coggan and Wolfgang 2012, Nadal-Jimenez et al. 2012). *P. aeruginosa* is a Gram-negative, metabolically versatile bacterium that exists in a wide range of environmental habitats, including aquatic sediments, water exposed surfaces, soil, plant roots and leaves and human and animal sewage. In humans, *P. aeruginosa* is associated with both acute and chronic infections in immunocompromised individuals and is capable of infecting a multitude of tissues, including lungs, burns, wounds, eyes, ears and indwelling medical devices (i.e., catheters and ventilators) (Driscoll et al. 2007). Acute *P. aeruginosa* infections are invasive and cytotoxic and frequently result in substantial tissue damage, systemic spread, sepsis and mortality. The pathogenesis of acute infections relies upon the expression of many surface-exposed and secreted virulence factors, including toxins, proteases (delivered by a type II secretion system; T2SS), type IV pili (Tfp), flagella and a type III secretion system (T3SS) that can inject a set of eukaryote specific effectors across the plasma membrane of target cells. On the other hand, chronic infections are minimally invasive and noncytotoxic and involve the formation of biofilms, which protect against the host immune system and provide resistance to antibiotics (Donlan and Costerston 2002, Ryder et al. 2007). Thus, chronic infections rarely result in systemic spread, but instead lead to unrelenting non-productive host inflammation that contributes to the resulting morbidity and mortality.

Recent work in the field has led to the hypothesis that the regulatory pathways that control the transition from free living (planktonic) to surface-adherent community (sessile) lifestyles in the environment are likely to play a similar role in dictating infection type (acute *vs*. chronic) in human disease (reviewed by Coggan and Wolfgang 2012, Nadal-Jimenez et al. 2012). The key regulatory pathways of *P. aeruginosa* lifestyle revealed up to now are cAMP/Vfr and c-di-GMP signalling, quorum sensing (QS), and the Gac/Rsm pathway. Cyclic AMP is now recognized as a major mediator of acute virulence gene expression and planktonic lifestyle phenotypes in *P. aeruginosa*. In contrast, c-di-GMP appears to play a critical role in controlling chronic infection-related phenotypes and the biofilm lifestyle. The coordination of population behaviour is also essential in determining the commitment to a particular lifestyle and QS is an important regulator of acute virulence phenotypes; however, it also plays a significant role in the transition from the planktonic to sessile lifestyle and in the establishment of chronic infection. Finally, the Gac/Rsm pathway appears to act as a global regulator of lifestyle by inversely regulating acute virulence factors and planktonic behaviour with chronic infection phenotypes and biofilm formation. Furthermore, the inverse transition from a sessile to planktonic lifestyle can occur in the environment enabling dissemination to new environmental niches. Similarly, this sessile to planktonic switch also occurs during chronic human diseases, although the mechanism and implications are less clear (McDougald et al. 2011).

Role of Motility and Chemotaxis in the *Rhizobium*-Legume Symbiosis

Like in many other motile bacteria, chemotaxis allows rhizobia to sense their external environment and respond appropriately by moving towards increasing concentrations of nutrients and away from increasing concentrations of toxic compounds. This process is achieved by signal transduction pathways that ultimately control the direction (or speed) of flagellar rotation. In the model bacterium *E. coli*, for which the chemotaxis signal transduction pathway is best understood, environmental signals sensed by transmembrane receptors known as methyl-accepting chemotaxis proteins (MCPs) regulate the activity of the histidine kinase, CheA. Upon chemostimulation, CheA becomes autophosphorylated at a conserved histidine residue and then transfers the phosphate to its cognate response regulator, the CheY protein. Phospho-CheY interacts directly with the flagellar motor and controls the bacterial swimming paths by switching the motor from counterclockwise to clockwise rotation (Wadhams and Armitage 2004).

Among rhizobia, the molecular mechanism of chemotaxis has been deeply investigated in *Sinorhizobium meliloti* (Sourjik and Schmitt 1996, Schmitt 2002). Certain features of motility and chemotaxis in this bacterium are clearly distinct to those observed in *E. coli* (Schmitt 2002). Whereas *E. coli* flagellar filaments comprise one type of flagellin, the rigid complex flagellar filaments of *S. meliloti* consist of four related flagellin subunits, and the interflagellin bonds lock the filaments into a right-handed helical conformation (Scharf et al. 2001). Flagella in *S. meliloti* rotate exclusively clockwise, and swimming cells respond to tactic stimuli by modulating their rotary speed (Sourjik and Schmitt 1996). Flagellar filaments of *R. leguminosarum* bv. viciae, which also rotate unidirectionally, are assembled from up to seven flagellins (Miller et al. 2007, Tambalo et al. 2010a). *Bradyrhizobium japonicum*, the soybean symbiont, has two types of flagella:

one formed by thin filaments consisting of two flagellins, and the other formed by thick filaments consisting of four flagellins (Althabegoiti et al. 2011). Furthermore, the chemotaxis signal transduction cascade of *S. meliloti* comprises nine chemoreceptors (MCPs), a CheA histidine kinase and two different response regulators, CheY1 and CheY2, but no CheZ phosphatase (Meier et al. 2007 and references therein). Both CheY1 and CheY2 can be phosphorylated by CheA-P. However, only CheY2-P interacts with the flagellar motor to decrease the rotary speed of the unidirectional clockwise-rotating flagellar motor, thereby mediating the tactic response (Sourjik and Schmitt 1996). Another distinct feature of the rhizobial motility system is its regulation. In *S. meliloti,* the 56-kb chromosomal region, named the flagellar region, encompasses the majority of genes associated with motility, which are expressed in a hierarchical manner and categorized into three main classes: class IA comprises the master regulatory genes, *visN* and *visR* which encode the LuxR-type subunits of the global transcriptional activator VisNR; class IB includes the *rem* gene, coding for the growth-dependent transcriptional activator intermediate between the master regulator VisNR and the flagellar and motility genes; class II, controlled by VisNR and Rem, consists of flagellar assembly and motility genes; and class III comprises flagellin and chemotaxis genes (Sourjik et al. 2000, Rotter et al. 2006). In *R. leguminosarum,* the VisNR-Rem cascade is also involved in the regulation of most of flagellar, motility and chemotaxis genes (*che1, motA, flaABCD, motB* and *mcpD*) (Tambalo et al. 2010b). All these genes except *mcpD,* are located within the main motility and chemotaxis gene cluster. However, other chemotaxis and motility genes (*che2* operon, *flaH, flaG, flaE* and *mcpC*), which are found outside of this cluster, are not regulated by VisNR and Rem activators, suggesting that an additional regulatory system participates in this process.

Remarkably, although molecular mechanisms governing chemotaxis and motility in *S. meliloti* are the best described in alpha-proteobacteria, their contribution to the establishment of a specific association with its legume host has not been extensively investigated using genetically defined bacterial mutants. Motility and rhizobial chemotaxis towards legume root exudates may facilitate the establishment of symbiosis by directing bacteria to the proper infection sites. In agreement with this hypothesis, different rhizobia have been shown to display chemotaxis towards plant root exudates (Caetano-Anollés et al. 1988a, Dharmatilake and Bauer 1992). Moreover, old studies performed with non-motile or non-chemotactic rhizobial strains, which were not genetically characterized, showed that these bacteria were less competitive and less efficient in nodulation, suggesting that motility and chemotaxis are important in the initial steps of the plant-bacterium interaction but dispensable for nodulation and nitrogen fixation (Ames and Bergman 1981, Soby and Bergman 1983, Caetano-Anollés et al. 1988b). Later, experiments performed with genetically characterized mutants confirmed these initial observations. Thus, nodulation competition experiments in *R. leguminosarum* bv. *viciae* have demonstrated that two MCPs and the chemotaxis cluster *che1* were important for efficient pea nodulation and competitive nodule formation (Yost et al. 1998, Miller et al. 2007). Also, *S. meliloti fliP* and *flgH* mutants (non-flagellated, non-motile) exhibited a delay in nodule development (Fujishige et al. 2006).

Once nodulation has been initiated, motility is no longer required for bacteroid development or effective nitrogen fixation. Consistent with this, the down-regulation of flagellar and chemotaxis genes has been observed in expression analyses of nodules induced by *S. meliloti* and *R. leguminosarum* (Becker et al. 2004, Yost et al. 2004). In fact, repression of *S. meliloti* motility genes mediated by the ExpR/Sin QS system is essential for effective nodule invasion (Gurich and González 2009). At low cell densities ExpR is required for the activation of motility-related genes, whereas at high population densities ExpR, in conjunction with acyl-homoserine lactones (AHLs) produced by the autoinducer synthase SinI, inhibits transcription of the *visNR* operon, resulting in the repression of genes belonging to the flagellar regulon. A *sinI* mutant induces fewer nodules per plant and leads to a lower pink/white nodule ratio compared to plants inoculated with the wild type, the *expR* or the *expRsinI* mutant strains. The inability to repress flagellar synthesis was found to be the reason for the symbiotic defect, since a *sinI* mutant unable to produce flagella regained the full capacity to establish symbiosis. At the moment it is not known how flagella can interfere with the development of the symbiotic interaction. Possible explanations are that flagella might impede bacterial movement through infection threads or induce plant defence responses (Gurich and González 2009). It is worth mentioning here that, in contrast to the highly conserved flagellin-derived peptide flg22 from pathogenic bacteria, the corresponding peptide from *S. meliloti* is inactive as elicitor of immune defences in *Arabidopsis* plants (Gómez-Gómez and Boller 2002).

Like in many other bacteria, expression of flagellar genes in rhizobia seems to be subjected to strict control in order to avoid undesirable costs. Different genome wide expression profiles performed in *S. meliloti* have shown that flagellar and motility genes are rapidly down-regulated in response to several environmental stresses (e.g., osmotic stress, acidic pH, iron and phosphate starvation) (Krol and Becker 2004, Domínguez-Ferreras et al. 2006, Hellweg et al. 2009). Another remarkable feature of motility in rhizobia is the inversed co-regulation of this behaviour with EPS production which is mediated by several regulatory systems. The ExpR/Sin QS system controls the expression of a large number of genes among which are those involved in the production of the symbiotically important EPSs succinoglycan (EPS I) and galactoglucan (EPS II) and, as already mentioned, motility genes which are regulated in a population-density manner via *visNR* (Hoang et al. 2004, Gurich and González 2009). The regulatory system ExoR/ExoS/ChvI, which is involved in the regulation of EPS I synthesis, also suppresses motility gene expression through VisNR and Rem (Hoang et al. 2008). Different mutants in this system that overproduce EPS I, are devoid of flagella. Mutants lacking CbrA, a putative two-component histidine kinase associated with a PAS sensory domain and initially described as a regulator of symbiosis, overproduce EPS I while repressing several motility genes, including *visNR* (Gibson et al. 2007). The regulatory protein MucR which plays an essential role in EPS production, represses the expression of *rem* (Bahlawane et al. 2008). Finally, the EmmABC system comprising the small regulatory protein EmmA and the putative two component system EmmBC, also regulates EPS production and motility (Morris and González 2009). Co-regulation of EPS production and motility has been reported for other host-associated bacteria. In rhizobia, the co-regulation of these two processes in response to different environmental cues might be essential for developmental switches required at some stages during the plant-bacteria interaction.

Data on chemotaxis and different aspects of rhizobial motility have been obtained by analyzing swimming motility, a flagella-dependent mode of bacterial translocation that takes place in liquid environments. Recent findings have unveiled the existence in rhizobia of additional types of translocation that occur in bacteria growing on surfaces (Fig. 3.1).

S. meliloti can move over semisolid surfaces using at least two different types of motility. One type is flagellum-independent surface spreading or sliding, which is positively influenced by a functional *expR* gene mainly through the production of EPS II (Nogales et al. 2012). The second type of surface translocation shown by *S. meliloti* is swarming, which is greatly

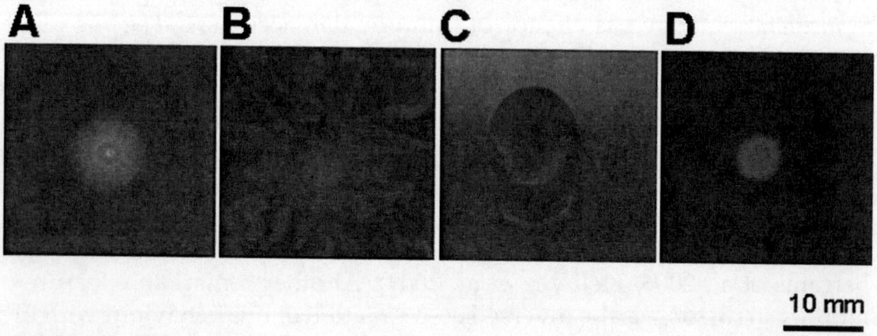

Figure 3.1 Macroscopic appearance of different types of motility in *Sinorhizobium meliloti*. A) Swimming of the wild-type strain Rm1021. B) Swarming and flagella-independent surface translocation of an *exoX* derivative mutant of Rm1021. C) Sliding shown by an *expR*⁺ derivative strain of Rm1021. D) No surface motility of a Rm1021 *rhbA* mutant. (For more details see Nogales et al. 2012).

dependent on flagella and synthesis of the siderophore rhizobactin 1021, is promoted by EPS I overproduction, but does not require ExpR (Nogales et al. 2010, 2012).

Swarming is defined as a flagella-dependent motility characterized by the rapid and co-ordinated population migration across solid surfaces. In contrast to movement in liquid medium (swimming), swarming is a multicellular process that requires a physiological and morphological differentiation which is characterized by an increase in the number of flagella and by cell elongation. The close connection between swarming and virulence in some pathogenic bacteria is well known. Swarm cell differentiation is accompanied by the expression of virulence factors and mutations affecting this multicellular migration significantly reduce the bacterial invasion ability. Moreover, several studies have revealed the existence of a link between swarming and biofilm formation (Verstraeten et al. 2008): i) both are surface-associated multicellular processes in which cell-cell communication and quorum sensing play important roles; ii) in both processes, the participation of the same cell surface-associated structures such as flagella, type IV pili, polysaccharidic matrix and biosurfactants has been reported; iii) swarming bacteria, like bacteria in biofilms, show increased resistance to several antimicrobial agents.

Swarming has been extensively studied in pathogenic bacteria. However, scarce information is available about this type of surface motility in rhizobia. Within this group, multicellular surface migration was first described for a *fadD* mutant of *S. meliloti* (Soto et al. 2002). In this bacterium, the lack of the *fadD* gene results in conditional swarming and also in defects in nodulation. A transcriptomic study of rhizobium growth on surfaces including swarming inducing conditions revealed major changes in the physiology of *S. meliloti* cells grown on a surface relative to liquid cultures (Nogales et al. 2010). Furthermore, this study also demonstrated that iron and genes involved in rhizobactin 1021 synthesis play a role in the surface motility shown by *S. meliloti* which can be circumvented in a *fadD* mutant. Swarming has also been described for the wild type strain CNPAF512 of *R. etli* (Daniels et al. 2004), and the *R. leguminosarum* bv. viciae VF39 strain (Tambalo et al. 2010c). The finding that mutants in the *cinIR* QS system of *R. etli* were no longer able to move over semisolid surfaces, led to the discovery that AHL carrying a long-chain fatty acid moiety have a dual role in swarming in this *Rhizobium*, as QS signals and as biosurfactants which promote surface translocation (Daniels et al. 2006). The characterization of several *R. etli* mutants defective in swarming has allowed the identification of additional genetic determinants which seem to play a role in this multicellular behaviour, including genes involved in polysaccharide synthesis or export, motility and amino acid and polyamine metabolism (Braeken et al. 2008). Interestingly, half of the mutants defective in swarming displayed defects in nodulation and/or symbiotic nitrogen fixation activity. The fact that some mutations which alter *S. meliloti* and *R. etli* swarming behaviour result in impairing the establishment of symbiosis suggests either that essential components for this multicellular motility and/or factors that are co-regulated during swarmer cell differentiation may play a role in the interaction with the host plant.

The discovery of surface translocation in rhizobia adds another level of complexity to the study of the yet poorly unexplored role of motility in the different lifestyles of these soil bacteria. Whereas the role of swarming motility in the establishment of the *Rhizobium*-legume symbiosis is still unclear, sliding motility may allow rhizobia to colonize surfaces under conditions where flagellar expression is down-regulated, for instance, at high cell population densities and during the invasion process. In line with this, a collective sliding movement of bacteria toward the infection thread tip has been proposed to contribute to colonization of the thread (Fournier et al. 2008). The biological significance of the different types of rhizobial motility in the establishment of the *Rhizobium*-legume symbiosis remains to be elucidated.

Biofilm and the Establishment of the *Rhizobium*-Legume Symbiosis

Although studies on biofilm formation have mostly focused on animal and human pathogens for obvious reasons, it has been demonstrated that rhizobia form biofilms on both abiotic and biotic (plant) surfaces. A modification of the microtiter plate assay initially developed by O'Toole and associates (1999) is being used as a first approach to investigate how environmental cues and genetic determinants quantitatively affect rhizobial biofilm abilities *in vitro*. Additionally, confocal lasser scanning microscopy (CLM) of cells tagged with autofluorescent proteins allows monitoring the process of biofilm formation on glass and plant surfaces and analyzing the three-dimensional structure of the biofilm.

Biofilm formation on both abiotic surfaces and roots was first reported in *S. meliloti* (Fujishige et al. 2006). Although this bacterium produces poor biofilms on abiotic surfaces, the use of the microtiter plate assay revealed that a nutritionally limiting environment increases the transition from planktonic to a sessile mode of life. Likewise, nutrients such as sucrose, phosphate and calcium enhance biofilm formation as their concentrations increase, whereas suboptimal temperatures or pHs and osmotic stress negatively affect biofilm formation on plastic surfaces (Rinaudi et al. 2006). *R. leguminosarum* bv. viciae seems to be better than *S. meliloti* for biofilm formation since it is able to attach to and develop thicker biofilms both on plastic and glass, where it develops three-dimensional and highly organized honeycomb-like structures (Russo et al. 2006, Williams et al. 2008). Moreover, unlike *S. meliloti*, it forms thick biofilm rings at the air liquid interface of shaken cultures in minimal medium. Like in *S. meliloti*, a nutrient-limited environment also promotes biofilm formation by *R. leguminosarum* bv. viciae. However, both rhizobia are able to develop biofilms on plant surfaces, a nutrient-rich environment, which has led to the proposal that plant factors may overcome the inhibitory effect of nutrients on rhizobial biofilm.

Several data indicate that cellular components and regulatory mechanisms involved in biofilm formation in other bacteria also influence rhizobial biofilm formation. The role of QS in biofilm formation will be mentioned later in this chapter. Like in many other bacteria, rhizobial EPSs and flagella have a role on biofilm formation. EPS defective mutants generally, display reduced biofilm phenotypes, whereas EPS overproducers exhibit more adherence and develop thicker although less stable biofilms than the wild type strains (Fujishige et al. 2006, Wells et al. 2007, Rinaudi and González 2009, Rinaudi et al. 2010). Thus, *S. meliloti exoY* mutants defective in EPS I, an EPS involved in successful bacterial invasion of plant roots, do not develop mature biofilms. It is also known that the production of low molecular weight fractions of EPS II or galactoglucan, another symbiotically important EPS whose synthesis requires a functional ExpR/Sin QS system, allows *S. meliloti* to develop highly structured biofilms on plant root surfaces. On the other hand, the *exoR95* and *exoS96* mutants, that overproduce EPS I, were found to be significantly more successful than the wild-type forming biofilms on plastic surfaces, although these were easily detached from the wells after washing (Fujishige et al. 2006, Wells et al. 2007). However, the increased biofilm ability exhibited by these mutants cannot solely be attributed to EPS I overproduction because they show pleiotropic phenotypes including the lack of flagella, another well-known structural element in biofilm formation. Nevertheless, results obtained in our group with different flagellated *S. meliloti exoX* mutants indicate that overproduction of EPS I alone confers increased biofilm ability on glass and plastic surfaces (Nogales et al. 2012, M.J.Soto, unpublished data). EPSs have also been involved in biofilm formation in *R. leguminosarum* bv. viciae, in *Mesorhizobium tianshanense* the symbiont of *Glycyrrhiza uralensis*, and in *B. japonicum* (Russo et al. 2006, Wang et al. 2008, Pérez-Giménez et al. 2009). As proposed for other bacteria, rhizobial EPS might be important for stabilizing the three dimensional structure of the biofilm or as a scaffold for other molecules that hold the biofilm together.

The role of flagella or motility in general on biofilm formation in rhizobia has not been yet extensively investigated. To our knowledge, a clear connection between flagella and biofilm formation has been only shown in one report from A. Hirsch's laboratory in which two nonflagellated mutants of *S. meliloti* (*fliP* and *flgH*) exhibited 50 percent reduction in biofilm capability in microtiter plates compared to the wild type, probably because of an impairment in bacterial adherence to surfaces (Fujishige et al. 2006). These mutants are able to form nitrogen-fixing nodules but show a delayed nodule formation phenotype, as mentioned earlier. However, it is difficult to conclude if their symbiotic defect is due to the absence of motility, impaired biofilm formation or both. Alterations in motility and biofilming abilities have been observed in several other rhizobial mutants; however, the pleiotropic phenotypes displayed by most of them, makes it difficult to establish a direct connection between motility and biofilm formation. This is the case for the already mentioned *exoR95* and *exoS96 S. meliloti* mutants. In *R. leguminosarum* bv. *trifolii* the lack of a functional *rosR* gene leads to a significant decrease in motility and an altered ability to form biofilm on plastic surfaces (Janczarek et al. 2010). These mutants form nodules on clover but they are unable to fix nitrogen and are outcompeted by the wild type in mixed inoculations, phenotypes that might be explained by the reduced EPS production associated to the *rosR* mutation. Likewise, *fabF2/F1* mutants of *R. leguminosarum* bv. viciae that show an altered lipopolysaccharide and different phenotypes associated with modifications in the cell envelope, are nonmotile and biofilm cells show different organization (Vanderlinde et al. 2009). More recently, increased motility has been correlated with increased biofilm dispersal in *S. meliloti* after reducing the intracellular c-di-GMP levels through the expression of *E. coli* BdcA, a protein that binds c-di-GMP (Ma et al. 2011a). However, changes in c-di-GMP levels might be affecting several other targets in the cell and therefore the impact on biofilm cannot solely be attributed to increased motility.

As mentioned previously , swarming motility can play an important role in biofilm formation in different bacteria (Verstraeten et al. 2008). Several genetic studies performed mostly in *P. aeruginosa* have shown an inverse relationship between swarming motility and biofilm formation; that is, mutations causing hyperswarmer phenotypes lead to defects in biofilm formation (Merritt et al. 2007, Kuchma et al. 2007, Caiazza et al. 2007). In this bacterium, c-di-GMP-dependent signalling seems to be part of a regulatory network controlling the two surface-associated group behaviour, in response to yet unknown environmental cues (Merritt et al. 2010). In *Rhizobium,* investigations on swarming motility are still in its infancy and, consequently, no information about its connection with biofilm formation is available. Nevertheless, several studies performed by our group indicate that genes and environmental factors that affect swarming motility in *S. meliloti* also influence biofilm formation (M.J. Soto, unpublished data).

Besides QS, flagella/motility and EPSs, rhizobial specific factors have also been reported to affect biofilm formation in those bacteria. The microtiter plate assay revealed that mutants of *R. leguminosarum* bv.

viciae lacking the symbiotic plasmid (pSym) exhibit a reduced ability for biofilm formation, suggesting that factors encoded by this plasmid play a role in regulating the transition from the planktonic to the sessile lifestyle (Fujishige et al. 2006). More interestingly, in *S. meliloti* the chitin backbone of Nod Factor, an essential molecule for the nodulation process, is critical for root colonization and biofilm formation (Fujishige et al. 2008). It has been proposed that, in addition to its role as morphogen, the core Nod factor can function as an adhesin holding bacteria together what confers higher resistance against desiccation and/or plant defences. Common *nod* genes responsible for the synthesis of the chitin moiety are expressed in the biofilm lifestyle of *S. meliloti* even in the absence of their plant inducers which for a long time have been considered to be essential for *nod*-gene expression. Perhaps, growth on a surface might be a signal used by rhizobia to switch on the expression of bacterial factors important for biofilm development. Therefore, the identification of such factors might be hindered in experiments performed using cells grown in liquid culture, and by contrast, facilitated in genetic, transcriptomic, proteomic or metabolomic analyses performed on cells grown on surfaces. A recent transcriptomic study of *Rhizobium* growth on surfaces revealed major changes in the physiology of rhizobial cells compared to those grown in liquid cultures, with more than a thousand genes showing differential expression (Nogales et al. 2010). Ongoing research in our laboratory aims to identify genes within this group with a role in biofilm development.

Despite increasing knowledge in the field, the role of rhizobial biofilms in establishing a successful symbiotic interaction is still not clear (reviewed in Rinaudi and Giordano 2010). In some cases, bacterial components involved in biofilm formation also have crucial roles in different stages of the plant-bacteria interaction, like EPSs of different rhizobia or the common *nod* genes of *S. meliloti*. In other instances, defects in biofilm ability and establishment of symbiosis are only part of the pleiotropic phenotypes displayed by some mutants, making it difficult to demonstrate a link between biofilm formation and the ability to establish symbiosis with the host. Finally, in many cases, genes reported to affect biofilm formation on abiotic surfaces do not have a major impact on nitrogen-fixing nodule formation. It is possible that components required for biofilm formation on root surfaces might not be involved in biofilm formation on abiotic surfaces and vice versa. In this sense, the rhizobial adhesion protein RapA1 from *R. leguminosarum* enhances bacterial attachment and colonization on plant root surfaces but not on abiotic surfaces. These plant-specific adhesins, however, do not affect nodulation (Mongiardini et al. 2008). Likewise, mutants defective in cellulose and glucomannan are defective in root colonization but form biofilms indistinguishable from those of the wild type (Williams et al. 2008). The development of new quantitative high throughput assays to determine rhizobial biofilm formation on natural surfaces, such as soil particles or different plant roots, will facilitate the identification of environmental cues and genetic determinants that really have an impact under more natural conditions. In addition, nodulation experiments are usually performed under optimal conditions in which high density inocula are applied to plants grown under the best environmental conditions and in the absence of other potential microbial competitors. The actual contribution of rhizobial biofilms to the *Rhizobium*-legume symbiosis will most probably be unveiled by performing nodulation experiments under adverse environmental conditions and/or in the presence of other microorganisms which reflect more realistic conditions.

Regulation of Life-Style Switch in Rhizobia

As stated above, rhizobia may adopt two very different lifestyles, as a single bacterium in soil or, in intimate association with a legume plant. The successful interaction with the eukaryotic host usually requires coordination to form a surface-adherent biofilm in root hairs and initiate a concerted infection and nodule morphogenesis. Regardless of the adaptive advantages that these two different ecological niches provide, rhizobia require a rapid adaptive response and thus finely-tuned regulatory systems to enable and control the transition between these two different habitats. It has been proposed that the ability to carry out this transition efficiently and accurately plays a crucial role among bacteria interacting with plants (Danhorn and Fuqua 2007).

Rhizobia should interpret the diverse environmental and endogenous signals they receive to obtain information about both their extracellular environment and their intracellular physiological status and act accordingly. The integration of the extra- and intracellular signals is a very fine-tuned process that relies on two fundamental regulatory pathways: extracellular QS signalling and intracellular nucleotide-based second messenger signalling. From our point of view, these two regulatory pathways may allow rhizobia to control the complex processes that take place in the transition between motile (free living) to a surface-adherent lifestyle associated with the host plant (Fig. 3.2A).

Intracellular signalling comprises diverse signal transduction mechanisms by which the perception of a primary environmental cue alters the level of a second intracellular signal also known as second messenger. Subsequently, the second messenger transduces this signal eliciting a physiological response in the cell. Among the different second messengers described up to date (recently reviewed by Pesavento and Hengge 2009) the cyclic di-GMP (also called cyclic diguanylate, 3′,5′-cyclic diguanylic acid or c-di-GMP) has turned out to be one of the most important and ubiquitous

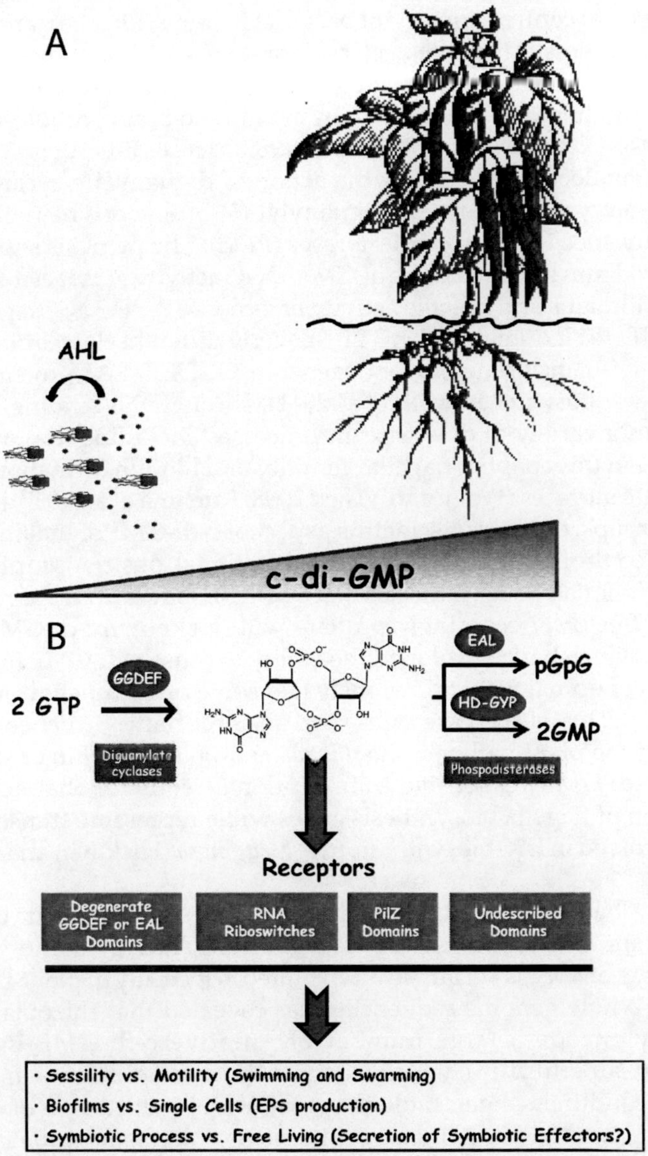

Figure 3.2 Role of bacterial second messenger c-di-GMP and Quorum Sensing in the regulation of rhizobial life-style. A) Proposed model where AHLs and c-di-GMP regulate the rhizobial transition from free-living to a symbiotic associative state with legume host plant. B) c-di-GMP is produced by diguanilate cyclases (DGCs) with GGDEF domains and degraded by phosphodiesterases (PDEs) bearing the EAL or HD-GYP domains. c-di-GMP binding receptors can be highly diverse and include proteins with a variety of domains as well as RNA motifs. Receptors exert its action downstream to promote different cellular functions (see text for more details).

molecules that controls important bacterial features, like the transition from planktonic to sessile lifestyles, cell cycle and virulence, among others (Mills et al. 2011). Cyclic di-GMP was discovered by Benziman and colleagues as an allosteric modulator that activated the membrane-bound cellulose synthase in *Gluconacetobacter xylinus* (Ross et al. 1987). The c-di-GMP is synthesized from two molecules of GTP by the action of diguanylate cyclases (DGC) and is hydrolyzed to 5′-phosphoguanylyl-(3′-5′)-guanosine (pGpG) and/or GMP by specific phosphodiesterases (PDE). The pGpG is subsequently hydrolyzed into two molecules of GMP. DGC activity is associated with the GGDEF domains and specific activity of c-di-GMP-PDE is associated with EAL or HD-GYP domains (Fig. 3.2B). Several different classes of molecules bind c-di-GMP including PilZ or degenerate GGDEF/EAL protein domains and mRNA riboswitches (Mills et al. 2011). Through this binding, c-di-GMP modulates a variety of downstream processes, including the two that are discussed in this chapter: flagellar motility and biofilm formation on biotic and abiotic surfaces (Wolfe and Visick 2008, Kuchma et al. 2007, Jonas et al. 2009). Perhaps, the most fascinating aspect of c-di-GMP signalling is that it exerts its control at multiple levels: transcriptional, posttranscriptional, and posttranslational, which indicates that c-di-GMP binds to an unprecedented range of effectors. Accordingly, proteins without known c-di-GMP binding domains, which have developed the ability to sense c-di-GMP through the evolution of domains that previously had some other function, are arising (Fig. 3.2B). This is the case with Clp, an important virulence regulator protein of the plant pathogen *Xanthomonas campestris* (Chin et al. 2010) or FleQ, a c-di-GMP-responsive transcriptional regulator that activate the expression of flagella biosynthesis genes while repressing transcription of genes involved in EPS biosynthesis in *P. aeruginosa* (Hickman and Harwood 2008).

Many bacteria contain multiple enzymes involved in c-di-GMP metabolism (i.e., synthesis or degradation), however the c-di-GMP regulatory network is yet largely unexplored, especially in plant-interacting bacteria. Whole-genome sequencing has revealed that rhizobia genomes generally encode a large number of putative c-di-GMP-interacting proteins (Galperin 2010), which is one of the main problems in the study of this second messenger. Little is yet known about the role of c-di-GMP in rhizobia, and so far, cellulose synthesis in *R. leguminosarum* is the only example of a function controlled by c-di-GMP levels (Ausmees et al. 1999). Also, a recent report showed that predicted GGDEF and EAL proteins from *S. meliloti* are involved in the control of motility, growth, EPS accumulation and competition for nodulation on alfalfa plants (Wang et al. 2010). However, the implication of c-di-GMP turnover has to be experimentally demonstrated in this latter case. In our laboratory, preliminary results have shown that c-di-GMP levels control cellular behaviour related with motility and biofilm formation in different symbiotic strains. In *S. meliloti*, *R. etli*, and *R. leguminosarum*, high intracellular levels of c-di-GMP repress swimming motility while promoting bacterial aggregation through the hyperproduction of EPSs (D. Pérez-Mendoza, unpublished data). On the other hand, inverse transition from a sessile to free-living lifestyle can also occur in the environment enabling bacterial dissemination to new environmental niches. A novel study suggests that c-di-GMP may be also a crucial signal in the regulation of rhizobia biofilm dispersion (Ma et al. 2011a). *E. coli* BdcA protein controls biofilm dispersal by binding c-di-GMP. Upon reducing the concentration of this second messenger, *E. coli* shifts to the planktonic state by increasing motility, decreasing aggregation and production of biofilm adhesins (Ma et al. 2011b). The introduction of *bdcA* via a broad-host range plasmid into *S. meliloti* increases motility as well as biofilm dispersal through a decrease in EPS production (Ma et al. 2011a).

Many strains of rhizobia use AHL to regulate different aspects of their physiology in a population density dependent manner. Detailed information regarding the components involved in rhizobia QS signalling is beyond the scope of this chapter and can be found in other reviews (González and Marketon 2003, Sánchez-Contreras et al. 2007, Downie 2010). However, a brief overview of the QS implication in the regulation of the transition from motile planktonic to settled rhizobial cells is provided here to aid the reader in understanding the key issues. Frequently, rhizobial settled microcolonies will divide and differentiate to form three-dimensional shapes that characterize a mature biofilm on the rhizoplane (Danhorn and Fuqua 2007). Chemotaxis and motility are required for the establishment of plant associated biofilm communities (Turnbull et al. 2001) and therefore, motility and biofilm formation are frequently related processes controlled by the same QS systems. It has already been mentioned that AHL (produced by SinI and accumulated under high population density conditions) down-regulate the expression of *S. meliloti* motility and chemotaxis genes through the action of the LuxR-type transcriptional regulator ExpR (Hoang et al. 2008, Bahlawane et al. 2008). Interestingly, the same ExpR/Sin QS system positively regulates biofilm formation in this bacterium through the production of EPS II, which provides the matrix for the development of structured and highly organized biofilms (Rinaudi and González 2009). However, the role of QS in motility and biofilm formation should be carefully evaluated in rhizobia since general considerations are not always applicable to all the species. For example, unlike what has been described in *S. meliloti*, either the disruption of one of the *R. leguminosarum* QS systems (CinIR) or its *expR* copy, led to an increase in biofilm formation (Edwards et al. 2009). Implication of QS systems in motility and/or biofilm formation have also been reported for other rhizobia. In *Mesorhizobium*, two different studies suggest that QS positively regulates biofilm formation in *M. huakuii*

(Wang et al. 2004) and *M. tianshanense* (Wang et al. 2008). In a similar way, QS signal molecules are involved in the control of *R. etli* swarming motility, where the dispersion of existing biofilms also helps (Daniels et al. 2004, Daniels et al. 2006).

Cyclic-di-GMP and QS have been proposed here as two central rhizobial "lifestyle-switch regulators" (Fig. 3.2). However, the complexity and the fined-tuned control required to govern the transition between these two very different lifestyles makes it reasonable to presume that other less-known, or even unknown, second messengers [e.g., cAMP-based, (p)ppGpp-based] or AHL-independent extracellular signalling molecules (e.g., plant-released compounds) may also have important roles. Furthermore, very interesting connections among these and other regulatory systems have been recently unveiled in other bacteria (Ueda and Wood 2009, Trimble and McCarter 2011) anticipating a complex but intriguing and fascinating future in this field.

Ackowledgements

The authors are grateful for the continued funding by grants from "Dirección General de Investigación" of the Spanish Ministry for Science and Innovation (MICINN) (BIO2010-18005, BIO2011-23032), grants from "Programa de Incentivos, Consejería de Innovación, Ciencia y Empresa" of the Andalusian Government (P08-CVI-3541, P10-CVI-5800), and FEDER funds. We also thank Dr. J. Sanjuán for reading the manuscript.

References

Althabegoiti, M.J., J.M. Covelli, J. Pérez-Giménez, J.I. Quelas, E.J. Mongiardini, M.F. López, S.L. López-García and A.R. Lodeiro. 2011. Analysis of the role of the two flagella of *Bradyrhizobium japonicum* in competition for nodulation of soybean. FEMS Microbiol. Lett. 319: 133–139.

Ames, P. and K. Bergman. 1981. Competitive advantage provided by bacterial motility in the formation of nodules by *Rhizobium meliloti*. J. Bacteriol. 148: 728–908.

Ausmees, N., H. Jonsson, S. Hoglund, H. Ljunggren and M. Lindberg. 1999. Structural and putative regulatory genes involved in cellulose synthesis in *Rhizobium leguminosarum* bv. *trifolii*. Microbiology. 145 (Pt 5): 1253–1262.

Bahlawane, C., M. McIntosh, E. Krol and A. Becker. 2008. *Sinorhizobium meliloti* MucR couples exopolysaccharide synthesis and motility. Mol. Plant Microbe Interact. 21: 1498–1509.

Becker, A., H. Bergés, E. Krol, C. Bruand, S. Rüberg, D. Capela, E. Lauber, E. Meilhoc, F. Ampe, F.J. de Brujin, J. Fourment, A. Francez-Charlot, D. Kahn, H. Küster, C. Liebe, A. Pühler, S. Weidner and J. Batut. 2004. Global changes in gene expression in *Sinorhizobium meliloti* 1021 under microoxic and symbiotic conditions. Mol. Plant Microbe Interact. 17: 292–303.

Braeken, K., M. Moris, R. Daniels, J. Vanderleyden and J. Michiels. 2008. Genetic determinants of swarming in *Rhizobium etli*. Microb. Ecol. 55: 54–64.

Caetano-Anollés, G., D.K. Cris-Estes and W.D. Bauer. 1988a. Chemotaxis of *Rhizobium meliloti* to the plant flavone luteolin requires functional nodulation genes. J. Bacteriol. 170: 3164–3169.

Caetano-Anollés, G., L.G. Wall, A.T. De Michelli, E.M. Macchi, W.D. Bauer and G. Favelukes. 1988b. Role of motility and chemotaxis in efficiency of nodulation by *Rhizobium meliloti*. Plant Physiol. 86: 1228–1235.

Caiazza, N.C., J.H. Merritt, K.M. Brothers and G.A. O'Toole. 2007. Inverse regulation of biofilm formation and swarming motility by *Pseudomonas aeruginosa* PA14. J. Bacteriol. 189: 3603–3612.

Chin, K.H., Y.C. Lee, Z.L. Tu, C.H. Chen, Y.H. Tseng, J.M. Yang, R.P. Ryan, Y. McCarthy, J.M. Dow, A.H. Wang and S.H. Chou. 2010. The cAMP receptor-like protein CLP is a novel c-di-GMP receptor linking cell-cell signaling to virulence gene expression in *Xanthomonas campestris*. J. Mol. Biol. 396: 646–662.

Coggan, K.A. and M.C. Wolfgang. 2012. Global regulatory pathways and cross-talk control *Pseudomonas aeruginosa* environmental lifestyle and virulence phenotype. Curr. Issues Mol. Biol. 14: 47–70.

Danhorn, T. and C. Fuqua. 2007. Biofilm formation by plant-associated bacteria. Annu. Rev. Microbiol. 61: 401–422.

Daniels, R., J. Vanderleyden and J. Michiels. 2004. Quorum sensing and swarming migration in bacteria. FEMS Microbiol. Rev. 28: 261–289.

Daniels, R., S. Reynaert, H. Hoekstra, C. Verreth, J. Janssens, K. Braeken, M. Fauvart, S. Beullens, C. Heusdens, I. Lambrichts, D.E. De Vos, J. Vanderleyden, J. Vermant and J. Michiels. 2006. Quorum signal molecules as biosurfactants affecting swarming in *Rhizobium etli*. Proc. Natl. Acad. Sci. U.S.A. 103: 14965–14970.

Dharmatilake, A.J. and W.D. Bauer. 1992. Chemotaxis of *Rhizobium meliloti* towards nodulation gene-inducing compounds from alfalfa roots. App. Environ. Microbiol. 58: 1153–1158.

Domínguez-Ferreras, A., R. Pérez-Arnedo, A. Becker, J. Olivares, M.J. Soto and J. Sanjuán. 2006. Transcriptome profiling reveals the importance of plasmid pSymB for osmoadaptation of *Sinorhizobium meliloti*. J. Bacteriol. 188: 7617–7625.

Donlan, R.M. and J.W. Costerston. 2002. Biofilms: survival mechanisms of clinically relevant microorganisms. Clin. Microbiol. Rev. 15: 167–193.

Downie, J.A. 2010. The roles of extracellular proteins, polysaccharides and signals in the interactions of rhizobia with legume roots. FEMS Microbiol. Rev. 34: 150–170.

Driscoll, J.A., S.L. Brody and M.H. Kollef. 2007. The epidemiology, pathogenesis and treatment of *Pseudomonas aeruginosa* infections. Drugs. 67: 351–368.

Duan, Q., M. Zhou, L. Zhu and G. Zhu. 2012. Flagella and bacterial pathogenicity. J. Basic Microbiol. 52: 1–8.

Edwards, A., M. Frederix, F. Wisniewski-Dye, J. Jones, A. Zorreguieta and J.A. Downie. 2009. The *cin* and *rai* quorum-sensing regulatory systems in *Rhizobium leguminosarum* are coordinated by ExpR and CinS, a small regulatory protein coexpressed with CinI. J. Bacteriol. 191: 3059–3067.

Felix, G., J.D. Duran, S. Volko and T. Boller. 1999. Plants have a sensitive perception system for the most conserved domain of bacterial flagellin. Plant J. 18: 265–276.

Flemming, H.C. and J. Wingender. 2010. The biofilm matrix. Nature Rev. Microbiol. 8: 623–633.

Fournier, J., A.C. Timmers, B.L. Sieberer, A. Jauneau, M. Chabaud and D.G. Barker. 2008. Mechanism of infection thread elongation in root hairs of *Medicago truncatula* and dynamic interplay with associated rhizobial colonization. Plant Physiol. 148: 1985–1995.

Fujishige, N.A., N.N. Kapadia, P.L. De Hoff and A.M. Hirsch. 2006. Investigations of *Rhizobium* biofilm formation. FEMS Microbiol. Ecol. 56: 195–206.

Fujishige, N.A., M.R. Lum, P.L. De Hoff, J.P. Whitelegge, K.F. Faull and A.M. Hirsch. 2008. *Rhizobium* common *nod* genes are required for biofilm formation. Mol. Microbiol. 67: 504–515.

Galperin, M.Y. 2010. Diversity of structure and function of response regulator output domains. Curr. Opin. Microbiol. 13: 150–159.

Gibson, K.E., M.J. Barnett, C.J. Toman, S.R. Long and G.C. Walker. 2007. The symbiosis regulator CbrA modulates a complex regulatory network affecting the flagellar apparatus and cell envelope proteins. J. Bacteriol. 189: 3591–3602.

Gómez-Gómez, L. and T. Boller. 2002. Flagellin perception: a paradigm for innate immunity. Trends Plant Sci. 7: 251–256.

González, J.E. and M.M. Marketon. 2003. Quorum sensing in nitrogen-fixing rhizobia. Microbiol. Mol. Biol. Rev. 67: 574–592.

Gurich, N. and J.E. González. 2009. Role of quorum sensing in *Sinorhizobium meliloti*-Alfalfa symbiosis. 191: 4372–4382.

Hall-Stoodley, L., J.W. Costerton and P. Stoodley. 2004. Bacterial biofilms: from the natural environment to infectious diseases. Nature Rev. Microbiol. 2: 95–108.

Hayashi, F., K.D. Smith, A. Ozinsky, T.R. Hawn, E.C. Yi, D.R. Goodlett, J.K. Eng, S. Akira, D.M. Underhill and A. Aderem. 2001. The innate immune response to bacterial flagellin is mediated by Toll-like receptor 5. Nature 410: 1099–1103.

Hellweg, C., A. Pühler and S. Weidner. 2009. The time course of the transcriptomic response of *Sinorhizobium meliloti* 1021 following a shift to acidic pH. BMC Microbiol. 15: 37.

Hickman, J.W. and C.S. Harwood. 2008. Identification of FleQ from *Pseudomonas aeruginosa* as a c-di-GMP-responsive transcription factor. Mol. Microbiol. 69: 376–389.

Hoang, H.H., A. Becker and J.E. González. 2004. The LuxR homolg ExpR, in combination with the Sin quorum sensing system, plays a central role in *Sinorhizobium meliloti* gene expression. J. Bacteriol. 186: 5460–5472.

Hoang, H.H., N. Gurich and J.E. González. 2008. Regulation of motility by the ExpR/Sin quorum-sensing system in *Sinorhizobium meliloti*. J. Bacteriol. 190: 861–871.

Janczarek, M., J. Kutkowska, T. Piersiak and A. Skorupska. 2010. *Rhizobium leguminosarum* bv. trifolii *rosR* is required for interaction with clover, biofilm formation and adaptation to the environment. BMC. Microbiol. 10: 284.

Jonas, K., O. Melefors and U. Römling. 2009. Regulation of c-di-GMP metabolism in biofilms. Future Microbiol. 4: 341–358.

Krol, E. and A. Becker. 2004. Global transcriptional analysis of the phosphate starvation response in *Sinorhizobium meliloti* strains 1021 and 2011. Mol. Genet. Genomics. 272: 1–17.

Kuchma, S.L., K.M. Brothers, J.H. Merritt, N.T. Liberati, F.M. Ausubel and G.A. O'Toole. 2007. BifA, a cyclic-Di-GMP phosphodiesterase, inversely regulates biofilm formation and swarming motility by *Pseudomonas aeruginosa* PA14. J. Bacteriol. 189: 8165–8178.

Ma, Q., G. Zhang and T.K. Wood. 2011a. *Escherichia coli* BdcA controls biofilm dispersal in *Pseudomonas aeruginosa* and *Rhizobium meliloti*. BMC. Res. Notes. 4: 447.

Ma, Q., Z. Yang, M. Pu, W. Peti and T.K. Wood. 2011b. Engineering a novel c-di-GMP-binding protein for biofilm dispersal. Environ. Microbiol. 13: 631–642.

Macnab, R.M. 2003. How bacteria assemble flagella. Annu. Rev. Microbiol. 57: 77–100.

McDougald, D., S.A. Rice, N. Barraud, P.D. Steinberg and S. Kjelleberg. 2011. Should we stay or should we go: mechanisms and ecological consequences for biofilm dispersal. Nat. Rev. Microbiol. 10: 39–50.

Meier, V.M., P. Muschler and B.E. Scharf. 2007. Functional analysis of nine putative chemoreceptor proteins in *Sinorhizobium meliloti*. J. Bacteriol. 189: 1816–1826.

Merritt, J.H., K.M. Brothers, S.L. Kuchma and G.A. O'Toole. 2007. SadC reciprocally influences biofilm formation and swarming motility via modulation of exopolysaccharide production and flagellar function. J. Bacteriol. 189: 8154–8164.

Merritt, J.H., D.G. Ha, K.N. Cowles, W. Lu, D.K. Morales, J. Rabinowitz, Z. Gitai and G.A. O'Toole. 2010. Specific control of *Pseudomonas aeruginosa* surface-associated behaviors by two c-di-GMP diguanylate cyclases. MBio. 1: e00183-10.

Miller, L.D., C.K. Yost, M.F. Hynes and G. Alexandre. 2007. The major chemotaxis gene cluster of *Rhizobium leguminosarum* bv. *viciae* is essential for competitive nodulation. Mol. Microbiol. 63: 348–362.

Mills, E., I.S. Pultz, H.D. Kulasekara and S.I. Miller. 2011. The bacterial second messenger c-di-GMP: mechanisms of signalling. Cell Microbiol.

Mongiardini, E.J., N. Ausmees, J. Perez-Gimenez, Althabegoiti M. Julia, Quelas J. Ignacio, S.L. Lopez-Garcia and A.R. Lodeiro. 2008. The rhizobial adhesion protein RapA1 is involved in adsorption of rhizobia to plant roots but not in nodulation. FEMS Microbiol. Ecol. 65: 279–288.

Morris, J. and J.E. González. 2009. The novel genes *emmABC* are associated with exopolysaccharide production, motility, stress adaptation, and symbiosis in *Sinorhizobium meliloti*. J. Bacteriol. 191: 5890–5900.

Nadal-Jimenez, P., G. Koch, J.A. Thompson, K.B. Xavier, R.H. Cool and W.J. Quax. 2012. The multiple signaling systems regulating virulence in *Pseudomonas aeruginosa*. Microbiol. Mol. Biol. Rev. 76: 46–65.

Nogales, J., A. Domínguez-Ferreras, C.V. Amaya-Gómez, P. van Dillewijn, V. Cuéllar, J. Sanjuán, J. Olivares and M.J. Soto. 2010. Transcriptome profiling of a *Sinorhizobium meliloti fadD* mutante reveals the role of rhizobactin 1021 biosynthesis and regulation genes in the control of swarming. BMC Genomics. 11: 157.

Nogales, J., L. Bernabéu-Roda, V. Cuéllar and M.J. Soto. 2012. ExpR is not required for swarming but promotes sliding in *Sinorhizobium meliloti*. J. Bacteriol. 194: 2027–2035.

O'Toole, G.A., L.A. Pratt, P.I. Watnick, D.K. Newman, V.B. Weaver and R. Kolter. 1999. Genetic approaches to study of biofilms. Methods Enzymol. 310: 91–109.

Pérez-Giménez, J., E.J. Mongiardini, M.J. Althabegoiti, J. Covelli, J.I. Quelas, S.L. Lopez-Garcia and A.R. Lodeiro. 2009. Soybean Lectin Enhances Biofilm Formation by *Bradyrhizobium japonicum* in the Absence of Plants. Int. J. Microbiol. 2009: 719367.

Pesavento, C. and R. Hengge. 2009. Bacterial nucleotide-based second messengers. Curr. Opin. Microbiol. 12: 170-176.

Rinaudi, L. V. and J. E. González. 2009. The low-molecular-weight fraction of exopolysaccharide II from *Sinorhizobium meliloti* is a crucial determinant of biofilm formation. J. Bacteriol. 191: 7216—7224.

Rinaudi, L.V. and W. Giordano. 2010. An integrated view of biofilm formation in rhizobia. FEMS Microbiol. Lett. 304: 1—11.

Rinaudi, L., N.A. Fujishige, A.M. Hirsch, E. Banchio, A. Zorreguieta and W. Giordano. 2006. Effects of nutritional and environmental conditions on *Sinorhizobium meliloti* biofilm formation. Res. Microbiol. 157: 867–875.

Rinaudi, L.V., F. Sorroche, A. Zorreguieta and W. Giordano. 2010. Analysis of the *mucR* gene regulating biosynthesis of exopolysaccharides: implications for biofilm formation in *Sinorhizobium meliloti* Rm1021. FEMS Microbiol. Lett. 302: 15–21.

Ross, P., H. Weinhouse, Y. Aloni, D. Michaeli, P. Weinbergerohana, R. Mayer, S. Braun, E. Devroom, G.A. Vandermarel, J.H. Vanboom and M. Benziman. 1987. Regulation of Cellulose Synthesis in *Acetobacter xylinum* by Cyclic Diguanylic Acid. Nature. 325: 279–281.

Rotter, C., S. Muhlbacher, D. Salamon, R. Schmitt and B. Scharf. 2006. Rem, a new transcriptional activator of motility and chemotaxis in *Sinorhizobium meliloti*. J. Bacteriol. 188: 6932–6942.

Russo, D.M., A. Williams, A. Edwards, D.M. Posadas, C. Finnie, M. Dankert, J.A. Downie and A. Zorreguieta. 2006. Proteins exported via the PrsD-PrsE type I secretion system and the acidic exopolysaccharide are involved in biofilm formation by *Rhizobium leguminosarum*. J. Bacteriol. 188: 4474–4486.

Ryder, C., M. Byrd and D.J. Wozniak. 2007. Role of polysaccharides in *Pseudomonas aeruginosa* biofilm development. Curr. Opin. Microbiol. 10: 6444–6448.

Sauer, K., A.K. Camper, G.D. Ehrlich, J.W Costerton and D.G. Davies. 2002. *Pseudomonas aeruginosa* displays multiple phenotypes during development as a biofilm. J. Bacteriol. 184: 1140–1154.

Sánchez-Contreras, M., W.D. Bauer, M. Gao, J.B. Robinson and D.J. Allan. 2007. Quorum-sensing regulation in rhizobia and its role in symbiotic interactions with legumes. Philos. Trans. R. Soc. Lond. B. Biol. Sci. 362: 1149–1163.

Scharf, B., H. Schuster-Wolff-Bühring, R. Rachel and R. Schmitt. 2001. Mutational analysis of the *Rhizobium lupini* H13-3 and *Sinorhizobium meliloti* flagellin genes: Importance of flagellin A for flagellar filament structure and transcriptional regulation. J. Bacteriol. 183: 5334–5342.

Schmitt, R. 2002. Sinorhizobial chemotaxis: a departure from the enterobacterial paradigm. Microbiology. 148: 627–631.

Soby, S. and K. Bergman. 1983. Motility and chemotaxis of *Rhizobium meliloti* in soil. Appl. Environ. Microbiol. 46: 995–998.

Soto, M.J., M. Fernández-Pascual, J. Sanjuán and J. Olivares. 2002. A *fadD* mutant of *Sinorhizobium meliloti* shows multicellular swarming migration and is impaired in nodulation efficiency on alfalfa roots. Mol. Microbiol. 43: 371–382.

Soto, M.J., A. Domínguez-Ferreras, D. Pérez-Mendoza, J. Sanjuán and J. Olivares. 2009. Mutualism versus pathogenesis: the give-and-take in plant-bacteria interactions. Cell Microbiol. 11: 381–388.

Sourjik, V. and R. Schmitt. 1996. Different roles of CheY1 and CheY2 in the chemotaxis of *Rhizobium meliloti*. Mol. Microbiol. 22: 427–436.

Sourjik, V., P. Muschler, B. Scharf and R. Schmitt. 2000. VisN and VisR are global regulators of chemotaxis, flagellar, and motility genes in *Sinorhizobium* (*Rhizobium*) *meliloti*. J. Bacteriol. 182: 782–788.

Tambalo, D.D., D.E. Bustard, K.L. Del Bel, S.F. Koval, M.F. Khan and M.F. Hynes. 2010a. Characterization and functional analysis of seven flagellin genes in *Rhizobium leguminosarum* bv. *viciae*. Characterization of *R. leguminosarum* flagellins. BMC Microbiology. 10: 219.

Tambalo, D.D., K.L. Del Bel, D.E. Bustard, P.R. Greenwood, A.E. Steedman and M.F. Hynes. 2010b. Regulation of flagellar, motility and chemotaxis genes in *Rhizobium leguminosarum* by the VisN/R-Rem cascade. Microbiology. 156: 1673–1685.

Tambalo, D.D., C.K. Yost and M.F. Hynes. 2010c. Characterization of swarming motility in *Rhizobium leguminosarum* bv. viciae. FEMS Microbiol. Lett. 307: 165–174.

Tolker-Nielsen, T., U.C. Brinch, P.C. Ragas, J.B. Andersen, C.S. Jacobsen and S. Molin. 2000. Development and dynamics of *Pseudomonas* sp. biofilms. J. Bacteriol. 182: 6482–6489.

Trimble, M.J. and L.L. McCarter. 2011. Bis-(3'-5')-cyclic dimeric GMP-linked quorum sensing controls swarming in *Vibrio parahaemolyticus*. Proc. Natl. Acad. Sci. U.S.A.

Turnbull, G.A., J.A. Morgan, J.M. Whipps and J.R. Saunders. 2001. The role of bacterial motility in the survival and spread of *Pseudomonas fluorescens* in soil and in the attachment and colonisation of wheat roots. FEMS Microbiol. Ecol. 36: 21–31.

Ueda, A. and T.K. Wood. 2009. Connecting quorum sensing, c-di-GMP, *pel* polysaccharide, and biofilm formation in *Pseudomonas aeruginosa* through tyrosine phosphatase TpbA (PA3885). PLoS Pathog. 5: e1000483.

Vanderlinde, E.M., A. Muszynski, J.J. Harrison, S.F. Koval, D.L. Foreman, H. Ceri, E.L. Kannenberg, R.W. Carlson and C.K. Yost. 2009. *Rhizobium leguminosarum* biovar viciae 3841, deficient in 27-hydroxyoctacosanoate-modified lipopolysaccharide, is impaired in desiccation tolerance, biofilm formation and motility. Microbiology. 155: 3055–3069.

Verstraeten, N., K. Breken, B. Debkumari, M. Fauvart, J. Fransaer., J. Vermant and J. Michiels. 2008. Living on a surface: swarming and biofilm formation. Trends Microbiol. 16: 496–506.

Wadhams, G.H. and J.P. Armitage. 2004. Making sense of it all: bacterial chemotaxis. Nat. Rev. Mol. Cell Biol. 5: 1024–1037.

Wang, H., Z. Zhong, T. Cai, S. Li and J. Zhu. 2004. Heterologous overexpression of quorum-sensing regulators to study cell-density-dependent phenotypes in a symbiotic plant bacterium *Mesorhizobium huakuii*. Arch. Microbiol. 182: 520–525.

Wang, P., Z. Zhong, J. Zhou, T. Cai, and J. Zhu. 2008. Exopolysaccharide biosynthesis is important for *Mesorhizobium tianshanense*: plant host interaction. Arch. Microbiol. 189: 525–530.

Wang, Y., J. Xu, A. Chen, Y. Wang, J. Zhu, G. Yu, L. Xu and L. Luo. 2010. GGDEF and EAL proteins play different roles in the control of *Sinorhizobium meliloti* growth, motility, exopolysaccharide production, and competitive nodulation on host alfalfa. Acta Biochim. Biophys. Sin. 42: 410–417.

Wells, D.H., E.J. Chen, R.F. Fisher and S.R. Long. 2007. ExoR is genetically coupled to the ExoS-ChvI two-component system and located in the periplasm of *Sinorhizobium meliloti*. Mol. Microbiol. 64: 647–664.

Williams, A., A. Wilkinson, M. Krehenbrink, D.M. Russo, A. Zorreguieta and J.A. Downie. 2008. Glucomannan-mediated attachment of *Rhizobium leguminosarum* to pea root hairs is required for competitive nodule infection. J. Bacteriol. 190: 4706–4715.

Wolfe, A.J. and K.L. Visick. 2008. Get the message out: cyclic-di-GMP regulates multiple levels of flagellum-based motility. J. Bacteriol. 190: 463–475.

Yost, C.K., P. Rochepeau and M.F. Hynes. 1998. *Rhizobium leguminosarum* contains a group of genes that appear to code for methyl-accepting chemotaxis proteins. Microbiology 144: 1945–1956.

Yost, C.K., K.L. Del Bel, J. Quandt and M.F. Hynes. 2004. *Rhizobium leguminosarum* methyl-accepting chemotaxis protein genes are down-regulated in the pea nodule. Arch. Microbiol. 182: 505–513.

4

Nod Factor Production and Abiotic Stress in *Rhizobium*

B. Guasch-Vidal,[1] *A.A.N. van Brussel,*[2] *J. Estévez,*[3] *R. Bellogín,*[1] *F.J. Ollero,*[1] *M.R. Espuny*[1] and *M. Megías*[3,*]

Introduction

Rhizobium bacteria are soil microorganisms which are able to induce root nodules in leguminous plants. In the nodules, rhizobium functions as a nitrogen fixing organel so that leguminous plants can grow without externally supplied fixed nitrogen. This gives legumes a selective advantage to non-leguminous plants in an environment without or with low concentrations of fixed nitrogen. This property can also be used in agriculture to grow (leguminous) plants without nitrogen fertilizer. Two organisms, a prokaryote (rhizobium) and a eukaryote (leguminous plant) together form a completely new organ, the symbiotic root nodule, which means that they have to communicate with one another. A complex series of events, coordinated by host and bacterial signalling, underlie and lead to the development of this symbiotic interaction. Rhizobia respond to flavonoids

[1]Departamento de Microbiología, Facultad de Biología, Universidad de Sevilla. Avd. de Reina Mercedes 6, 41012 Sevilla, Spain.
[2]Sylvius Laboratory, Institute of Biology, Leiden University. Sylviusweg. 70, 2333 BE, Leiden, The Netherlands.
[3]Departamento de Microbiología y Parasitología, Facultad de Farmacia, Universidad de Sevilla, Profesor García González, 41012 Sevilla, Spain.
*Corresponding author: megiasg@us.es

exuded by the leguminous plant roots by secreting Nod factors, which are the key molecular signals to induce nodule morphogenesis and allow the bacteria to initiate the symbiosis with the host plant.

An important feature of symbiotic root nodule formation is the so-called host specificity. Groups of rhizobium bacteria and plants can form root nodules with one another, however only within the group, and not with plants/rhizobia outside this group. The term used for such a group is cross-inoculation group. Exceptions to this rule of host specificity are the so-called promiscuous rhizobia which can form root nodules with many very different plants. It must be mentioned that host specificity is mainly determined by the structure of the Nod factor(s).

Apart from complicated signalling as indicated above, both symbiotic partners are constantly challenged by a variety of stresses in their natural environments, including nutrient limitation and exposure to different abiotic stresses, such as elevated temperature, acidity, salinity, high osmolarity and oxidative shock (Zahran 1999). That is, not only in the rhizosphere, but also during the different steps of infection leading to the formation of nitrogen-fixing nodules, rhizobia have to face different stresses. Bacteroids, the nitrogen-fixing form of the rhizobia inside the nodules, may perceive the symbiosome compartment as a stressful environment and have to adapt to these conditions. For example, the *bacA* gene in the rhizobia of inverted repeat-lacking clade is required to stand these conditions (Haag et al. 2011).

In conclusion, the ability of rhizobia to adapt to and/or resist several adverse conditions is a complex phenomenon that involves many physiological and biochemical processes and changes in gene expression. Thus, in order to establish an efficient symbiosis, rhizobia have evolved mechanisms to alter gene expression patterns and the activity of the enzymes and transport proteins.

It has recently been reported that different adverse environmental conditions (stress) can exert a great influence on the amount and chemical composition of the population of Nod factors produced by *Rhizobium tropici* CIAT899 (Morón et al. 2005, Estévez et al. 2009). These findings open up new avenues in research focused on the regulation of Nod factor production and on the mechanisms of adaptation of this strain to the environment.

The Rhizobial Signalling Molecule: Nod Factors

Chemical Structure of Nod Factors

For more than 20 yr, Nod factors have been recognized as essential signalling molecules for the development of nitrogen-fixing nodules on legume roots (Lerouge et al. 1990). The Nod factors produced by various rhizobium bacteria are similar to one another with regard to their general chemical structure (Fig. 4.1). All the bacterial Nod factors characterized up to date share a core structure consisting of a backbone from two to six β-1,4-linked *N*-acetyl glucosamine (GlcNAc) residues. The non-reducing terminal residue of these (chitin) fragments is substituted with a long-chain fatty acid moiety, ranging from 16 to 20 carbon units, which can be saturated or unsaturated in one or more of its bonds. The length as well as the number and positions of the double bonds in the fatty acyl chain are important for the efficient formation of nodule primordia. Due to their structural similarity to chitin molecules, which are β-1,4 glucosamine polymers present in fungi and in the external skeleton of insects, Nod factors are also known as lipo-chitin oligosaccharides (LCO). These LCO structures are further elaborated by the addition of reducing and/or to the non-reducing terminal GlcNAc residues of one or more of a range of different small organic or other monosaccharide substituents (Spaink 2000). Often, *N*-methyl, *O*-acetyl and *O*-carbamoyl groups are commonly found on the non-reducing terminal residue of the molecules and *L*-fucosyl, 2-*O*-methyl-fucosyl, 4-*O*-acetyl-fucosyl, acetyl substituents or a sulphate ester on the reducing-terminal residue (D'Haeze and Holsters 2002, Downie 2010).

When induced by plant exudates in the rhizosphere, the majority of rhizobia synthesize and secrete populations of Nod factors that consist of a mixture with a variable number of different individual structures. For instance, only two different structures in the Nod factors population

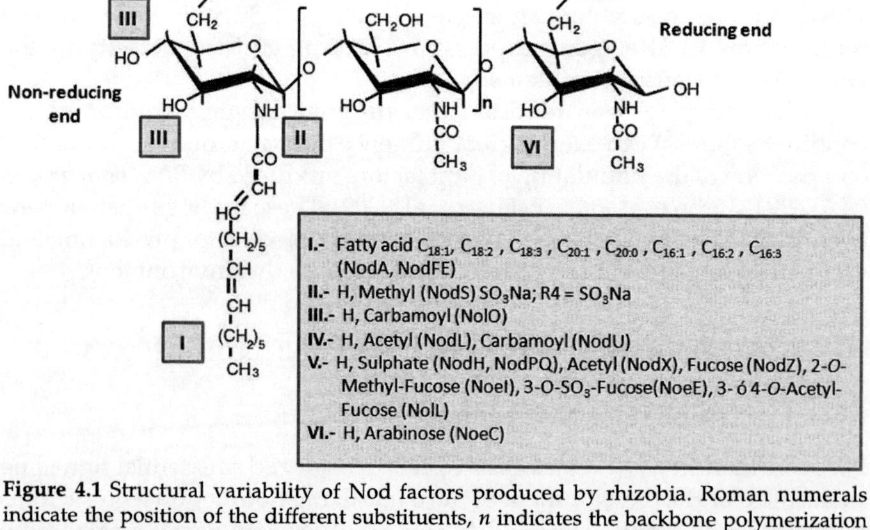

Figure 4.1 Structural variability of Nod factors produced by rhizobia. Roman numerals indicate the position of the different substituents, *n* indicates the backbone polymerization degree. Enzymes which synthesize or couple the most common substituents to the backbone are indicated between brackets.

produced by *Rhizobium etli* CFN42 have been identified (Poupot et al. 1995), but 60 different molecules were described to *R. galegae* HAMBI207 (Yang et al. 1999). In one way or another, these individual structures act cooperatively to induce the right response in the specific host plant (Minami et al. 1996). Qualitative and quantitative aspects of these Nod factor populations are strain-specific. This means that changes in the chemical composition and in the relative number of the different Nod factors produced by a strain can change the host range of the bacterial symbiont itself (Perret et al. 2000). Not only the number of *N*-acetylglucosamine residues and the length and saturation of the acyl chain, but also the nature of modifications on the basic backbone are important for the biological activity of the Nod factors and define much of the host specificity observed in each symbiotic relationship. For instance, the lack of a sulphated group in the Nod factors produced by *Sinorhizobium* (*Ensifer*) *meliloti* means a loss in the capacity of this strain to nodulate alfalfa plants (Schultze et al. 1992). In the same way, the loss of the fucosyl group in the Nod factors of *E. fredii* reduces the host range of the strain (Lamrabet et al. 1999).

Moreover, there is evidence that some Nod factors exist which have different structures from the one just described above, that is, they have different molecular backbones and/or carry different kinds of substituents. This is the case for the pentameric Nod factors from *E. fredii* USDA191, which carry a glucosyl group instead of the GlcNAc residue which is located in the centre of the backbone (Berc-Ferté et al. 1996). Some other Nod factors harbour a wide range of residues which are not often present at the reducing terminal of these molecules. Thus, glycerol and mannose groups have been identified as common substituents in the C_1 position of the reducing end of the Nod factors of *Bradyrhizobium elkanii* USDA61 (Stokkermans et al. 1996) and *R. tropici* CIAT899 (Folch-Mallol et al. 1996), respectively. The Nod factors produced by *R. gallicum* bv. gallicum R592 also differ from the general structure of Nod factors because they bear a glucosaminitol group instead of the glucosamine at the reducing end of the molecule (Soria-Díaz et al. 2006).

The Nodulation Genes

The biosynthesis and secretion of the Nod factors is mediated by a great number of bacterial proteins which are codified by the nodulation genes (Table 4.1). The nodulation genes were initially called *nod* genes, but the constant description of new nodulation genes made it necessary to use the *nol* and *noe* designations too. The *nod*, *nol* and *noe* genes are usually located in a large rhizobium plasmid called the symbiotic plasmid (pSym) in almost all rhizobium. But in some species, such as *B. japonicum* and *Mesorhizobium loti*, the nodulation genes are grouped as symbiotic islands

Table 4.1 Some nodulation genes (Downie 1998, Perret et al. 2000, Vance 2002).

Gene	Function of the protein
nodA	Acyltransferase
nodB	Deacetylase
nodC	Oligochitin-synthase
nodD	Transcriptional regulator of nodulation genes
nodEF	Synthesis of polyunsaturated fatty acids
nodH	Sulphotransferase
nodIJ, nodT	Nod factor secretion system
nodL, nodX, nolL	*O*-acetyltransferase
nodM	Glucosaminesynthase
nodPQ	Synthesis of sulphate in active form
nodS	*N*-methyltransferase
nodU	6-*O*-carbamoyltransferase
nodZ	Fucosyltransferase
noeE	Sulphuryltransferase
noeI	2-*O*-methyltransferase
noeK	Phosphomannomutase
nolK	Fucose synthase
nolR	Nodulation gene repressor

on the chromosome (Downie 1998). Moreover, the nodulation genes are usually grouped as genetic operons which are widely conserved among the genomes of different genera of rhizobium. Thus, for example, the *nodABCIJnolOnoeI* operon from *Rhizobium* sp. NGR234 is also present in the genome of *E. fredii* strains (Schlaman et al. 1998).

The different rhizobia harbour different nodulation genes, and so this heterogeneity is responsible for the different Nod factor structures each rhizobium is able to produce. Several studies made it possible to figure out the roles of the different nodulation proteins and the order in which they act in the Nod factor biosynthesis pathway. It has been proposed that the Nod factor biosynthesis takes place in enzymatic complexes which appear in the interphase between the cytoplasmic membrane and the cytosol (Dénairé et al. 1996). The first step in the synthesis of the basic structure of the Nod factors is catalyzed by the *N*-acetylglucosamine transferase codified by the *nodC* gene. The NodC protein is located in the cytoplasmic membrane and catalyzes the association between *N*-acetylglucosamine molecules which were previously synthesized by the NodM protein or by homologous proteins involved in the general metabolism of the bacterium. The elongation takes place by the addition of *N*-acetylglucosamine molecules to the non-reducing end of the new oligomer (Geremia et al. 1994). Then, the NodB protein removes the *N*-acetyl residue from the non-reducing end of the molecule, and finally, the NodA protein transfers an acyl group to that position (Perret et al. 2000).

The nodulation genes have been traditionally divided into different groups according to their function. Initially, the *nodA*, *nodB* and *nodC* genes were called the common nodulation genes as they were thought to be functionally interchangeable between different rhizobium species without disturbing the symbiotic properties of the receptor strain. Nevertheless, these days it is known that this is not true as it was reported that the NodA protein has different specificity for different fatty acids and the NodC protein determines the backbone length, so they both contribute to the host specificity as well (Roche et al. 1996). In this way, in *E. meliloti* and *R. leguminosarum* the acyl transferase NodA is specialized in the transference of α and β-unsaturated fatty acids to the new molecules (Ritsema et al. 1996, Debellé et al. 1996), respectively. Furthermore, the length of the *N*-acetylglucosamine molecules determined by the NodC protein is different in *E. meliloti*, *M. loti* and *R. leguminosarum* and specific of each strain (Kamst et al. 1995).

Another classical group of nodulation genes are those known as the *hsn* genes (host-specific nod genes). These genes encode the proteins responsible for the incorporation of the different specific residues on the basic backbone of the Nod factors in each rhizobium strain. Given that the Nod factor structures produced by the bacteria are specific of the strain, these genes and their function in each strain are also specific. Thus, the proteins codified by the *nodHPQ* genes are involved in the sulphation of the Nod factors of *R. tropici* (Folch-Mallol et al. 1996), and the NodZ protein synthesizes and/or incorporates fucosyl groups to the Nod factor backbones of *E. fredii* HH103 (Lamrabet et al. 1999).

Once they have been synthesized, Nod factors are secreted by the bacteria due to the activity of the NodI, NodJ and NodT transport proteins. The secretion rate depends on the environmental conditions the bacteria encounter such as the temperature or the pH (MacKay and Djordjevic 1993, Downie 1998). Mutations in these transport genes only cause a little decrease on nodulation capacity of the strain because their function can be taken over by other transport genes in the cell, even if they are not implied in symbiosis (Higgins et al. 1990). For instance, the Nod factors could cross the cytoplasmic membrane helped by the Ndv transport system, which is usually implied in cyclic glucan transport in several strains, or by the proteins codified by the *kpsT*/*kpsM* genes, which are usually implied in the secretion of the capsular polysaccharide in several rhizobia (Stanfield et al. 1988, Whitfield and Valvano 1993).

Regulation of Nodulation Gene Expression by NodD

Since the Nod factor biosynthesis is a process closely linked to the establishment of the symbiosis, the nodulation gene expression takes place when the bacteria detect a compatible host plant through the root exudates it produces. The expression of the majority of the nodulation genes requires the presence of flavonoids, which are phenolic compounds exuded by the plant roots, and their association with the NodD protein in the inner membrane of the bacteria (Schell 1993). The carboxy-terminal end of the NodD protein takes part in the specific recognition of the flavonoid structure (Brencic and Winans 2005). Thus, the NodD protein has a dual nature as the key regulator in nodulation gene expression and also as a sensor of the plant signals to the bacteria (Schlaman et al. 1992).

The NodD protein belongs to the LysR family of transcriptional regulators and is constitutively transcribed in rhizobia. It binds to the plant flavonoids or derived compounds with different specificity depending on both the symbiotic partners (Broughton et al. 2000). When bound to its carboxy-terminal region, flavonoids induce conformational changes in the NodD protein. As a result, NodD is able to bind to conserved DNA regions of 49 pb located in the promoter regions of the majority of the nodulation genes and thus altering their expression patterns. These regions, called the *nod*-boxes, contain two or three palindrome inverted sequences of the ATC-N9-GAT motif where the NodD protein associates as a tetramer in the same way as has been described for many others LysR regulators (Schell 1993). In some species, such as *R. leguminosarum* bv. viciae and *E. meliloti*, the NodD protein also binds to the *nod*-boxes in the absence of inducers, but the presence of flavonoids has been described as fundamental to the nodulation genes induction (Perret et al. 2000).

NodD together with the plant roots exudates control which plants are going to be nodulated by the bacteria. When transferring the *nodD1* gene from *Rhizobium* sp. NGR234 to the narrow-host range rhizobium strain *R. leguminosarum* bv. trifolii it is possible to extend its symbiotic capacity to even the non-leguminous plant *Parasponia andersonii* (Bender et al. 1988). However, the precise mechanism by which NodD proteins respond to the inductor and initiate the nodulation gene transcription still remains unknown. In *R. leguminosarum* bv. viciae the NodD protein is located in the cytoplasmic membrane in order to facilitate the association with flavonoids. However, in *E. meliloti* a large amount of the protein is located in the cytosol and migrates towards the membrane where NodD seems to contact the flavonoids (Kondorosi et al. 1989). Also the presence of some chaperone-like proteins, homologous to the GroEL protein from *Escherichia coli*, have been described in *E. meliloti* which appear to be essential for the activation of the *nod* genes transcription depending on NodD (Bittner et al. 2007). It has been proposed that in this strain those proteins participate in the translocation of NodD towards the membrane and its storage in a soluble conformation (Long et al. 1991). It has also been suggested that the flavonoid-NodD complex is formed on the *nod*-boxes. Although the existence of this complex has not been demonstrated yet, there is some indirect evidence that indicates the interaction between these two molecules (Peck et al. 2006).

The *nod*D gene is conserved among all *Rhizobium* genera described up to now and is normally located on the symbiotic plasmid, either as a unique-copy gene, as in *R. leguminosarum* bv. viciae and *A. caulinodans*, or with more than one *nodD* copy (Sousa et al. 1993, van Rhijn et al. 1993, Fellay et al. 1995, Schlaman et al. 2006) which is more common. These species with more than one *nodD* copy have developed specific responses to different groups of flavonoids. Therefore it has been speculated that the *nodD* gene divergence might be the result of an evolutionary process adapting the Nod factor chemical structure to be able to interact with multiple hosts which secrete different kinds of flavonoids (Demont et al. 1994). Whereas *E. meliloti* has three *nodD* copies and it is necessary to mutate all of them to suppress the nodulation capacity (Honma et al. 1990), in *Rhizobium* sp. NGR234 the mutation of only one of the two *nodD* copies removes the nodulation ability of this strain (Relic et al. 1994). There are also two *nodD* copies in *E. fredii*, but only the *nodD1* is implied in the regulation of the *nod* genes expression and this expression is auto-regulated negatively (Vinardell et al. 2004). Strain *R. tropici* CIAT899 has five NodD regulator proteins (van Rhijn et al. 1993). The *nodD1* is essential for nodulation in *Leucaena leucocephala* and in *Macroptilium atropurpureum*; while the *nodD1* mutation greatly reduces nodulation of *R. tropici* CIAT899 in *Phaseolus vulgaris* plants (van Rhijn et al. 1994).

Besides flavonoids, sometimes the NodD protein can also exert a repressive activity on the nodulation gene expression (Spaink et al. 1989) In this sense, the NodD2 protein in *Rhizobium* sp. NGR234, *B. japonicum* and *E. fredii*, which lacks the classical *nod*-box, acts as a repressor of the *nod* gene expression and is also involved in the exopolysaccharide production. Different mutants in the *nodD2* gene show up to a five-fold increase in their Nod factor production compared to that of the wild type strain (Fellay et al. 1998, Machado and Krishnan 2003).

Other Regulatory Genes

Besides NodD, in some rhizobium species other proteins are also involved in regulation of the expression of nodulation genes, which is a really complex process in these bacteria. One of these proteins is the SyrM protein, which belongs to the LysR family of transcriptional regulators and appears to co-regulate the Nod factor synthesis in a flavonoid independent manner

(Barnett et al. 1998). Normally, the SyrM proteins are also implied in the biosynthesis of the extracellular polysaccharides which, together with Nod factors and flavonoids, are key signals to the specificity of the symbiosis (Etzler et al. 1999). In *E. meliloti*, the SyrM protein activates the expression of the *nodD3* gene which, in turn, activates the expression of the *syrM* gene, and thus in a multicopy state SyrM leads to the activation of the *nod* gene transcription in the absence of either flavonoids or *nodD1* (Kondorosi et al. 1991). Moreover, in *E. meliloti* SyrM also controls the exopolysaccharide biosynthesis through the regulation of the *syrA* espression (Barnett and Long 1990). In *Rhizobium* sp. NGR234 the *syrM* gene is located in the symbiotic plasmid and has two copies. The SyrM1 may need other regulators to act, as occurs in *E. meliloti*, but SyrM2 has a *nod*-box and so its expression is induced by flavonoids (Hanin et al. 1998). The *syrM* and *nodD3* genes are implied in the control of the *nod* gene expression together with the nitrogen metabolism regulation system *ntr*, which responds to the level of ammonia, the first product of the nitrogen fixation in the cell. This regulatory mechanism might explain the repressive effect that the excess of ammonia exerts on nodulation (Dusha and Kondorosi 1993).

There are also other Nod proteins implied in the regulation of the Nod factor production. Thus, in the genus *Bradyrhizobium*, the two-component regulatory system composed by NodV and NodW, responds to the isoflavonoid genistein and is essential for nodulation in *M. atropurpureum*, *Vigna radiata* and *V. unguiculata* (Loh et al. 1997). NodV is located on the membrane and has a sensor and kinase function which phosphorylates NodW. NodW is a transcriptional regulator which in turn activates the nodulation genes. It is known that NodW together with NodD1 stimulate the common *nod* genes expression, but the precise induction mechanism is still unclear. Besides its positive control on the induction of some *nod* genes, it is possible that NodW could modify plant signals involved in NodD activation (Sanjuán et al. 1994).

The nodulation gene expression is also subjected to a negative control. After the initial induction by flavonoids, a range of the nodulation genes must be repressed in order to get a successful and optimal nodulation. Thus, the *nolR* repressor gene has been identified in the chromosome of *Rhizobium* and *Sinorhizobium* (*Ensifer*) genera (Kiss et al. 1998). NolR activation and, thus, the repression of many nodulation genes is essential for an optimal nodulation of *E. meliloti* in *Medicago sativa* (Kondorosi et al. 1989) and of *R. leguminosarum* bv. viciae TOM in *Pisum sativum* cv. Wisconsin Perfection (Kiss et al. 1998). It has been described that in *E. meliloti* as well as in *E. fredii* HH103 the NolR protein binds to the promoter regions of *nodD1*, *nodD2* and *nodABC* and represses their transcription (Vinardell et al. 2004).

Another *nod* gene repressor, known as *nolA*, was first identified in the *Bradyrhizobium* genera. The *nolA* gene lacks the classical *nod*-box, so its expression does not depend on flavonoids, but it requires the presence of NodD2 to exert its repressive function. It has been postulated that *nolA* could act by inhibiting the nodulation signals to the bacteria (García et al. 1996, Schlaman et al. 1998).

In the same way, there is some evidence that the regulation of the Nod factor production takes place at the post-transcriptional level. This is the case of NodX protein, which acetylates the glucosamine backbone of Nod factors in the R4 position depending on the growth temperature the bacteria encounter (Olsthoorn et al. 2000).

There is little evidence about the regulation of the Nod factor production in the soil or in the plant after the initial steps of the infection process. Although it has been demonstrated that the essential *nod* genes are repressed in later steps of the symbiotic relationship (Schlaman et al. 1991), there is no information about the precise mechanisms implied in this negative regulation process.

Environmental Stresses which Most Commonly Affect the Rhizobium-Legume Symbiosis

Rhizobium-legume symbiosis represents the most important nitrogen-fixing mechanism, which may have the potential to increase nitrogen input in arid and semi-arid ecosystems, which are often located in developing countries. Leguminous plants will play an important role in the development of sustainable agricultural systems, particularly in those areas. However, few agriculturalists have the luxury of production under ideal conditions. The great majority of abiotic stresses, such as salinity, osmotic stress, soil acidity or temperatures, are integral parts of the crop cycle and limit legume crop production. In root nodule formation both partners may encounter stress and also the root nodule morphogenesis itself is sensitive to different stresses.

It is known that soil acidity, salinity and temperature affect the persistence of rhizobium in the soil and in the rhizosphere of plants, as well as their efficiency of nodulation, especially in tropical areas (Anyango et al. 1995, Bohlool et al. 1992, Hungría et al. 1993). Therefore strategies to improve the symbiosis efficiency and legume production under these conditions should target both symbiotic partners, together with appropriate crop and soil management (Zahran 1999, Drevon et al. 2001, Yamaguchi and Blumwald 2005).

Salinity

Among all the adverse conditions affecting agricultural soils, salinity is the one which has influenced human population settlement the most. Lands affected by salt stress spread all over the climate regions of the Earth, from the tropics to the poles, and at different altitudes, from the sea level to high mountains (Singh and Chatrath 2001). Although these areas are not well-defined, it has been estimated that a third part of the cultivable lands in the world are affected by salt stress. So salinity is one of the main restrictions for agricultural production (Zahran 1999).

High salinity in the soil can result from naturally high salt levels or from local salt accumulation due to irrigation or the application of chemical fertilizers. Arid and semiarid soils often suffer from elevated salinity due to the high evaporation rates, which is a serious threat to agriculture, particularly in the third world. Thus, nitrogen-fixing legumes tolerant to salinity represent an important alternative to improve fertility in those areas. These days, more than 20 percent of the cultivated soils and about 50 percent of the irrigated lands in the world are listed as potentially saline, and most of them are located in the Mediterranean basin (Singh and Chatrath 2001).

Common salt (NaCl) is an inorganic compound that exerts both an ionic and osmotic stress effects on the cell, due to the ionic strength it causes in each aqueous solution and its high ability to retain water molecules. When water deficiency causes a decrease in soil or plant water potential, plants develop defensive responses in order to overcome the adverse situation. In fact, the most common effect of salinity on plants is a reduction in their growth due to a decrease of the osmotic potential in the soil, a specific toxicity which is often associated with an excessive absorption of Na^+ and Cl^- ions, a nutritional disorder due to the interference of saline ions with essential nutrients, and the combination of the effects just mentioned. As a result of these primary effects, secondary effects, such as an oxidative stress, often arise in the plant.

To overcome these adverse effects, plants can maintain turgor or reduce the rate of turgor loss by actively accumulating inorganic solutes. Under salt stress osmotic adjustment can occur through compartmentalization of toxic ions from the cytoplasm into the vacuole (Hasegawa et al. 2000). Plants can also protect themselves from drought and salt stress by accumulating organic compounds called compatible solutes. Compatible solutes are non-toxic, highly soluble and uncharged molecules which do not inhibit normal metabolic reactions but stabilize proteins and membranes (McNeil et al. 1999). Compatible solutes also protect enzymes from heat inactivation (Paleg et al. 1984).

Leguminous plants are less tolerant to salinity than other vegetables (Greenway and Munns 1980) but some variability exits among the different species. Tolerance of rhizobium to salinity is also very important for the symbiosis. For an extended review on the mechanisms of adaptation of the rhizobium-legume symbiosis to salinity, see Chapter 5.

Soil Acidity

Together with salinity, soil acidity constitutes one of the main problems in the agricultural production worldwide, because they both cause a decrease in the crop productivity (Bordeleau and Prevost 1994, Graham 1992). Low pH values can affect all the steps in the *Rhizobium*-legume symbiosis, ranging from the survival of the bacteria in the soil, root hairs infection, nodule development and nitrogen fixation (Wood et al. 1984). This reduction in the nodulation rate is especially evident when legumes grow in pH values under 5.0, due to the inability of the rhizobial inoculum to survive in these severely adverse conditions (Graham et al. 1982, Carter et al. 1994). Nevertheless, there are some exceptions as it has been reported that the nodulation capacity of *R. tropici* CIAT899, a highly resistant bacterium, is as good at pH 4.5 as at pH 7.0 (Morón et al. 2005).

Acidity also becomes a challenge in nodule development, affecting both the plant and bacteria, as well as their interaction. The mechanisms bacteria have to overcome acidity are known as acid resistance (AR), which has mainly been investigated in *E. coli*. In rhizobium, several genes have been characterized as essential for AR (Tiwari et al. 1996a,b, Riccillo et al. 2000, Vinuesa et al. 2003, Rojas-Jiménez et al. 2005, Reeve et al. 2006). By using a transcriptome approach, Tiwari et al. (2004) have reported that genes of *Ensifer medicae* are induced about 2.25-fold higher at shifting the pH from 7.0 to 5.7. It has been reported that glutathione (γ-glutamylcysteinylglycine) participates in acid resistance in *R. tropici* and is essential for it to grow at low pH. Without glutathione, either produced by its own biosynthetic metabolism or supplied externally, *R. tropici* CIAT899 is unable to grow under acidic conditions. Glutathione plays keys roles in the physiology of bacterial cells. Glutathione ensures correct folding, synthesis, regulation and degradation of proteins, protects against oxidative stress and participates in the detoxification of numerous xenobiotics. It is the major low-molecular-mass thiol found in most organisms, where levels can reach concentrations exceeding 10 mM (Noctor et al. 2002, Neumann et al. 2003). It has been reported that the expression of acid tolerance in *R. tropici* CIAT899 involves the transcriptional activation of the *gshB* gene in which protein(s) binds the 5′-upstream region, a feature in which inverted repeated sequences appear to be important (Muglia et al. 2007).

Osmotic Stress

Both drought and salinity impose osmotic stress, as a result of high concentrations of either salt or non-ionic solutes in the surrounding medium (Galinski 1995). Rhizobia in the rhizosphere are subjected to frequent fluctuations in the osmolarity of their environment due to the succession of drought and rain periods, the exclusion of salts like NaCl from root tissues, the release of plant exudates, or the production of exopolymers by plant roots and other rhizobacteria. In addition, rhizobia must also adapt to the osmotic situation during the infection process and in the nodule when exchanging nutrients with the host plant (Miller and Wood 1996). Therefore, besides symbiotic efficiency, osmotolerance may constitute a competitive trait for native and inoculant rhizobia and may play a role in the colonization and/or infection process. One of the main mechanisms of bacterial adaptation to hyperosmotic conditions is the intracytoplasmic accumulation of low molecular-weight organic osmolytes. These molecules are called compatible solutes because they do not interact with macromolecules in detrimental ways (Brown 1976). Compatible solutes are accumulated either by uptake from the environment, as exogenous compatible solutes, or by de novo biosynthesis as endogenous compatible solutes. The diversity of compatible solutes is large but falls into a few major chemical categories, such as sugars (i.e., sucrose, trehalose), polyols (i.e., sorbitol, mannitol), amino acids and derivatives (i.e., proline, glutamate, glutamine), betaines and ectoines (da Costa et al. 1998). It is very common for microorganisms to use a cocktail of compatible solutes, a strategy that allows the cell to adapt the compatible solute pool to different environmental injuries. Indeed, the role of compatible solutes goes beyond osmotic adjustment alone, they also give protection to cells and cell components against freezing, desiccation, high temperature and oxygen radicals (da Costa et al. 1998, Welsh 2000, Santos and da Costa 2002). Still another way of hypoosmotic adaptation in Gram-negative bacteria, including the *Rhizobiaceae,* is the accumulation of periplasmic cyclic glucans, which appear to contribute substantially to periplasmic osmolarity (Miller and Wood 1996, Breedveld and Miller 1994).

Extreme Temperatures

Plants must adopt the temperature of the environment in which they live because they do not have internal mechanisms to regulate their own temperature. If external temperatures fluctuate out of the optimal growth range of the plant, the vegetable enters into a stressful situation which can affect its development and then may have a serious impact on agricultural yields.

High soil temperatures in tropical areas are a major problem for biological nitrogen fixation by legume crops, especially in the hotter seasons, when the high temperatures hamper the establishment of the symbiosis between legumes and the inoculated rhizobia or even with the endemic ones (Michiels et al. 1994). The influence of temperature on nodulation and symbiotic nitrogen fixation has been shown to depend on the plant cultivar and the nodulating bacterial strain (La Favre and Eaglesham 1986, Arayankoon et al. 1990). As a general rule, an optimal nodulation takes place when temperatures are under 30°C. For instance, nodule functioning in the common bean is optimal between 25 and 30°C, and is hampered by root temperatures greater than 33°C (Piha and Munns 1987). However, there are some exceptions such as the rhizobial strains isolated from roots of tropical legumes by Surange et al. (1997) and by Kulkarni and Nautiyal (2000), when free-living tolerate temperatures up to 50°C.

Not only at high temperatures but also at temperatures which are lower than the optimal temperature for plant growth, the symbiotic activity is usually shut off (Flower et al. 2001, Ercoli et al. 2004). In these cases, plant survival depends on the speed of the temperature fall and on the existence of protective mechanisms against low temperatures in the plant (Pearce 1999).

Although resistence to temperature stress may constitute a prerequisite for symbiotic nitrogen fixation in tropical areas, heat may also affect other properties. The loss of *Rhizobium* symbiotic effectiveness at elevated temperature was attributed to deletions or to the curing of the symbiotic plasmid (Zurkowski 1982, Toro and Olivares 1986). In *R. tropici* strains, the symbiotic plasmid seems to be particularly stable, even after long incubation at high temperatures (Michiels et al. 1994). The symbiotic performance of *R. tropici* CIAT899 on a high temperatures regime is due not to its ability to fix atmospheric nitrogen at high temperature but probably to its capacity to survive the thermal stress and to recover afterwards. So, in addition to the resistance of *R. tropici* CIAT899 to thermal stress, the stability of its symbiotic properties may constitute a second advantage in tropical and subtropical areas.

The phenomenon whereby cells can adapt to high temperatures when they are first exposed to a sub lethal dose of heat is called adquired thermotolerance. In several organisms, including bacteria, plants, and animals, the acquired thermotolerance appeared to be dependent on the protein synthesis. All organisms so far examined respond to a sudden increase in growth temperature by inducing the synthesis of a small number of heat-shock molecular chaperones Hsps (Neidhardt et al. 1984), some of them highly conserved throughout evolution, which may be indicative of an important cellular function of these proteins (Lindquist 1986).

Several Hsps or homologs thereof have been shown to be involved in the processes of nodulation and nitrogen fixation. GroEL may assist in the correct folding or assembly of *E. meliloti* NodD into an active form (Long et al. 1991). In *Klebsiella pneumoniae,* GroEL interacts with several Nif polypeptides and determines their rates of synthesis and levels of accumulation (Govezensky et al. 1991). In *B. japonicum,* the anaerobic expression of a GroEL homolog is dependent on symbiotic nitrogen fixation genes such as NifA and NtrA (Fischer et al. 1993). The *E. fredii* NodC protein, which is involved in cultivar-specific nodulation of soybean, is homologous to the *E. coli* Hsp DnaJ (Krishnan and Pueppke 1990), which is involved in protein folding, protein targetting to membranes and protein renaturation after stress. In addition to its role in salt tolerance, DnaJ is also important for growth of *R. tropici* CIAT899 at high temperatures, and thus involved in at least two different stresses, namely heat and osmotic stress (Nogales et al. 2002).

Rhizobium tropici CIAT899: a Highly Stress Resistant Bacterium

Rhizobium tropici, first classified as *Rhizobium leguminosarum* bv. phaseoli type II, is present in the soils of America, France and different regions in Africa (Martínez-Romero et al. 1991, Amarger et al. 1994, Hungría et al. 2000, Priefer et al. 2001, Zurdo-Piñeiro et al. 2004). In general, strains belonging to this species are able to nodulate in acid soils and seem to be adapted to living in sandy soils. Based on their 16S RNA sequences, DNA-DNA hybridizations and their phenotype characterization it was possible to distinguish two types within this species, called A and B, which seem to diverge but contain a common symbiotic plasmid (Martínez-Romero and Caballero-Mellado 1996). There are several *R. tropici* strains with intermediate features between types A and B (Mostasso et al. 2002). *R. tropici* type A strains associate with *P. vulgaris, Leucaena* spp. and *Amorpha fruticosa,* whereas type B strains establish a symbiotic association with *P. vulgaris, L. leucocephala* and *M. atropurpureum* (Hernández-Lucas et al. 1995, Sawada et al. 2003). Only the type B strains can grow at 37°C, tolerate the presence of heavy metals and high concentrations of NaCl in the medium (Martínez-Romero, 2003).

Rhizobium tropici CIAT899 is the reference strain belonging to type B and has been well characterized. This strain is highly tolerant to diverse adverse conditions, and was chosen as a model organism to study its symbiotic properties under severe environmental conditions (Martínez-Romero et al. 1991, Nogales et al. 2002, Vinuesa et al. 2003). Furthermore, strain CIAT899 is a good competitor for bean nodule occupancy in the different conditions (Vlassak et al. 1996, Morón et al. 2005).

R. tropici CIAT899 genome consist of a chromosomic structure and three plasmids, which are called pRtCIAT899a (218 Kb), also present in other *Rhizobium* strains that tolerate NaCl stress, pRtCIAT899b or symbiotic plasmid (540 Kb), which harbours the nodulation genes, and pRtCIAT899c or megaplasmid (2.1 Mb), also present on other bean nodulating strains such as *R. etli* (Acosta-Durán and Martínez-Romero 2002, Shamseldin 2007, Ormeño-Orrillo and Martínez-Romero 2009). In contrast to other bean nodulating strains, *R. tropici* CIAT899 maintains its symbiotic properties under several stressful conditions, which add up to its genetic stability (Hungría et al. 2000). This strain tolerates up to 100 µM of Al^{3+} and 200 µM of Mn in the medium, and is able to grow at pH 4.25 and with high NaCl concentrations and very high osmotic pressures (Vargas and Graham 1988, Nogales et al. 2002). Osmoadaptation responses of this strain are: an increase in the intracellular levels of potassium, regulation of the cytoplasmic pH values, accumulation of compatible solutes as glutamate or glutathione, production of chaperone-like proteins and modifications of the membrane composition (Graham 1992, Ricillo et al. 2000, Muglia et al. 2007, Vences-Guzmán et al. 2011).

Production of Nod Factors by *R. tropici* CIAT899

R. tropici CIAT899 nodulation genes were first studied by Vargas et al. (1990). Up to date, the *nodABCSUIJHPQ* operon has been identified and localized on the symbiotic plasmid of this strain, where it is preceded by a *nod*-box and the *nodD1* regulator gene, where transcription occurs in the opposite direction of the genes in this operon. van Rhijn et al. (1993) described the existence of four more different copies of the *nodD* regulator gene, which are apparently located, but without grouping, on the symbiotic plasmid of *R. tropici* CIAT899 (Fig. 4.2). Several mutants of this strain were obtained that emphasize the symbiotic importance of all these genes and *nodD1* in the association with *P. vulgaris* plants (Sousa et al. 1993, van Rhijn et al. 1994, Folch-Mallol et al. 1998, Manyani et al. 2001).

The population of Nod factors produced by *R. tropici* CIAT899 was first characterized by Folch-Mallol et al. (1996). This work reported that their backbone structure consisted of three to five β-1,4-linked *N*-acetyl glucosamine residues and that the more frequent decorations were a sulphated group at the reducing end and carbamoyl, methyl and acyl groups attached to the non-reducing end of the molecules. According to the nature of these substituents, the Nod factors produced by *R. tropici* CIAT899 have been classified in two families: a hydrophilic one, which contains sulphated compounds, and a hydrophobic family which contains the non-sulphated

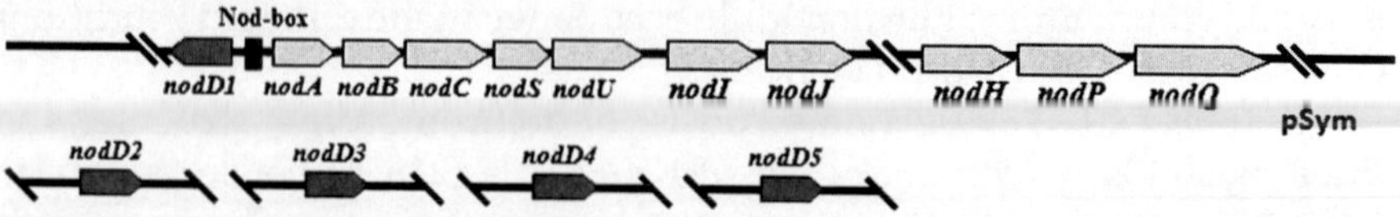

Figure 4.2 Genetic map of the nodulation genes in *R. tropici* CIAT899. Arrows indicate the sense of transcription. pSym: symbiotic plasmid.

Nod factors (Morón 2004). The structural variabily of these Nod factors is related to the broad host range this strain is able to nodulate. Thus, while a Nod factor carrying a fatty acid chain at the non-reducing end of the molecule is sufficient for *R. tropici* CIAT899 to nodulate common bean (Laeremans and Vanderleyden 1998), a sulphate group attached to the reducing end of the molecule is an essential feature for the nodulation of the additional plant host *Leucaena* (Folch-Mallol et al. 1996). However, the functions of only a fraction of the observed Nod factors have been explained in this way. The biological significance of the larger portion of the various Nod factor structures is as yet unexplained.

Biotic and Abiotic Factors Affecting Nod Factor Production

Laeremans and Vanderleyden (1998) indicated that biotic and environmental stresses could affect the Nod factor production in rhizobial strains. Oltshoorn et al. (2000) published data showing temperature dependence of *nodX*- related production of Nod factors of *Rhizobium leguminosarum* bv. viciae and bv. trifolii. At 12°C there was a relative larger amount of NodX-mediated acetylated Nod factors than at 28°C. So the fact that biotic and abiotic factors affect Nod factor production is not new, but has, to date, not been investigated systematically.

Nod Factor Production under Acid Stress Condition

Recently it was reported that the production of Nod factors of *R. tropici* CIAT899 can vary a great deal depending on the environmental and growth conditions the bacteria encounter. Morón et al. (2005) provided the first report on this phenomenon showing that acidic conditions increased the Nod factor production of *R. tropici* CIAT899. A wide variety of Nod factors were produced by this strain when cultured at pH 4.5. *R. tropici* CIAT899 produced 52 different Nod factors when growing with apigenin under acidic culture conditions (pH 4.5), while at neutral pH with apigenin the bacteria produced 29 different Nod factors, and only 15 of those Nod factors were produced under both conditions. Among the 37 Nod factors that were exclusively produced under acidic conditions, was a hitherto unknown structure which has two sulphate groups in its backbone [V($C_{18:1}$, NMe,

S, S)]. Additionally, novel *R. tropici* CIAT899 Nod factor structures were produced at neutral pH, such as molecules containing a carbamoyl group or molecules containing mannose or hexoses as part of the molecular backbone among others [(V($C_{10:0}$-OH, Cb); V($C_{10:0}$-OH, NMe, Cb); IV-Hex($C_{10:0}$-OH, Cb); IV-Hex($C_{12:0}$, NMe); IV-Hex($C_{18:1}$, NMe)]. These results show that nodulation gene expression is affected at acid pH, not only allowing a higher induction, but also activating the expression of other genes, allowing new "decorations" on the signal molecule produced by *R. tropici* CIAT899. This was the first report of the carbamoyl group in *R. tropici* Nod factors. This group appears only in molecules bearing C_{10} hydroxylated fatty acyl chains, suggesting that the *nodU*, which is present in the symbiotic plasmid of *R. tropici* CIAT899 (Waelkens et al. 1995), appears to be functional under stress and requires the presence of a $C_{10:0\text{-}OH}$ fatty acid in this strain.

Based on the relatively intact nodulation abilities of *R. tropici* CIAT899 under acidic plant growth conditions, these authors speculated that the increased production of Nod factors detected in *R. tropici* CIAT899 following acidification of the medium could be a response developed by the bacteria in an attempt to diminish, as much as possible, the adverse effects of soil acidity on the establishment of the symbiosis, and to be effective in nodulation under such conditions, or even to contribute in making Nod factors more stable under these conditions. However, experiments confirming these speculations have not yet been performed with *R. tropici* CIAT899. Duzan et al. (2004) reported a similar effect of acidity in the *Glycine max-Bradyrhizobium japonicum* symbiosis; here acidity also dramatically reduces the root hair response to Nod factors, and the effect is reversed by increasing the concentration of these molecules.

Nod Factor Production under Conditions of NaCl Stress

In order to gain more knowledge about Nod factor production under stress conditions, the influence of high NaCl concentrations (up to 300 mM) on the Nod factor production of *R. tropici* CIAT899 has been also studied (Estévez et al. 2009). This strain produces 46 different Nod factors when growing with apigenin under salt stress (300 mM NaCl), and 15 of them are common to those produced with apigenin but without NaCl in the medium. These 46 Nod factors that CIAT899 produces with NaCl were classified into seven types according to the nature of their substituents (Fig. 4.3).

Under NaCl stress, the Nod factor of *R. tropici* CIAT899 include modifications of the non-reducing and reducing terrminal residues, like the production of new saturated, unsaturated, and hydroxylated fatty acids (ranging from 10 to 22 carbon units), substituents thought to be essential for bean-nodulating rhizobia (Laeremans and Vanderleyden 1998). There

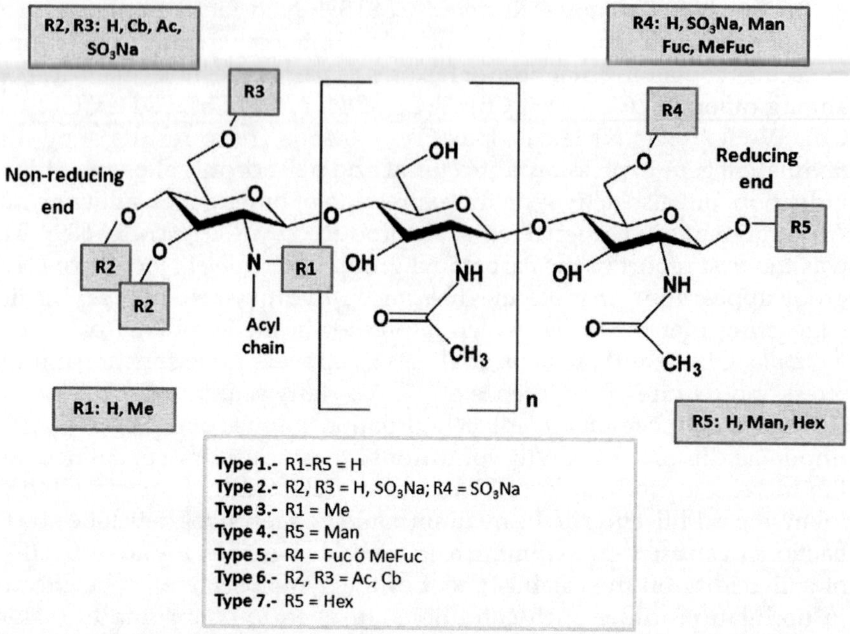

Figure 4.3 Structural variability of *R. tropici* CIAT899 Nod factors produced under NaCl stress (Estévez et al. 2009). Ac: acetyl group; Cb: carbamoyl group; Fuc: fucose; H: hydrogen; Hex: hexose; Man: mannose; Me: methyl group; MeFuc: methylfucose; n: from 2 to 3 N-acetyl-D-glucosamine molecules and SO_3Na: sulphated group.

are several differences in fatty acid production patterns found between acidic- and NaCl stress conditions. For instance, the fatty acid characterized as $C_{22:0}$ was synthesized by *R. tropici* CIAT899 under NaCl stress conditions as well as under control conditions, but it was not detected under acidic conditions. Also, the shorter-chain fatty acids were predominantly found under acidic conditions (i.e., C_{16}) while the longer ones (i.e., C_{20}) were prevalent under saline conditions, with C_{18} being the most common fatty acid produced under control, acid and saline conditions. The most common Nod factor structure produced under all three conditions is a pentameric structure with a $C_{18:0}$ fatty acid, *N*-methylated at the non-reducing terminal residue and with a sulphate group attached to the reducing terminal residue [V($C_{18:0}$, NMe, S)].

Moreover, a new doubly sulphated Nod factor produced by *R. tropici* CIAT899 grown under NaCl stress was detected. In this case, the additional sulphate group is located on the non-reducing terminal residue [V($C_{16:1}$, NMe, S, S)]. This molecule is different from the double sulphated species [V($C_{18:1}$, NMe, S, S)] described in the study of Morón et al. (2005).

Nod factors were found to be fucosylated/methylfucosylated on the reducing terminal end of those bearing a C_{18} fatty acyl chain [V($C_{18:1}$, MeFuc); IV($C_{18:1}$, MeFuc) and IV($C_{18:1}$, Fuc)]. As no sulphate group was identified in this class of Nod factors, it is possible that these two substituents compete for the C6 position of the backbone. As far as it is known, this is the first time that a fucosyl group, which for instance is very common among Nod factors from bacteria of the genus *Sinorhizobium* (*Ensifer*) (Lamrabet et al. 1999), has been described as a Nod factor substituent in a *R. tropici* strain. Nodulation genes in the symbiotic plasmid related to fucosyl synthesis, transport or attachment, such as *nodZ*, have not been described in *R. tropici* CIAT899, so it is quite possible that the chromosomal copies of these genes are responsible for the incorporation of the fucosyl group. Another possibility is that the fucosylated Nod factors are related to *noeJ* gene activity (Nogales et al. 2002), a gene that is also involved in *R. tropici* CIAT899 NaCl tolerance. In fact, preliminary MS analysis failed to identify fucosyl groups in any of the Nod factors produced by a *R. tropici* CIAT899 *noeJ* mutant. The fucosyl substituent is essential for determining a broad host range in various plant species belonging to the family of *Phaseoleae* (López-Lara et al.1995), so we can speculate that the production of these fucosylated Nod factors by *R. tropici* CIAT899 under NaCl stress could possibly expand the host range of the stressed bacteria.

Nod factors of *R. tropici* CIAT899 produced under NaCl stress conditions had biological activity as they induced the development of nodule-like structures, called pseudonodules, in common bean plants. However, these pseudonodules were only produced in absence of NaCl stress, with Nod factors extracted with butanol from 300 mM NaCl containing media. Therefore, it is difficult to explain the biological significance of the production of different Nod factors structures under conditions of a high NaCl concentration, at least for bean nodulation. Most bean plants, like the common bean plants, are inhibited in their growth at 50 mM NaCl (Saadallah et al. 1998) and die at the concentration of NaCl required for these Nod factors to be produced. However, other legumes can tolerate quite high salt concentrations in the growth medium (Teakle et al. 2007, Zahran et al. 2007), and thus it can not be excluded that the higher qualitative and quantitative production of Nod factors has a role in the induction of root nodules on such plants under NaCl stress. Another possibility might be the following: high concentrations of salts will be present in periods of drought, which will be followed by periods with sufficient amounts of water and lower concentrations of salts of such a level that beans are able to germinate and grow and then can be induced for nodulation by the Nod factors already produced in the preceding period of drought.

Possible Explanations for Stress Induced Variability in Nod Factor Production and Future Perspectives

The common *nodABC* genes have a known relaxed substrate specificity. Furthermore, the acquired stress tolerance on *R. tropici* CIAT899 was in some cases found to be dependent on protein biosynthesis (Michiels et al. 1994). Hence, it can be conceived that the diversity of Nod factors observed under stress could have been a consequence of a different availability of enzymes and intracellular biosynthetic precursors due to the effects of abiotic stress. Advanced genomics could help to identify additional genes and gene regulation mechanisms involved in this environmentally regulated Nod factor biosynthesis.

The fact that stress has a huge influence on Nod factor biosynthesis indicates that new very interesting regulatory mechanisms, worth investigating, are involved in Nod factor biosynthesis in *R. tropici* CIAT899. Both studies, Morón et al. (2005) and Estévez et al. (2009), are largely descriptive, and they do not address the molecular or regulatory mechanisms underlying the biosynthesis of the large variety of Nod factors or their functional aspects, but they do raise many important issues about Nod factor structural complexity. They emphasize that rhizobium physiology and ecology are underappreciated aspects of Nod factor biology. Taking all the results together, *R. tropici* CIAT899 can produce 80 different Nod factors; such a very broad variety of chemical substituents combination has never been described before for any *Rhizobium* strain. These results do suggest that the biologically relevant Nod factors for a particular *Rhizobium*-legume host symbiosis may be habitat-dependent and more complex than currently thought. This implies that examination of Nod factor function should include consideration of soil and rhizosphere environmental conditions.

References

Acosta-Durán, C. and E. Martínez-Romero. 2002. Diversity of rhizobia from nodules of the leguminous tree *Gliricidia sepium*, a natural host of *Rhizobium tropici*. Arch. Microbiol. 178: 161–164.

Amarger, N., M. Bours, F. Revoy, M.R. Allard and G. Laguerre. 1994. *Rhizobium tropici* nodulates field-grown *Phaseolus vulgaris* in France. Plant Soil. 2: 147–156.

Anyango, B., K.J. Wilson, J.L. Beynon and K.E. Giller. 1995. Diversity of rhizobia nodulating *Phaseolus vulgaris* L. in two Kenyan soils with contrasting pHs. Appl. Environ. Microbiol. 61: 4016–4021.

Arayankoon, T., H.H. Schomberg and R.W. Weaver. 1990. Nodulation and N_2 fixation of guar at high room temperature. Plant Soil. 126: 209–213.

Barnett, M.J. and S.R. Long. 1990. DNA sequence and translational product of a new nodulation-regulatory locus: *syrM* has sequence similarity to NodD proteins. J. Bacteriol. 172: 3695–3700.

Barnett, M.J., J.A. Swanson and S.R. Long. 1998. Multiple genetic controls on *Rhizobium meliloti syrA*, a regulator of exopolysaccharide abundance. Genetics. 148: 19–32.

Bender, G.L., M. Nayudu, K.K. Le Strange and B.G. Rolfe. 1988. The *nodD1* gene from *Rhizobium* strain NGR234 is a key determinant in the expression of host range to the nonlegume *Parasponia*. Mol. Plant-Microbe Interact. 1: 259–266.

Berc-Ferté, M.P., H.B. Krishnan, A. Savagnac, S.G. Pueppke and J.C. Promé. 1996. *Rhizobium fredii* synthesizes an array of lipooligosaccharides, including a novel compound with glucose inserted into backbone of the molecule. FEBS. Lett. 393: 273–279.

Bittner, A.N., A. Foltz and V. Oke. 2007. Only one of five *groEL* genes is required for viability and successful symbiosis in *Sinorhizobium meliloti*. J. Bacteriol. 189: 1884–1889.

Bohlool, B.B., J.K. Ladha, D.P. Garrity and T. George. 1992. Biological nitrogen fixation for sustainable agriculture: a perspective. Plant Soil. 141: 1–11.

Bordeleau, L.M. and D. Prevost. 1994. Nodulation and nitrogen fixation in extreme environments. Plant Soil. 161: 115–124.

Breedveld, M.W. and K.J. Miller. 1994. Cyclic beta-glucans of members of the family *Rhizobiaceae*. Microbiol. Rev. 58: 145–161.

Brencic, A. and S.C. Winans. 2005. Detection of and response to signals involved in host-microbe interactions by plant-associated bacteria. Microbial. Mol. Boil. Rev. 69: 155–194.

Broughton, W.J., S. Jabbouri and X. Perret. 2000. Keys to symbiotic harmony. J. Bacteriol. 182: 5641–5652.

Brown, A.D. 1976. Microbial water stress. Bacterial. Rev. 40: 803–846.

Carter, J.M., W.K. Gardner and A.H. Gibson. 1994. Improved growth and yield of faba beans (*Vicia faba* cv. Fiord) by inoculation with strains of *Rhizobium leguminosarum* biovar viciae in acid soils in South-West Victoria. Aust. J. Agric. Res. 45: 613–623.

da Costa, M.S., H. Santos and E.A. Galinski. 1998. An overview of the role and diversity of compatible solutes in Bacteria and Archaea. Adv. Biochem. Eng. Biotechnol. 61: 117–153.

Debellé, F., C. Planazet, P. Roche, C. Pujol, A. Savagnac, C. Rosenberg, J.C. Promé and J. Denairé. 1996. The NodA proteins of *Rhizobium meliloti* and *Rhizobium tropici* specify the N-acylation of Nod factors by different fatty acids. Mol. Microbiol. 22: 303–314.

Demont, N., M. Ardourel, F. Maillet, D. Promé and M. Ferro. 1994. The *Rhizobium mellloti* regulatory *nodD3* and *syrM* genes control the synthesis of a particular class of nodulation factors N-acylated by (omega-1)-hydroxylated fatty acids. EMBO J. 13: 2139–2149.

Dénairé, J., F. Debellé and J.C. Promé. 1996. *Rhizobium* lipo-chitooligosaccharide nodulation factors: signalling molecules mediating recognition and morphogenesis. Ann. Rev. Biochem. 65: 503–535.

D'Haeze, W. and M. Holsters. 2002. Nod factor structures, responses, and perception during initiation of nodule development. Glycobiology. 12: 79–105.

Downie, J.A. 1998. Functions of rhizobial nodulation genes. *In*: H.P. Spaink, A. Kondorosi and P.J.J. Hooykaas [eds.]. The Rhizobiaceae: Molecular Biology of Model Plant Associated Bacteria. Kluwer Academic Publishers. Dordrecht. The Netherlands. pp. 387–402.

Downie, J.A. 2010. The roles of extracellular proteins, polysaccharides and signals in the interactions of rhizobia with legume roots. FEMS Microbiol. Rev. 34: 150–170.

Drevon, J.J., C. Abdelly, N. Amarger, E.A. Aouani, J. Aurag, H. Gherbi, M. Jebara, C. Lluch, H. Payre, O. Schump, M. Soussi, B. Sifi and M. Trabelsi. 2001. An interdisciplinary research strategy to improve symbiotic nitrogen fixation and yield of common bean (*Phaseolus vulgaris*) in salinised areas of the Mediterranean basin. J. Biotechnol. 91: 257–268.

Dusha, I. and A. Kondorosi. 1993. Genes at different regulatory levels are required for the ammonia control of nodulation in *Rhizobium meliloti*. Mol Gen. Genet. 240: 435–444.

Duzan, H.M., X. Zhou, A. Souleimanov and D.L. Smith. 2004. Perception of *Bradyrhizobium japonicum* Nod factor by soybean [*Glycine max* (L.) Merr.] root hairs under abiotic stress conditions. J. Exp. Bot. 55: 2641–2646.

Ercoli, L., M. Mariotti, A. Masoni and I. Arduini. 2004. Growth responses of sorghum plants to chilling temperature and duration of exposure. Eur. J. Agro. 21: 93–103.

Estévez, J., M.E. Soria-Díaz, F. Fernández de Córdoba, B. Morón, H. Manyani, A. Gil, J. Thomas-Oates, A.A.N. van Brussel, M.S. Dardanelli, C. Sousa and M. Megías. 2009. Different and new Nod factors produced by *Rhizobium tropici* CIAT899 following Na^+ stress. FEMS Microbiol. Lett. 293: 220–231.

Etzler, M.E., G. Kalsi, N.N. Ewing, N.J. Roberts, R.B. Day and J.B. Murphy. 1999. A Nod factor binding lectin with apyrase activity from legume roots. Proc. Natl. Acad. Sci. U.S.A. 96: 5856–5861.

Fellay, R., X. Perret, V. Viprey, W.J. Broughton and S. Brenner. 1995. Organisation of host-inducible transcripts on the symbiotic plasmid of *Rhizobium* sp. NGR234. Mol. Microbiol. 16: 657–667.

Fellay, R., M. Hanin, G. Montorzi, J. Frey, C. Freiberg, W. Golinowski, C. Staehelin, W.J. Broughton and S. Jabbouri. 1998. *nodD2* of *Rhizobium* sp. NGR234 is involved in the repression of the *nodABC* operon. Mol. Microbiol. 27: 1039–1050.

Fischer, H.M., M. Babst, T. Kaspar, G. Acuña, F. Arigoni and H. Hennecke. 1993. One member of a *groESL*-like chaperonin multigene family in *Bradyrhizobium japonicum* is coregulated with symbiotic nitrogen fixation genes. EMBO J. 12: 2901–2916.

Flower, D.B., G. Breton, A.E. Limin, S. Mahfoozi and F. Sarhan. 2001. Photoperiod and temperature interactions regulate low-temperature-induced gene expression in barley. Plant Physiol. 127: 1676–1681.

Folch-Mallol, J.L., S. Marroquí, C. Sousa, H. Manyani, I.M. López-Lara, K.M.G.M. van der Drift, J. Haverkamp, C. Quinto, A. Gil-Serrano, J. Thomas-Oates, H.P. Spaink and M. Megías. 1996. Characterization of *Rhizobium tropici* CIAT899 nodulation factors: the role of *nodH* and *nodPQ* genes in their sulfation. Mol. Plant-Microbe Interact. 9: 151–163.

Folch-Mallol, J.L., H. Manyani, S. Marroquí, C. Sousa, C. Vargas, N. Nava, C. Colmenero-Flores, C. Quinto and M. Megías. 1998. Sulfation of Nod factors via *nodHPQ* is *nodD* independent in *Rhizobium tropici* CIAT899. Mol. Plant-Microbe Interact. 11: 979–987.

Galinski, E.A. 1995. Osmoadaptation in bacteria. Adv. Microb. Physiol. 37: 272– 328.

García, M.L., J. Dunlap, J. Loh and G. Stacey. 1996. Phenotypic characterization and regulation of the *nolA* gene of *Bradyrhizobium japonicum*. Mol. Plant-Microbe Interact. 16: 827–834.

Geremia, R.A., P. Mergaert, D. Geelen, M. van Montagu and M. Holsters. 1994. The NodC protein of *Azospirillum caulinodans* is an *N*-acetylglucosaminyltransferase. Proc. Natl. Acad. Sci. U.S.A. 91: 2669–2673.

Govezensky, D., T. Greener, G. Segal and A. Zamir. 1991. Involvement of GroEL in *nif* gene regulation and nitrogenase assembly. J. Bacteriol. 173: 6339–6346.

Graham, P.H. 1992. Stress tolerance in *Rhizobium* and *Bradyrhizobium* and nodulation under adverse soil conditions. Can. J. Microbiol. 38: 475–484.

Graham, P.H., S.E. Viteri, F. Mackie, A.T. Vargas and R. Palacios. 1982. Variation in acid soil tolerance among strains of *Rhizobium phaseoli*. Field Crops Res. 5: 121–128.

Greenway, H. and R. Munns. 1980. Mechanisms of salt tolerance in non halo-phytes. Annu. Rev. Plant Physiol. 31: 149–190.

Haag, A.F., M. Baloban, M. Sani, B. Kerscher, O. Pierre, A. Farkas, R. Longhi, E. Boncompagni, D. Herouart, S. Dall'Angelo, E. Kondorosi, M. Zanda, P. Mergaert and G.P. Ferguson. 2011. Protection of *Sinorhizobium* against host cysteine-rich antimicrobial peptides is critical for symbiosis. PLoS Biology. 9: 1–10.

Hanin, M., S. Jabbouri, W.J. Broughton and R. Fellay. 1998. SyrM1 of *Rhizobium* sp. NGR234 activates transcription of symbiotic loci and controls the level of sulphated Nod factors. Mol. Plant-Microbe Interact. 11: 343–350.

Hasegawa, P.M., R.A. Bressan, J.-K. Zhu and H.J. Bohnert. 2000. Plant cellular and molecular responses to high salinity. Annu. Rev. Plant Physiol. Plant. Mol. Biol. 51: 463–499.

Hernández Lucas, I., L. Segovia, E. Martínez-Romero and S.G. Pueppke. 1995. Phylogenetic relationships and host range of *Rhizobium* spp. that nodulate *Phaseolus vulgaris* L. Appl. Environ. Microbiol. 61: 2775–2779.

Higgins, C.F., S.C. Hyde, M.M. Mimmack, U. Gileadi, D.R. Gill and M.P. Gallagher. 1990. Binding protein-dependent transport systems. J. Bioenerg. Biomembr. 22: 571–592.

Honma, M.A., M. Asomaning and F.M. Ausubel. 1990. *Rhizobium meliloti nodD* genes mediate host-specific activation of *nodABC*. J. Bacteriol. 172: 901–911.

Hungría, M., A.A. Franco and J.I. Sprent. 1993. New sources of high-temperature tolerant rhizobia for *Phaseolus vulgaris* L. Plant Soil. 149: 103–109.

Hungría, M., D.S. Andrade, L.M.O. Chueire, A. Probanza, F.J. Gutiérrez-Mañero and M. Megías. 2000. Isolation and characterization of new efficient and competitive bean (*Phaseolus vulgaris* L.) rhizobia from Brazil. Soil Biol. Biochem. 32: 1515–1528.

Kamst, E., K.M. van der Drift, J. Thomas-Oates, B.J.J. Lugtenberg and H.P. Spaink. 1995. Mass spectrometric analysis of chitin oligosaccharides produced by *Rhizobium* NodC protein in *Escherichia coli*. J. Bacteriol. 21: 6282–6285.

Kiss, E., P. Mergaert, B. Oloh, A. Kereszt and C. Staehelin. 1998. Conservation of *nolR* in the *Sinorhizobium* and *Rhizobium* genera of the *Rhizobiaceae* family. Mol. Plant-Microbe Interact. 12: 1186–1195.

Kondorosi, E., J. Gyuris, M. John, E. Duda, B. Hoffman, J. Schell and A. Kondorosi. 1989. Positive and negative control of *nod* gene expression in *Rhizobium meliloti* is required for optimal nodulation. EMBO J. 8: 1331–1341.

Kondorosi, E., M. Buiré, M. Cren, N. Iyer, B. Hoffmann and A. Kondorosi. 1991. Involvement of the *syrM* and *nodD3* genes of *Rhizobium meliloti* in *nod* gene activation and in optimal nodulation of the plant host. Mol. Microbiol. 5: 3035–3048.

Krishnan, H.B. and S.G. Pueppke. 1990. *nolC*, a *Rhizobium fredii* gene involved in cultivar-specific nodulation of soybean, shares homology with a heat-shock gene. Mol. Microbiol. 5: 737–745.

Kulkarni, S. and C.S. Nautiyal. 2000. Effects of salt and pH stress on temperature-tolerant *Rhizobium* sp. NBRI330 nodulating *Prosopis juliflora*. Curr. Microbiol. 40: 221–226.

La Favre, A.K. and A.R.J. Eaglesham. 1986. The effects of high temperatures on soybean nodulation and growth with different strains of bradyrhizobia. Can. J. Microbiol. 32: 22–27.

Laeremans, T. and J. Vanderleyden. 1998. Review: infection and nodulation signaling in *Rhizobium-Phaseolus vulgaris* symbiosis. World J. Microbiol. Biotechnol. 14: 787–808.

Lamrabet, Y., R.A. Bellogín, T. Cubo, R. Espuny, A. Gil, H.B. Krishnan, M. Megías, F.J. Ollero, S.G. Pueppke, J.E. Ruíz-Sainz, H.P. Spaink, P. Tejero-Mateo, J. Thomas-Oates and J.M. Vinardell. 1999. Mutation in GDP-fucose synthesis genes of *Sinorhizobium fredii* alters Nod factors and significantly decreases competitiveness to nodulate soybeans. Mol. Plant-Microbe Interact. 12: 207–217.

Lerouge, P., P. Roche, C. Faucher, F. Maillet, G. Truchet, J. Promé and J. Dénairé. 1990. Symbiotic host specificity of *Rhizobium meliloti* is determined by a sulphated and acylated glucosamine oligosaccharide signal. Nature. 344: 781–784.

Lindquist, S. 1986. The heat-shock response. Annu. Rev. Biochem. 55: 1151–1191.

Loh, J., M. García and G. Stacey. 1997. NodV and NodW, a second flavonoid recognition system regulating *nod* gene expression of the *Bradyrhizobium japonicum* nodulation genes. J. Bacteriol. 184: 1759–1766.

Long, S.R., R.F. Fisher, J. Ogawa, J.J. Swanson, D.W. Erhardt, E.M. Atkinson and J. Schwedock. 1991. *Rhizobium meliloti* nodulation gene regulation and molecular signals. *In*: H. Hennecke and D.P.S. Verma [eds.]. Advances in Molecular Genetics of Plant-Microbe Interactions. Kluwer Academic Publishers. Dordrecht. The Netherlands. pp. 127–133.

López-Lara, I.M., K.M. Van Der Drift, A.A.N. van Brussel, J. Haverkamp, B.J.J. Lugtenberg, J. Thomas-Oates and H.P. Spaink. 1995. Induction of nodule primordia on *Phaseolus* and *Acacia* by lipo-chitin oligosaccharide nodulation signals from broad-host-range *Rhizobium* strain GRH2. Plant Mol. Biol. 29: 465–477.

Machado, D. and H.B. Krishnan. 2003. *nodD* alleles of *Sinorhizobium fredii* USDA 191 differentially influence soybean nodulation, *nodC* expression, and production of exopolysaccharides. Curr. Microbiol. 47: 134–137.

MacKay, T. and M.A. Djordjevic. 1993. Production and excretion of Nod metabolites by *Rhizobium leguminosarum* bv. trifolii are disrupted by the same environmental factors that reduce nodulation in the field. Appl. Environ. Microbiol. 59: 3385–3392.

Manyani, H., C. Sousa, M.E. Soria-Díaz, A.M. Gil-Serrano and M. Megías. 2001. Regulation of Nod factor sulphation genes in *Rhizobium tropici* CIAT899. Can. J. Microbiol. 47: 574–579.

Martínez-Romero, E. 2003. Diversity of *Rhizobium-Phaseolus vulgaris* symbiosis: overview and perspectives. Plant Soil. 252: 11–23.

Martínez-Romero, E. and J. Caballero-Mellado. 1996. *Rhizobium* phylogenies and bacterial genetic diversity. Crit. Rev. Plant Sci. 15: 113–140.

Martínez-Romero, E., L. Segovia, F. Martins Mercante, A.A. Franco, P. Graham and M.A. Pardo. 1991. *Rhizobium tropici*, a novel species nodulating *Phaseolus vulgaris* L. beans and *Leucaena* sp.trees. Int. J. System. Bacteriol. 41: 417–426.

McNeil, S.D., M.L. Nuccio and A.D. Hanson. 1999. Betaines and related osmoprotectants. Targets for metabolic engineering of stress resistance. Plant Physiol. 120: 945–949.

Michiels, J., C. Verreth and J. Vanderleyden. 1994. Effects of temperature stress on bean nodulating *Rhizobium* strains. Appl. Environ. Microbiol. 60: 1206–1212.

Miller, K.J. and J.M. Wood. 1996. Osmoadaptation by rhizosphere bacteria. Annu. Rev. Microbiol. 50: 101–136.

Minami, E., H. Kouchi, J.R. Cohn, T. Ogawa and G. Stacey. 1996. Expression of the early nodulin, ENOD40, in soybean roots in response to various lipo-chitin signal molecules. Plant J. 10: 23–32.

Morón, B. 2004. Influencia del esters abiótico en el establecimiento de la simbiosis *Phaseolus vulgaris-Rhizobium* spp. Thesis, University of Seville, Spain.

Moron, B., M.E. Soria-Díaz, J. Ault, G. Verroios, N. Sadaf, D.N. Rodríguez-Navarro, A. Gil-Serrano, J. Thomas-Oates, M. Megías and C. Sousa. 2005. Low pH changes the profile of nodulation factors produced by *Rhizobium tropici* CIAT899. Chem. Biol. 12: 1029–1040.

Mostasso, L., F.L. Mostasso, B.G. Dias, M.A.T. Vargas and M. Hungría. 2002. Selection of bean (*Phaseolus vulgaris* L.) rhizobial strains for Brazilian Cerrados. Field Crops Res. 73: 121–132.

Muglia, C.I., D.H. Grasso and O.M. Aguilar. 2007. *Rhizobium tropici* response to acidity involves activation of glutathione synthesis. Microbiology. 153: 1286–1296.

Neidhardt, F.C., R.A. VanBogelen and V. Vaughn. 1984. The genetics and regulation of the heat-shock proteins. Annu. Rev. Genet. 18: 295–329.

Neumann, C., Boubakari, R. Grunert and P.J. Bednarski. 2003. Nicotinamide adenine dinucleotide phosphate-regenerating system coupled to a glutathione-reductase microtiter method for determination of total glutthione concentrations in adherent growing cancer cell lines. Anal. Biochem. 320: 170–178.

Noctor, G., L.D. Gómez, H. Vanacker and C.H. Foyer. 2002. Interactions between biosynthesis, compartmentation and transport in the control of glutathione homeostasis and signaling. J. Exp. Bot. 53: 1283–1304.

Nogales, J., R. Campos, H. BenAbdelkhalek, J. Olivares, C. Lluch and J. Sanjuán. 2002. *Rhizobium tropici* genes involved in free-living salt tolerance are required for the establishment of efficient nitrogen-fixing symbiosis with *Phaseolus vulgaris*. Mol. Plant-Microbe Interact. 15: 225–232.

Olsthoorn, M.A.M., E. Stokvis, J. Haverkamp, H.P. Spaink and J.E. Thomas-Oates. 2000. Growth temperature regulation of host-specific modifications of rhizobial lipo-chitin oligosaccharides. the function of *nodX* is temperature regulated. Mol. Plant-Microbe Interact. 13: 808–813.

Ormeño-Orrillo, E. and E. Martínez-Romero. 2009. Genómica de *Rhizobium tropici* CIAT899. pp. 94. *In:* Memorias de la XXIV RELAR y I IBEMPA. La Habana. Cuba.

Paleg, L.G., G.R. Stewart and J.W. Bradbeer. 1984. Proline and glycine betaine influence protein salvation. Plant Physiol. 75: 974–978.

Pearce, R.S. 1999. Molecular analisys of acclimation to cold. Plant Growth Regul. 29: 47–76.

Peck, M.C., R.F. Fisher and S.R. Long. 2006. Diverse flavonoids stimulate NodD1 binding to *nod* gene promoters in *Sinorhizobium meliloti*. J. Bacteriol. 188: 5417–5427.

Perret, X., C. Staehelin and W.J. Broughton. 2000. Molecular basis of symbiotic promiscuity. Microbiol. Mol. Biol. Rev. 64: 180–201.

Piha, M.I. and D.N. Munns. 1987. Sensitivity of the common bean (*Phaseolus vulgaris* L.) symbiosis to high soil temperature. Plant Soil. 98: 183–194.

Poupot, R., E. Martínez-Romero, N. Gautier and J.C. Promé. 1995. Wild type *Rhizobium etli*, a bean symbiont, produces acetyl-fucosylated, *N*-methylated and carbomoylated nodulation factors. J. Biol. Chem. 270: 6050–6055.

Priefer, U.B., J. Aurag, B. Boesten, I. Bouchmouch, R. Defez, A. Filali-Maltouf, M. Miklis, H. moawad, B. Mouhsine, J. Prell, A. Schluter and B. Senatore. 2001. Characterization of *Phaseolus* symbionts isolated from Mediterraneam soils and analysis of genetic factors related to pH tolerance. J. Biotechnol. 91: 223–236.

Reeve, W.G., L. Bräu, J. Castelli, G. Garau, C. Sohlenkamp, O. Geiger, M.J. Dilworth, A.R. Glenn, J.G. Howieson and R.P. Tiwari. 2006. The *Sinorhizobium medicae* WSM419 *lpiA* gene is transcriptionally activated by FsrR and required to enhance survival in lethal acid conditions. Microbiology. 152: 3049–3059.

Relic, B., X. Perret, M.T. Estrada-García, J. Kopcinska, W. Grolinowski, H. Krishnan, S.G. Pueppke and W.J. Broughton. 1994. Nod factors of *Rhizobium* are a key to the legume door. Mol. Microbiol. 13: 171–178.

Ricillo, P.M., C. Muglia, F.J. De Bruijn, A. Roe, I.R. Booth and O.M. Aguilar. 2000. Glutathione is involved in environmental stress responses in *Rhizobium tropici*, including acid tolerance. J. Bacteriol. 182: 1748–1753.

Ritsema, T., A.H.M. Wijfjes, B.J.J. Lugtenberg and H.P. Spaink. 1996. *Rhizobium* nodulation protein NodA is a host-specific determinant of the transfer of fatty acids in Nod factor biosynthesis. Mol. Gen. Genet. 251: 44–51.

Roche, P., F. Maillet, C. Plazanet, F. Debellé, M. Ferro, G. Truchet, J.C. Promé and J. Dénairé. 1996. The common *nodABC* genes of *Rhizobium meliloti* are host-range determinants. Proc. Natl. Acad. Sci. U.S.A. 93: 15305–15310.

Rojas-Jiménez, K., K. Sohlenkamp, O. Geiger, E. Martínez-Romero, D. Werner and P. Vinuesa. 2005. A ClC chloride channel homolog and ornithine-containing membrane lipids of *Rhizobium tropici* CIAT899 are involved in symbiotic efficiency and acid tolerance. Mol. Plant-Microbe Interact. 18: 1175–1185.

Saadallah, K., C. Abdelly, M. Lachaal, A. Soltani and M. Hajji. 1998. Effects of N source and P nutrition on the limitation by naCl of the common bean growth (*Phaseolus vulgaris*). *In*: C. Elmerich, A. Kondorosi and W.E. Newton [eds.]. Biological Nitrogen Fixation for the 21st Century. Kluwer Academic. Dordrecht, The Netherlands. pp. 515.

Sanjuán, J., P. Grob, H.H. Gröttfert and G. Stacey. 1994. NodW is essential for full expression of the common nodulation genes in *Bradyrhizobium japonicum*. Mol. Plant-Microbe Interact. 7: 364–369.

Santos, H. and M.S. da Costa. 2002. Compatible solutes of organisms that live in hot saline environments. Environ. Microbiol. 4: 501–509.

Sawada, H., L.D. Kuykendall and J.M. Young. 2003. Changing concepts in the systematic of bacterial nitrogen-fixing legume symbionts. J. Gen. Appl. Microbiol. 49: 155–179.

Schell, M.A. 1993. Molecular biology of the LysR family of transcriptional regulators. Annu. Rev. Microbiol. 47: 597–626.

Schlaman, H.R.M., B. Horvath, E. Vijgenboom, R.J.H. Okker and B.J.J. Lugtenberg. 1991. Supression of nodulation gene expression in bacteroids of *Rhizobium leguminosarum* biovar viciae. J. Bacteriol. 173: 4277–4287.

Schlaman, H.R.M., R.J. Okker and B.J.J. Lugtenberg. 1992. Regulation of nodulation gene expression by NodD in rhizobia. J. Bacteriol. 174: 5177–5182.

Schlaman, H.R.M., D.A. Phillips and E. Kondorosi. 1998. Genetic organization and transcriptional regulation of rhizobial nodulation genes. *In*: H.P. Spaink, A. Kondorosi and P.J.J. Hooykaas [eds.]. The Rhizobiaceae: Molecular Biology of Model Plant Associated Bacteria. Kluwer Academic Publishers. Dordrecht, The Netherlands. pp. 371–380.

Schlaman, H.R.M., M.M.A. Olsthoorn, M. Harteveld, L. Dörner, M.A. Djordjevic, J.E. Thomas-Oates and H.P. Spaink. 2006. The production of species-specific highly unsaturated fatty acid acyl-containing LCOs from *Rhizobium leguminosarum* bv. trifolii is stringently regulated by *nodD* and involves the *nodRL* genes. Mol. Plant-Microbe Interact. 19: 215–226.

Schultze, M., B. Quiclet-Sire, E. Kondorosi, H. Virelizier, J.N. Glushka, G. Endre, S.D. Géro and A. Kondorosi. 1992. *Rhizobium meliloti* produces a family of sulphated lipooligosaccharides exhibiting different degrees of plant host specificity. Proc. Natl. Acad. Sci. U.S.A. 89: 192–196.

Shamseldin, A. 2007. Plasmid content of salt-stress tolerant *Rhizobium* strains from Egyptian soils nodulating common bean (*Phaseolus vulgaris* L.). World J. Microbiol. Biotechnol. 24: 1603–1606.

Singh, K.N. and R. Chatrath. 2001. Distribution of saline soils. *In*: M.P. Reynolds, J.I. Ortíz-Monasterio and A. McNab [eds.]. Application of Physiology in Wheat Breeding. CIMMYT. México D.F. pp. 101–110.

Soria-Díaz, M.E., M.A. Rodríguez-Carvajal, P. Tejero-Mateo, J.L. Espartero, B. Morón, C. Sousa, M. Megías, J. Thomas-Oates and A.M. Gil-Serrano. 2006. Structural determination of the Nod factors produced by *Rhizobium gallicum* bv. gallicum R592. FEMS. Microbiol. Let. 255: 164–173.

Sousa, C., J.L. Folch, P. Boloix, M. Megías, N. Nava and C. Quinto. 1993. A *Rhizobium tropici* DNA region carrying the amino-terminal half of a *nod* gene and a *nod*-box-like sequence confers host range extension. Mol. Microbiol. 9: 1157–1168.

Spaink, H.P. 2000. Root nodulation and infection factors produced by rhizobial bacteria. Annu. Rev. Microbiol. 54: 257–288.

Spaink, H.P., C.A. Wijffelman, E. Pees, R.J.H. Okker and B.J.J. Lugtenberg. 1989. Localisation of the functional regions of the *Rhizobium nodD* product using hybrid nod genes. Plant Mol. Biol. 12: 59–73.

Standfield, S.W., L. Ielpi, D. O'Brochta, D.R. Helinski and G.S. Ditta. 1988. The *ndvA* gene product of *Rhizobium meliloti* is required for beta-(1→2) glucan production and has homology to the ATP-binding export protein HlyB. J. Bacteriol. 170: 3523–3530.

Stokkermans, T.J.W., R. Orlando, V.S. Kumar Kolli, R.W. Carlson and N.K. Peters. 1996. Biological activities and structures of *Bradyrhizobium elkanii* low abundance lipo-chitin oligosaccharides. Mol. Plant-Microbe Interact. 9: 298–304.

Surange, S., A.G. Wollum, N. Kumar and S.C. Nautiyal. 1997. Characterization of *Rhizobium* from root nodules of leguminous trees growing in alkaline soils. Can. J. Microbiol. 43: 891–894.

Teakle, N.L., T.J. Flowers, D. Real and T.D. Colmer. 2007. *Lotus tenuis* tolerates the interactive effects of salinity and waterlogging by "excluding" Na^+ and Cl^- from the xylem. J. Exp. Bot. 58: 2169–2180.

Tiwari, R.P., W.G. Reeve, M.J. Dilworth and A.R. Glenn. 1996a. An essential role for *actA* in acid tolerance in *Rhizobium meliloti*. Microbiology. 142: 601–602.

Tiwari, R.P., W.G. Reeve, M.J. Dilworth and A.R. Glenn. 1996b. Acid tolerance in *Rhizobium meliloti* strain WSM419 involves a two-component sensor-regulator system. Microbiology. 142: 1693–1701.

Tiwari, R.P., W.G. Reeve, B.J. Fenner, A.R. Glenn and J.G. Howieson. 2004. Probing for pH regulated genes in *Sinorhizobium medicae* using transcriptional analysis. J. Mol. Microbiol. Biotechnol. 7: 133–139.

Toro, N. and J. Olivares. 1986. Analysis of *Rhizobium meliloti* Sym mutants obtained by heat treatment. Appl. Environ. Microbiol. 51: 1148–1150.

van Rhijn, P.J., B. Feys, C. Verreth and J. Vanderleyden. 1993. Multiple copies of *nodD* in *Rhizobium tropici* CIAT899 and BR816. J. Bacteriol. 175: 438–447.

van Rhijn, P.J., J. Desair, K. Vlassak and J. Vanderleyden. 1994. Functional analysis of *nodD* genes of *Rhizobium tropici* CIAT899. Mol. Plant-Microbe Interact. 7: 666–677.

Vance, C.P. 2002. Root-bacteria interactions: symbiotic nitrogen fixation. *In*: Y. Waisel, A. Eshel and U. Kafkati [eds.]. Plant Roots: the Hidden Half. Marcel Dekker Publishers. Dordrecht, The Netherlands. pp. 839–867.

Vargas, A.A.T. and P.H. Graham. 1988. Cultivars and pH effects on competition for nodule sites between isolates of *Rhizobium* in beans. Plant Soil. 117: 195–200.

Vargas, C., L.J. Martínez, M. Megías and C. Quinto. 1990. Identification and cloning of nodulation genes and host specificity determinants of the broad host-range *Rhizobium leguminosarum* biovar phaseoli strain CIAT899. Mol. Microbiol. 4: 1899–1910.

Vences-Guzmán, M.A., Z. Guan, E. Ormeño-Orrillo, N. González-Silva, I.M. López-Lara, E. Martínez-Romero, O. Geiger and C. Sohlenkamp. 2011. Hydroxylated ornithine lipids increase stress tolerance in *Rhizobium tropici* CIAT899. Mol. Microbiol. 79: 1496–1514.

Vinardell, J.M., F.J. López-Baena, A. Hidalgo, F.J. Ollero, R. Bellogín, M.R. Espuny, F. Temprano, F. Romero, H.B. Krishnan, S.G. Pueppke and J.E. Ruiz-Sainz. 2004. The effect of FITA mutations on the symbiotic properties of *S. fredii* varies in a chromosomal-background-dependent manner. Arch. Microbiol. 181: 144–154.

Vinuesa, P., F. Neumann-Silkow, C. Pacios-Bras, H.P. Spaink, E. Martínez-Romero and D. Werner. 2003. Genetic analysis of a pH-regulated operon in *Rhizobium tropici* CIAT899 involved in acid tolerance and nodulation competitiveness. Mol. Plant-Microbe Interact. 16: 159–168.

Vlassak, K., J. Vanderleyden and A. Franco. 1996. Competition and persistence of *Rhizobium tropici* and *Rhizobium etli* in tropical soil during successive bean (*Phaseolus vulgaris* L.) cultures. Biol. Fertil. Soils. 21: 61–68.

Waelkens, F., T. Voets, K. Vlassak, J. Vanderleyden and P. van Rhijn. 1995. The *nodS* gene of *Rhizobium tropici* CIAT899 is necessary for nodulation on *Phaseolus vulgaris* and on *Leucaena leucocephala*. Mol. Plant-Microbe Interact. 8: 147–154.

Welsh, D.T. 2000. Ecological significance of compatible solute accumulation by microorganisms: from single cells to global climate. FEMS Microbiol. Rev. 24: 263–290.

Whitfield, C. and M.A. Valvano. 1993. Biosynthesis and expression of cell-surface polysaccharides in Gram-negative bacteria. Adv. Microb. Physiol. 35: 135–246.

Wood, M., J.E. Cooper and A.J. Holding. 1984. Soil acidity factors and nodulation of *Trifolium repens*. Plant Soil. 78: 367–380.

Yamaguchi, T. and E. Blumwald. 2005. Developing salt-tolerant crop plants: challenges and opportunities. Trends Plant Sci. 10: 615–620.

Yang, G.P., F. Debellé, A. Savagnac, M. Ferro, O. Schiltz, F. Maillet, D. Promé, M. Treilhou, C. Vialas, K. Lindstrom, J. Denairé and J.C. Promé. 1999. Structure of the *Mesorhizobium huakuii* and *Rhizobium galegae* Nod factors: a cluster of phylogenetically related legumes are nodulated by rhizobia producing Nod factors with α,β-unsaturated *N*-acyl substitutions. Mol. Microbiol. 34: 227–237.

Zahran, H.H. 1999. *Rhizobium*-legume symbiosis and nitrogen fixation under severe conditions and in an arid climate. Microbiol. Mol. Biol. Rev. 63: 968–989.

Zahran, H.H., M.C. Marin-Manzano, A.J. Sánchez-Raya, E.J. Bedmar, K. Venema and M.P. Rodríguez-Rosales. 2007. Effect of salt stress on the expression of NHX-type ion transporters in *Medicago intertexta* and *Melilotus indicus* plants. Physiol. Plantarum. 131: 122–130.

Zurdo-Piñeiro, J.L., E. Velázquez, M.J. Lorite, G. Brelles-Mariño, E.C. Schröder, E.J. Bedmar, P.F. Mateos and E. Martínez-Molina. 2004. Identification of fast-growing rhizobia nodulating tropical legumes from Puerto Rico as *Rhizobium gallicum* and *Rhizobium tropici*. Syst. Appl. Microbiol. 27: 469–477.

Zurkowski, W. 1982. Molecular mechanism for loss of nodulation properties of *Rhizobium trifolii*. J. Bacteriol. 150: 999–1007.

5

Strategies of Salt Tolerance in the Rhizobia-Legume Symbiosis

M. López-Gómez, F. Palma and *C. Lluch**

Introduction: Origin of Soil Salinity in the World

The sustainability of irrigated agriculture in many arid and semi arid areas of the world is at risk because of a combination of several interrelated factors, including the lack of fresh water, lack of drainage, the presence of high water tables and salinization of soil and ground water resources. Soil salinity often leads to the development of other problems in soils such as sodicity and alkalinity soil, sodicity being the result of the binding of Na^+ to the negatively charged clay particles. Hydrolysis of the Na-clay complex results in soil alkalinity (Manchanda and Garg 2008) thus soil salinity is a major factor limiting sustainable agriculture and crop productivity (Munns 2002). More than 800 millions ha of land throughout the world are salt-affected (FAO 2008; http://www.fao.org). Also, it is estimated that worldwide about 20 percent of total cultivated lands and 33 percent of irrigated agricultural lands are affected by high salinity (Flowers 2004), most of them confined to the tropics and the Mediterranean regions where

Departamento de Fisiología Vegetal, Facultad de Ciencias, Universidad de Granada, Campus de Fuentenueva s/n, Granada 18071, Spain.
*Corresponding author: clluch@ugr.es

evapo-transpiration exceeds annual precipitation and irrigation is therefore necessary to meet crops water needs. Furthermore, the salinized areas are increasing at a rate of 10 percent annually for various reasons, including the accumulation of salts over long periods of time, low precipitation, high surface evaporation, irrigation with saline water and poor agricultural practices (Rengasamy 2002). Watering of saline rocks releases soluble salts of various types, mainly chlorides of sodium, calcium and magnesium and a lesser extend sulphates and carbonates (Szabolcs 1994).

Characteristics of a Saline Soil

Salinity is a soil condition characterized by a high concentration of soluble salts. A saline soil is generally defined as one in which the electrical conductivity (EC) of the saturation extract (ECe) in the root zone exceeds 4 dS m-1 (~40mM NaCl) at 25°C, has an exchangeable sodium of 15 percent and generates and osmotic pressure of approximately 0.2 MPa. This definition of salinity originates from the ECe that significantly reduces the yield of most crops, although many of them exhibit yield reduction at lower ECs according to USDA Laboratory Salinity (USDA 2008).

Consequences of Soil Salinity in Plants

Salinity imposes at least three primary stresses on plants, the first is a hyperosmotic stress caused by the reduction of water potential of the external soil solution that can exceed the osmotic potential of the plant cell, reducing the water availability for the plant. The second consists in a hyperionic stress, related to the toxic effects of the accumulated ions, especially Na^+ and Cl^- that produce a nutritional imbalance (Munns and Tester 2008). The third effect stems from the generation of reactive oxygen species (ROS) . The combination of all of these factors causes adverse effects on plant growth and development at physiological, biochemical (Munns 2002, Munns and Janus 2003) and at molecular levels (Tester and Darenport 2003). The identification of genes and/or salt tolerance indicators can be used by plant breeders to improve the crops yield under stressful conditions. Intense research on the molecular basis of plant response to salinity has been carried out over the last few years, but this is outside the scope of this chapter and can be found elsewhere (Yamaguchi and Blumwald 2005, Pereira et al. 2007, Jamil et al. 2011, Mudgal et al. 2010).

Importance of Legumes in Soil Fertility

Any increase in the agricultural productivity of saline lands which are often grouped under marginal or waste lands, can contribute to the socioeconomic improvement of millions of inhabitants who sustain their livelihood from those areas. Legumes have been suggested as appropriate crops for enhancement of bioproductivity and for restoration of marginal soils because they are relevant for sustainable cropping systems and could offer many economic and environmental benefits if they were grown more widely in crop rotation worldwide. In addition, legumes can improve saline soils fertility and help to reintroduce agriculture to these lands due to their capacity to grow on nitrogen-poor soils (Crespi and Galvez 2000). In this sense, considerable variability in salinity tolerance among crop legumes has been reported (Läuchi 1984). Among cultivated legumes, *Sesbania cannabina* and *Lupinus luteus* are particularly tolerant to salinity (Keating and Fisher 1985). Some of the tree legumes such as *Prosopis* and *Accacia* spp. are highly tolerant to salinity as well, with their tolerance levels approaching that of sea water (Rhodes and Felker 1988). On the other hand, grain legumes such as *Phaseolus vulgaris*, *Cicer arietinum* and *Vigna radiata* are highly sensitive to salinity with 50 percent growth reduction at 3–4 dSm^{-1} salinity levels. (Subbarao and Johansen 1991) meanwhile *Pisum sativum*, *Vicia faba*, and *Medicago sativa* are moderately sensitive to salinity.

Increasing demand for agricultural production results in greater nitrogen-fertilizer application; however, with the current technology for fertilizers production and the inefficient methods employed, fertilizer usage will eventually become prohibitive (Flechard et al. 2007). Rather than relying only on applications of nitrogen fertilizer, alternative nitrogen sources are needed to help the development of more sustainable farming systems. Legumes have the potential to fulfil these requirements due to their unique ability to fix nitrogen biologically from the atmosphere. The ability to fix atmospheric nitrogen comes from the symbiotic relationship between legumes and a diverse group of soil bacteria known as rhizobia which provides the legume with nitrogen in the form of ammonium and ureides in exchange of energy and carbon molecules product of the photosynthesis. The symbiosis between legumes and rhizobia is a cheaper and usually more effective agronomic practice for ensuring an adequate supply of nitrogen to legume based crop and pasture production than the application of fertilizers (Andres et al. 2012).

Rhizobia-Legume Symbiosis

Successful establishment of the symbiotic interaction involve chemotaxis of the bacteria toward the root followed by root colonization, root hair deformation and simultaneously, the division of the cortical cells leading to the formation of nodule primordia which finally transform into a mature nodule where the bacteria differentiate into bacteroids. Early events of nodule formation require expression of bacterial nodulation genes which are induced by plant flavonoids, a large group of structurally related compounds (Stougaard 2000). The rhizobia synthesize lipoquitooligosaccharides known as Nod Factors that allow the bacteria to enter and induce nodule gene expression (see Chapter 4). The nitrogen fixation process relies on the nitrogenase enzyme that catalyzes the reaction to split the nitrogen molecule and reduce it to ammonia with the plant energy and electron supply (Postgate 1982). Oldroyd and Downie 2008 reviewed the signalling involved in the symbiotic establishment at molecular level.

This kind of symbiosis exists in many types of legumes including grain and forage legumes and some leguminous trees. The benefit to soil nitrogen fertility from grain/forage legumes is positively correlated with nitrogen fixation (Evans et al. 2001) assuming that successful rhizobia-legume symbiosis will definitely increase the incorporation of nitrogen into soil ecosystems after the harvesting of crops. Therefore, grain and forage legumes are used to build up soil nitrogen fertilizer in sustainable farming systems.

Response of the Symbiosis to Salt Stress

The rhizobia-legume symbiosis and root nodule function on legumes are more sensitive to salt or osmotic stress than the free living rhizobia (Rao et al. 2002). The symbiosis establishment involves a complex interaction between the host root, rhizobial strain and environmental conditions. Salt stress may differently affect any step of the rhizobia-legume symbiosis such as: a) rhizobia survival and growth in bulk soil or rhizosphere of the host legume, b) rhizobial infection of host root tissue, c) nodule initiation and development and d) nodule functioning and growth of the host legume. It is difficult to distinguish which phase is primordially affected because of the independence of these phases (Borucki and Sujkowska 2008).

The effects of salt stress on nodulation and nitrogen fixation processes have been examined in several studies. Soybean root hairs showed little curling or deformation when inoculated with *Bradyrhizobium japonicum* in the presence of 170 mM NaCl and nodulation was completely suppressed by 210 mM NaCl (Tu 1981). Serraj et al. (1994) reported 80 percent decrease in nitrogenase activity of soybean plants subjected to 0.1 M NaCl. Cordovilla et al. (1995) compared faba bean, pea, soybean and the common bean under NaCl and reported that the depressive effect of salinity on nitrogenase activity of nodules was directly related to the salt-induced decline in dry weight and nitrogen content in shoots. Tejera et al. (2005) found that common bean plants responded to salinity by decreasing nitrogen fixation activity and nodule number per plant, a similar behaviour was also detected in pea (Bolaños et al. 2006) and chickpea (Soussi et al. 2001). The reduction of nitrogen-fixing activity by salinity is usually attributed to a reduction in nodular respiration (Delgado et al. 1994, Soussi et al. 2001) and in the leghaemoglobine content (Delgado et al. 1993). In *Pisum sativum* nodular structure, the salt induced some disturbances including loss of turgor of the nodule peripheral cells, changed nodule zonation, stimulated infection thread expansion, and disorders in the bacterial release from the infection threads (Borucki and Sujkowska 2008). All these alterations could also be the reason for the decline in the nitrogen fixation activity in legumes subjected to salt stress (Requena et al 1997).

Implication of the Rhizobial Bacteria

Although the root nodule-colonizing bacteria, originally classified in the genus *Rhizobium*, are more salt tolerant than their host legumes, they show a marked variation in salt tolerance since many reports have shown that highly tolerant strains are symbiotically more efficient than salt-sensitive ones under salt stress (Zahran 1999, Saxena et al. 1996, Pereira et al. 2007). Growth of a number of rhizobial species was inhibited by 100 mM NaCl while some rhizobia such as *Sinorhizobium meliloti* were tolerant to 300 mM (Mohammad et al. 1991, Embalomatis et al. 1994). It is not clear if salt-tolerant rhizobia are better nitrogen fixers than salt-sensitive ones in saline environments. A major criticism of these reports is that the differences in symbiotic efficiency could have been due to genotypic variations not related to salt tolerance (Zahran 1999). Indeed, few reports have used isogenic or near-isogenic strains with regard to salt tolerance and symbiotic effectiveness. For instance, Chen et al. (1992) described that spontaneous mutants of salt-sensitive *Rhizobium leguminosarum* strains with enhanced salt tolerance induce nodules on *Vicia faba* that lacked nitrogenase activity. In contrast, the work by Ohwada et al. (1998) showed that a transposon-induced mutant of *Sinorhizobium fredii* with reduced tolerance to NaCl formed ineffective symbiosis with soybean. In neither of these two cases were the genes characterized.

Mechanisms of Tolerance to Salinity of the Rhizobial Bacteria

Rhizobia may use distinct mechanisms for osmotic adaptation on salt stress such as the accumulation of low molecular weight organic solutes (osmolites), including amino acids as glutamate (Botsford and Lewis 1990), *N*-acetylglutaminyl-glutamine (Smith et al. 1994) sugars and polyamines (Fujihara and Yoneyama 1994) or the accumulation of ions (e. K^+) (Miller and Wood 1996). Another osmoprotectant identified in rhizobia is the intracellular accumulation of glycine-betaine (Smith et al. 1988), and ectoine was as effective as glycine-betaine in improving the growth of *Rhizobium meliloti* under 0.5 M NaCl conditions (Talibart et al. 1997).

Trehalose. Trehalose is a non-reducing disaccharide that has been shown to be accumulated in all 27 strains of rhizobia tested and that by far was the major carbohydrate found in cultured *Bradyrhizobium* strains, regardless of the carbon source supplied in the medium (Streeter 1985). Also, it has been reported that, among mono- and disaccharides in nodules, only trehalose is present at higher concentrations in bacteroids than in the nodule cytosol (Streeter 1987). In combination, these results strongly suggest that trehalose is synthesized in soybean nodules by bacteroids. In addition, the presence of three totally independent mechanisms for the synthesis of trehalose by *Bradyrhizobium* species suggests that this disaccharide is important in the function of this organism both in the free-living state and in symbiosis (Streeter and Gómez 2006).

Extracellular Polysaccharides (EPS) and Lipopolysaccharide (LPS). EPS and LPS constitute other changes in the bacterial response to salinity (Soussi et al. 2001) which may have an impact on symbiosis because of the implication of these molecules in the initial steps of the symbiotic interaction.

Changes in the Gene Expression. Eight different gene *loci* involved in *Rhizobium tropici* CIAT899 to salinity (NaCl) stress were identified by Nogales et al. (2002). Most of these genes are also required for adaptation to hyperosmotic media. Interestingly, all the genes appear to be important for the bacteroid nitrogen-fixing function into the nodules.

Morphological Modifications. The rhizobial cells appear as spiral or filamentous-like structures and the cell size expands greatly . The cell ultra-structure is severely affected as well and the cellular envelope undergoes modifications in its composition. Nevertheless, the response and adaptation to environmental stresses is probably a complex phenomenon involving many physiological and biochemical processes (Fernández Pascual et al. 1996).

Plasmid-Mediated Resistance. Pereira et al. (2007) suggested that this mechanism may be ecologically important since salt resistance can be rapidly transferred from tolerant to sensitive bacteria, thus extra chromosomal genes can contribute to survival in saline soils. However, Soussi et al. (2001) reported that changes in the protein profile induced by salt are sufficient to substantiate the presumption that proteins have a role in salt tolerance.

Carbon Metabolism

Nodule nitrogen fixation depends on the photosynthate delivered from the phloem in form of sucrose in a similar way to any other sink tissue in the plant (Hawker 1985). Sucrose supports nodule growth, provides energy for the nitrogen fixation, maintains a functional population of rhizobia and allows the synthesis of amino compounds produced from nitrogen fixation (Vance 2008). The cost of C per unit of fixed N varies widely with species, growth stage and environmental conditions, and ranges from 1.4 to 8.5 (Minchin and Pate 1974, Schulze et al 1999).

Once in the nodule, sucrose is cleaved by either sucrose synthase (SS) or alkaline invertase (AI). Sucrose synthase has been reported to be essential for nitrogen fixation in legume nodules (Gordon et al. 1999); however, less is known about the role of AI, although it has been reported as a variation in the relative activity of both enzymes with the developmental stage (Ocaña et al. 1998) and salt stress (Ben Salah et al. 2009).

Hexose phosphates derived from sucrose cleavage are metabolized through the glycolytic pathways to render phosphoenolpyruvate, which is converted to malate via the combined action of phosphoenolpyruvate carboxylase (PEPC) and malate dehydrogenase (MDH) (Lodwig and Poole 2003). The carboxilase activity is particularly abundant in nodules with 10 to 15 higher expression levels than in other organs and comprise up to 2 percent of the soluble protein in nodules of *Medicago sativa* (Vance et al. 1994) and *Vicia faba* (Ocaña et al. 1996). Indeed, PEPC plays a significant role in supporting nitrogen fixation by anaplerotic CO_2 fixation for recycling carbon in nodules to produce oxaloacetate, as demonstrated by Nomura et al. (2006) in *Lotus japonicus*. Some nodule enhanced forms of carbonic anhydrase, PEPC and MDH have been identified from legume nodules (Coba de la Peña et al. 1997, Miller et al. 1998). In nodules of model legumes *Medicago truncatula* and *Lotus japonicus*, PEPC and MDH were inhibited by salinity showing a correlation with nitrogen fixation activity (López et al. 2008).

The decrease in nitrogen fixation in root nodules under saline stress may be related to the limitation in the energetic substrate shortage to bacteroids following photosynthetic activity and nodule sucrose breakdown reduction (Soussi et al. 1998, 1999, López and Lluch 2008). Oxaloacetate is rapidly converted into malate which is used as a source of carbon and energy for bacteroid consumption to support nitrogen fixation in nodules (Lodwig and Poole 2003). In this sense, Fischinger and Schulze (2010) suggested that an improvement of nodule capability to channel assimilates into oxaloacetate and malate formation through CO_2 fixation might prolong intensive nitrogen fixation in grain legumes into the later stages of ontogeny.

Under salt stress, malate concentration was depressed in bacteroids and citosol (Delgado et al. 1993), which suggested that the inhibitory effect of NaCl on nitrogen fixation activity of bacteroids from pea nodules may be related to the decrease in bacteroid respiration. The activation of fermentative metabolism by salinity suggests some reduction in the O_2 availability within the nodule. In addition, the salinity also alters their potential to transfer and metabolize different compounds such as proline or lactate (Ferri et al.

2000). *Mesorhizobium cicery* bacteroids could utilize glucose as respiratory substrate under normal and stress conditions while malate did not appear to be the preferred substrate under saline conditions.

Nitrogen Assimilation

The presence of a high concentration of combined nitrogen in the rooting medium is inhibitory to the establishment and function of the nitrogen-fixing nodules (Rigaud 1981). However, it has been reported that a certain concentration of mineral nitrogen in the root zone, defined as "starter N", stimulated nodule establishment, although the optimal concentration varied widely with cultivar and growth conditions (Gan et al. 2004), and with the plant age (Hervás et al. 1991). There is however strong evidence of variations among isolates of rhizobia and legumes species in their tolerance to combined nitrogen (Gibson and Harper 1985). In effect, a number of grain legumes including *Pisum sativum* may benefit from the complementary effects of simultaneous nitrogen fixation and nitrate reduction (Ligero et al 1987), which may be due to the fact that nitrogen deficiency in the initial stages of development during the formation of effective nodules can cause a severe decline in plant growth from which the plant recovers slowly when symbiotically-fixed nitrogen become available (Postgate 1982).

Root nodules have the capacity to reduce nitrate as demonstrated by Ligero et al. (1987). However, the inhibitory effect of exogenous nitrate on nitrogen fixation has been attributed to a direct competition between nitrate reductase and nitrogenase for reducing power (Stephen and Neyra 1983) or to the fact that nitrite, a product of nitrate reductase activity, inhibits the function of nitrogenase activity and leghemoglobine (Becana and Spent 1989). Streeter (1988) concluded that neither explanation was correct and that other explanations should also be considered. Nitrate and nitrite reduction are carried out by inducible enzymes in roots of *Vicia faba* and by constitutive enzymes in the nodule, which indicate that there may be different forms of these enzymes in each tissues (Caba et al. 1990).

Nitrogen fixation as well as nitrate assimilation in root nodules of legumes generates ammonia which is also processed by the GS/GOGAT cycle after being excreted into the plant citosol (Schubert 1981). The first step is catalyzed by glutamine synthetase (GS) which forms glutamine, subsequently glutamine is converted to glutamate by glutamate synthase (GOGAT). Caba et al. (1993) suggested that assays of nodule glutamine synthase (GS) and glutamato sintasa might be included in *Vicia faba* breeding programs to indicate the capacity for symbiotic nitrogen fixation, particularly in the presence of nitrate. However, in *Pisum sativum* nitrate accumulation in leaves, stems and roots can be principally associated with the nitrate treatment, nodulation by rhizobia strains also affected the nitrate concentration of tissues in roots and stems, especially when the nitrate concentration in the medium ranged from 2 to 4 mM (Hervas et al. 1991). Ligero et al. (1987) and Lluch and Gavilan (1989) found that the rhizobia strain can affect the nitrate concentration and the expression of nitrate reductase activity.

The nitrogen fixation and plant growth inhibition by salinity are consistent with the significant sensitivity to salt stress detected in the enzymes involved in ammonium assimilation in root nodules of *Vicia faba* (Cordovilla et al. 1995a), *Pisum sativum* (Cordovilla et al. 1999a) and *Cicer arietinum* (Soussi et al. 1998), although in pea nodules, nitrogenase and glutamato synthase proved to be more sensitive to salinity than glutamine synthetase (Cordovilla et al. 1999b). Therefore, the activity of these enzymes should be considered as a reliable criteria for selection of salt tolerance. However, the inhibition of GS/GOGAT pathway promotes glutamato dehydrogenase (GDH) and aspartate amino transferase (AAT), an alternative pathway for ammonia assimilation in nodules. The inhibition of enzymes related to purine catabolism xanthine dehydrogenase and uricasa decreased the ureide content and alternatively the amino acid content increased in the nodule of the common bean (Khadri et al. 2001).

Nodular Mechanisms of Tolerance to Salinity

Osmoprotectants

Almost all organisms ranging from microbes to animals and plants synthesize and accumulate compounds termed osmoprotectants (or compatible solutes) in response to osmotic stress. These are small, electrically neutral molecules that are non-toxic at molar concentrations, and stabilize proteins and membranes against the denaturing effect of high concentrations of salts and other harmful solutes. In dry or saline environments osmoprotectants can therefore serve both to raise cellular osmotic pressure and to protect cell constituents (Bartels and Sunkar 2005). Chemically, osmoprotectants are of three types: betaines and related compounds, certain amino acids like proline and ectoine, and polyols and nonreducing sugars such as trehalose. Not all of these occur in crop plants. However, an important feature of osmoprotectants is that their beneficial effects are generally not species-specific, so that alien osmoprotectants can be engineered into plants and protect their new host (Rontein et al. 2002). In plant cells, osmoprotectants are typically confined mainly to the cytosol, chloroplasts, and other cytoplasmic compartments that together occupy 20 percent or less of the volume of mature cells (the other 80 percent is the large central vacuole). Natural osmoprotectant concentrations in cytoplasmic compartments can therefore reach or exceed 200 mM (Rhodes et al. 2002). Identification of intracellular solutes and their modification under salt stress conditions could be of potential use for the selection of salt-adapted cultivars.

Studies of metabolite concentrations in root nodules, and especially in bacteroids, are essential to understand the consequence of salt stress on the functioning of legume nodules at the molecular level. Salinity is known to boost the nodular carbohydrate and amino acid pools. Among carbohydrates, **sucrose** is predominant under salinity associated with the inhibition of the sucrose hydrolytic enzyme sucrose synthase (Arrese-Igor et al. 1999, López et al. 2008). **Pinitol** proved to be another carbohydrate that accumulates in nodules of *Lotus japonicus* plants submitted to salt stress (López-Gómez et al. 2012). Previously, this polyol was described as a compatible solute in plants (Obendorf et al. 2008) and in soybean as a major carbohydrate with up to 50–60 percent of soluble sugars under water stress conditions (Streeter et al. 2001). A strong correlation between sugar accumulation and osmotic stress tolerance has been widely reported (Streeter et al. 2001).

Trehalose was first described as a major carbohydrate in soybean root nodules in 1980 (Streeter 1980). In later studies, trehalose appeared to be a common carbohydrate in almost all nodules tested (Phillips et al. 1984). Initially, the presence of trehalose in uninfected higher plants was not known, which led to the prediction that it was synthesized by bacteroids. This prediction was supported later by the fact that trehalose was synthesized in bacteroids isolated from soybean nodules (Streeter 1985), trehalose concentration did not decline in soybean senescing nodules while other carbohydrates declined (Müller et al. 2001), and concentration of trehalose in bacteroids was related to the rhizobial strain (Streeter 1985). The role of trehalose in legume root nodules under salt stress conditions has been studied by the addition of the trehalase inhibitor validamycin A. Validamycin A caused an increase in the amount of trehalose that improved the response to salinity of *Medicago truncatula* (López et al. 2009) and *Lotus japonicus* (López et al. 2006) by increasing the biomass production under stress conditions. In addition, salt stress increased the nodular trehalose concentration in the model legumes *Lotus japonicus* and *Medicago truncatula* about 40 and 100 percent respectively (López et al. 2008). Similarly, in *Medicago sativa* plants infected with *Rhizobium meliloti*, maltose and trehalose concentrations were significantly enhanced upon 0.15 M sodium chloride stress, especially in roots and bacteroids (Fougère et al. 1991). All these data together support a role for trehalose as osmoprotectant under stress conditions in the rhizobia-legume symbiosis.

Within the amino acids, **proline** shows the largest increase in the nodular tissue of salt-stressed *Medicago truncatula* and *Lotus japonicas* plants (López-Gómez et al. 2012). In soybean bacteroids subjected to salt stress an enhanced proline dehydrogenase activity (ProDH) has been reported, the main proline catabolic enzyme, suggesting that this amino acid could be used as an energy source for nitrogen fixation (Zhu et al. 1992, Kohl et al. 1994). Other studies have attributed an antioxidant feature to proline due to its reactive oxygen species scavenging activity (Szabados and Savouré 2009). Transgenic plants that accumulate high levels of proline are reported to display increased tolerance to salt and osmotic stress, as well as to cold and frost (Kishor et al. 2005, Han and Hwang 2003). One example would be transgenic *Medicago truncatula* plants over-expressing pyroline-5-carboxylate synthase (P5CS), in those plants, nitrogen fixation was significantly less affected by salt treatment compared to wild type plants (Verdoy et al. 2006).

Ionic Balance

Osmotic stress is caused by ions (mainly Na^+ and Cl^-) in the soil solution decreasing the availability of water to the root. Ion toxicity occurs when plants take up Na^+ and Cl^- and these ions are accumulated to detrimental levels in leaves, and consequently affect the absorption of other nutrients (Greenway and Munns 1980). Ion transport (influx and efflux) and maintenance of ionic homeostasis employing transporters, exclusion, long distance transport and compartmentalization, have also emerged as a crucial input in salt stress response (Serrano and Rodríguez 2002, Yamaguchi and Blumwald 2005). Under this constraint, potassium and calcium content in nodules decreased, probably following competition with sodium on the absorption site. This decrease may alter the physiological activity of nodules. In fact, it was proposed by Minchin et al. (1994) that calcium and magnesium content are involved in the operation of the cortical diffusion. In parallel, increased levels of Na^+ and Cl^- in the cytoplasm could affect a variety of metabolic activities, reducing the nodular efficiency probably by a toxic effect of those ions or as consequence of changes in nodule permeability to O_2 (Aydi et al. 2004).

Fernández-Pascual et al. (1996) observed that following a 6-d exposure to 150 mol m^{-3} NaCl, Na^+ ions were largely excluded from the infected zone, whilst only Cl^- ions penetrated into this region. Differential distribution of these ions across nodule cell layers (infected and uninfected cells) of two lines of faba bean has been described as well (Abd-Alla et al. 2001), which supports the fact that salt tolerance of the nodules involves ion exclusion from the central zone, possibly involving a K^+/Na^+ interchange within the cortex. Such a substitution of K^+ by Na^+ is a common adaptation to salt stress and has been proposed as criteria for salt tolerance in chickpea plants growing under symbiotic conditions and salinity (Tejera et al. 2006). The capacity to accumulate Na^+ in the root might be related to a protection of the photosynthetic apparatus from sodium damage (Soussi et al. 1998). In this regard, Baalbaky et al. (2000) postulated the involvement of two physiological mechanisms to reduce the impact of salinity: sodium compartmentalization in roots and the K^+/Na^+ selectivity in reducing damage associated with excessive sodium levels in soil.

Oxidative Stress and Antioxidant Mechanisms

All aerobic organisms, while having the ability to use oxygen, have the associated risk of oxidative stress caused by reactive oxygen species (ROS), which includes singlet oxygen ($^1O_2$), superoxide anion (O_2^-), hydrogen peroxide (H_2O_2), and hydroxyl radical (OH^-) species. ROS cause oxidative damage to different cellular components including membrane lipids, proteins and nucleic acids. In addition to a consequence of the aerobic metabolism, ROS are constantly produced by plants during their development or in response to biotic and abiotic stresses, including saline stress (Rubio et al. 2009). In this context, ROS are described as powerful signalling molecules and are considered as secondary messengers in the regulation of many pathways associated with plant development and environmental stress responses (Foyer and Noctor 2009).

Nodules have a high potential for ROS production due to, among other factors, the elevated rates of bacteroid respiration, the highly reducing conditions required for nitrogen fixation, and the presence of many electron transfer components including ferredoxin, uricase and hydrogenase susceptible to auto-oxidation which results in superoxide formation (Dalton et al. 1998). However, ROS production has been shown to play a key role in the rhizobia-legume interaction and maintenance of the symbiosis between the plant and the bacteria (D'Haeze et al. 2003). The accumulation of ROS observed in legume roots during its interaction with rhizobia, together with the impairment of root hair curling and infection threads formation by the inhibition of ROS production (Peleg-Grossman et al. 2007), indicates that the production of ROS may not be a plant defence response to the microbe, but rather a process needed for the development of a proper interaction. The importance of ROS production has been confirmed in a *Sinorhizobium meliloti* strain overexpressing catalase activity, this bacterium acting as a H_2O_2 sink, provoked a delayed nodulation and enlargement of infection threads (Jamet et al. 2007).

Root nodules possess a pool of antioxidant metabolites and enzymes to prevent accumulation of ROS and hence the damage of lipids, proteins, and DNA despite of the role of ROS as signalling molecules in stressful situations and in activating defence genes (Mittler et al. 2004).

Different studies suggest that redox balance is involved in the regulation of nodule metabolism; for instance, the diminution of nitrogen fixation under abiotic stress is correlated with a modification of the redox balance and a strong decline in the antioxidant defence (Jebara et al. 2005). In this context, a strong antioxidant defence system may be crucial to allow an efficient nodule functioning. Under salt stress, nitrogen fixation and antioxidant enzyme activities in nodules were affected. In *Phaseolus vulgaris* catalase (CAT) and ascorbate peroxidase (APX) were inhibited by salt stress (Tejera et al. 2004), whereas superoxide dismutase (SOD) and peroxidase were activated (Jebara et al. 2010). Another study in *Phaseolus vulgaris* and *Medicago sativa* nodules indicated an inhibition of the nodule CAT activity when plants were exposed to NaCl (Tejera et al. 2007), this decrease was proportional to the nitrogen fixation, confirming the correlation between CAT activity and the functioning of symbiosis (Mhadhbi et al. 2004). CAT are common enzymes that catalyze the decomposition of hydrogen peroxide to water and oxygen, they are mostly located in nodular peroxisomes and glyoxisomes, although bacteroids also contain high levels of catalase (Becana et al. 2000). On the other hand, peroxidases are present as multiple isozymes in plant tissues and several of them are present in nodules although their functions and subcellular localization are unknown (Becana et al. 2000). Superoxide dismutases are enzymes that catalyze the dismutation of superoxide into oxygen and hydrogen peroxide. They are metalloenzymes with a key role in protection against oxidative stress (Moran et al. 2003). Bacterial SODs have a critical protective role for efficient nodulation, nitrogen fixation and delayed senescence (Santos et al. 2000). Several enzymes involved in the biosynthesis of ascorbate (Matamoros et al. 2006) are expressed in nodules reaching up to 1 percent of the total soluble protein of nodules (Dalton et al. 1998) which indicates its importance in the rhizobia-legume symbiosis.

Ascorbate and glutathione are molecular antioxidants that play important functions in growth and stress responses, both of them can be found in nodules at concentrations ranging from 0.5 to 2 mM, which is consistent with their multiple and essential functions (Dalton et al. 1998), and they are a potent water-soluble antioxidant, acting as a direct ROS-scavenger and as a metabolite of the ascorbate-glutathione pathway for H_2O_2 detoxification. Although nodule host cells synthesize their own glutathion, some of this critical antioxidant needs to be produced by the bacterial partner to achieve optimal nitrogen fixation, as evidenced by the observation that rhizobia deficient in GSH formed nodules with early senescence and diminished symbiotic performance (Muglia et al. 2008). Homoglutathione is another thiol tripeptide present in nodules and other organs of some legume species and may share antioxidant and regulatory properties (Frendo et al. 2005). Homoglutathione may partially or completely replace GSH (Matamoros et al. 2003).

Phytohormones

The protective response of the rhizobia-legume symbiosis to biotic and abiotic stresses is regulated by phytohormones and plant growth regulators that include: salicylic acid (SA), abscisic acid (ABA) and a group of polycationic compounds known as polyamines (PAs) (Lluch et al. 2011). These endogenous low molecular weight molecules regulate, via synergistic and antagonistic actions, the expression of different but overlapping suites of genes, which is referred to as signalling cross-talk (Kraft et al. 2007).

Salicylic acid is one of plant hormones whose function in biotic stress response has been better studied. It has been reported that SA is involved in nodule formation, as shown in plants of *Vigna mungo* where the number of nodules, the leghaemoglobin content and the nitrogenase activity were reduced by SA (Ramanujam et al. 1998). Exogenous SA inhibited the growth of rhizobia and the production of nod factors by them and also delayed the nodule formation, thereby decreasing the number of nodules per plant (Mabood and Smith 2007). However, *Medicago sativa* plants when inoculated with an incompatible strain of rhizobia, resulted in a marked accumulation of SA in the roots of host plant. It was therefore concluded that the compatible strains of rhizobia produce certain signals (specific nod factors) which are perceived by the host plant that suppress the accumulation of SA in the roots (Martínez-Abarca et al. 1998). The *nahG* gene, which encodes a SA hydroxylase that degrades SA to catechol, when expressed in *Lotus japonicus* resulted in a marked reduction of SA levels, and this reduction correlated with an increase in the number of infections and mean nodule number when compared to controls (Stacey et al. 2006). This data points to an important role for SA-mediated plant defence pathways in controlling nodule formation. Salicylic acid is involved in the response to abiotic stresses; however, its role remains unresolved. It has been reported in the literature that SA mitigate the adverse effects of drought and salinity in legumes (Palma et al. 2009, Nazar et al. 2011). On the contrary, SA treatment negatively affected the response of pea (cv. Lincoln) to 70 mm NaCl, and this response was correlated with a decrease in the antioxidant enzymes APX and GR and an increase in SOD, as well as in H_2O_2 content (Barba-Espin et al. 2011). Treatment of bean with 0.5 mM SA resulted in a decrease in the content of Na^+, Cl^-, H_2O_2, and electrolyte leakage under saline conditions compared to the control, while it increased antioxidant metabolism (Khan et al. 2010). Other reports demonstrate that application of SA may promote the formation of ROS during salt and osmotic stresses (Palma et al. 2009). The mechanisms by which SA enhances H_2O_2 content are not clear, but could involve inactivation of H_2O_2-removing enzymes by SA as described by Palma et al. (2009) in bean and/or activation of SOD as reported by Khan et al. (2010). These results demonstrate that the effect of SA vary depending on plant species, growth conditions, as well as SA and stress application.

The abscisic acid is the key hormone in plant responses to abiotic stresses (drought, salinity and cold) and during specific developmental stages. ABA has been known to be a negative regulator of legume root nodule formation and nitrogen fixation (Tominaga et al. 2010). However, under salinity stress ABA content increases in various tissues, including nodules of some legumes such as *Phaseolus vulgaris* (Khadri et al. 2006) and *Medicago ciliaris* (Ben Salah et al. 2012) where ABA alleviates the damages caused by salinity. Exogenous ABA application induced accumulation of proline in *Phaseolus vulgaris* nodules (Khadri et al. 2007). Interestingly, ABA has been reported to have regulatory effects on the pathway involved in proline metabolism (Khadri et al. 2007) exerting beneficial effects in plant tissues, especially via its involvement in osmotic adjustment. Besides, in *Phaseolus vulgaris* ABA treatment limited the sodium translocation to the shoot resulting in the maintenance of a high K^+/Na^+ ratio (Khadri et al. 2007). Several reports have demonstrated the relation between the accumulation of ABA and the biosynthesis of polyamines under stress conditions, suggesting a role for ABA as a signal molecule involved in the maintenance of the cellular pool of PAs under salinity conditions. In addition, ABA contributes to the conversion from the bound and conjugated to the free soluble forms of polyamine (Ben Hassine et al. 2009).

Polyamines are polycationic compounds widespread in many organisms and particularly in plants, where they have been implicated in the regulation of many physiological processes and a variety of stress responses. Spermidine (Spd), spermine (Spm) and putrescine (Put) accumulate under abiotic stress conditions. The enhanced level of PAs plays an important role in the protective response of plants to various abiotic stresses. It has been suggested that PAs involvement in abiotic stress adaptation could be due to their roles in osmotic adjustment, membrane stability, free-radical scavenging and regulation of stomatal movements (Liu et al. 2007). Together with the above mentioned roles of PAs, they have been also found to be involved in biochemical and physiological processes related to plant growth and development, including root nodules development in legumes (Fortes et al. 2011). Polyamine levels are generally five to 10 times higher in legume nodules than in the root, leaf and stem (Fujihara et al. 1994). In nodules of *Lotus japonicus*, the expression of genes involved in the synthesis of PAs are induced early in nodule development and declines with aging, whereas PAs accumulate steadily during nodule maturation, suggesting that they are involved in nodule cell division and expansion, but also in other functions related to nitrogen fixation (Efrose et al. 2008). Indeed, significant linear correlations between the total concentrations of free polyamines in nodules and nitrogenase activity and leghaemoglobin content have been reported in field-grown bean (Lahiri et al. 2004). The accumulation of PAs is associated with salinity tolerance (Duan et al. 2008) and it has been reported that the exogenous applications of PAs increases K^+ level under salinity, maintaining high K^+/Na^+ ratio.

References

Abd Alla, M.H., A.E. El-Enany, A.E. Hamada and A.M. Abdel Wahab. 2001. Element distribution in fababean root nodules under salinity and its effects on growth, nodulation and nitrogen fixation. *Rostlina-Vyroba*. UZPI. 47(9): 399–404.

Andrés, J.A., M. Rovera, L.B. Guiñazú, N.A. Pastor and S.B. Rosas. 2012. Interactions between legumes and Rhizobia under stress conditions. *In:* D.K. Maheshwari [ed.]. Bacteria in Agrobiology: Stress Management. Springer-Verlag, Berlin-Heidelberg. pp. 77–94.

Arrese-Igor, C., E.M. González, A.J. Gordon, F.R. Minchin, L. Gálvez, M. Royuela, P.M. Cabrerizo and P.M. Aparicio-Tejo. 1999. Sucrose synthase and nodule nitrogen fixation under drought and other environmental stresses. Symbiosis. 27: 189–212.

Aydi, D., J.J. Drevon and C. Abdelly. 2004. Effect of salinity on root-nodule conductance to oxygen the *Medicago truncatula-Sinorhizobium meliloti* symbiosis. Plant Physiol. Biochem. 42(10): 833–840.

Baalbaki, R.Z., R.A. Zurayak, M.A.M. Adlan and C.M. Saxena. 2000. Effect of nitrogen source and salinity levels on salt accumulation of two chick-pea genotypes. J. Plant Nutr. 23: 805–814.

Barba-Espin, G., M.J. Clemente-Moreno, S. Alvarez, M.F. Garcia-Legaz, J.A. Hernandez and P. Diaz-Vivancos. 2011. Salicylic acid negatively affects the response to salt stress in pea plants. Plant Biol. 13: 909–917.

Bartels, D. and R. Sunkar. 2005. Drought and salt tolerance in plants. Crit. Rev. Plant Sci. 24: 23–58.

Becana, M., F.R. Minchin and J.I. Sprent. 1989. Short-term inhibition of legume N_2 fixation by nitrate. Nitrate effects on nitrate reductase activities of bacteroids and nodule cytosol. Planta. 180: 40–45.

Becana, M., D.A. Dalton, J.F. Moran, I. Iturbe-Ormaetxe, M.A. Matamoros and M.C. Rubio. 2000. Reactive oxygen species and antioxidants in legume nodules. Physiol. Plantarum. 109: 372–381.

Ben Hassine, A., M.E. Ghanem, S. Bouzid and S. Lutts. 2009. Abscisic acid has contrasting effects on salt excretion and polyamine concentrations of an inland and a coastal population of the Mediterranean xero-halophyte species *Atriplex halimus*. Ann. Bot. 104: 925–936.

Ben Salah, I., A. Albacete, C. Martínez Andújar, R. Haouala, N. Labidi, F. Zribi, V. Martinez, F. Pérez-Alfocea and C. Abdelly. 2009. Response of nitrogen fixation in relation to nodule carbohydrate metabolism in *Medicago ciliaris* lines subjected to salt stress. J. Plant Physiol. 166: 477–488.

Ben Salah, I., N. Jelali, T. Slatni, M. Gruber, A. Albacete, C. Martinez Andujar, V. Martinez, F. Perez-Alfocea and C. Abdelly. 2012. Involvement of source-sink relationship and hormonal control in the response of *Medicago ciliaris-Sinorhizobium medicae* symbiosis to salt stress. Acta Biol. Hung. 63: 97–112.

Bolaños, L., M. Martín, A. El-Handaoui, R. Rivilla and I. Bonilla. 2006. Nitrogenase inhibition in nodules from pea plants grown under salt stress occurs at the physiological level and can be alleviated by B and Ca. Plant Soil. 280: 135–142.

Borucki, W. and M. Sujkowska. 2008. The effects of sodium chloride-salinity upon growth, nodulation, and root nodule structure of pea (*Pisum sativum* L.) plants. Acta Physiol. Plant. 30: 293–301.

Botsford, J.L. and T.A. Lewis. 1990. Osmoregulation in *Rhizobium meliloti*: production of glutamic acid in response to osmotic stress. Appl. Environ. Microbiol. 56: 488–494.

Caba, J.M., C. Lluch, A. Hervás and F. Ligero. 1990. Nitrate metabolism in roots and nodules of *Vicia faba* in response to exogenous nitrate. Physiol. Plant. 79: 531–539.

Caba, J.M., C. Lluch and F. Ligero. 1993. Genotypic differences in nitrogen assimilation in *Vicia faba:* Effect of nitrate. Plant and Soil. 151: 167–174.

Chen, C.T., J. Maundu, J. Cavaness, L.M. Dandurand and C.S. Orser. 1992. Characterization of salt tolerant and salt sensitive mutants of *Rhizobium leguminosarum* biovar. *viciae* strain C12046. FEMS Microbiol. Lett. 90: 135–140.

Coba de la Peña, T., F. Frugier, H.I. McKhann, P. Bauer, S. Brown, A. Kondorosi and M. Crespi. 1997. A carbonic anhydrase gene is induced in the nodule primordium and its cell-specific expression in controlled by the presence of *Rhizobium* during development. Plant J. 11: 407–420.

Cordovilla, M.P., A. Ocaña, F. Ligero and C. Lluch. 1995a. Growth and macronutrient contents of faba bean plants: effects of salinity and nitrate nutrition. J. Plant Nutrition. 18(8): 1611–1628.

Cordovilla, M.P., A. Ocaña, F. Ligero and C. Lluch. 1995b. Salinity effects on growth analysis and nutrient composition in four grain legumes-*Rhizobium* symbiosis. J. Plant Nutrition. 18: 1595–1609.

Cordovilla, M.P., F. Ligero and C. Lluch. 1999a. Effect of NaCl on growth and nitrogen fixation and assimilation of inoculated and KNO_3 fertilized *Vicia faba* L. and *Pisum sativum* L. plants. Plant Sci. 140: 127–136.

Cordovilla, M.P., S.I. Berrido, F. Ligero and C. Lluch. 1999b. *Rhizobium* strain effects on the growth and nitrogen assimilation in *Pisum sativum* and *Vicia faba* growth under salt stress. J. Plant Physiol. 154: 127–131.

Crespi, M. and S. Galvez. 2000. Molecular mechanisms in root nodule development. J. Plant Grow. Regul. 19: 155–166.

Dalton, D.A., S.L. Joyner, M. Becana, I. Iturbe-Ormaetxe and J.M. Chatfield. 1998. Antioxidant defenses in the peripheral cell layers of legume root nodules. Plant Physiol. 116: 37–43.

Delgado, M.J., J.M. Garrido, F. Ligero and C. Lluch. 1993. Nitrogen fixation and carbon metabolism by nodules and bacteroids of pea plants under sodium chloride. Physiol. Plant. 89: 824–829.

Delgado, M.J., F. Ligero and C. Lluch. 1994. Effects of salt stress on growth and nitrogen fixation by pea, faba-bean, common bean and soybean plants. Soil Biol. Biochem. 26(3): 371–376.

D'Haeze, W., R. De Rycke, R. Mathis, S. Goormachtig, S. Pagnotta, C. Verplancke, W. Capoen and M. Holsters. 2003. Reactive oxygen species and ethylene play a positive role in lateral root base nodulation of a semiaquatic legume. PNAS. 100: 11789–11794.

Duan, J.J., J. Li, S.R. Guo and Y.Y. Kang. 2008. Exogenous spermidine affects polyamine metabolism in salinity-stressed *Cucumis sativus* roots and enhances short-term salinity tolerance. J. Plant Physiol. 165: 1620–1635.

Efrose, R.C., E. Flemetakis, L. Sfichi, C. Stedel, E.D. Kouri, M.K. Udvardi, K. Kotzabasis and P. Katinakis. 2008. Characterization of spermidine and spermine synthases in *Lotus japonicus*: induction and spatial organization of polyamine biosynthesis in nitrogen-fixing nodules. Planta. 228: 37–49.

Elbein, A.D., Y.T. Pan, I. Pastuszak and D. Carroll. 2003. New insights on trehalose: a multifunctional molecule. Glycobiology. 13: 17–27.

Embalomatis, A., D.K. Papacosta and P. Katinakis. 1994. Evaluation of *Rhizobium meliloti* strains isolated from indigenous populations northern Greece. J. Agric. Crop Sci. 172: 73–80.

Evans, J, A.M. McNeill, M.J. Unkovich, N.A. Fetell and D.P. Heeman. 2001. Net nitrogen balances for cool season grain legume crops and contributions to wheat nitrogen uptake: a review. Aust. J. Exp. Agric. 41: 347–359.

FAO. 2008. FAO Land and Plant Nutrition Management Service. http://www.fao.org/ag/agl/agll/spush.

Fernández-Pascual, M., C. de Lorenzo, M.R. de Felipe, S. Rajalakshmi, A.J. Gordon, B.J. Thomas and F.R. Minchin. 1996. Possible reason for relative salt stress tolerance in nodules of white lupin cv. Multolupa. J. Exp. Bot. 47: 1709–1716.

Ferri, A., C. Lluch and A. Ocaña. 2000. Effect of salt stress on carbon metabolism and bacteroid respiration in root nodules of common bean (*Phaseolus vulgaris* L.). Plant Biol. 2: 396–402.

Fischinger, S.A. and J. Schulze. 2010. The importance of nodule CO_2 fixation for the efficiency of symbiotic nitrogen fixation in pea at vegetative growth and during pod formation. J. Exp. Bot. 61: 2281–2291.

Flechard, C.R., P. Ambus, U. Skiba, R.M. Rees, A. Hensen, A. van Amstel, A. van den Pol-van Dasselaar, J.F. Soussana, M. Jones, J. Clifton-Brown, A. Raschi, L. Horvath, A. Neftel, M. Jocher, C. Ammann, J. Leifeld, J. Fuhrer, P. Calanca, E. Thalman, K. Pilegaard, C. Di Marco, C. Campbell, E. Nemitz, K.J. Hargreaves, P.E. Levy, B.C. Ball, S.K. Jones, W.C.M. van de Bulk, T. Groot, M. Blom, R. Domingues, G. Kasper, V. Allard, E. Ceschia, P. Cellier, P. Laville, C. Henault, F. Bizouard, M. Abdalla, M. Williams, S. Baronti, F. Berretti, B. Grosz. 2007. Effects of climate and management intensity on nitrous oxide emissions in grassland systems across Europe Agric. Ecosyst. Environ. 121: 135–152.

Flowers, T.J. 2004. Improving crop salt tolerance. J. Exp. Bot. 55: 307–319.

Fortes, A.M., J. Costa, F. Santos, J.M. Segui-Simarro, K. Palme, T. Altabella, A.F. Tiburcio and M.S. Pais. 2011. Arginine decarboxylase expression, polyamines biosynthesis and reactive oxygen species during organogenic nodule formation in hop. Plant Sign. Behav. 6: 258–269.

Fougère, F., D. Rudulier and J.G. Streeter. 1991. Effects of salt stress on amino acid, organic acid, and carbohydrate composition of roots, bacteroids, and cytosol of alfalfa (*Medicago sativa L.*). Plant Physiol. 96: 1228–1236.

Foyer, C.H. and G. Noctor. 2009. Redox regulation in photosynthetic organisms: signaling, acclimation and practical implications. Antioxid. Redox Sign. 11: 1–45.

Frendo, P., J. Harrison, C. Norman, M.J.H. Jimenez, G. Van de Sype, A. Gilabert and A. Puppo. 2005. Glutathione and homoglutathione play a critical role in the nodulation process of *Medicago truncatula*. Mol. Plant Microbe In. 18: 254–259.

Fujihara, S and T. Yoneyama. 1994. Response of *Rhizobium fredii* P220 to osmotic shock: interactions between K^+, Mg^{2+}, glutamate and homospermidine. Microbiol. 140: 1909–1916.

Fujihara, S., H. Abe, Y. Minakawa, S. Akao and T. Yoneyama. 1994. Polyamines in nodules from various plant-microbe symbiotic associations. Plant Cell Physiol. 35: 1127–1134.

Gan, Y., I. Stulen, H. van Keulen and P.J.C. Kuiper. 2004. Low concentration of nitrate and ammonium stimulate nodulation and N_2 fixation while inhibiting specific nodulation (nodule DW g^{-1} root dry weight) and specific N_2 fixation (N_2 fixed g^{-1} root dry weight) in soybean. Plant Soil. 258: 281–292.

Gibson, A.H. and J.E. Harper. 1985. Nitrate effect on nodulation of soybean by *Bradyrhizobium japonicum*. Crop. Sci. 25: 497–501.

Gordon, A.J., F.R. Minchin, L.C. James and O. Komina. 1999 Sucrose synthase in legume nodules is essential for nitrogen fixation. Plant Physiol. 120: 867–878.

Greenway, H. and R. Munns. 1980. Mechanisms of salt tolerance in non-halophytes. Annu. Rev. Plant Physiol. 31: 149–190.

Han, K.H. and C.H. Hwang. 2003. Salt tolerance enhanced by transformation of a *P5CS* gene in carrot. J. Plant Biotechnol. 5: 149–153.

Hawker, J.S. 1985. Sucrose. *In*: P.M. Dey and R.A. Dixon [eds.]. Biochemistry of Storage Carbohydrates in Green Plants. Academic Press, London. pp. 1–51.

Hervás, A., J.M. Caba, C. Lluch and F. Ligero. 1991. Effect of combined nitrogen on dinitrogen fixation and productivity in *Pisum sativum L.* inoculated with different strains of *Rhizobium*. Chemosphere. 22: 1153–1160.

Jamet, A., K. Mandon, A. Puppo and D. Herouart. 2007. H_2O_2 is required for optimal establishment of the *Medicago sativa/Sinorhizobium meliloti* symbiosis. J. Bacteriol. 189: 8741–8745.

Jamil, A., S. Riaz, M. Ashraf and R. Foolad. 2011. Gene expression profiling in plants under salt stress. Crit. Rev. Plant Sci. 30: 435–458.

Jebara, S., M. Jebara, F. Limam and M.E. Aouani. 2005. Changes in ascorbate peroxidase, catalase, guaiacol peroxidase and superoxide dismutase activities in common bean (*Phaseolus vulgaris*) nodules under salt stress. J. Plant Physiol. 162: 929—936.

Jebara, S., J.J. Drevon and M. Jebara. 2010. Modulation of symbiotic efficiency and nodular antioxidant enzyme activities in two *Phaseolus vulgaris* genotypes under salinity. Acta Physiol. Plant. 32: 925–932.

Keating, B.A. and M.J. Fisher. 1985. Comparative tolerance of tropical grain legumes to salinity. Aut. J. Agric. Res. 36: 373–383.

Khadri, M., M. Soussi, L. Pliego, C. Lluch and A. Ocaña. 2001. Ammonium assimilation and ureide metabolism in common bean (*Phaseolus vulgaris* L.) nodules. Agronomie 21 (6–7): 635–644.

Khadri, M., N.A. Tejera and C. Lluch. 2006. Alleviation of salt stress in common bean (*Phaseolus vulgaris*) by exogenous abscisic acid supply. J. Plant Growth Regul. 25: 110–119.

Khadri, M., N.A. Tejera and C. Lluch. 2007. Sodium chloride-ABA interaction in two common bean (*Phaseolus vulgaris*) cultivars differing in salinity tolerance. Environ. Exp. Bot. 60: 211–218.

Khan, N.A., S. Syeed, A. Masood, R. Nazar and N. Iqbal. 2010. Application of salicylic acid increases contents of nutrients and antioxidative metabolism in mungbean and alleviates adverse effects of salinity stress. Int. J. Plant Biol. 1: 1–8.

Kishor, P.B.K., S. Sangam, R.N. Amrutha, P.S. Laxmi, K.R. Naidu, K.R.S.S. Rao, S. Rao, K.J. Reddy, P. Theriappan and N. Sreenivasulu. 2005. Regulation of proline biosynthesis, degradation, uptake and transport in higher plants: its implications in plant growth and abiotic stress tolerance. Curr. Sci. 88: 424–438.

Kohl, D.H., P.F. Straub and G. Shearer. 1994. Does proline play a special role in bacteroid metabolism? Plant Cell Environ. 17: 1257–1262.

Kraft, M., R. Kuglitsch, J. Kwiatkowski, M. Frank and K. Grossmann. 2007. Indole-3-acetic acid and auxin herbicides up-regulate 9-cis-epoxycarotenoid dioxygenase gene expression and abscisic acid accumulation in cleavers (*Galium aparine*): interaction with ethylene. J. Exp. Bot. 58: 1497–1503.

Lahiri, K., S. Chattopadhyay and B. Ghosh. 2004. Correlation of endogenous free polyamine levels with root nodule senescence in different genotypes in *Vigna mungo* L. J. Plant Physiol. 161: 563–571.

Läuchli, A. 1984. Salt exclusion: an adaptation of legume for crops and pastures under saline conditions. *In*: R.C. Staples and G.H. Toenniessen [eds.]. Salinity Tolerance in Plants. Strategies for Crop Improvement. John Wiley and Sons, New York. pp. 171–187.

Ligero, F., C. Lluch, J. Olivares and E.J. Bedmar. 1987. Nitrate reductase activity in nodules of pea inoculated with hydrogenase positive and hydrogenase negative strains of *Rhizobium leguminosarum*. Physiol. Plant. 69: 313–316.

Liu, X.G., Y.L. Yue, B. Li, Y.L. Nie, W. Li, W.H. Wu and L.G. Ma. 2007. A G protein coupled receptor is a plasma membrane receptor for the plant hormone abscisic acid. Science. 315: 1712–1716.

Lluch, C. and P. Gavilan. 1989. Fijación biológica de nitrógeno. Naturalia Baetica. 1: 73–82.

Lluch, C., C. Iribarne, F. Palma, M. López and N. Tejera. 2011. Integración de las hormonas vegetales en el establecimiento de la simbiosis Rhizobium leguminosa. *In*: M. Megías, R. Rivilla, M.J. Soto, M.J. Delgado, E. González, P.F. Mateos, M. León, B. Rodelas and E.J. Bedmar [eds.]. Fundamentos y Aplicaciones Agroambientales de las Interacciones Beneficiosas Plantas-Microorganismos. Sociedad Española de Fijación de Nitrógeno, España. pp. 61–73.

Lodwig, E. and P. Poole. 2003. Metabolism of *rhizobium* bacteroids. Crit. Rev. Plant Sci. 22: 37–78.

López, M. and C. Lluch. 2008. Nitrogen fixation is synchronized with carbon metabolism in *Lotus japonicus* and *Medicago truncatula* nodules under salt stress. J. Plant Interact. 3: 137–144.

López, M., J.A. Herrera-Cervera, C. Lluch and N.A. Tejera. 2006. Trehalose metabolism in root nodules of the model legume *Lotus japonicus* in response to salt stress. Physiol. Plant. 128 (4): 701–709.

López, M., J.A. Herrera-Cervera, C. Iribarne, N.A. Tejera and C. Lluch. 2008. Growth and nitrogen fixation in *Lotus japonicus* and *Medicago truncatula* under NaCl stress: Nodule carbon metabolism. J. Plant Physiol. 165: 641–650.

López, M., N.A. Tejera and C. Lluch. 2009. Validamycin A improves the response to salt stress of *Medicago truncatula* by inducing trehalose accumulation in the root nodules. J. Plant Physiol. 166: 1218–1222.

López-Gómez, M., N.A. Tejera, C. Iribarne, J.A. Herrera-Cervera and C. Lluch. 2012. Different strategies for salt tolerance in determined and indeterminate nodules of *Lotus japonicus* and *Medicago truncatula*. Arch. Agron. Soil Sci. (in press).

Mabood, F. and D. Smith. 2007. The role of salicylates in *Rhizobium*-legume symbiosis and abiotic stresses in higher plants. *In*: S. Hayat and A. Ahmad [eds.]. Salicylic Acid, A Plant Hormone. Springer Publishers. Dordrechts, The Netherlands. pp. 151–162.

Manchanda, G. and N. Garg. 2008. Salinity and its effects on the functional biology of legumes. Acta Physiol. Plant. 30 (5): 595–618.

Martínez-Abarca, F., J.A. Herrera-Cervera, P. Bueno, J. Sanjuan, T. Bisseling and J. Olivares. 1998. Involvement of salicylic acid in the establishment of the *Rhizobium meliloti*-alfalfa symbiosis. Mol. Plant Microbe In. 11: 153–155.

Matamoros, M.A., D.A. Dalton, J. Ramos, M.R. Clemente, M.C. Rubio and M. Becana. 2003. Biochemistry and molecular biology of antioxidants in the rhizobia legume symbiosis. Plant Physiol. 133: 499–509.

Matamoros, M.A., J. Loscos, M.J. Coronado, J. Ramos, S. Sato, P.S. Testillano, S. Tabata and M. Becana. 2006. Biosynthesis of ascorbic acid in legume root nodules. Plant Physiol. 141: 1068–1077.

Mhadhbi, H., M. Jebara, F. Limam and M.E. Aouani. 2004. Rhizobial strain involvement in plant growth, nodule protein composition and antioxidant enzyme activities of chickpea rhizobia symbioses: modulation by salt stress. Plant Physiol. Bioch. 42: 717–722.

Miller, K.J. and J.M. Wood. 1996. Osmoadaptation by rhizosfere bacteria. Annu. Rev. Microbiol. 50: 101–136.

Miller, S.S., B.T. Driscoll, R.G. Gregerson and C.P. Vance. 1998. Alfalfa malate dehydrogenase (MDH): molecular cloning and characterization of five different forms reveals an unique nodule-enhanced MDH. Plant J. 15: 173–184.

Minchin, F.R. and J.S. Pate. 1974. Diurnal functioning of the legume root nodule. J. Exp. Bot. 25: 295–308.

Minchin, F.R., B.J. Thomas and L.R. Mytton. 1994. Ion distribution across the cortex of soybean nodules: possible involvement in control of oxygen diffusion. Ann. Bot. 74: 613–617.

Mittler, R., S. Vanderauwera, M. Gollery and F. Van Breusegem. 2004. Reactive oxygen gene network of plants. Trends Plant Sci. 9: 490–498.

Mohammad, R.M., M. Akhavan-Kharazian, W.F. Campbell and M.D. Rumbaugh. 1991. Identification of salt and drought-tolerant *Rhizobium meliloti* L. strains. Plant Soil. 134: 271–276.

Moran, J.F., E.K. James, M.C. Rubio, G. Sarath, R.V. Klucas and M. Becana. 2003. Functional characterization and expression of a cytosolic iron-superoxide dismutase from cowpea root nodules. Plant Physiol. 133: 773–782.

Mudgal, V., N. Madaan and A. Mudgal. 2010. Biochemical mechanisms of salt tolerance in plants: a review. Int. J. Bot. 6(2): 136–143.

Muglia, C., G. Comai, E. Spegazzini, P.M. Riccillo and O.M. Aguilar. 2008. Glutathione produced by *Rhizobium tropici* is important to prevent early senescence in common bean nodules. Fems Microbiol. Lett. 286: 191–198.

Müller, J., T. Boller and A. Wiemken. 2001. Trehalose becomes the most abundant non-structural carbohydrate during senescence of soybean nodules. J. Exp. Bot. 52: 943–947.

Munns, R. 2002. Comparative physiology of salt and water stress. Plant Cell Environ. 25: 239–250.

Munns, R. and R.A. James. 2003. Screening methods for salinity tolerance: a case study with tetraploid wheat. Plant Soil. 253: 201—218.

Munns, R. and M. Tester. 2008. Mechanisms of salinity tolerance. Annu. Rev. Plant Biol. 59: 651–681.

Nazar, R., N. Iqbal, S. Syeed and N.A. Khan. 2011. Salicylic acid alleviates decreases in photosynthesis under salt stress by enhancing nitrogen and sulfur assimilation and antioxidant metabolism differentially in two mungbean cultivars. J. Plant Physiol. 168: [illegible].

Nogales, J., R. Campos, H. BenAbdelhalek, J. Olivares, C. Lluch and J. Sanjuán. 2002. *Rhizobium tropicci* genes involved in free living salt tolerance are required for the establishment of efficient nitrogen-fixing symbiosis with *Phaseolus vulgaris*. Mol. Plant-Microbe Interact. 15: 225–232.

Nomura, M., H.T. Mai, M. Fujii, S. Hata, K. Izui and S. Tajima. 2006. Phosphoenolpyruvate carboxylase plays a crucial role in limiting nitrogen fixation in *Lotus japonicus* nodules. Plant Cell Physiol. 47: 613–621.

Obendorf, R.L., E.M. Sensenig, J. Wu, M. Ohashi, T.E. O'Sullivan, S.M. Kosina and S.R. Schnebly. 2008. Soluble carbohydrates in mature soybean seeds after feeding D-chiro inositol, myo-inositol, or D-pinitol to stem-leaf-pod explants of low raffinose, low stachyose lines. Plant Sci. 175: 650–655.

Ocaña, A., M.P. Cordovilla, F. Ligero and C. Lluch. 1996. Phosphoenolpyruvate carboxylase in root nodules of *Vicia faba*: partial purification and properties. Physiol. Plant. 97: 724–730.

Ocaña, A., J.M. Caba, F. Ligero and C. Lluch. 1998. Nitrogen fixation in *Vicia faba* root nodules over ontogeny. Isr. J. Plant Sci. 47: 35–41.

Ohwada, T., Y. Sasaki, H. Koike, K. Igawa and T. Sato. 1998. Correlation between NaCl sensitivity of *Rhizobium* bacteria and ineffective nodulation of leguminous plants. Biosci. Biotechnol. Biochem. 62: 2086–2090.

Oldroyd, G.E. and J.A. Downie. 2008. Coordinating nodule morphogenesis with rhizobial infection in legumes. Annu. Rev. Plant Biol. 59: 519–46.

Palma, F., C. Lluch, C. Iribarne, J.M. Garcia-Garrido and N.A. Tejera Garcia. 2009. Combined effect of salicylic acid and salinity on some antioxidant activities, oxidative stress and metabolite accumulation in *Phaseolus vulgaris*. Plant Growth Regul. 58: 307–316.

Peleg-Grossman, S., H. Volpin and A. Levine. 2007. Root hair curling and *Rhizobium* infection in *Medicago truncatula* are mediated by phosphatidylinositide-regulated endocytosis and reactive oxygen species. J. Exp. Bot. 58: 1637–1649.

Pereira, S.I.A., A.I.G. Lima and E.M.A.P. Figueira. 2007. *Rhizobium leguminosarum* isolated from agricultural ecosystems subjected to different climatic influences: the relation between genetic diversity, salt tolerance and nodulation efficiency. *In*: L. Tian-Xiao [ed.]. Soil Ecology Research Developments. Nova Science Publishers Inc. New York, pp. 1–17.

Phillips, D.V., D.O. Wilson and D.E. Dougherty. 1984. Soluble carbohydrates in legumes and nodulated nonlegumes. J. Agri. Food Chem. 32: 1289–1291.

Postgate, J.R. 1982. The Fundamentals of Nitrogen Fixation. Cambridge University Press, New York.

Ramanujam, M.P., V.A. Jaleel and G. Kumaravelu. 1998. Effect of salicylic acid on nodulation, nitrogenous compounds and related enzymes of *Vigna mungo*. Biol. Plantarum. 41: 307–311.

Rao, D.N.L., K.E. Giller, A.R. Yeo and T.J. Flowers. 2002. The effect of salinity and sodicity upon nodulation and nitrogen fixation in chickpea (*Cicer arietinum*). Ann. Bot. 89: 563–570.

Rengasamy, P. 2002. Transient salinity and subsoil constraints to dryland farming in Australian sodic soil: an overview. Aust. J. Exp. Agric. 42: 351–361.

Requena, N., I. Jimenezm, M. Toro and J.M. Barea. 1997. Interaction between plant-growth-promoting rhizobacteria (PGPR), arbuscular mycorrhizal fungi and *Rhizobium* spp. In the rhizosphere of *Anthyllis cytisoides*, amodel legume for revegetation in Mediteranean semi-arid ecosystems. New Phytol. 136: 667–677.

Rhodes, D. and P. Felker. 1988. Mass screening of Prosopis (mesquite) seedlings for growth at seawater salinity concentrations. Forest. Ecol. Manage. 24: 169–176.

Rhodes, D, A. Nadolska-Orczyk and P.J. Rich. 2002. Salinity, osmolytes and compatible solutes. *In*: A. Läuchli and U. Lüttge [eds.]. 2002. Salinity: Environment Plant Molecules. Kluwer Academic Publishers, The Netherlands. pp. 181–204.

Rigaus, J. 1981. Comparison of the efficiency of nitrate and nitrogen fixation in crop yield. *In:* J.D. Bewley [ed.]. Nitrogen and Carbon Metabolism. Martinus Nijhoff, The Hague, Amsterdam. pp. 17–48.

Rontein, D., G. Basset and A.D. Hanson. 2002. Metabolic engineering of osmoprotectant accumulation in plants. Met. Engineering. 4: 49–56.

Rubio, M.C., P. Bustos-Sanmamed, M.R. Clemente and M. Becana. 2009. Effects of salt stress on the expression of antioxidant genes and proteins in the model legume *Lotus japonicus*. New Phytol. 181: 851–859.

Santos, R., D. Herouart, A. Puppo and D. Touati. 2000. Critical protective role of bacterial superoxide dismutase in *Rhizobium*-legume symbiosis. Mol. Microbiol. 38: 750–759.

Saxena, D., M. Amin and S. Khanna. 1996. Modulation of protein profiles in *Rhizobium* sp. under salt stress. Can. J. Microbiol. 42: 617–620.

Schubert, K.R. 1981. Enzymes of purine biosynthesis and catabolism in *Glycine max*. Plant Physiol. 68: 1115–1122.

Schulze, J., E. Adgo and W. Merbach. 1999. Carbon costs associated with N_2 fixation in *Vicia faba* L. *Pisum sativum* over a 14-day period. Plant Biol. 1: 625–631.

Serraj, R., G. Roy and J.J. Devron. 1994. Salt stress induces a decrease in the oxygen uptake of soybean nodules and in their permeability to oxygen diffusion. Physiol. Plant. 91: 161–168.

Serrano, R. and P. Rodríguez. 2002. Plants, genes and ions. EMBO rep. 3: 116–119.

Shamseldin, A., J. Nyalwidhe and D. Werner. 2006. A proteomic approach toward the analysis of salt tolerance in *Rhizobium etli* and *Sinorhizobium meliloti* strains. Cur. Microbiol. 52: 333–339.

Smith, L.T., J.A. Pocard, T. Bernard and D. Le Ruddulier. 1988. Osmotic control of glycine betaine biosynthesis and degradation in *Rhizobium meliloti*. J. Bacteriol. 170: 3142–3149.

Smith, L.T., G.B. Smith, M. D'Souza, J.-A. Pocard, D. Le Rudulier and M.A. Madkour. 1994. Osmoregulation in Rhizobium meliloti: mechanism and control by other environmental signals. J. Exp. Zool. 268: 162–165.

Soussi, M., A. Ocaña and C. Lluch. 1998. Effects of salt stress on growth, photosynthesis and nitrogen fixation in chick-pea (*Cicer arietinum* L.). J. Exp. Bot. 49: 1329–1337.

Soussi, M., M. Khadri, C. Lluch and A. Ocaña. 2001. Carbon metabolism and bacteroid respiration in nodules of chick-pea (*Cicer arietinum* L.) plants grown under saline conditions. Plant Biosystems. 135: 157–164.

Stacey, G., M. Libault, L. Brechenmacher, J.R. Wan and G.D. May. 2006. Genetics and functional genomics of legume nodulation. Curr. Opin. Plant Biol. 9: 110–121.

Stephens, B.D. and C.A. Neyra. 1983. Nitrate and nitrite reduction in relation to nitrogenase activity in soybean nodules and *Rhizobium japonicum* bacteroids. Plant Physiol. 71: 731–735.

Stougaard, J. 2000. Regulators and regulation of legume root nodule development. Plant Physiol. 124: 531–540.

Streeter, J.G. 1980. Carbohydrates in soybean nodules. Distribution of compounds in seedling during the onset of nitrogen fixation. Plant Physiol. 66: 471–476.

Streeter, J.G. 1985. Accumulation of α α-trehalose by *Rhizobium* bacteria and bacteroids. J. Bacteriol. 164: 78–84.

Streeter, J.G. 1987. Carbohydrate, organic acid, and amino acid composition of bacteroids and cytosol from soybean nodules. Plant Physiol. 85: 768–773.

Streeter, J.G. 1988. Inhibition of legume nodule formation and nitrogen fixation by nitrate. Crit. Rev. Plant Sci. 7: 1–23.

Streeter, J.G. and M.L. Gómez. 2006. Three enzymes for trehalose synthesis in Bradyrhizobium cultured bacteria and bacteroids from soybean nodules. App. Environ. Microbiol. 72: 4250–4255.

Streeter, J.G., D.G. Lohnes and R.J. Fioritto. 2001. Pattern of pinitol accumulation in soybean plants and relationships to drought tolerance. Plant Cell Environ. 24: 429–438.

Subbarao, G.V. and C. Johansen. 1991. Potential for genetic improvement in salinity tolerance in legumes: pigeon pea. *In*: M. Pessarakli [ed.]. Hand Book of Plant and Crop Stress. Marcel Dekker Inc. New York, USA.

Szabados, L. and A. Savouré. 2009. Proline: a multifunctional amino acid. Trends Plant Sci. 15(2): 89–97.

Szabolcs, I. 1994. Soil salinization. *In*: M. Pessarakli [ed.]. Handbook of Plant Crop Stress. Marcel Dekker, New York. pp. 3–11.

Talibart, R., M. Jebbar, K. Gouffi, V. Pichereau, G. Gouesbet, C. Blanco, T. Bernard and J.A. Pocard. 1997. Transient accumulation of glycine betaine and dynamics of endogenous osmolytes in salt-stressed cultures of *Sinorhizobium meliloti*. App. Env. Microbiol. 63: 4657–4663.

Tejera, N.A., R. Campos, J. Sanjuan and C. Lluch. 2004. Nitrogenase and antioxidant enzyme activities in *Phaseolus vulgaris* nodules formed by *Rhizobium tropici* isogenic strains with varying tolerance to salt stress. J. Plant Physiol. 161: 329–338.

Tejera, N.A., R. Campos, J. Sanjuán and C. Lluch. 2005. Effect of sodium chloride on growth, nutrient accumulation, and nitrogen fixation of common bean plants in symbiosis with isogenic strains. J. Plant Nutr. 28: 1907–1921.

Tejera, N.A., M. Soussi and C. Lluch. 2006. Physiological and nutritional indicators of tolerance to salinity in chickpea plants growing under symbiotic conditions. Environ. Exp. Bot. 58: 17–24.

Tejera, N.A., C. Iribarne, F. Palma and C. Lluch. 2007. Inhibition of the catalase activity from *Phaseolus vulgaris* and *Medicago sativa* by sodium chloride. Plant Physiol. Bioch. 45: 535—541.

Tester, M. and R. Davenport. 2003. Na^+ tolerance transport in higher plants. Ann. Bot. 91: 503–527.

Tominaga, A., M. Nagata, K. Futsuki, H. Abe, T. Uchiumi, M. Abe, K.-I. Kucho, M. Hashiguchi, R. Akashi, A. Hirsch, S. Arima and A. Suzuki. 2010. Effect of abscisic acid on symbiotic nitrogen fixation activity in the root nodules of *Lotus japonicus*. Plant Sign. Behav. 5: 440–443.

Tu, J.C. 1981. Effect of salinity on *Rhizobium* root hair interaction, nodulation and growth of soybean. Can. J. Plant Sci. 61: 231–239.

USDA-ARS. 2008. Research databases. Bibliography on salt tolerance. Salinity Lab. US dep. Agric. Agric Res. Serv. Riverside, CA. http://www.ars.usda.gov/Services/docs.htm?docid=8908.

Vance, C.P. 2008. Carbon and nitrogen metabolism in legume nodules. *In*: M.J. Dilwoth, E.K. James, J.I. Sprent and W.E. Newton [eds.]. 2008. Nitrogen Fixing Leguminous Symbiosies. Springer, Dordrecht. pp. 293–330.

Vance, C.P., R.G. Gregerson, D.L. Robinson, S.S. Miller and J.S. Gantt. 1994. Primary assimilation of nitrogen in alfalfa nodules: molecular features of the enzymes involved. Plant Sci. 101: 51–64.

Verdoy, D., T. Coba de la Peña, F.J. Redondo, M.M. Lucas and J.J. Pueyo. 2006. Transgenic *Medicago truncatula* plants that accumulate proline display nitrogen-fixing activity with enhanced tolerance to osmotic stress. Plant Cell Environ. 29: 1913–1923.

Yamaguchi, T and E. Blumwald. 2005. Developing salt-tolerant crop plants: challenges and opportunities. Trends Plant Sci. 12: 615–620.

Zahran, H.H. 1999. *Rhizobium*-legume symbiosis and nitrogen fixation under severe conditions and in an arid climate. Microbiol. Mol. Biol. Rev. 63: 968–989.

Zhu, Y, G. Shearer and D.H. Kohl. 1992. Proline fed to intact soybean plants influences acetylene reducing activity and content and metabolism of proline in bacteroids. Plant Physiol. 98: 1020–1028.

6

Mineral Nutrition in the Legume-Rhizobia Nitrogen Fixing Symbiosis

I. Abreu,[1] M. Reguera,[1] A. Bonilla,[2] L. Bolaños[1] and *I. Bonilla[1,*]*

Introduction

The increase in world population means a higher demand of food resources which only can be met with an exponential increase in agricultural production. It is important to consider that human nutrition is mainly held by 20 plant crops belonging to cereals, vegetables (which include legumes) and fruits. To increase crop yield, improving-programs have tried to select cultivars with enhanced efficiency between fertilizer application and productivity. This has been relatively successful in cereals such as wheat or barley, used mainly as an energy source, but they have a low concentration of essential amino acids lysine and threonine, and a total protein content of only about 12 percent. Legumes such as soybean, bean or pea, have a more complex mineral nutrition, but are good and cheap sources of protein

[1]Department of Biology, Faculty of Sciences, University Autónoma de Madrid. c/Darwin 2, Campus de Cantoblanco. 28049-Madrid. Spain.
[2]Department of Environmental Sciences and Natural Resources. Faculty of Pharmacy, University of San Pablo CEU. Ctra. Boadilla del Monte Km 5.3. 28668-Boadilla del Monte. Madrid, Spain.
*Corresponding author: ildefonso.bonilla@uam.es

(around 25 percent), and with a lysine and threonine content higher than cereals. FAO recommends a mixed diet, which includes cereals and legumes in an adequate proportion not only to get a balanced protein nutrition and essential amino acids , but also an equilibrated content of sugars, vitamins and minerals.

From the Green Revolution in the mid-20th century, cereal production has depended on global consumption of chemical fertilizers, which has been dramatically increased. For example, the application of nitrogen fertilizer in U.S.A, ranged from 8 kg nitrate/Ha/year before World War II to an average of more than 250 kg/Ha/year at the end of the last century. While global consumption of fertilizers in the early 21st century was 136 million tons (International Fertilizer Development Centre—IFDC), experts of the International Fertilizers Association (IFA) and FAO have estimated an increasing demand of 196 to 216 million tons for the period 20011/12. Its production will rise to 241 million tons for the same period, so production will exceed consumption. Fertilizer use has stabilized in highly populated countries like the USA (about 19 million tons per year since 1984) and India (16 million tons per year since 1998). Today China has the highest consumption of chemical fertilizers (over 40 million tons in 2004 and an estimate of almost 70 million by 2013), including N, P and K. Statistics from the European Union show excessive nitrogen fertilization and application of pesticide active ingredients (2.0 kg/Ha in 2001). This indiscriminate use of fertilizers and pesticides affects health of both farmers and residents of rural areas, and increases pollution of natural ecosystems associated with agribusiness. Human and biosphere survival are compromised by the increasing pollution of inland waters (nitrate and phosphate concentration are reaching toxic levels) and by the eutrophication of lakes and rivers. These negative effects cause an additional economic problem. Only in the USA, the cost of the environment and health problems due to pesticide use is estimated to be US$10,000 million per year, US$2,500 million per year is the cost of losses due to pollution by an excessive use of fertilizers, and US$45,000 million per year is the cost associated with soil erosion by conventional modern agriculture. Many problems arising from these practices have resulted in a social claim for sustainable agriculture based on ecological practices that produce healthy food. However, the fertilizer industry is a sector growing in a continuously increasing population. Despite repeated attempts to control it, according to FAO it is estimated that by 2050 there will be 9,000 million people on the planet.

The increase of productivity cannot be addressed only by increasing nutrient addition, as cited above for cereals, or by improving plant cultivars (i.e., cold, drought or pathogen resistance), because this approach has often resulted in a huge and sometimes indiscriminate use of pesticides and nitrogen fertilizers, as the only response to the lack of plant-available nitrogen in arable soils.

Both aerial plant organs and plant roots are able to interact with many microorganisms, and several of these plant-microbe interactions, which are described in this book, can increase crop yield in a healthy way. Of special relevance are the beneficial interactions that take place in the rhizosphere involving roots and soil microbes, three of which are susceptible of being used for the above mentioned objective. These are the beneficial effects of the Plant Growth-Promoting Rhizobacteria (PGPR) on plant fitness, the improved acquisition of phosphate mediated by the vesicular-arbuscular mycorrhizae (VAM) which establish symbiosis with almost all plants, and the rhizobia-legume symbiosis which allow fixed atmospheric nitrogen under N-deficient conditions. The research on these topics is largely focused to the molecular and physiological dissection of the interaction, or to the selection of bacteria and plants with the best symbiotic properties. However, with the exception of the rhizobia-legume symbiosis, there is a lack of knowledge of how different availability of soil nutrients affects these interactions. Despite the benefits in agriculture and in sustainability, in the last decade only two reviews (Bonilla and Bolaños 2009, O'Hara 2001) synthesize the state-of-the-art of mineral nutrition and symbiosis. Therefore, the goal of this chapter is to summarize what is the essential element, those functions which justify its essentiality for either symbiotic partner or for the rhizobia-legume symbiosis, and the major worldwide nutritional constraints. In connection with this chapter, metal transport mechanisms in the *Rhizobium*-legume symbiosis are reviewed in Chapter 7.

Some Details about the Rhizobia-Legume Symbiosis

As mentioned above, rhizobia-legume symbiosis is one of the most studied interactions in biology because of its importance in agriculture, due to its capacity to fix atmospheric nitrogen to ammonium, which diminishes the addition of fertilizer-nitrogen to the fields. The studies on this topic started with an agricultural focus but have evolved to different approaches which cover different aspects of the relation between about 18000 species of legumes and an indeterminate number of rhizobia strains, belonging both to α- and β-proteobacteria (Masson-Boivin et al. 2009). Today, the "hottest" topic is the genetic dissection of the dialogue between the legume and rhizobia (Popp and Ott 2011) contrasting with past molecular and biochemical studies which described the steps of the symbiosis development, from the early interaction between the free-living bacteria and the legume and the infection process, to the transport of metabolites once the rhizobia reside endosymbiotically in the plant cell (for reviews see Brewin 1991, 2004, Gage 2004, Jones et al. 2007).

Mineral nutrition can affect all the stages of development of an effective symbiosis (summarized in Fig. 6.1), so it is essential to to briefly describe

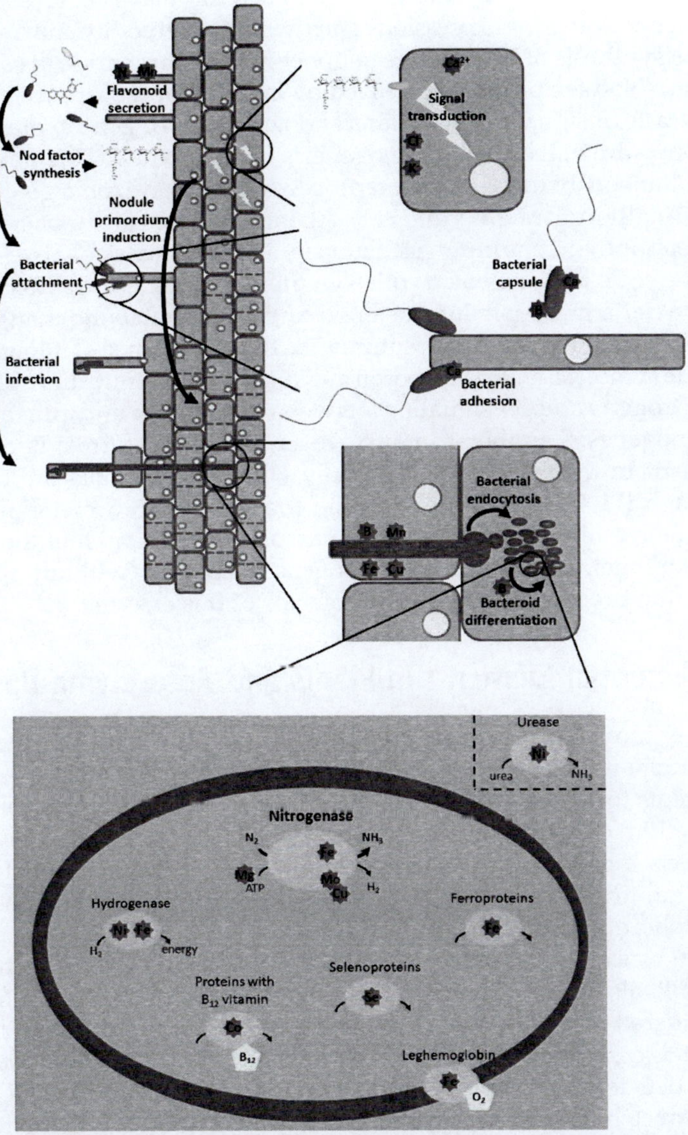

Figure 6.1 Effects of different elements on nodulation and enzymatic activities inside the symbiosome. The drawing represents the symbiosis development in an indeterminate nodule, where the elements with a key role during the symbiosis are schematized with a red star, although the functions can be extended to determinate nodule-forming plants.

Color image of this figure appears in the color plate section at the end of the book.

the symbiotic events (for more information see the references above). The interaction starts when bacteria sense the plant, usually responding to flavonoids secreted by the root. The bacteria, that is colonizing the root surface activates the expression of *nod* genes which codify enzymes involved in the synthesis of Nod factors (a lipochitooligosaccharide specific for each strain). Nod factors are recognized by LysM receptor kinases, followed by a cascade of signal transductions and activation of gene expression which finally culminate in the organogenesis of the nodule. This process occurs simultaneously to the infection process whereby rhizobia go into the root, and finally into the nodule cell. At the end, when the bacteria reaches the endophytic environment, late signaling processes between both the prokaryotic and the eukaryotic symbionts mediate the differentiation of the bacteria into N_2-fixing bacteroid and plant metabolic changes.

Just three examples can summarize the importance of mineral nutrition for the symbiosis: nitrogen, boron and molybdenum nutrition. The presence of nitrogen inhibits signaling between symbionts and prevents nodule organogenesis, as they can survive as free-living forms. Boron is a key element in the symbiosis, especially when the rhizobia are progressing through the infection thread in the root, although a metabolic function in any symbiont has not yet been associated. Meanwhile the deficiency of molybdenum inhibits the nitrogenase activity (without affecting the infection process) because it is a cofactor of this enzyme.

Essential Elements in Plants and Free-Living Bacteria

The organisms, based on their physiology, have different nutritional requirements. There was an early interest in the establishment of essential elements for plants, mainly centered in increasing the crop yield. As early as the 19th century, von Liebig enunciated the Law of the minimum which predicts increases in plant growth following the increase of the amount of a limiting nutrient. At that time, several nutrients such as nitrogen, phosphorous, and potassium were included as beneficial for plants. The actual concept of an essential element was proposed by Arnon and Stout in 1939, also in plant nutrition. Citing those authors: "an element is not considered essential unless (a) a deficiency of it makes it impossible for the plant to complete the vegetative or reproductive stage of its life cycle; (b) such deficiency is specific to the element in question, and can be prevented or corrected only by supplying this element; and (c) the element is directly involved in the nutrition of the plant quite apart from its possible effects in correcting some unfavorable microbiological or chemical condition of the soil or other culture medium". Following these criteria, 17 elements were described as essential in plants (reviewed in Bonilla et al. 2010) and classified by its required concentrations in macronutrients: calcium (Ca), carbon (C), hydrogen (H), magnesium (Mg), nitrogen (N), oxygen (O), phosphorous (P), potassium (K) and sulfur (S), and micronutrients: boron (B), chlorine (Cl), copper (Cu), iron (Fe), manganese (Mn), molybdenum (Mo), nickel (Ni) and zinc (Zn) (Table 6.1).

As far as we know, the first studies in mineral nutrition of rhizobia started with Clarck (1936) and Steinberg (1938). After an increase in the purity of mineral salts, some authors like Bergensen and Sherwood developed a synthetic media as a prelude to the analysis of some micronutrients as cobalt (in the 60s) manganese and zinc (in the 70s) in the free-living bacteria. In accordance with O'Hara (1988, 2001) there were 17 elements considered essential to rhizobia, 15 of them matching with the plant list (Ca, C, Cu, Fe, H, K, Mg, Mn, Mo, N, Ni, O, P, S, Zn), plus cobalt (Co) and selenium (Se), that were described as specific for rhizobia. Boron, as will be discussed below, is essential for the symbiosis and some recent results indicate that it also affects cell surface polysaccharides of free-living rhizobia (Abreu et al. 2012). As in plants, the essential elements for bacteria are classified in macronutrient (the same in plants) and micronutrients too (Table 6.1).

Macroelements

Nitrogen

Nitrogen is an essential element as a basic constituent of biomolecules like proteins, nucleic acids or several polysaccharides. In the atmosphere, nitrogen is present in high amounts as molecular nitrogen (N_2) but this molecular form is inaccessible for plants or free-living-rhizobia because of the absence of nitrogenase. The main objective of the symbiosis is to create inside the nodule conditions required for the expression and function of the nitrogenase encoded in the genome of the bacteria. The symbiotic interaction is not mandatory to any partner, and nitrogen availability actively controls all the processes, including the symbiosis establishment and also the maintenance of an effective nodule. In this sense, it has been reported that nitrogen controls the adhesion of the rhizobia to the root, infection rate and the expression of nitrogenase and leghemoglobin as later processes.

As a field recommendation, N_2 fixation depends on the number of effective nodules, which justifies a reduction of the nitrogen fertilization to levels that ensure a correct development of the plant without inhibiting the process of the symbiosis. If the symbiosis does not progress correctly or the nitrogen fixation rate is insufficient, it can be added as a nitrogen source. The nitrogen specie and the added amount must depend of the type of soil.

Table 6.1 Essential elements in free living organisms and principal functions in rhizobia-legume symbiosis.

Element	Bioavailable ion	Dry Weight Concn (μmol·g⁻¹)	Essentiality			Function					
			Rhizobia	Legumes	Rhizobia-legume symbiosis	Structural (cellular components)	Structural (Membranes, cell wall)	Energy (Chemical or potential energy)	Osmolyte	Enzymatic (Cofactor)	Signaling
Macronutrients											
Hydrogen (H)	H^+, H_2O	60,000	+	+	+	+	+	+	+		
Carbon (C)	CO_2	40,000	+	+	+	+	+	+			
Oxygen (O)	O_2, H_2O	30,000	+	+	+	+	+	+			
Nitrogen (N)	$\mathbf{NO_3^-}$, NH_4^+	1,000	+	+	+	+	+	+			
Potassium (K)	K^+	250	+	+	+				+	+	+
Calcium (Ca)	Ca^{2+}	125	+	+	+		+		+	+	+
Magnesium (Mg)	Mg^{2+}	80	+	+	+		+	+		+	
Phosphorous (P)	$\mathbf{H_2PO_4^-}$, HPO_4^{2-}	60	+	+	+	+	+	+			+
Sulfur (S)	SO_4^{2-}	30	+	+	+	+	+			+	
Micronutrients											
Chlorine (Cl)	Cl^-	3	+	+	+				+	+	+
Boron (B)	H_3BO_3	2	+[a]	+	+		+				+
Iron (Fe)	Fe^{2+}, Fe^{3+}	2	+	+	+			+		+	
Manganese (Mn)	Mn^{5+}	1	+	+	+					+	
Zinc (Zn)	Zn^{2+}	0.3	+	+	+					+	

Cobalt (Co)	Co^{2+}, Co^{3+}	0,002	+	+[b]	+					+	
Nickel (Ni)	Ni^{2+}	0,001	+	+	+					+	
Molybdenum (Mo)	MoO_4^{2-}	0,001	+	+	+					+	
Selenium (Se)	SeO_4^{2-}	(nd)	+		+	+				+	

[a]The element affects cell surface polysaccharide production and structure in some species.
[b]The element is necessary in some species.

Phosphorous

No element has as many functions in the cell as phosphorous, and maybe it is one of the elements whose distribution and availability is more capricious. Phosphorous is taken by the plant as one of the phosphate anion species and remains in the cell in the oxidized form of phosphate, or covalently bound to a carbon skeleton in sugars, in energy storage nucleotides, in nucleic acids (storing and managing the genetic information), in membrane phospholipids, and also in proteins (as activation/deactivation tag) (Maathuis 2009).

With these features and properties, it is obvious that phosphorus plays several important roles in the symbiosis. Moreover, the nitrogen fixation process itself requires a correct balance in phosphorous nutrition because it is very expensive in energy terms. It is calculated that the energy required per N_2-fixed mol is higher than 16–18 moles of ATP. Some laboratory studies show a higher phosphorous requirement in nodulated than in fertilized soybean, and some studies recommend an application of 80 to 120 kg/Ha of P_2O_5 for a correct development of this legume (O'Hara et al. 1988, O'Hara 2001, Bonilla and Bolaños 2009). However, the final supply of phosphorous depends on soil pH or calcium concentration among other factors that have to be considered before fertilization. P deficiency is widespread, and both the plant and the bacteria have developed mechanisms to chelate and solubilize phosphorous. Among them, other plant-microbe interaction has a special relevance: the symbiosis between plants and vesicular-arbuscular mycorrhizae. It is undergone by 90 percent of plant species, and is compatible with the rhizobia symbiosis (see Chapter 15 in this volume).

Potassium

Potassium is the ion which preferentially controls osmoregulation in plants. Also, potassium cell content regulates activation of several cytosolic enzymes (Maathuis 2009). A complex system of transporters of high and low affinity ensures the concentration of this element in the plant cell, both in the cytosol and in the vacuole.

Although potassium deficiency is uncommon, frequently this element is applied as a fertilizer to increase biomass. This practice is particularly relevant in saline soils because of the role of potassium to maintain the ionic balance and cell turgor to protect the activity of several enzymes. As an example, a supply of 50–80 kg K_2O/Ha is recommended to ensure correct nutrition of soybean.

Calcium

Calcium is found in plants mainly as divalent cation Ca^{2+}, which is relatively immobile because of its high reactivity with negatively-charged groups, such as phosphates or nitrates in soils, or as phosphates and carboxyls of sugars, lipids and proteins. This property supports a major structural role of calcium in plants and in rhizobia. In plants, Ca^{2+} is linked to the structure of the cell wall and membranes. The cation in the cell wall is coordinated to the carboxyl groups of opposing pectins, but Ca^{2+} is also determined for the stability of the cell membrane (Maathuis 2009). In bacteria, Ca^{2+} is also an important element for the integrity of the cell envelope and plays an important role in the formation of biofilms (O'Hara 2001). Besides structural roles, recent work has focused on the particularly important function of a small presence of free cytosolic Ca^{2+} (around 100 nM) as one of the most important messengers involved in signal transduction. External stimuli cause changes in the concentration of cytosolic Ca^{2+}, which acts as a second messenger in signaling environmental changes to induce organism responses, in a process known as "Ca signature" (Bonilla et al. 2010).

Early in 1960, Greenwood and Hallsworth recorded for the first time a role of Ca^{2+} in symbiotic N_2 fixation in legumes, and some years later, in 1968, the necessity to supply high Ca^{2+} concentrations to induce a high nodulation rate was described. These studies indicate an important role of Ca^{2+} signaling mechanisms for the recognition between the plant and bacteria. An external supply of calcium enhances the activity of the phenylalanine ammonia-lyase (PAL), a key enzyme in the flavonoid pathway synthesis, one of the major components of root exudates. In accordance with this increase of root exudates caused by calcium, the activity of the *nod* genes is enhanced when Ca^{2+} concentration increases. Calcium is also necessary for optimum root colonization, because several components of the plant and bacteria are capable of using Ca^{2+} as a ligand to enhance adhesion. Of special importance are plant lectins, or some rhizobial Ca^{2+}-dependent ricadhesins. Furthermore, bacterial exopolysaccharide (EPS) can form a gel in the presence of Ca^{2+}, which is a non-specific mechanism for bacterial attachment.

As we mentioned above, calcium plays an important role in signal transduction during the symbiosis. An influx of Ca^{2+} at the tip of the root hair is the first detectable effect following application of Nod factor. This determines an efflux of Cl^- and membrane depolarization, causing an increase in cytosolic Ca^{2+} at the root hair tip in a few minutes. As a second messenger, calcium interacts with other signal transduction pathways and messengers, as inositol triphosphate (IP3), which is a signal molecule produced in the G protein-phospholipase C signaling pathway, which is also induced by Nod factors and Ca^{2+} dependent. Both signaling mechanisms regulate some actin-binding proteins which mediate the rearrangement of the actin filaments, and that could be the basis of several other pre infection events, including root hair deformation, vacuolization, alignment of the endoplasmic reticulum with the plasma membrane, nuclear movement toward the swollen root hair tip, and the inward growth of the cell wall to initiate infection channel. As "Ca signature", a first wave of calcium occurs about 3 min after the application of Nod factors, when the mentioned reorganization of the actin cytoskeleton begins. Others Ca^{2+} waves are later formulated in the perinuclear region at the root tip, approximately 9 min after application of Nod factor, and are extended for a minimum of 60 min to 3 hr. Although the role of these waves is still unclear, there is some information about changes in gene expression and cell cycle regulation during nodule organogenesis. For instance, mutations in some Ca^{2+}/calmodulin-dependent protein kinases induce the formation of nodules without rhizobia.

Magnesium

Magnesium is not usually a limiting factor for plant growth except in very acid or some sandy soils. As divalent cation (Mg^{2+}) has an important role in the maintenance of turgor pressure in the plant cell vacuole. Other ubiquitous roles of Mg^{2+} are related to the handling of phosphoester compounds as phosphosugars, nucleotides or nucleic acids. Mg^{2+} can stabilize the negative charges of compounds such as pyrophosphate (PPi) or ATP in phosphorylation/dephosphorylation processes with an energetic function; also the same charge-stabilizing role is applicable in nucleic acid metabolism, where the presence of magnesium is required to replicate, transcribe and translate the genetic information. Moreover, in plants, magnesium is present in the chlorophyll molecule and needs pigment photosynthetic functionality (Maathuis 2009).

Sulfur

Sulfur is present in two essential amino acids, cysteine and methionine, and several vitamins such as biotin or thiamine. This justifies its essentiality for any organism, including bacteria and plants. Furthermore sulfur is an important element to prevent oxidative damage and heavy metal toxicity, as it is present as tiol group (–SH) in molecules as glutation and phytochelatins. This protective role is based in the reactivity of tiols with any reactive oxygen species (ROS) and its affinity to coordinate heavy metals (Maathuis 2009).

Recently, sulfur deficiency in nodulated legumes have been reported. The consequences are low nodulation, incorrect development of the nodule, and a general decrease in enzyme activity which results in whole metabolism failure.

Microelements

Chlorine

Chlorine is usually present in its anionic form (Cl^-). Its negative charge is used by living organisms to maintain the electric charge balance, and in that sense is involved in membrane depolarization mediated signaling mechanisms (Hänsch and Mendel 2009). This function is important during the first step of symbiosis because, after Nod factor perception by the root hair, a Ca^{2+} influx takes place (the first "Ca signature") followed by an efflux of Cl^- (and after that by a K^+ efflux) that ensures the propagation of the depolarization wave along the root hair.

Chlorine also has metabolic functions: it is a cofactor in the oxygen evolving complex of photosystem II, and also is an activator of proton-pumping ATPase at the tonoplast.

Iron

Iron acts as a cofactor in several proteins with functions that imply redox reactions such as respiration (cytochromes, globins), nitrogen assimilation (nitrogenase, nitrate reductase), nucleic acid synthesis and regulation or ROS detoxification (catalases, peroxidases). These iron functions can be generalized to all organisms, and in photosynthetic organisms, Fe is also involved in photosynthesis. This ubiquitous presence in life explains why this microelement is required in higher amounts than others by all organisms, and is a possible explanation of why autotrophic organisms have developed several strategies to increase the acquisition of iron, especially because the element is unavailable at alkaline pH (Hänsch and Mendel 2009).

In symbiosis, iron is required in the highest quantities because it is present in key enzymes mentioned above, highly abundant in the nodule and in the nitrogenase complex. Plant and bacterial peroxidases are induced during the development of the infection thread, because thread matrix polymerization is driven by the generation of ROS and peroxide cross-linking. Despite other important roles of iron, the effects of Fe deficiency seem to be not as clear in other symbiotic processes as in the N_2 fixation stage. Fe is a cofactor of both the nitrogenase reductase and the dinitrogenase components of the nitrogenase complex. Furthermore, the element is a cofactor of the leghemoglobin, which maintains the anaerobic environment inside the nodule at the time that maintains the flux of oxygen required by the bacteroids to oxidize the organic acids used as an energy source. Together nitrogenase and leghemoglobin are about 30 percent of the total protein content in the nodule (O'Hara 1988, 2001, Bonilla and Bolaños 2009).

There are particular chemical conditions associated with the physiological pH and the presence of several lipid layers in each cell inside the nodule. The symbiotic partners, as in its free-living form, have developed some strategies to solubilize and chelate iron ions from the soil to either the plant or the bacteria cell, and, in an analogous way, from the host plant cell to the bacteroid. These strategies include the acidification of the soil and the peribacteroid space; the redox function of the Fe (III)-chelate reductase; the chelation of iron with phytosiderophores, bacterial siderophores, and organic acids as citrate (which are also used to obtain energy), or the introduction of Fe (II) into the bacteroid with specific protein carriers (Slatni et al. 2012). For these reasons, regardless other factors which can alter its availability, an application of 0.30 kg/Ha during growth of nodulated soybean is recommended.

Boron

Boron was described as essential for plants, heterocyst-forming cyanobacteria, diatoms, and recently for some animal physiology. In plants, where the essentiality has been more deeply investigated, boron stabilizes the cell wall structure, maintains membrane function, and supports sugar, RNA or flavonoid metabolisms. However, the appearance of alterations in these metabolic processes at low B concentrations is often considered a secondary effect, and primary functions of boron other than maintaining cell wall architecture remain still elusive. The anion does not act as an osmolyte and to date no enzymes containing boron as a cofactor have been described. *In vivo*, only rhamnogalacturonan II (RGII) binding to borate as a crosslinking agent for the dimerization of that pectin has been convincingly demonstrated in plants. However, this does not seem sufficient to explain all the postulated roles of the micronutrient. Recently, some studies have described others *in vivo* or *in vitro* boron ligands, like AI-2 (a quorum sensing signal molecule in some bacteria), sorbitol or ribonucleotides. These findings could relate to boron primary functions with signaling mechanisms.

Our group has been working on the effects of deficiency of boron in several organisms, and also in the N_2-fixing symbioses, where the effects of the micronutrient deficiency are very severe. Moreover, the element has recently been described as necessary for the correct synthesis of the rhizobial polysaccharide capsule, although not for bacterial survival (Abreu et al. 2012). This capsule is important for the correct recognition of the bacteria by the legume, preventing the induction of the plant defense response which is elicited in interactions with some capsule mutants or under boron deficiency. Boron also plays a role in the direct plant-microbe interaction which precedes the infection step (Bolaños et al. 1996, Reguera et al. 2010). Inside the infection thread, boron is mediating the interaction between a matrix glycoprotein called AGPE (Arabino Galactan Protein like-Extensin) and the RGII. This AGPE-RGII binding both prevents the bacteria trapped by the infection thread matrix, and promotes infection thread apical growth (Reguera et al. 2010). Finally, the element plays a role in signaling during nodule organogenesis, as judged by the inhibition of differentiation of infected cells (Reguera et al. 2009), or a broad alteration in the bacteroid differentiation (Bolaños et al. 2001). Boron requirements vary among legumes but in soybean and pea an addition of 0.5 to 1 kg/Ha of boric acid (experimental data) is recommended. Because boron deficiency causes high damage in the legume crop yield, it is important to ensure correct nutrition, although sometimes it is difficult because of the narrow optimal range of the element. Nevertheless, boron deficiency can be often alleviated by a supplement of calcium due to the synergic effects of both elements.

Manganese

Manganese is a cofactor or activator of certain enzymes, although it can be replaced in most of the activated enzymes by magnesium. In plants, manganese is present as a cofactor in the manganese-containing water splitting system of photosystem II or in the superoxide dismutase (SOD) whose function is to scavenge superoxide radicals. This role can be relevant during symbiosis in order to ensure the survival of the interaction under stressing conditions. Manganese also activates some enzymes of the Krebs's cycle or enzymes involved in the shikimic acid (substrate of other enzymes involved in the synthesis of aromatic amino acids, lignins and flavonoids) pathway (Hänsch and Mendel 2009). Given that manganese concentration can affect both flavonoid synthesis and the level of carboxylic acids by altering the Krebs's cycle, this element could be important in both the first signaling events between the symbionts, and in the maintenance of the symbiosis. However, no evidence has been reported .

Zinc

Zinc is an element with a prominent role in genetic regulation because it is involved in DNA transcription, RNA processing and translation. The element forms the zinc-finger motif involved in protein-protein interactions. The element is also a cofactor of several metalloproteases, nucleases and aminopeptidases, and CuZn-SOD isoform (Hänsch and Mendel 2009).

Zinc is as well essential for symbiosis, but it is not associated with a specific role during the development or maintenance of the interaction. Its function in the genetic regulation can be an interesting target of future studies, because symbiosis involves several coordinated changes in gene expression in both symbionts.

Copper

Copper plays a major role associated with energetic processes, as it is a cofactor in cytochrome c oxidase (catalyzing the terminal oxidation in the mitochondria), or in plastocyanin and other electron carrier proteins in the chloroplast. The element is further involved in nitrogen and carbon metabolism, cell wall synthesis, and ROS detoxification, as will be mentioned below. Like iron, copper has two oxidation states, which determines its use as redox agent, but its potential toxicity make regulation of copper content extremely important. Because of this reason, copper free levels are reduced by two proteins, the metallothioneins and the copper chaperones, highly abundant in plants (Hänsch and Mendel 2009).

During symbiosis copper is essential as the cofactor of Cu and CuZn superoxide dismutases (Cu-SOD or the CuZn-SOD respectively), responsible of superoxide radical detoxification. Also, copper is important to inhibit membrane lipid peroxidation by controlling polyphenoloxidase (PPO) activity. The maintenance of the membrane integrity and permeability is critical to ensure ion transport (Bonilla and Bolaños 2009). Finally, copper has a crucial role in the assembly of molybdenum with the pterin molecule, where it forms an intermediate complex (Hänsch and Mendel 2009).

Molybdenum

Molybdenum and nickel are the elements less abundant in plant tissues, but often the amount stored in the seed is enough to ensure the entire life cycle of the organism. Molybdenum is present in few enzymes, but is very important in metabolism. These are nitrate reductase (catalyzing the first step of nitrate assimilation), aldehide oxidase (enzyme of the absicic acid –ABA–synthesis), and sulfite oxidase (involved in sulfite detoxification) (Hänsch and Mendel 2009). Molybdenum is essential during symbiosis, with a specific role in the nitrogen fixation process, because the element is the cofactor of the dinitrogenase component of nitrogenase. The last enzyme to which the micronutrient is a cofactor is the xanthine dehydrogenase/ oxidase, which is of special interest in plants with determinate nodule development, because it is involved in purine catabolism and ureid biosynthesis, the sources of mobile nitrogen in these plants (Bonilla and Bolaños 2009).

Molybdenum availability can be affected by low pH in the soil or in the nodule, dropping the nitrogenase synthesis and nitrogen fixation. As molybdenum has to be complexed with pterin to be a functional cofactor (called molybdenum cofactor or Moco), and for this process copper has to form an intermediate complex, a controlled addition of both elements is recommended in nodulated legumes. In soybean, the recommended fertilizer includes around 0.02 kg/Ha of molybdate.

Cobalt

This micronutrient is not necessary in plants, although it is considered a beneficial element. However, in rhizobia and other bacteria such as cyanobacteria, cobalt is essential for the homocysteine methyltransferase enzyme involved in the synthesis of methionine and for the ribonucleotide reductase which catalyzes the synthesis of deoxyribonucleotides. These enzymes are dependent of vitamin B_{12}, where Co is present. Another pathway in which vitamin B_{12} is essential is fatty acids metabolism and synthesis of heme groups of cytochromes and leghemoglobins, acting through the methylmalonil coenzyme A mutase. The mentioned functions, which determine the correct metabolism and differentiation of the rhizobia, explain the essentiality of the element during the symbiosis (Pilon-Smits et al. 2009).

Nickel

In bacteria, this element was described as essential in the 1960s. Since then, nickel has been described as a cofactor in several enzymes: ureases, hydrogenases, dehydrogenases, decarbonylases or methyl reductases (Mulrooney and Hausinger 2003). In plants, nickel was the last element included in the list of essential nutrients, because only 200 µg are enough to complete the life cycle. To date, nickel has been reported to be present in plant ureases and in a Ni-metallochaperone probably related with the binding of Ni to the urease (Hänsch and Mendel 2009).

Ureides are the nitrogen-transporting molecules used in plants such as soybean or bean, which develop determinate nodules in symbiosis. One product of the degradation pathway of ureides is urea, which is toxic to the cell at high concentrations. Ureases degrade urea to carbon dioxide and ammonia, which easily assimilates in organic molecules. As mentioned, nickel is also present in bacterial NiFe-hydrogenases, which degrade H_2 to obtain energy. Some rhizobia strains synthesize NiFe-hydrogenases that can use the hydrogen gas produced by nitrogenase activity, improving the nitrogen fixed and crop yield (Brito et al. 2000). To ensure nitrogen metabolism in determinate nodule-forming legumes and to ensure the correct synthesis of NiFe-nitrogenase, it is important to control nickel availability in soil, and the suitable additions of this nutrient.

Selenium

Selenium is chemically similar to sulfur, and in several organisms the element has a metabolic significance because it forms amino acids selenomethionine and selenocysteine. Selenocysteine presents a higher reactivity than its sulfur analogous, and it is usually present in the active site of some enzyme families such as glutathione peroxidases, deiodinases, thioredoxin reductases, etc. (Copeland 2005).

Several symbiotic plasmids of rhizobia carry the *sel* genes, involved in the synthesis of a selenocysteine tRNA or the selenocysteine synthase. In the same plasmids three members of formate dehydrogenases family are encoded, and at least one of them has a selenocysteine residue in its sequence. The formate respiration could be important in nodule metabolism as a carbon source, although as yet there is no direct data confirming this hypothesis.

Conclusions

Rhizosphere is a complex environment which includes plant roots, invertebrates, bacteria, fungi, viruses, and the soil composed of both organic and inorganic fraction. Improvement of crop yield is mainly based on improving growth and productivity in plants, making harvesting easier, making plants resistant to abiotic and biotic stresses, or ensuring beneficial plant-microbe interactions, through the addition of strains with improved colonization abilities as bacterial adhesion or biofilm formation. However, the effects of a proper and optimal mineral nutrition are underestimated, and usually field practices include an abusive use of fertilizers. Sustainable agriculture based in the symbiotic potential of beneficial interactions cannot bypass the correct supplement of these nutrients essential for a symbiotic process, as summarized here for rhizobia-legume symbiosis. Future works should help to complete the understanding of mineral requirements and roles of nutrients in other plant-microbe interactions, and must be complemented with a translational effort to transfer this knowledge to farmers.

Acknowledgments

This work was supported by MICROAMBIENTECM Program from Comunidad de Madrid. Isidro Abreu is the recipient of a FPU fellowship from the Spanish Ministry of Education.

References

Abreu, I., M.E. Cerda, M. Pérez de Nanclares, I. Baena, I. Lloret, I. Bonilla, L. Bolaños and M. Reguera. 2012. Boron deficiency affects rhizobia cell surface polysaccharides important for suppression of plant defense mechanisms during legume recognition and for development of nitrogen-fixing symbiosis. Plant Soil. 361: 385–395.

Arnon, D.I. and P.R. Stout. 1939. The essentiality of certain elements in minute quantity for plants with special reference to copper. Plant Physiol. 14: 371–375.

Bolaños, L., N.J. Brewin and I. Bonilla. 1996. Effects of boron on *Rhizobium*-legume cell-surface interactions and nodule development. Plant Physiol. 110: 1249–1256.

Bolaños, L., A. Cebrián, M. Redondo-Nieto, R. Rivilla and I. Bonilla. 2001. Lectin-like glycoprotein PsNLEC-1 is not correctly glycosylated and targeted in boron deficient pea nodules. Mol Plant-Microbe Interact. 14: 663–670.

Bolaños, L., K. Lukaszewsky, I. Bonilla and D. Blevins. 2004. Why boron? Plant Physiol. Biochem. 42: 907–912.

Bonilla, I. and L. Bolaños. 2009. Mineral nutrition for legume-rhizobia symbiosis: B, Ca, N, P, S, K, Fe, Mo, Co and Ni. A review. *In:* E. Lichtfouse [ed.]. Organic Farming, Pest Control and Remediation of Soil Pollutants. Sustainable Agriculture Reviews 1. Springer, New York. pp. 253–274.

Bonilla, I., J. Abadía and L. Bolaños. 2010. Introduction to mineral nutrition of plants. *In:* A. González-Fontes, A. Gárate, I. Bonilla [eds.]. Agricultural Sciences: Topics in Modern Agriculture. Studium Press LCC, Houston. pp. 145–171.

Brewin, N.J. 1991. Development of the legume root nodule. Ann. Rev. Cell Dev. Biology. 7: 191–226.

Brewin, N.J. 2004. Plant cell wall remodeling in the *Rhizobium*-legume symbiosis. Crit. Rev. Plant Sci. 23: 293–316.

Brito, B., A.J. Monza, J. Imperial, T. Ruiz-Argüeso and J.M. Palacios. 2000. Nickel availability and *hupSL* activation by heterologous regulators limit symbiotic expression of the *Rhizobium leguminosarum* bv. *viciae* hydrogenase system in Hup^- rhizobia. Appl. Environ. Microbiol. 66: 937–942.

Clarck, D.G. 1936. Physiological studies on *Rhizobium* species. New York Agr. Exp. Sta. Ithaca Mem. 196.

Copeland, P.R. 2005. Making sense of nonsense: the evolution of selenocysteine usage in proteins. Genome Biol. 6: 221–225.

Gage, D.J. 2004. Infection and invasion of roots by symbiotic, nitrogen-fixing rhizobia during nodulation of temperate legumes. Microbiol. Mol. Biol. Rev. 68: 280–300.

Hänsch, R. and R.R. Mendel. 2009. Physiological functions of mineral micronutrients (Cu, Zn, Mn, Fe, Ni, Mo, B, Cl). Curr. Opin. Plant Biol. 12: 259–266.

Jones, K.M., H. Kobayashi, B.W. Davies, M.E. Taga and G.C. Walker. 2007. How rhizobial symbionts invade plants: the *Shinorhizobium-Medicago* model. Nature Rev. Microbiol. 5: 619–633.

Maathuis, F.J.M. 2009. Physiological functions of mineral macronutrients. Curr. Opin. Plant Biol. 12: 250–258.

Masson-Boivin, C., E. Giraud, X. Perret and J. Batut. 2009. Establishing nitrogen-fixing symbiosis with legumes: how many Rhizobium recipes? Trends Microbiol. 17: 458–466.

Mulrooney, S.B. and R.P. Hausinger. 2003. Nickel uptake and utilization by microorganisms. FEMS Microbiol. Rev. 27: 239–261.

O'Hara, G.W. 2001. Nutritional constraints on root nodule bacteria affecting symbiotic nitrogen fixation: a review. Aust. J. Exper. Agriculture. 41: 417–433.

O'Hara, G.W., N. Boonkerd and M.J. Dilworth. 1988. Mineral constrains to nitrogen fixation. Plant Soil. 108: 93–110.

Pilon-Smits, E.A.H., C.F. Quinn, W. Tapken, M. Malagoli and M. Schiavon. 2009. Physiological functions of beneficial elements. Curr. Opin. Plant Biol. 12: 267–274.

Popp, C. and T. Ott. 2011. Regulation of signal transduction and bacterial infection during root nodule symbiosis. Curr. Opin. Plant Biol. 14: 458–467.

Reguera, M., I. Abreu, N.J. Brewin, I. Bonilla and L. Bolaños. 2010. Borate promotes the formation of a complex between legume AGP-extensin and Rhamnogalacturonan II and enhances production of *Rhizobium* capsular polysaccharide during infection thread development in *Pisum sativum* symbiotic root nodules. Plant, Cell & Environ. 33: 2112–2120.

Reguera, M., A. Espí, L. Bolaños, I. Bonilla, and M. Redondo-Nieto. 2009. Endoreduplication before cell differentiation fails in boron deficient legume nodules. Is boron involved in signaling during cell cycle regulation? New Phytol. 183: 8–12.

Slatni, T., M. Dell'Orto, I.B. Salah, G. Vigani, A. Smaoui, H. Gouia, G. Zocchi and C. Abdelly. 2012. Immunolocalization of H^+-ATPase and IRT1 enzymes in N_2-fixing common bean nodules subjected to iron deficiency. J. Plant Physiol. 160: 242–248.

Steinberg, R.A. 1938. Applicability of nutrient-solution purification to the study of trace-element requirements of *Rhizobium* and *Azotobacter*. J. Agr. Res. 57: 461–476.

7

Metal Transport in the *Rhizobium*-Legume Symbiosis

M. González-Guerrero, L. Rubio-Sanz, B. Rodríguez-Haas, M. Albareda, M. Menéndez-Cerón, B. Brito and *J.M. Palacios**

Introduction

Iron, zinc, copper, molybdenum and some other transition metals are essential nutrients (Fraustro da Silva and Williams 2001). They are constitutive elements of around half of the proteins of a typical cell (Andreini et al. 2008), where they may act as structural elements, such as zinc in the zinc finger domain, but mostly as key elements in the active site of enzymes involved in almost every physiological process, from oxidative respiration to photosynthesis. These elements are growth-limiting nutrients for autotrophic organisms, since they form very stable complexes, resulting in low solubility and making their uptake difficult (Ruel and Bouis 1998, Fung et al. 2000, Grotz and Guerinot 2006). Moreover, living beings cannot accumulate high amounts of essential transition metals, since these elements can catalyze the production of free radicals in Fenton-style reactions or compete with other metals for the active site of metalloenzymes (Goldstein et al. 1993, Ranquet et al. 2007, Macomber and Imlay 2009). This is also the basis of the toxic effect of non-biogenic metals such as cadmium, lead or mercury.

Centro de Biotecnología y Genómica de Plantas. Universidad Politécnica de Madrid. Carretera M-40 k. 38. 28223 Pozuelo de Alarcón (Madrid)-Spain.
*Corresponding author: jose.palacios@upm.es

Consequently, a concerted set of systems must be in place to ensure high affinity metal uptake while simultaneously avoiding the noxious effects derived from metal accumulation. These systems include a number of metal transporter families, several types of organic molecules that can bind metals with high affinity, soluble proteins that shuttle metal in the cytosol (metallochaperones) or that simply bind excess metals (metallothioneins), and transcription factors that regulate the process (Wandersman and Delepelaire 2004, Waldron and Robinson 2009, Blindauer and Leszczyszyn 2010, Reyes-Caballero et al. 2011, Argüello et al. 2012). Overall, the mechanisms governing metal homeostasis are so efficient that the "free", hydrated, cytosolic metal concentration of a typical bacteria is in the pM-fM range (Outten and O'Halloran 2001), which is less than one free ion per cell. This has two important consequences for how metals are handled by the cell: one of them is the high metal affinity of all the elements involved in metal homeostasis; the other is that metal transport involves either the cotransport with a metal ligand (nicotianamine, glutathione, and other molecules) or the physical interaction between a metal delivery metallochaperone and the transporter itself.

In this chapter we focus on how metals are important for nitrogen fixation, and how metal uptake is carried out in the *Rhizobium*-legume system.

Metals and Symbiotic Nitrogen Fixation

Metal levels in the host legume are critical for the establishment and functionality of the symbiosis with rhizobia. For instance, the effect of low iron levels ranges from the inhibition of the nodulation to the loss of the capability to fix nitrogen (Tang et al. 1990, Tang et al. 1992, O'Hara 2001). This is due to the high number and relevance of metalloproteins involved in symbiotic nitrogen fixation.

Nitrogenase is one of the most abundant proteins in the symbiosome (around 10 percent) and arguably the most important enzyme of this symbiosis, since it catalyzes the conversion of N_2 into NH_4^+ (Miller et al. 1993). This enzyme is a multimeric protein that contains metallic cofactors, 2 FeMoCo, 2 P-clusters (Fe_8S_7), and one Fe_4S_4 cluster, totalling 34 iron and two molybdenum atoms. These metal cofactors direct the reducing electrons from a ferredoxin donor (also an iron-sulfur protein) through the nitrogenase Fe_4S_4 clusters to the P-clusters reaching the FeMoCo that would finally use them to break the triple bond in N_2 (Miller et al. 1993). This is an energetically inefficient process, in which reducing power is wasted in the production of H_2. To recover some of it, some rhizobia express the enzyme hydrogenase, a metalloprotein that carries a Ni-Fe cluster (O'Brian and Maier 1989, Palacios et al. 2005).

Leghemoglobin is also expressed at high levels in the nodule (around 20 percent of the total nodular protein) (Appleby 1984). This is a heme-carrying metalloprotein, whose iron confers the characteristic reddish color to functional nodules. Leghemoglobin is responsible of sequestering O_2, creating a microaerobic environment in which nitrogenase can function. In spite of the oxygen sensitivity of nitrogenase, the bacteroids use O_2-dependent respiration to obtain most of their energy. In order to be able to function in this microaerobic environment, the bacteroid expresses a high affinity cytochrome oxidase (*cbb*$_3$-type), whose synthesis requires a steady supply of copper (Preisig et al. 1996a, b).

Metalloproteins, such as catalase or superoxide dismutase (SOD), also play an important role in controlling reactive oxygen species (ROS) (Fridovich 1998). In the nodule, ROS are created by respiration in mitochondria and bacteroids, the direct reduction of O_2 by nitrogenase, hydrogenase and ferredoxin, and the autooxidation of leghemoglobin (Puppo et al. 1981, Dalton et al. 1991). Manganese and ferric-SOD are expressed in the nitrogen fixing areas of the nodule, where they seem to be controlling ROS (Rubio et al. 2004, 2007). The relative levels of Mn-SOD or of Fe-SOD appear to be controlled by iron availability in the nodule, in such a way that as the nodule senesces and heminic iron from the leghemoglobin is released, the relative levels of Fe-SOD increase, as a means to both protect against ROS, and to sequester free iron (Rubio et al. 2007). SODs transform superoxide anion to H_2O_2, which is still toxic. This peroxide is further detoxified by iron-containing catalases (Nicholls et al. 2001), which are critical for nodulation (Jamet et al. 2003, Hanyu et al. 2008). However, low levels of H_2O_2 are necessary for infection thread development and for nod signal transduction (Jamet et al. 2007, Cárdenas et al. 2008). Consequently, catalase overexpression negatively affects the symbiosis (Jamet et al. 2007). It has been hypothesized that a Cu, Zn-SOD is responsible for producing this H_2O_2 (Rubio et al. 2004, 2007).

Therefore, metals are essential nutrients for symbiotic nitrogen fixation, playing relevant roles from the transmission of the nod factor signal to the nitrogen fixation itself. Given that they are endosymbionts, and that many of these proteins are newly synthesized in the nodule, this means that for the metal to reach the symbiosome it has to cross several barriers. First, it has to be incorporated from the soil into the epidermal cells, where they symplastically reach the endodermis. Metals are transported to the vasculature and translocated to other parts of the plant, including the nodule. Then, it has to cross back the endodermis into the nodule cortex where the plant cells will uptake them to synthesize plant-derived nodular metalloproteins. Another part, the one required for bacteroid metalloproteins synthesis, has to cross not only the plasma membrane, but also the peribacteroid membrane and the rhizobial outer and inner membranes. Given that biological membranes are impermeable to ions, metal transporters play an important role in delivering and controlling which and how metals reach the nodule.

Metal Transporters

Metal transport is as ancient as the first cell, and consequently many metal transporter families (CDF, ZIP or P_{IB}-ATPases) are conserved in all three domains of life. Attending to their substrate, we can classify them in two groups: those transporting metal ions and those transporting metal complexes. Similarly, the direction of transport seems to be conserved, and those families involved in loading the cytosol with metals (either from the cell surface or from an organelle lumen) do not normally export metal out of the cell and vice versa. These are the most common families of metal transporters:

Metal importers

- ZIP (Zrt1-Irt1 like Protein). They typically transport divalent metals (Fe^{2+}, Zn^{2+} and Mn^{2+}) (Eide 2004). They are mainly responsible for iron and zinc uptake in dicotyledonous plants (Vert et al. 2002).
- Nramp (Natural Resistance-Associated Macrophage Protein). Although first discovered in macrophages, these transporters are present in all domains of life (Forbes and Gros 2001, Nevo and Nelson 2006). Their role is complementary to ZIP transporters in many instances, and share similar substrates (Fe^{2+}, Mn^{2+}, Zn^{2+}) (Curie et al. 2000, Cailliatte et al. 2010).
- YSL (Yellow stripe-like). These transporters are present only in plants. Their substrate is a nicotianamine or related molecule complexed with metals (DiDonato et al. 2004). They are responsible for metal uptake from soil in monocotyledonous plants, as well as for long distance metal trafficking in both monocotyledonous and dicotyledonous plants (Curie et al. 2001, Jean et al. 2005).
- Ctr (Cu transporter). This is a homotrimeric protein present only in eukaryotes (Aller and Unger 2006, Dumay et al. 2006, De Feo et al. 2009). They transport Cu^{+} towards the cytosol, being the main copper uptake system in plants (Burkhead et al. 2009).
- MOT (Molybdenum Transporter). They are an extremely high affinity (nM k_M) system of molybdate import in eukaryotes (Tejada-Jiménez et al. 2007, Tomatsu et al. 2007, Tejada-Jiménez et al. 2011). In some cases, their function may be carried out by sulfate transporters (Fitzpatrick et al. 2008).

- ABC (ATP-Binding Cassette-type). These proteins transport a wide range of substrates, including mono- and oligosaccharides, organic and inorganic ions, amino acids, peptides, iron siderophores, metals, polyamine cations, opines and vitamins (Davidson et al. 2008). The minimal structure of the ABC systems comprises two ATP binding domains for energy transduction, and two membrane-embedded domains forming the channel across the cytoplasmic membrane, which can be encoded in just one gene or by four of them (Rea 2007, Davidson et al. 2008). Most bacterial ABC uptake systems also contain a periplasmic solute-binding protein (SBP) that provides high-affinity binding to the corresponding substrate (Imperial et al. 1998). Extensive reviews on the mechanism and structure of ATP-driven transition metal transporters have been published recently (Cui and Davidson 2011, Klein and Lewinson 2011). While in bacteria they are mostly associated with metal import, in eukaryotes they are involved in metal detoxification (Rea 2007).
- TBDT (TonB-dependent transporter). They are the outer membrane proteins present only in bacteria. Most TBDT-like proteins are known to transport iron compounds (siderophores or heme) (Wandersman and Stojiljkovic 2000).
- HupE/UreJ. These bacterial transporters are involved in Ni^{2+} uptake (Brito et al. 2010).
- NiCoT (Ni^{2+}-Co^{2+} transporter). These transporters are involved in nanomolar Ni^{2+} and Co^{2+} uptake in bacteria, archaea and eukaryotes (Eitinger and Friedrich 1991, Eitinger et al. 2000).

Metal exporters

- P_{IB}-ATPases. These are members of the bigger P-type ATPase family that includes the Ca^{2+}, the Na^+/K^+ and the H^+ pumps (Palmgren and Nissen 2011). They are an ancient family of transporters present in all domains of life (Argüello 2003). They export Cu^+, Cu^{2+}, Zn^{2+}, Mn^{2+}, or Co^{2+}, either with detoxification purposes or to synthesize extracytoplasmic metalloproteins (González-Guerrero et al. 2010, Argüello et al. 2011, Raimunda et al. 2011).
- CDF (Cation Diffusion Facilitator). The homodimer of these transporters extrudes divalent cations (Zn^{2+}, Fe^{2+}, or Mn^{2+}) either to the cell exterior or to organelles to synthesize metalloproteins (Blaudez et al. 2003, Anton et al. 2004, Wei et al. 2004, Lu and Fu 2007).
- Ferroportin. These transporters have been found in animals and plants (McKie et al. 2000, Morrissey et al. 2009). They are involved in Fe^{2+} and Co^{2+} delivery to the vasculature and, consequently, in long distance metal transport.

- CCC1/VIT1. Members of this family have been found only in eukaryotes. They are involved in iron and manganese transport to the vacuole (Li et al. 2001, Kim et al. 2006).
- RND (Resistance-Nodulation-Division). These transporters are present only in Gram negative bacteria, since they form a protein complex that spans both cytoplasmic and outer membranes. The complex is formed by a trimeric inner membrane element and a trimeric outer membrane one that meet in the periplasm. This interaction is stabilized by a hexameric complex of a third protein (Long et al. 2010, Kim et al. 2011, Kulathila et al. 2011). They seem to be involved with metal detoxification from the periplasm (Anton et al. 1999, Nies 2003).

While not directly transporting metals, other transporters are also essential for metal homeostasis. Among them, probably the best characterized is the FRD3 citrate transporter in *Arabidopsis thaliana* (Rogers and Guerinot 2002, Durrett et al. 2007). FRD3 and its homologues are critical for iron transport across symplastically disconnected tissues (pericycle-xylem, embryo-seed envelope or pollen-anther tissue) (Roschzttardtz et al. 2011).

Plant Metal Uptake Transport and Symbiotic Nitrogen Fixation

As previously stated, bioavailable metal levels in soils are very low, especially in basic soils (Grotz et al. 1998). As a consequence, plants are often in a state close to metal deficiency, affecting crop production worldwide (Grotz et al. 1998, Ruel and Bouis 1998). This is particularly evident in the case of legumes since symbiotic nitrogen fixation requires relatively high amounts of metals. Consequently, as the nodulation process is being initiated, it also triggers the "metal deficiency response" (Terry et al. 1991). In dicotyledonous plants, this response involves an acidification of the surrounding soil to increase metal solubility, and the induction of metal reductases to reduce metals which are subsequently incorporated by epidermal metal transporters and also up-regulated (Puig et al. 2007, Andaluz et al. 2009, Bernal et al. 2012). Concomitant to this, and in order to ensure that the nodule needs are satisfied, the legume should down-regulate metal transporters directing metals to non-essential roles, as it seems to be the case of MtMTP1 (Chen et al. 2009).

To date, only three plant transporters have been associated with metal transfer to the nodule. The best characterized of them is *Lotus japonicus* SEN1, identified in a screen for *fix*⁻ mutants (Hakoyama et al. 2011). SEN1 is a member of the CCC1/VIT1 family. It probably transports iron, since the nodules of *sen1* plants have lower iron concentrations. Histochemical and transcriptional data indicate that SEN1 is nodule-specific. All these observations suggest that SEN1 is involved in iron delivery to the symbiosome. However, the authors of this work were not able to show the precise location of SEN1 in the cell, nor verify the metal transported. This was achieved in the other two identified nodule-specific metal transporters: the Nramp transporter GmDMT1 and the ZIP GmZIP1, both from *Glycine max* (Moreau et al. 2002, Kaiser et al. 2003). These transporters are located in the peribacteroid membrane. Yeast complementation assays indicate that they transport Fe^{2+} and Zn^{2+}, respectively. Their precise direction of transport is not clear. The complementation data and biochemical analysis of homologous transporters indicate that they introduce metal towards the cytosol (Eide 2004, Nevo and Nelson 2006). However, in contradiction with this, an antibody raised against GmZIP1 inhibited Zn uptake in isolated symbiosomes (Moreau et al. 2002). Unfortunately, at that time it was not possible to study mutant lines of either gene, and consequently there is no data on the actual importance and role of each transporter in the symbiosis.

The nodule senesces a few weeks post infection. This is a genetically programmed process, often times coupled to flowering, in which the plant stops sending nutrients to the nodule and directs them instead to the seed (Fedorova et al. 2002, Puppo et al. 2005). It is estimated that around 50 percent of the nodular metal is transferred back to the seed (Burton et al. 1998), where it will be used for embryogenesis and germination (Sancenon et al. 2004, Kim et al. 2006, Roschzttardtz et al. 2011). Consequently, from a metal point of view, nodule senescence involves not only stopping metal delivery to the nodule but recycling the metal "stored" in the nodule. In this process, YSL transporters would play a role. As stated earlier, YSLs are involved in long-distance metal trafficking in which the metal substrate forms complexes with nicotianamine (DiDonato et al. 2004). Although no YSL transporter has as yet been identified in nodules, the presence of the nodule-specific senescence-induced nicotianamine synthase LjNAS2 indicates that this mechanism should be in place (Hakoyama et al. 2009). In this process of recycling metals, ferritins would also participate. Ferritins are plastidial iron storage proteins that form a shell-like structure that contains the iron in a quasi-crystalline form (Briat et al. 2010). In functional nodules, ferritins are highly expressed in the infected cells, but they are down-regulated as the nodule senesces or in the senescence areas of indeterminate nodules (Lucas et al. 1998). This is consistent with a role of ferritin in the protection against ROS (Briat et al. 2010). However, in *Lupinus* nodules, as the nodule senesces, ferritins are up-regulated in the nodular cortex, which might be responsible for protecting the rest of the plant against free radicals produced in the degradation of the nodule, as well as storing the iron before its translocation to the vasculature via YSL transporters.

Metal Transport Mechanisms in the Bacterial Side

As discussed above, the endosymbiotic state of rhizobia requires the uptake of high amounts of metals for the synthesis of metalloproteins. This has an additional layer of complexity, since these bacteria live surrounded by a plant-derived peribacteroid membrane immersed in the plant cytoplasm. Given the complexity of the symbiotic situation, most metal transport studies have been performed with free-living cultures, with further analysis of the effect that mutations in the different systems may have in symbiotic performance. We present here the main recent advances reported on rhizobial mechanisms for uptake of metals through outer and inner membranes. Metal efflux mechanisms, mainly involved in resistance against high levels of metals, have been covered in other reviews (Nies 2003, Macomber and Hausinger 2011, Raimunda et al. 2011).

Metallo-organic Complexes and Metal Uptake in Rhizobia

The extremely low solubility of iron under aerobic conditions is a problem that many organisms, including bacteria and plants, solve through the production of siderophores, organic compounds of low molecular weight (200–2000 Da) with high affinity for Fe^{3+} that allow the bacteria to scavenge the environment for this metal. Siderophores have a variety of chemical structures and form a family of at least 500 different compounds (Budzikiewicz 2010, Hider and Kong 2010). So far, the structures of three types of rhizobial siderophores have been characterized: cyclic trihydroxamates (vicibactin), α-hydroxycarboxylates (rhizobactin) and citrate-derivative hydroxamates (rhizobactin 1021 and schizokinen). Genes involved in the synthesis of these siderophores have been identified for different rhizobia (see O'Brian and Fabiano 2010 for a review).

Metallophores specific for other cations could be used for metal uptake. The analysis of a novel nickel uptake system through the outer membrane of *Helicobacter pylori* led to the hypothesis of the existence of a nickel-complexing compound (nickelophore) required to uptake this element when present at very low concentrations (Schauer et al. 2007). The genome of *Bradyrhizobium japonicum* encodes a similar outer membrane receptor for nickel (see below) suggesting that this metal could also be taken as a metallophore-complex in this bacterium. On the other hand, the existence of a chalkophore (the copper metallophore methanobactin) has been described in the alphaproteobacterium *Methylosinus trichosporium* and other methanotrophic bacteria (Balasubramanian et al. 2011), although the presence of such systems has not been studied in rhizobia.

In addition to bacterial metallophores, rhizobia can also use other complexes such as heme and citrate for uptake of iron or other metals.

Heme is a natural iron source for pathogenic bacteria, and it can be also used by rhizobia. Heme acquisition mediated by the *hmu* system has been described in different rhizobia, and the use of such a system for iron uptake from leghemoglobin in senescing nodules has been proposed (Balasubramanian et al. 2011). Citrate production and uptake of iron-citrate complexes have been described in *B. japonicum* free-living cells (Guerinot et al. 1990). Furthermore, Moreau et al. (1995) showed that isolated soybean bacteroids were able to transport iron citrate, and proposed this chemical species as the main supply for iron in symbiotic conditions. Besides this report , information on the "chemical landscape" of metal species available to the bacteroids within the nodules is scarce. For instance, while nickel is present as malate and citrate complexes in pea nodules (Cacho et al. 2010), other legumes show a different speciation for this metal (our unpublished results). This might affect the availability of nickel for the synthesis of Ni-enzymes such as hydrogenase, and might account for the marked host effect observed on expression of *R. leguminosarum* hydrogenase metalloenzyme (Brito et al. 2008).

Transporters for Metal Uptake through the Outer Membrane

Once the metallo-organic complexes are formed outside the cell, they are internalized through specific receptors located in the outer membrane (OM). Although transport of divalent cations through OM can proceed through general porins (Zeth and Thein 2010), the low availability of metal ions in the free form implies the requirement of receptors to import metallo-organic complexes. The OM does not maintain a proton gradient nor ATP synthesis, and the energy for transport through these receptors is collected from the cytoplasmic membrane by ExbB/ExbD membrane protein complexes and transduced to the OM by the periplasm-spanning inner-membrane protein TonB (Postle and Larsen 2007, Noinaj et al. 2010). For this reason these receptors have been designated as TonB-dependent transporters (TBDTs). The mechanism of uptake for ferric-siderophore complexes by a TBDT has been described in detail in the *E. coli* model system (Chakraborty et al. 2007). Most TBDT-like proteins are known to transport iron compounds (siderophores or heme) or cobalamine (Noinaj et al. 2010); recent reports, however, indicate that these kind of receptors are also involved in the uptake of nickel (Schauer et al. 2007) and zinc (Stork et al. 2010).

In the case of rhizobia, Fe^{3+}-siderophore complexes are recognized by different OM receptors depending on the siderophore. *R. leguminosarum* contains a receptor for vicibactin (FhuA) whose expression is induced in free-living cultures under low Fe conditions (Yeoman et al. 2000). Interestingly, *fhuA* expression is also induced in the meristematic zone of pea nodules, but not in mature bacteroids. Mutations in *fhuA*, however, do not show significant effects on symbiotic performance (Yeoman et al. 2000). The TBDT for Fe-rhizobactin complex is encoded by *Sinorhizobium meliloti rhtA*. This gene is regulated by iron, and encodes a protein highly similar to enterobactin receptor LutA. *S. meliloti rhtA*⁻ mutants have no significant defects in symbiotic performance (Lynch et al. 2001).

Genomic searches have unveiled the existence of other potential TonB-dependent receptors for which there is no experimental evidence. Analysis of the genome of 13 selected *Rhizobiaceae* strains revealed that the number of TBDT-like genes ranged from one in *Mesorhizobium loti* MAFF303099 to 14 in *Azorhizobium caulinodans* ORS571 (Lim 2010). The majority of the 54 rhizobial TBDTs identified in that study are predicted to be involved in the uptake of iron-siderophore (26) or heme (13). Also, *Mesorhizobium loti* TBDT-like encoded by *meso2063* is known to transport cobalamine, and *B. japonicum bll6948* might transport nickel. The latter assumption is based on the similarity of the *bll6948* gene product to a TBDT involved in nickel uptake in *Helicobacter pylori* (Schauer et al. 2007), and on the genomic context of *bll6948*, located between the gene encoding a nickel transporter of the NiCoT family (*hupN*) and a cluster of genes for the nickel-containing enzyme hydrogenase (Schauer et al. 2008). Although there are no direct studies on the relevance of this potential Ni-specific TBDT, data from transcriptomic analysis indicate that *bll6948* is induced, although at low level, in *B. japonicum* bacteroids (Chang et al. 2007).

Some rhizobia do not produce siderophores, but do synthesize OM receptors able to bind siderophore-metal complexes, allowing the uptake of iron chelates (Small et al. 2009). Such siderophores used by one organism but synthesized and secreted by other organisms are referred to as xenosiderophores. Examples of this type are ferrichrome and rhodotorulic acid, siderophores of fungal origin used by *B. japonicum* as a source of iron (Plessner et al. 1993). The use of xenosiderophores in rhizobia is extensively discussed in another review (O'Brian and Fabiano 2010). A recent report demonstrated that the expression of siderophore receptors from *Pseudomonas* in rhizobia isolated from pigeon pea resulted in better rhizosphere colonization and improved legume plant growth (Arif et al. 2012). This data suggests that iron availability is a major factor limiting the rhizobium-legume symbiosis, and that expression of heterologous siderophore receptors might constitute a potential tool to improve competitiveness and nitrogen fixation.

Although TBDTs appear to be the major OM metal receptors, other types of OM transporters might also participate in metal uptake. Recently, an outer membrane protein with a β-barrel structure (MnoP) has been described as essential for manganese uptake in *B. japonicum* (Hohle et al. 2011). This protein is similar to specific porins OmpA and OprB, and unrelated to the TBDTs described above, and might represent a distinct model for metal uptake through the outer membrane. The fact that the expression of this protein is co-regulated with that of the inner membrane Mn^{2+} transporter MntH strongly suggests the existence of a two step transport system for crossing both membranes in this bacterium (Hohle et al. 2011).

Mechanisms for Metal Uptake through Inner Membrane

Once in the periplasm, metal ions must cross the cytoplasmic membrane, the major permeability barrier for metal uptake. As described above, bacteria have developed a number of mechanisms to overcome this barrier.

ABC Transporters

ABC transporters are the prevalent transport systems in rhizobia (Young et al. 2006). *S. meliloti* contains 146 such systems, and similar numbers are present in other *Rhizobiaceae* (Galibert et al. 2001). The role of ABC transporters in the uptake of iron, manganese, cobalt, molybdenum and zinc has been demonstrated in free-living cultures of different rhizobia. Examples of iron-siderophore uptake systems include the *fhuBCD* system described in *Rhizobium leguminosarum* for uptake of vicibactin-Fe, and the *hmuTUV* system described in several rhizobia for heme-Fe complexes (O'Brian and Fabiano 2010).

A manganese-specific ABC importer (*sitABCD*) has been described in *S. meliloti*. Analysis of *S. meliloti* mutants affected in this operon indicates that this system is required for growth under manganese-limiting conditions. The same mutants display a normal symbiotic phenotype (Platero et al. 2003). A similar situation was found for the cobalt transport system (*cbtJKL*) recently described also in *S. meliloti* (Cheng et al. 2011). Expression of this system is controlled by a cobalamin-binding riboswitch located upstream of the operon. Mutants affected in this system are unable to grow on LB medium unless it is supplemented with cobalt. Interestingly, these mutants show a residual level of cobalt uptake, potentially due to an alternate, unspecific transport system. Such a system might be responsible for the normal symbiotic phenotype showed by the *cbt* mutants in symbiosis with alfalfa (Cheng et al. 2011).

The ModABC system mediates high affinity molybdate uptake in *B. japonicum* and probably also in other rhizobia, since it is encoded in most rhizobial genomes (http://genome.kazusa.or.jp/rhizobase/). *B. japonicum* mutants affected in the *mod* system lack nitrate reductase activity in free living cultures, and show reduced levels of nitrogen fixation in symbiosis with soybean. These deficiencies were corrected by addition of molybdenum to the plants (Delgado et al. 2006). Such complementation has a strong dependence on the presence of sulfate, suggesting that, in addition to being mainly incorporated through the Mod system, molybdate is also taken up through sulfate transporters under symbiotic conditions, as it has been shown for a plant sulfate transporter (Fitzpatrick et al. 2008).

ABC transporters for zinc uptake homologous to the ZnuABC system described in *E. coli* and other bacteria (Hantke 2005) are present in the genome of most rhizobial strains. However, no specific work on zinc transport by endosymbiotic bacteria is available. The only mention of relevance of these transport systems in rhizobia is the induction of expression of *S. meliloti* Znu system in response to zinc limitation (Mauchline et al. 2006). Interestingly, Znu-deficient mutants of the phylogenetically relative animal pathogen *Brucella abortus* show a reduced intracellular survival and virulence (Kim et al. 2004).

There is not much information about the uptake of copper in rhizobia. Copper is required in bacteria mainly for proteins located either in the cytoplasmic membrane or in the periplasmic space. Based on that, some authors maintain that bacteria may not have a general requirement for cytoplasmic copper (Solioz et al. 2010). However, genes for two ATP-dependent systems (*nosDFYL* and *fixI*) have been linked to copper metabolism in rhizobia. The *nosDFYL* operon encodes an ABC uptake system. Since these genes are located in the nitrous oxide reductase gene clusters from *S. meliloti* and *B. japonicum* (Holloway et al. 1996, Velasco et al. 2004), their participation in the uptake of copper for the synthesis of this copper-metalloenzyme has been proposed (Holloway et al. 1996). Mutations in these genes do not affect symbiotic performance, suggesting that such a system is not critical for copper uptake under symbiotic conditions. On the other hand, *fixI* encodes a P-type Cu-ATPase known to participate in the synthesis of cbb_3-type heme-copper cytochrome *c* oxidase (cbb_3-Cox), an essential component of the symbiotic respiratory branch (Preisig et al. 1996a). These authors suggested the possibility that this ATPase imports Cu into the cytoplasm; however, this possibility has been contested by biochemical data that indicates that FixI is actually an efflux system (González-Guerrero et al. 2010). Then, FixI should participate in the transport of Cu from cytoplasm to periplasm for cbb_3-Cox biosynthesis. Copper uptake for cbb_3-Cox synthesis in rhizobia might instead be carried out by CcoA, an MFS transporter recently described in *Rhodobacter capsulatus* (see below).

Permease Metal Transporters

These are secondary transporters, dependent on proton motive force, composed by single proteins with multiple transmembrane domains in a monomeric or dimeric form. In the case of endosymbiotic bacteria, this type of transport system mediates uptake of nickel, cobalt and manganese ions.

Three main classes of secondary transport mechanisms for metal uptake have been described in rhizobia:

- *NiCoTs.* NiCoTs expression in *E. coli* identified ion preferences ranging from strict selectivity for nickel to a strong preference for cobalt through unbiased transport of both ions (Eitinger et al. 2005). *B. japonicum* HupN was the first member identified in rhizobia (Fu et al. 1994). The corresponding gene is located adjacent to a DNA region encoding the nickel-containing enzyme hydrogenase. Analysis of mutants affected in this gene revealed a significant Ni-dependent decrease in hydrogenase activity in free-living cells (Fu et al. 1994).
- *HupE/UreJ.* Their encoding genes are associated to gene clusters encoding NiFe hidrogenases in different rhizobia like *R. leguminosarum* and *A. caulinodans*. *R. leguminosarum hupE* encodes an integral membrane protein with six known transmembrane domains. Mutant analysis demonstrated the essentiality of this protein for the synthesis of *R. leguminosarum* NiFe hydrogenase both in free-living culture and in symbiosis (Brito et al. 2010). Genes encoding members of this group of transporters in other bacteria are sometimes preceded by sequences for cobalamin riboswitch RNA regulatory elements, suggesting that they might transport this cobalt compound rather than nickel (Schauer et al. 2008). So far, no evidence for a cobalt transporter of this type has been reported in rhizobia.
- *Nramp.* A relevant member of this group is MntH, a proton symporter acting as the main manganese uptake system in free-living cultures of *B. japonicum* (Hohle and O'Brian 2009). This transporter is not essential for the development of nodules nor for nitrogen fixation activity in symbiosis with soybean, suggesting that either the host provides enough manganese to make high-affinity transport unnecessary, or the bacterium has another mechanism for manganese acquisition that is not expressed in free-living cells (Hohle and O'Brian 2009). Interestingly, MntH plays a critical role in Mn^{2+} transport in the close phylogenetic relative *B. abortus*, where the presence of this manganese transporter is required for wild-type virulence in mouse (Anderson et al. 2009).

An additional type of secondary metal transporter might be relevant for rhizobia. A recent report described a gene (*ccoA*) required to maintain normal amounts of intracellular Cu and synthesis of *cbb*$_3$-Cox (Ekici et al. 2012) in the alpha-proteobacterium *Rhodobacter capsulatus*, strongly suggesting that the corresponding protein is a copper uptake transporter. CcoA is a member of the major facilitator superfamily (MFS) with 12 putative transmembrane helices split into two subdomains of six helices each, separated by a large cytoplasmic loop. Interestingly, this protein has homologs (*ca.* 40 percent identity) encoded in the genome of *Bradyrhizobium*, *Mesorhizobium* and

Sinorhizobium species. Further studies are required to elucidate whether these homologs might function as copper transporters in rhizobia.

Global Analysis of Expression of Bacterial Metal Uptake Systems

The information on direct analysis of metal transport under symbiotic conditions is still scarce, due to the inherent difficulty of working with bacteroids. This information can be complemented by data arising form transcriptomic and proteomic analysis of genes expressed by rhizobia in the endosymbiotic state. A global study performed through transcriptomic analysis of symbiotic expression in *S. meliloti* revealed that genes involved in iron uptake were mostly repressed in the bacteroids (Becker et al. 2004). However, repression does not indicate that these genes are not expressed in the nodule bacteria. In fact, solute-binding proteins from ABC systems mediating uptake of iron, manganese and zinc were identified in the proteomic analysis of *S. meliloti* bacteroids isolated from alfalfa nodules (Djordjevic et al. 2003). In the case of *B. japonicum* soybean bacteroids, transcriptomic analysis (Pessi et al. 2007) revealed that components of the molybdenum uptake system ModABC were amongst the most strongly induced genes in bacteroids (over 120-fold as compared to free-living cells). This data is consistent with the relevant symbiotic role of this transporter deduced from the mutant analysis mentioned above (Delgado et al. 2006). Further proteomic analyses of *B. japonicum* bacteroids induced in soybean indicate the presence of proteins involved in iron transport, including the TonB-dependent ferrichrome receptors blr3904 and bll4920 (Delmotte et al. 2010). Finally, a recent report on the transcriptomics of *R. etli* bacteroids induced in *Phaseolus* indicates moderate induction of genes for inorganic ion transporters including *afuA3*, which encodes a solute binding protein from an ABC iron uptake system not expressed in free-living cells (Vercruysse et al. 2011).

Summary and Outlook

In the last two decades, our understanding of metal homeostasis processes in plants and bacteria has increased considerably, but it is still lagging in certain aspects such as the legume-rhizobium interaction. This is in spite of metals being essential nutrients for symbiotic nitrogen fixation and critical components of the key enzymes of this metabolic process. In this process of metal exchange between the symbionts, the peribacteroid membrane is critical. However, many transporters in this membrane remain to be discovered. With the development of genomic, transcriptomic, proteomic and metallomic tools we are filling this gap in knowledge. New elements involved in metal transfer to the symbiosome have been identified, as well as the mechanisms involved in bacteroid metal uptake and use (summarized in Fig. 7.1). Interestingly, not many of the rhizobial transporters involved in metal uptake in free living conditions have a relevant role in symbiosis, probably reflecting a more favorable environment for metal uptake than in soils. Nevertheless, there are still many aspects to study, such as how the overall plant metal homeostasis mechanisms are affected by nodulation, what role rhizobial metal efflux transporters play in the invasion stage, which transporters are directing metals from the soil to the bacteroid, how metal speciation affects metal delivery, and how metals are directed to the corresponding apoproteins in the bacteroid cytosol.

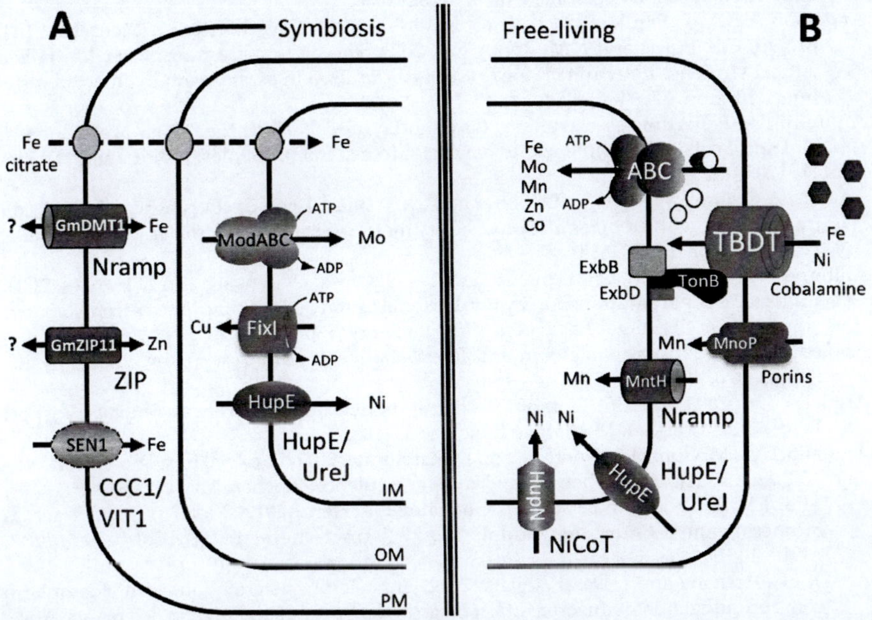

Figure 7.1 Metal uptake transporters described in *Rhizobium*-legume interactions. Transporters involved in metal uptake in symbiotic (A) and free-living (B) conditions are localized in the inner (IM), outer (OM) and plant membranes (PM). Siderophores are shown by hexagons whereas circles represent metal ions. Transport orientation for the plant Nramp and ZIP proteins remains elusive. The dotted arrow symbolizes an unknown mechanism for ferric citrate transfer across the plant and bacterial membranes. ABC: ATP-Binding Cassette transporters.

Acknowledgements

The work at the author's laboratories was funded by grants from Spain's Ministry of Science and Innovation (BIO2010-15301 to J.M.Palacios) and from the European Union Marie Curie International Reintegration program

(FP7-PEOPLE-2010-RG-276771 to M.González-Guerrero). M.González-Guerrero is a recipient of a Ramón y Cajal research contract. M.Menéndez-Cerón is a recipient of a grant from CONACYT (México).

References

Aller, S.G. and V.M. Unger. 2006. Projection structure of the human copper transporter CTR1 at 6-A resolution reveals a compact trimer with a novel channel-like architecture. Proc. Natl. Acad. Sci. U.S.A. 103: 3627–3632.

Andaluz, S., J. Rodríguez-Celma, A. Abadía, J. Abadía and A.F. López-Millán. 2009. Time course induction of several key enzymes in *Medicago truncatula* roots in response to Fe deficiency. Plant Physiol. Biochem. 47: 1082–1088.

Anderson, E.S., J.T. Paulley, J.M. Gaines, M.W. Valderas, D.W. Martin, E. Menscher, T.D. Brown, C.S. Burns and R.M. Roop, 2nd. 2009. The manganese transporter MntH is a critical virulence determinant for *Brucella abortus* 2308 in experimentally infected mice. Infect. Immun. 77: 3466–3474.

Andreini, C., I. Bertini, G. Cavallaro, G. Holliday and J. Thornton. 2008. Metal ions in biological catalysis: from enzyme databases to general principles. J. Biol. Inorg. Chem. 13: 1205–1218.

Anton, A., C. Groe, J. Reiman, T. Probyl and D.H. Nies. 1999. CzcD is a heavy metal ion transporter involved in regulation of heavy metal resistance in *Ralstonia* sp. strain CH34. J. Bacteriol. 181: 6876–6881.

Anton, A., A. Weltrowski, C.J. Haney, S. Franke, G. Grass, C. Rensing and D.H. Nies. 2004. Characteristics of zinc transport by two bacterial cation diffusion facilitators from *Ralstonia metallidurans* CH34 and *Escherichia coli*. J. Bacteriol. 186: 7499–7507.

Appleby, C.A. 1984. Leghemoglobin and *Rhizobium* respiration. Annu. Rev. Plant Physiol. 35: 443–478.

Argüello, J.M. 2003. Identification of ion-selectivity determinants in heavy-metal transport P-1B-type ATPases. J. Membrane Biol. 195: 93–108.

Argüello, J.M., M. González-Guerrero and D. Raimunda. 2011. Bacterial transition metal P1B-ATPases: Transport mechanism and roles in virulence. Biochemistry. 50: 9940–9949.

Argüello, J.M., D. Raimunda and M. González-Guerrero. 2012. Metal transport across biomembranes: Emerging models for a distinct chemistry. J. Biol. Chem. 287: 13510–13517.

Arif, K., G. Archana and J. Desai Anjana. 2012. Engineering heterologous iron siderophore complex utilization in rhizobia: Effect on growth of peanut and pigeon pea plants. Appl. Soil Ecol. 53: 65–73.

Balasubramanian, R., G.E. Kenney and A.C. Rosenzweig. 2011. Dual pathways for copper uptake by methanotrophic bacteria. J. Biol. Chem. 286: 37313–37319.

Becker, A., H. Berges, E. Krol, C. Bruand, S. Ruberg, D. Capela, E. Lauber, E. Meilhoc, F. Ampe, F.J. de Bruijn, J. Fourment, A. Francez-Charlot, D. Kahn, H. Kuster, C. Liebe, A. Puhler, S. Weidner and J. Batut. 2004. Global changes in gene expression in *Sinorhizobium meliloti* 1021 under microoxic and symbiotic conditions. Mol. Plant Microbe Interact. 17: 292–303.

Bernal, M., D. Casero, V. Singh, G.T. Wilson, A. Grande, H. Yang, S.C. Dodani, M. Pellegrini, P. Huijser, E.L. Connolly, S.S. Merchant and U. Krämer. 2012. Transcriptome sequencing identifies SPL7-regulated copper acquisition genes FRO4/FRO5 and the copper dependence of iron homeostasis in Arabidopsis. Plant Cell. 24: 738–761.

Blaudez, D., A. Kohler, F. Martin, D. Sanders and M. Chalot. 2003. Poplar metal tolerance protein 1 confers zinc tolerance and is an oligomeric vacuolar zinc transporter with an essential leucine zipper motif. Plant Cell. 15: 2911–2928.

Blindauer, C.A. and O.I. Leszczyszyn. 2010. Metallothioneins: unparalleled diversity in structures and functions for metal ion homeostasis and more. Nat. Prod. Rep. 27: 720–741.

Briat, J.F., C. Duc, K. Ravet and F. Gaymard. 2010. Ferritins and iron storage in plants. Biochim Biophys. Acta. 1800: 806–814.

Brito, B., A. Toffanin, R.-I. Prieto, J. Imperial, T. Ruiz-Argüeso and J.M. Palacios. 2008. Host-dependent expression of *Rhizobium leguminosarum* bv. viciae hydrogenase is controlled at transcriptional and post-transcriptional levels in legume nodules. Mol. Plant Microbe. Interact. 21: 597–604.

Brito, B., R.-I. Prieto, E. Cabrera, M.-A. Mandrand-Berthelot, J. Imperial, T. Ruiz-Argueso and J.-M. Palacios. 2010. *Rhizobium leguminosarum hupE* encodes a nickel transporter required for hydrogenase activity. J. Bacteriol. 192: 925–935.

Budzikiewicz, H. 2010. Microbial siderophores. Fortschr Chem. Org. Naturst. 92: 1–75.

Burkhead, J.L., K.A. Gogolin Reynolds, S.E. Abdel-Ghany, C.M. Cohu and M. Pilon. 2009. Copper homeostasis. New Phytol. 182: 799–816.

Burton, J.W., C. Harlow and E.C. Theil. 1998. Evidence for reutilization of nodule iron in soybean seed development. J. Plant Nutr. 5: 913–927.

Cacho, C., B. Brito, J. Palacios, C. Pérez-Conde and C. Cámara. 2010. Speciation of nickel by HPLC-UV/MS in pea nodules. Talanta. 83: 78–83.

Cailliatte, R., A. Schikora, J.F. Briat, S. Mari and C. Curie. 2010. High-affinity manganese uptake by the metal transporter NRAMP1 is essential for *Arabidopsis* growth in low manganese conditions. Plant Cell. 22: 904–917.

Cárdenas, L., A. Martínez, F. Sánchez and C. Quinto. 2008. Fast, transient and specific intracellular ROS changes in living root hair cells responding to Nod factors (NFs). Plant J. 56: 802–813.

Chakraborty, R., E. Storey and D. van der Helm. 2007. Molecular mechanism of ferric siderophore passage through the outer membrane receptor proteins of *Escherichia coli*. Biometals. 20: 263–274.

Chang, W.S., W.L. Franck, E. Cytryn, S. Jeong, T. Joshi, D.W. Emerich, M.J. Sadowsky, D. Xu and G. Stacey. 2007. An oligonucleotide microarray resource for transcriptional profiling of *Bradyrhizobium japonicum*. Mol. Plant Microbe Interact. 20: 1298–1307.

Chen, M., X. Shen, D. Li, L. Ma, J. Dong and T. Wang. 2009. Identification and characterization of MtMTP1, a Zn transporter of CDF family, in the *Medicago truncatula*. Plant Physiol. Biochem. 47: 1089–1094.

Cheng, J., B. Poduska, R.A. Morton and T.M. Finan. 2011. An ABC-type cobalt transport system is essential for growth of *Sinorhizobium meliloti* at trace metal concentrations. J. Bacteriol. 193: 4405–4416.

Cui, J. and A.L. Davidson. 2011. ABC solute importers in bacteria. Essays Biochem. 50: 85–99.

Curie, C., J.M. Alonso, M. Le Jean, J.R. Ecker and J.F. Briat. 2000. Involvement of NRAMP1 from *Arabidopsis thaliana* in iron transport. Biochem. J. 347: 749–755.

Curie, C., Z. Panaviene, C. Loulergue, S.L. Dellaporta, J.F. Briat and E.L. Walker. 2001. Maize yellow stripe1 encodes a membrane protein directly involved in Fe(III) uptake. Nature. 409: 346–349.

Dalton, D.A., C.J. Post and L. Langeberg. 1991. Effects of ambient oxygen and of fixed nitrogen on concentrations of glutathione, ascorbate and associated enzymes in soybean root nodules. Plant Physiol. 96: 812–818.

Davidson, A.L., E. Dassa, C. Orelle and J. Chen. 2008. Structure, function, and evolution of bacterial ATP-binding cassette systems. Microbiol. Mol. Biol. Rev. 72: 317–364.

De Feo, C.J., S.G. Aller, G.S. Siluvai, N.J. Blackburn and V.M. Unger. 2009. Three-dimensional structure of the human copper transporter hCTR1. Proc. Natl. Acad. Sci. U.S.A. 106: 4237–4242.

Delgado, M.J., A. Tresierra-Ayala, C. Talbi and E.J. Bedmar. 2006. Functional characterization of the *Bradyrhizobium japonicum modA* and *modB* genes involved in molybdenum transport. Microbiology. 152: 199–207.

Delmotte, N., C.H. Ahrens, C. Knief, E. Qeli, M. Koch, H.M. Fischer, J.A. Vorholt, H. Hennecke and G. Pessi. 2010. An integrated proteomics and transcriptomics reference data set provides new insights into the *Bradyrhizobium japonicum* bacteroid metabolism in soybean root nodules. Proteomics. 10: 1391–1400.

DiDonato, R.J., L.A. Roberts, T. Sanderson, R.B. Eisley and E.L. Walker. 2004. *Arabidopsis Yellow Stripe-Like2 (YSL2)* a metal-regulated gene encoding a plasma membrane transporter of nicotianamine-metal complexes. Plant J. 39: 403–414.

Djordjevic, M.A., H.C. Chen, S. Natera, G. Van Noorden, C. Menzel, S. Taylor, C. Renard, O. Geiger and G.F. Weiller. 2003. A global analysis of protein expression profiles in *Sinorhizobium meliloti*: discovery of new genes for nodule occupancy and stress adaptation. Mol. Plant Microbe Interact. 16: 508–524.

Dumay, Q.C., A.J. Debut, N.M. Mansour and M.H. Saier. 2006. The copper transporter (Ctr) family of Cu^+ uptake systems. J. Mol. Microbiol. Biotechnol. 11: 10–19.

Durrett, T.P., W. Gassmann and E.E. Rogers. 2007. The FRD3-Mediated efflux of citrate into the root vasculature is necessary for efficient iron translocation. Plant Physiol. 144: 197–205.

Eide, D. 2004. The ABC of solute carriers. The SLC39 family of metal ion transporters. Eur. J. Physiol. 447: 796–800.

Eitinger, T. and B. Friedrich. 1991. Cloning, nucleotide sequence, and heterologous expression of a high-affinity nickel transport gene from *Alcaligenes eutrophus*. J. Biol. Chem. 266: 3222–3227.

Eitinger, T., O. Degen, U. Böhnke and M. Müller. 2000. Nic1p, a relative of bacterial transition metal permeases in *Schizosaccharomyces pombe*, provides nickel ion for urease biosynthesis. J. Biol. Chem. 275: 18029–18033.

Eitinger, T., J. Suhr, L. Moore and J.A. Smith. 2005. Secondary transporters for nickel and cobalt ions: theme and variations. Biometals. 18: 399–405.

Ekici, S., H. Yang, H.G. Koch and F. Daldal. 2012. Novel transporter required for biogenesis of *cbb3*-type cytochrome c oxidase in *Rhodobacter capsulatus*. MBio 3: e00293-e00311.

Fedorova, M., J. van de Mortel, P.A. Matsumoto, J. Cho, C.D. Town, K.A. VandenBosch, J.S. Gantt and C.P. Vance. 2002. Genome-wide identification of nodule-specific transcripts in the model legume *Medicago truncatula*. Plant Physiol. 130: 519–537.

Fitzpatrick, K.L., S.D. Tyerman and B.N. Kaiser. 2008. Molybdate transport through the plant sulphate transporter SHST1. FEBS Lett. 582: 1508–1513.

Forbes, J.R. and P. Gros. 2001. Divalent-metal transport by NRAMP proteins at the interface of host-pathogen interactions. Trends Microbiol. 1 9: 397–403.

Fraustro da Silva, J.J.R. and R.J.P. Williams. 2001. The biological chemistry of the elements Oxford University Press, New York, U.S.A.

Fridovich, I. 1998. Oxygen toxicity: a radical explanation. J. Exp. Biol. 201: 1203–1209.

Fu, C., S. Javedan, F. Moshiri and R.J. Maier. 1994. Bacterial genes involved in incorporation of nickel into a hydrogenase enzyme. Proc. Natl. Acad. Sci. U.S.A. 91: 5099–5103.

Fung, I.Y., S.K. Meyn, I. Tegen, S.C. Doney, J.G. John and J.K.B. Bishop. 2000. Iron supply and demand in the upper ocean. Global Biogeochem. Cycles. 14: 281–295.

Galibert, F., T.M. Finan, S.R. Long, A. Puhler, P. Abola, F. Ampe, F. Barloy-Hubler, M.J. Barnett, A. Becker, P. Boistard, G. Bothe, M. Boutry, L. Bowser, J. Buhrmester, E. Cadieu, D. Capela, P. Chain, A. Cowie, R.W. Davis, S. Dreano, N.A. Federspiel, R.F. Fisher, S. Gloux, T. Godrie, A. Goffeau, B. Golding, J. Gouzy, M. Gurjal, I. Hernandez-Lucas, A. Hong, L. Huizar, R.W. Hyman, T. Jones, D. Kahn, M.L. Kahn, S. Kalman, D.H. Keating, E. Kiss, C. Komp, V. Lelaure, D. Masuy, C. Palm, M.C. Peck, T.M. Pohl, D. Portetelle, B. Purnelle, U. Ramsperger, R. Surzycki, P. Thebault, M. Vandenbol, F.J. Vorholter, S. Weidner, D.H. Wells, K. Wong, K.C. Yeh and J. Batut. 2001. The composite genome of the legume symbiont *Sinorhizobium meliloti*. Science. 293: 668–672.

Goldstein, S., D. Meyerstein and G. Czapski. 1993. The Fenton reagents. Free radical biology and medicine. 15: 435–445.

González-Guerrero, M., D. Raimunda, X. Cheng and J.M. Argüello. 2010. Distinct functional roles of homologous Cu^+ efflux ATPases in *Pseudomonas aeruginosa*. Mol. Microbiol. 78: 1246–1258.

Grotz, N. and M.L. Guerinot. 2006. Molecular aspects of Cu, Fe and Zn homeostasis in plants. Biochim. Biophys. Acta. 1763: 595–608.

Grotz, N., T. Fox, E. Connolly, W. Park, M.L. Guerinot and D. Eide. 1998. Identification of a family of zinc transporter genes from *Arabidopsis* that respond to zinc deficiency. Proc. Natl. Acad. Sci. U.S.A. 95: 7220–7224.

Guerinot, M.L., E.J. Meidl and O. Plessner. 1990. Citrate as a siderophore in *Bradyrhizobium japonicum*. J. Bacteriol. 172: 3298–3303.

Hakoyama, T., H. Watanabe, J. Tomita, A. Yamamoto, S. Sato, Y. Mori, H. Kouchi and N. Suganuma. 2009. Nicotianamine synthase specifically expressed in root nodules of *Lotus japonicus*. Planta. 230: 309–317.

Hakoyama, T., K. Niimi, T. Yamamoto, S. Isobe, S. Sato, Y. Nakamura, S. Tabata, H. Kumagai, Y. Umehara, K. Brossuleit, T.R. Petersen, N. Sandal, J. Stougaard, M.K. Udvardi, M. Tamaoki, M. Kawaguchi, H. Kouchi and N. Suganuma. 2011. The integral membrane protein SEN1 is required for symbiotic nitrogen fixation in *Lotus japonicus* nodules. Plant Cell Physiol. 53: 225–236.

Hantke, K. 2005. Bacterial zinc uptake and regulators. Curr. Opin. Microbiol. 8: 196–202.

Hanyu, M., H. Fujimoto, K. Tejima and K. Saeki. 2008. Functional differences of two distinct catalases in *Mesorhizobium loti* MAFF303099 under free-living and symbiotic conditions. J. Bacteriol. 191: 1463–1471.

Hider, R.C. and X. Kong. 2010. Chemistry and biology of siderophores. Nat. Prod. Rep. 27: 637–657.

Hohle, T.H. and M.R. O'Brian. 2009. The *mntH* gene encodes the major Mn(2+) transporter in *Bradyrhizobium japonicum* and is regulated by manganese via the Fur protein. Mol. Microbiol. 72: 399–409.

Hohle, T.H., W.L. Franck, G. Stacey and M.R. O'Brian. 2011. Bacterial outer membrane channel for divalent metal ion acquisition. Proc. Natl. Acad. Sci. U.S.A. 108: 15390–15395.

Holloway, P., W. McCormick, R.J. Watson and Y.K. Chan. 1996. Identification and analysis of the dissimilatory nitrous oxide reduction genes, *nosRZDFY*, of *Rhizobium meliloti*. J. Bacteriol. 178: 1505–1514.

Imperial, J., M. Hadi and N.K. Amy. 1998. Molybdate binding by ModA, the periplasmic component of the *Escherichia coli* mod molybdate transport system. Biochim. Biophys. Acta. 1370: 337–346.

Jamet, A., S. Sigaud, G. van de Sype, A. Puppo and D. Hérouart. 2003. Expression of the bacterial catalase genes during *Sinorhizobium meliloti-Medicago sativa* symbiosis and their crucial role during the infection process. Mol. Plant Microbe Interact. 16: 217–225.

Jamet, A., K. Mandon, A. Puppo and D. Hérouart. 2007. H_2O_2 is required for optimal establishment of the *Medicago sativa/Sinorhizobium meliloti* symbiosis. J. Bacteriol. 189: 8741–8745.

Jean, M.L., A. Schikora, S. Mari, J.F. Briat and C. Curie. 2005. A loss-of-function mutation in AtYSL1 reveals its role in iron and nicotianamine seed loading. Plant J. 44: 769–782.

Kaiser, B.N., S. Moreau, J. Castelli, R. Thomson, A. Lambert, S. Bogliolo, A. Puppo and D.A. Day. 2003. The soybean NRAMP homologue, GmDMT1, is a symbiotic divalent metal transporter capable of ferrous iron transport. Plant J. 35: 295–304.

Kim, E.-H., D.H. Nies, M.M. McEvoy and C. Rensing. 2011. Switch or funnel: How RND-type transport systems control periplasmic metal homeostasis. J. Bacteriol. 193: 2381–2387.

Kim, S., K. Watanabe, T. Shirahata and M. Watarai. 2004. Zinc uptake system (*znuA* locus) of *Brucella abortus* is essential for intracellular survival and virulence in mice. J. Vet. Med. Sci. 66: 1059–1063.

Kim, S.A., T. Punshon, A. Lanzirotti, L. Li, J.M. Alonso, J.R. Ecker, J. Kaplan and M.L. Guerinot. 2006. Localization of iron in *Arabidopsis* seed requires the vacuolar membrane transporter VIT1. Science. 314: 1295–1298.

Klein, J.S. and O. Lewinson. 2011. Bacterial ATP-driven transporters of transition metals: physiological roles, mechanisms of action, and roles in bacterial virulence. Metallomics. 3: 1098–1108.

Kulathila, R., R. Kulathila, M. Indic and B. van den Berg. 2011. Crystal structure of *Escherichia coli* CusC, the outer membrane component of a heavy metal efflux pump. PLoS ONE 6: e15610.

Li, L., O.S. Chen, D. McVey Ward and J. Kaplan. 2001. CCC1 is a transporter that mediates vacuolar iron storage in yeast. J. Biol. Chem. 276: 29515–29519.

Lim, B.L. 2010. TonB-dependent receptors in nitrogen-fixing nodulating bacteria. Microbes Environ. 25: 67–74.

Long, F., C.C. Su, M.T. Zimmermann, S.E. Boyken, K.R. Rajashankar, R.L. Jernigan and E.W. Yu. 2010. Crystal structures of the CusA efflux pump suggest methionine-mediated metal transport. Nature. 467: 484–488.

Lu, M. and D. Fu. 2007. Structure of the Zinc Transporter YiiP. Science. 317: 1746–1748.

Lucas, M.M., G. Van de Sype, D. Hérouart, M.J. Hernández, A. Puppo and M.R. de Felipe. 1998. Immunolocalization of ferritin in determinate and indeterminate legume root nodules. Protoplasma. 204: 61–70.

Lynch, D., J. O'Brien, T. Welch, P. Clarke, P.O. Cuiv, J.H. Crosa and M. O'Connell. 2001. Genetic organization of the region encoding regulation, biosynthesis, and transport of rhizobactin 1021, a siderophore produced by *Sinorhizobium meliloti*. J. Bacteriol. 183: 2576–2585.

Macomber, L. and J.A. Imlay. 2009. The iron-sulfur clusters of dehydratases are primary intracellular targets of copper toxicity. Proc. Natl. Acad. Sci. U.S.A. 106: 8344–8349.

Macomber, L. and R.P. Hausinger. 2011. Mechanisms of nickel toxicity in microorganisms. Metallomics. 3: 1153–1162.

Mauchline, T.H., J.E. Fowler, A.K. East, A.L. Sartor, R. Zaheer, A.H. Hosie, P.S. Poole and T.M. Finan. 2006. Mapping the *Sinorhizobium meliloti* 1021 solute-binding protein-dependent transportome. Proc. Natl. Acad. Sci. U.S.A. 103: 17933–17938.

McKie, A.T., P. Marciani, A. Rolfs, K. Brennan, K. Wehr, D. Barrow, S. Miret, A. Bomford, T.J. Peters, F. Farzaneh, M.A. Hediger, M.W. Hentze and R.J. Simpson. 2000. A novel duodenal iron-regulated transporter, IREG1, implicated in the basolateral transfer of iron to the circulation. Mol. Cell. 5: 299–309.

Miller, R.W., Z. Yu and C.G. Zarkadas. 1993. The nitrogenase proteins of *Rhizobium meliloti*: purification and properties of the MoFe and Fe components. Biochim. Biophys. Acta. 1163: 31–41.

Moreau, S., J.M. Meyer and A. Puppo. 1995. Uptake of iron by symbiosomes and bacteroids from soybean nodules. FEBS Lett. 361: 225–228.

Moreau, S., R.M. Thomson, B.N. Kaiser, B. Trevaskis, M.L. Guerinot, M.K. Udvardi, A. Puppo and B.A. Day. 2002. GmZIP1 encodes a symbiosis-specific zinc transporter in soybean. J. Biol. Chem. 277: 4738–4746.

Morrissey, J., I.R. Baxter, J. Lee, L. Li, B. Lahner, N. Grotz, J. Kaplan, D.E. Salt and M.L. Guerinot. 2009. The ferroportin metal efflux proteins function in iron and cobalt homeostasis in *Arabidopsis*. Plant Cell. 21: 3326–3338.

Nevo, Y. and N. Nelson. 2006. The NRAMP family of metal-ion transporters. Biochim. Biophys. Acta. 1763: 609–620.

Nicholls, P., I. Fita and P.C. Lowen. 2001. Enzymology and structure of catalases. Adv. Inorg. Chem. 51: 51–106.

Nies, D.H. 2003. Efflux-mediated heavy metal resistance in prokaryotes. FEMS Microbiol Rev. 27: 313–339.

Noinaj, N., M. Guillier, T.J. Barnard and S.K. Buchanan. 2010. TonB-dependent transporters: regulation, structure, and function. Annu. Rev. Microbiol. 64: 43–60.

O'Brian, M.R. and R.J. Maier. 1989. Hydrogen metabolism in *Rhizobium*: energetics, regulation, enzymology and genetics. Adv. Microb. Physiol. 1988: 1–52.

O'Brian, M.R. and E. Fabiano. 2010. Mechanisms and regulation of iron homeostasis in rhizobia. *In:* P. Cornelis and C. Andrews [eds.]. Iron uptake and homeostasis in microorganisms Caister Academic Press, Norfolk, U.K.

O'Hara, G.W. 2001. Nutritional constraints on root nodule bacteria affecting symbiotic nitrogen fixation: a review. Austr. J. Exp. Agric. 41: 417–433.

Outten, C.E. and T.V. O'Halloran. 2001. Femtomolar sensitivity of metalloregulatory proteins controlling zinc homeostasis. Science. 292: 2488–2492.

Palacios, J.M., H. Manyani, M. Martinez, A.C. Ureta, B. Brito, E. Bascones, L. Rey, J. Imperial and T. Ruiz-Argueso. 2005. Genetics and biotechnology of the H(2)-uptake [NiFe] hydrogenase from *Rhizobium leguminosarum* bv. *viciae*, a legume endosymbiotic bacterium. Biochem. Soc. Trans. 33: 94–96.

Palmgren, M.G. and P. Nissen. 2011. P-type ATPases. Annu. Rev. Biophys. 40: 243–266.

Pessi, G., C.H. Ahrens, H. Rehrauer, A. Lindemann, F. Hauser, H.M. Fischer and H. Hennecke. 2007. Genome-wide transcript analysis of *Bradyrhizobium japonicum* bacteroids in soybean root nodules. Mol. Plant Microbe Interact. 20: 1353–1363.

Platero, R.A., M. Jaureguy, F.J. Battistoni and E.R. Fabiano. 2003. Mutations in *sitB* and *sitD* genes affect manganese growth requirements in *Sinorhizobium meliloti*. FEMS Microbiol. Lett. 218: 65–70.

Plessner, O., T. Klapatch and M.L. Guerinot. 1993. Siderophore utilization by *Bradyrhizobium japonicum*. Appl. Environ. Microbiol. 59: 1688–1690.

Postle, K. and R.A. Larsen. 2007. TonB-dependent energy transduction between outer and cytoplasmic membranes. Biometals. 20: 453–465.

Preisig, O., R. Zufferey and H. Hennecke. 1996a. The *Bradyrhizobium japonicum fixGHIS* genes are required for the formation of the high-affinity cbb(3)-type cytochrome oxidase. Arch. Microbiol. 165: 297–305.

Preisig, O., R. Zufferey, L. Thony-Meyer, C. Appleby and H. Hennecke. 1996b. A high-affinity *cbb3*-type cytochrome oxidase terminates the symbiosis-specific respiratory chain of *Bradyrhizobium japonicum*. J. Bacteriol. 178: 1532–1538.

Puig, S., N. Andés-Colás, A. García-Molina and L. Penarrubia. 2007. Copper and iron homeostasis in *Arabidopsis*: responses to metal deficiencies, interactions and biotechnological applications. Plant Cell Environ. 30: 271–290.

Puppo, A., J. Rigaud and D. Job. 1981. Role of the superoxide anion in the leghemoglobin autoxidation. Plant Sci. Lett. 22: 353–360.

Puppo, A., K. Groten, F. Bastian, R. Carzaniga, M. Soussi, M.M. Lucas, M.R. de Felipe, J. Harrison, H. Vanacker and C.H. Foyer. 2005. Legume nodule senescence: roles for redox and hormone signalling in the orchestration of the natural aging process. New Phytol. 165: 683–701.

Raimunda, D., M. González-Guerrero, B.W. Leeber and J.M. Argüello. 2011. The transport mechanism of bacterial Cu(+)-ATPases: distinct efflux rates adapted to different function. Biometals. 24: 467–475.

Ranquet, C., S. de-Choudens, L. Loiseau, F. Barras and M. Fontecave. 2007. Cobalt stress in *Escherichia coli*. The effect on the iron-sulfur proteins. J. Biol. Chem. 282: 30442–30451.

Rea, P.A. 2007. Plant ATP-Binding Cassette transporters. Ann. Rev. Plant Biol. 58: 347–375.

Reyes-Caballero, H., G.C. Campanello and D.P. Giedroc. 2011. Metalloregulatory proteins: Metal selectivity and allosteric switching. Biophys. Chem. 156: 103–114.

Rogers, E.E. and M.L. Guerinot. 2002. FRD3, a member of the multidrug and toxin efflux family, controls iron deficiency responses in *Arabidopsis*. Plant Cell. 14: 1787–1799.

Roschzttardtz, H., M. Séguéla-Arnaud, J.-F. Briat, G. Vert and C. Curie. 2011. The FRD3 citrate effluxer promotes iron nutrition between symplastically disconnected tissues throughout *Arabidopsis* development. Plant Cell. 23: 2725–2737.

Rubio, M.C., E.K. James, M.R. Clemente, B. Bucciarelli, M. Fedorova, C.P. Vance and M. Becana. 2004. Localization of superoxide dismutases and hydrogen peroxide in legume root nodules. Mol. Plant Microbe Int. 17: 1294–1305.

Rubio, M.C., M. Becana, S. Sato, E.K. James, S. Tabata and H.P. Spaink. 2007. Characterization of genomic clones and expression analysis of the three types of superoxide dismutases during nodule development in *Lotus japonicus*. Mol. Plant Microbe Interact. 20: 262–275.

Ruel, M. and H. Bouis. 1998. Plant breeding: a long-term strategy for the control of zinc deficiency in vulnerable populations. Am. J. Clin. Nutr. 68: 488S–494S.

Sancenon, V., S. Puig, I. Mateu-Andres, E. Dorcey, D.J. Thiele and L. Penarrubia. 2004. The *Arabidopsis* copper transporter COPT1 functions in root elongation and pollen development. J. Biol. Chem. 279: 15348–15355.

Schauer, K., B. Gouget, M. Carriere, A. Labigne and H. de Reuse. 2007. Novel nickel transport mechanism across the bacterial outer membrane energized by the TonB/ExbB/ExbD machinery. Mol. Microbiol. 63: 1054–1068.

Schauer, K., D.A. Rodionov and H. de Reuse. 2008. New substrates for TonB-dependent transport: do we only see the 'tip of the iceberg'? Trends Biochem. Sci. 33: 330–338.

Small, S.K., S. Puri, I. Sangwan and M.R. O'Brian. 2009. Positive control of ferric siderophore receptor gene expression by the Irr protein in *Bradyrhizobium japonicum*. J. Bacteriol. 191: 1361–1368.

Solioz, M., H.K. Abicht, M. Mermod and S. Mancini. 2010. Response of gram-positive bacteria to copper stress. J. Biol. Inorg. Chem. 15: 3–14.

Stork, M., M.P. Bos, I. Jongerius, N. de Kok, I. Schilders, V.E. Weynants, J.T. Poolman and J. Tommassen. 2010. An outer membrane receptor of *Neisseria meningitidis* involved in zinc acquisition with vaccine potential. PLoS Pathog 6: e1000969.

Tang, C., A.D. Robson and M.J. Dilworth. 1990. The role of iron in nodulation and nitrogen fixation in *Lupinus angustifolius* L. New Phytol. 114: 173–182.

Tang, C.X., A.D. Robson, M.J. Dilworth and J. Kuo. 1992. Microscopic evidence on how iron-deficiency limits nodule initiation in *Lupinus angustifolius* l. New Phytol. 121: 457–467.

Tejada-Jiménez, M., A. Llamas, E. Sanz-Luque, A. Galván and E. Fernández. 2007. A high-affinity molybdate transporter in eukaryotes. Proc. Natl. Acad. Sci. U.S.A. 104: 20126–20130.

Tejada-Jiménez, M., A. Galván and E. Fernández. 2011. Algae and humans share a molybdate transporter. Proc. Natl. Acad. Sci. U.S.A. 108: 6420–6425.

Terry, R.E., K.U. Soerensen, V.D. Jolley and J.C. Brown. 1991. The role of active *Bradyrhizobium japonicum* in iron stress response of soy-beans. Plant Soil. 130: 225–230.

Tomatsu, H., J. Takana, H. Takahashi, A. Watanabe-Takahashi, N. Shibagaki and T. Fujiwara. 2007. An *Arabidopsis thaliana* high affinity molybdate transporter required for efficient uptake of molybdate from soil. Proc. Natl. Acad. Sci. U.S.A. 104: 18807–18812.

Velasco, L., S. Mesa, C.A. Xu, M.J. Delgado and E.J. Bedmar. 2004. Molecular characterization of *nosRZDFYLX* genes coding for denitrifying nitrous oxide reductase of *Bradyrhizobium japonicum*. Antonie Van Leeuwenhoek. 85: 229–235.

Vercruysse, M., M. Fauvart, S. Beullens, K. Braeken, L. Cloots, K. Engelen, K. Marchal and J. Michiels. 2011. A comparative transcriptome analysis of *Rhizobium etli* bacteroids: specific gene expression during symbiotic nongrowth. Mol. Plant Microbe Interact. 24: 1553–1561.

Vert, G., N. Grotz, F. Dedaldéchamp, F. Gaymard, M.L. Guerinot, J.F. Briat and C. Curie. 2002. IRT1, an *Arabidopsis* transporter essential for iron uptake from the soil and the plant growth. Plant Cell. 14: 1223–1233.

Waldron, K.J. and N.J. Robinson. 2009. How do bacterial cells ensure that metalloproteins get the correct metal? Nat. Rev. Microbiol. 7: 25–35.

Wandersman, C.C. and I. Stojiljkovic. 2000. Bacterial heme sources: the role of heme, hemoprotein receptors and hemophores. Curr. Opin. Microbiol. 3: 215–220.

Wandersman, C.C. and P. Delepelaire. 2004. Bacterial iron sources: From siderophores to hemophores. Annu. Rev. Microbiol. 58: 611–647.

Wei, Y., H. Li and D. Fu. 2004. Oligomeric state of the *Escherichia coli* metal transporter YiiP. J. Biol. Chem. 279: 39251–39259.

Yeoman, K.H., F. Wisniewski-Dye, C. Timony, J.B. Stevens, N.G. de Luca, J.A. Downie and A.W. Johnston. 2000. Analysis of the *Rhizobium leguminosarum* siderophore-uptake gene *fhuA*: differential expression in free-living bacteria and nitrogen-fixing bacteroids and distribution of an *fhuA* pseudogene in different strains. Microbiology. 146: 829–837.

Young, J.P., L.C. Crossman, A.W. Johnston, N.R. Thomson, Z.F. Ghazoui, K.H. Hull, M. Wexler, A.R. Curson, J.D. Todd, P.S. Poole, T.H. Mauchline, A.K. East, M.A. Quail, C. Churcher, C. Arrowsmith, I. Cherevach, T. Chillingworth, K. Clarke, A. Cronin, P. Davis, A. Fraser, Z. Hance, H. Hauser, K. Jagels, S. Moule, K. Mungall, H. Norbertczak, E. Rabbinowitsch, M. Sanders, M. Simmonds, S. Whitehead and J. Parkhill. 2006. The genome of *Rhizobium leguminosarum* has recognizable core and accessory components. Genome Biol. 7: R34.

Zeth, K. and M. Thein. 2010. Porins in prokaryotes and eukaryotes: common themes and variations. Biochem. J. 431: 13–22.

8

Ecology of Denitrification in Soils and Plant-Associated Bacteria

E.J. Bedmar, E. Bueno, D. Correa, M.J. Torres, M.J. Delgado* and *S. Mesa*

The Microbial N Cycle

After carbon (C), hydrogen (H) and oxygen (O_2), nitrogen (N) is the fourth most abundant element in the biomass, where it is part of essential compounds such as proteins, nucleic acids, hormones, etc. N is present in large amounts around the Earth's surface, making up to about 80 percent of the Earth's atmosphere. With the exception of water, N is also the most common limiting element in agriculture. Despite its abundance in the atmosphere, availability of N in a form suitable for plant and animal consumption is a major constraint to life on our planet. Most of the N in the atmosphere is found in the form of dinitrogen gas (N_2), which is inaccessible to eukaryotes and many bacteria. Diazotrophic microorganisms, mainly bacteria, contain the enzyme nitrogenase, which converts bio-unavailable N_2 gas to bio-available ammonia (NH_4^+). This process is called biological nitrogen fixation (BNF), and initiates the N cycle in the biosphere

Department of Soil Microbiology and Symbiotic Systems. Estación Experimental del Zaidín. Consejo Superior de Investigaciones Científicas, P. O. Box 419-Granada, Spain.
*Corresponding author: eulogio.bedmar@eez.csic.es

(Fig. 8.1). Ammonia is subsequently incorporated into cellular biomass mainly via the glutamine synthetase-glutamate synthase (GS-GOGAT) pathway. Alternatively, glutamate dehydrogenase (GDH) may also be involved in aerobic ammonium assimilation. In addition to its incorporation into organic nitrogen compounds, ammonia can be oxidized to nitrate (NO_3^-) by nitrifying bacteria in a two-step process called nitrification. During nitrification, the enzymes ammonia monooxygenase and nitrite oxidoreductase (nitrite oxidase) oxidize ammonia to nitrite (NO_2^-) and nitrite to nitrate, respectively. Some bacteria such as *Escherichia coli* or *Bacillus subtilis* are able to perform nitrate respiration, but they do not denitrify with N_2 as a product. Instead, they reduce nitrate to ammonium, so-called nitrate-ammonification. In many species of nitrate ammonifying bacteria, there are two biochemically distinct nitrate reductases, one membrane-bound with the active site located in the cytoplasm and the other in the periplasm. In the cytoplasm, nitrate is reduced to nitrite by a membrane-bound respiratory nitrate reductase system (NarGHI), and nitrite is further reduced to ammonium by a siroheme-containing Nir (NirB). A periplasmic nitrate reductase (NapA) reduces nitrate to nitrite in the periplasm and a cytochrome cNir further reduces nitrite to ammonium.

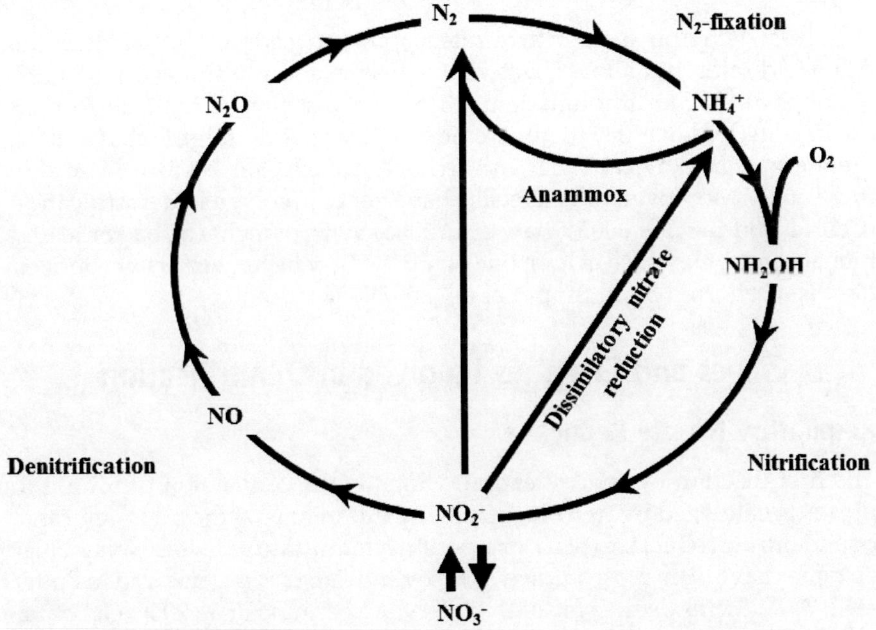

Figure 8.1 A scheme of the N-cycle.

Finally, denitrification is the process by which nitrate becomes N_2, which returns to the atmosphere, thus closing the N cycle in the biosphere. In addition to denitrification, anaerobic ammonia oxidation (anammox) converts nitrite and ammonium directly into N_2, thus largely contributing to production of N_2. Genome sequencing of several N-cycle organisms, the nitrite-dependent anaerobic methane oxidation (Raghoebarsin et al. 2006) and hypertermophilic N_2-fixing methane producing archaea (Mehta and Baross, 2006) are examples of the biodiversity and metabolic capacity of new nitrogen conversions within the N cycle (Jetten 2008). Together, these processes form the global N cycle and microorganisms are essential for maintaining the balance between reduced and oxidized forms of nitrogen (van Spanning 2011).

Denitrification

Denitrification is an alternative form of respiration in which bacteria sequentially reduce nitrate (NO_3^-) or nitrite (NO_2^-) to N_2 by the intermediates nitric oxide (NO) and nitrous oxide (N_2O) when oxygen concentrations are limiting, according to the following reaction:

$$NO_3^- \rightarrow NO_2^- \rightarrow NO \rightarrow N_2O \rightarrow N_2$$

The switch from oxygen to nitrate respiration leads to a reduction in the ATP yield rates, but allows bacteria to survive and multiply (Zumft 1997, Simon et al. 2008). Although denitrification was believed to be performed exclusively by bacteria, there is evidence that some fungi (Takaya 2002, Prendergast-Miller et al. 2011) and archaea (Treush et al. 2005) are also able to denitrify. Moreover, nitrifiers also have genes involved in denitrification (Cebron and Garnier 2005, Shaw et al. 2006). A list of archaeal, bacterial and fungal genera for which at least one denitrifying gene has been characterized has been reported by Philippot et al. (2007).

Genes and Enzymes Involved in Denitrification

Respiratory Nitrate Reductases

The first reaction of denitrification, which is the conversion of nitrate to nitrite, is catalyzed by two biochemically different enzymes, a membrane-bound nitrate reductase (Nar), or a periplasmic nitrate reductase (Nap). Nar enzymes have been most studied in *E. coli* and *Paracoccus* (reviewed in Potter et al. 2001, Richardson et al. 2001, 2007, van Spanning et al. 2005, González et al. 2006, Richardson 2011, van Spanning 2011). It is a 3-subunit enzyme composed of NarGHI (Bertero et al. 2003, Jormakka et al. 2004), where NarG is the catalytic subunit of about 140 kDa that contains a bismolybdopterin guanine dinucleotide cofactor and a [4Fe-4S] cluster. NarH, of about 60 kDa, contains one [3Fe-4S] and three [4Fe-4S]. NarG and NarH are located in the cytoplasm and associate with NarI, an integral membrane protein of about 25 kDa with five transmembrane helices and the N-terminus facing the periplasm. Nar proteins are encoded by genes of a *narGHJI* operon. Whereas *narGHI* encode the structural subunits, *narJ* codes for a cognate chaperone required for the proper maturation and membrane insertion of Nar (Blasco et al. 1992). The organization of this operon is conserved in most species that express Nar. *E. coli* has a functional duplicate of the *narGHJI* operon named *narZYWV*, which physiologically has a function during stress response rather than anaerobic respiration (Blasco et al. 1990, Spector et al. 1999). In some archaea and bacteria the NarGH subunits are on the outside rather than the inside of the cytoplasmic membrane. This enzyme is supposed to be the evolutionary precursor of the Nar system (Martínez-Espinosa et al. 2007). The Nar enzyme couples quinol oxidation with proton translocation and energy conservation. This respiratory function permits cell growth under oxygen-limiting conditions (Potter et al. 2001, Simon et al. 2008).

Nap is widespread in all classes of denitrifying and non-denitrifying proteobacteria (reviewed in Potter et al. 2001, Richardson et al. 2001, 2007, van Spanning et al. 2005, Gonzalez et al. 2006, Richardson 2011, van Spanning 2011). The best studied Nap enzymes were isolated from *Paracoccus pantotrophus*, *E. coli*, *Rhodobacter sphaeroides* and *Desulfovibrio desfuromonas* (Dias et al. 1999, Arnoux et al. 2003, Jepson et al. 2007). Nap is a 2-subunit enzyme composed of the NapAB complex located in the periplasm and a transmembrane NapC component. The catalytic subunit NapA contains the bis-MGD cofactor at its active site and an FeS center. NapB is diheme cytochrome c_{552}, and NapC is a *c*-type tetra-heme membrane-anchored protein that is involved in the electron transfer from the quinol pool to NapAB (Roldán et al. 1998, Cartron et al. 2002). Eight different genes have been identified as components for operons that encode Naps in different organisms (Richardson et al. 2001). Most bacteria studied thus far have the *napABC* genes in common. The remaining *napDEFKL* genes encode for different proteins that are not directly involved in the nitrate reduction. NapD is a cytoplasmic protein that acts as a chaperone. NapF is a cytoplasmic iron–sulfur containing protein with four loosely bound [4Fe-4S] clusters, and is thought to participate in the assembling of the iron–sulfur cluster of NapA (Olmo-Mira et al. 2004, Nilavongse et al. 2006). The *napEKL* genes encode for proteins with so far unknown functions. In *E. coli*, the *nap* operon includes *napGH* genes encoding a periplasmic and an integral membrane protein with [4Fe-4S] clusters. NapH and NapG interact, making an electron transfer supercomplex that can channel electrons from both menaquinol and ubiquinol to NapA (Brondijk et al. 2002, 2004). Although Nap is also linked to quinol oxidation, it does not synthesize ATP (Simon et al. 2008). Physiological functions for Nap systems include the disposal of reducing equivalents during aerobic growth on reduced carbon substrates and anaerobic nitrate respiration as a part of bacterial ammonification or denitrification pathways (Potter et al. 2001).

Respiratory Nitrite Reductases

Two types of respiratory nitrite reductases (Nir) have been described in denitrifying bacteria, NirS and NirK (Rinaldo and Cutruzzolà 2007, Rinaldo et al. 2008, van Spanning et al. 2005, 2007). They catalyze the one-electron reduction of nitrite to nitric oxide, however, neither of the enzymes is electrogenic. Both are located in the periplasmic space, and receive electrons from cytochrome *c* and/or a blue copper protein, pseudoazurin, via the cytochrome bc_1 complex (Pearson et al. 2003). The cd_1 NirS nitrite reductase is a homodimeric enzyme with hemes *c* and d_1. Electrons are transferred via the heme *c* of NirS to heme d_1, where nitrite binds and is reduced to nitric oxide (Rinaldo et al. 2008). The best-characterized *nirS* genes clusters are those from *Pseudomonas aeruginosa* (*nirSMCFDLGHJEN*) and *P. denitrificans* (*nirXISECFDLGHJN*). In the model denitrifier *P. stutzeri* there are two *nir* clusters (*nirSTBMCFDLGH* and *nirJEN*) which are separated by a part of the *nor* gene cluster encoding nitric oxide reductase.

The Cu-containing NirK enzymes are homotrimeric complexes harboring three type I copper centers, and three type II copper centers, which form the active site. Nitrite binds to the type II site where it is reduced to nitric oxide by electrons transferred from the type I copper site. In contrast to the complex organization of the genes encoding the NirS proteins, the Cu-NirK enzyme is encoded by the *nirK* gene (Rinaldo and Cutruzzolà 2007). Although both Nir enzymes are widespread among denitrifiers, there is no evidence that the same species could have both enzymes (Zumft 1997).

Respiratory Nitric Oxide Reductases

Three types of nitric oxide reductases (Nor) have been characterized, cNor, qNor, and qCuANor (reviewed in Zumft 2005, de Vries et al. 2007, van Spanning et al. 2005, 2007, van Spanning 2011). The cNor is an integral membrane enzyme composed of two subunits, the heme *c* containing-NorC, and NorB, which contains hemes *b* and a non-heme iron. Electron transfer to cNor is mediated by the cytochrome bc_1 complex and a soluble cytochrome *c* or pseudoazurin (van Spanning et al. 2007). Electrons are transferred to the heme *c* and then via the heme *b* to the active site. There, two molecules of nitric oxide are reduced to form nitrous oxide. The qNor uses quinol or menaquinol as electron donors. The enzyme has been found not only in denitrifying archaea and soil bacteria, but also in pathogenic microorganisms that do not denitrify (de Vries et al. 2003). The qCuANor has been described in the Gram-positive bacterium *Bacillus azotoformans* (Suharti et al. 2004). This enzyme is bifunctional using both menahydroquinone (MKH2) and a specific *c*-type cytochrome c_{551} as electron donor. It was suggested that the MKH2-linked activity of qCuANor serves detoxification and the c_{551} pathway has a bioenergetics function.

The cNor is encoded by the *norCBQD* operon. The *norC* and *norB* genes encode subunit II and subunit I, respectively. The *norQ* and *norD* genes encode proteins essential for activation of cNor. Some denitrifiers have additional *norEF* genes, the products of which are involved in maturation and/or stability of Nor activity (Hartsock and Shapleigh 2010). As a unique case, the Nor of *Roseobacter denitrificans* is similar to cNor, but differs in that it contains copper (Matsuda et al. 2002).

Respiratory Nitrous Oxide Reductase

The final step in denitrification consists of the two-electron reduction of nitrous oxide to N_2, a reaction catalyzed by the nitrous oxide reductase (Nos) (reviewed in van Spanning et al. 2005, 2007, van Spanning 2011, Zumft and Kroneck 2007). The enzyme is located in the periplasmic space and has been purified from some denitrifying strains, including *P. denitrificans*, *P. pantotrophus*, and *P. stutzeri*. Nos is a homo-dimer of a 65 kDa copper-containing subunit. Each monomer is made up of the Cu_A and CuZ domains. Electron input into Cu_A is usually via *c*-type cytochromes or cupredoxins (Berks et al. 1993). The *nos* gene clusters often comprise the *nosRZDFYLX* genes. The *nosZ* gene encodes the monomers of Nos. The *nosDFYL* genes encode proteins that are apparently required for copper assemblage into Nos, although their specific role still remains unknown. The NosRX proteins have roles in transcription regulation, activation, and Cu assemblage of Nos (Zumft and Kronek 2007).

Alterations in the N Cycle

It is estimated that BNF provides between 90 and 130 Tg N per year (Tg = Teragram = 1 billion grams). In addition to nitrate from BNF, industrial manufacture of ammonium by the Haber-Bosch process contributes to nitrate formation with some additional 140 Tg per year, which results in a considerable increase in soil nitrate concentration. This excess nitrate cannot be removed by denitrification, resulting in the accumulation of nitrate in soil, water and sediments. This large increase in N load in the environment, in turn, leads to serious alterations in the cycling of N and will likely cause severe damage to environmental services at local, regional and global scales (Galloway et al. 2008).

Denitrification and Health

Contamination of groundwater by nitrates is increasingly frequent and intense (Sutton et al. 2011). Although the origin of this contamination is diverse, that caused by N-fertilizers, septic systems, and manure storage or application is noteworthy. The intensive use of N-fertilizers during agricultural practices determines, after leaching by rain and irrigation waters, the progressive NO_3^- enrichment of aquifers. Nitrate pollution of groundwater represents a threat for human and animal health and also a social and economic problem as a shortage or water resources may prevent water utilization for public consumption. High concentrations of nitrates in drinking water are toxic, especially for children under 6 mon. Nitrate itself is not toxic unless it turns into nitrite by bacteria inhabiting the digestive tract. As nitrite enters the blood it reacts with hemoglobin to form methemoglobin, which blocks the transport of oxygen and cause asphyxia, a condition known as blue baby syndrome or methemoglobinemia. The World Health Organization (WHO) has estimated that the upper limit for nitrates in drinking water should not exceed 45 mg NO_3^-/liter.

Denitrification and Environment

Bacteria capable of achieving complete denitrification, this is the conversion of nitrate/nitrite to N_2, are scarce. Most of them do not possess, or do not express, the complete set of enzymes required to carry out each one of the reduction steps that comprise denitrification. It is therefore a paradox that being the only known biological process to remove excess nitrates that pollute soil and water ecosystems, denitrification is also a mechanism whose gaseous intermediate N_2O has an enormous impact on air pollution.

Although the human perturbation of the N cycle due to the increased production of N-fertilizer and of oxidized reactive N compounds originating from fossil fuel combustion have led to an unprecedented accumulation of nitrogen oxides (mainly N_2O) in the biosphere (Erisman et al. 2009), there is evidence that suggests that most N_2O originates from microbial denitrification activity in soil and water contaminated with nitrates. More than half the fertilizer applied to the soil ends up in rivers, lakes and seas, contributing to eutrophication and massive growth of algae on offshore continental platforms (Diaz and Rosenberg 2008). Hence, excess nitrates affects not only terrestrial and marine ecosystems, but also contributes to the release to the atmosphere of greenhouse gases involved in climate change (Sutton et al. 2011).

Among non-CO_2 greenhouse gases, the contribution of N_2O to climate forcing ranks second only to methane, with a warming potential nearly 300 times greater than CO_2 (IPCC 2007). Moreover, atmospheric N_2O is the leading contributor to the destruction of the ozone layer which protects the Earth from ultraviolet light (Ravishankara et al. 2007). In addition, N_2O can persist for up to 150 yr while it is slowly broken down in the stratosphere (Lassey and Harvey 2007). N_2O in the atmosphere can be converted to nitric oxide which can act as a precursor for other chemical compounds which fall to the Earth surface as constituents of the so-called acid rain.

N_2O is not only produced during denitrification, but nitrification and nitrate ammonification also result in N_2O production. These processes rarely occur in isolation, with possible competition for substrates under near limiting conditions, and the possibility of transfer of N_2O or intermediary products from one process to another depends on the prevailing environmental conditions, microbial community structure and location within soil matrix. This means that N_2O produced during several processes form one pool that might be reduced to N_2 during denitrification (Baggs 2011, Baggs and Philippot 2011).

There is also evidence for the occurrence of chemodenitrification, a name coined to include all non-biological processes leading to production of NO, N_2O or N_2 under specific conditions. Chemodenitrification occurs when N_2O accumulates and reacts with organic compounds to produce NO and N_2O (Bremner 1997). This may be a significant source of N_2O in acid soil where the N_2O-HNO_2 equilibrium shifts in favor of nitrous acid, which either by itself decomposes to form NO, HNO_3 and H_2O or reacts with constituents of the soil organic matter to form N_2, N_2O and CH_3ONO (Venterea and Ralston 2000).

Codenitrification is the process by which one N atom from NO or N_2O combines with one atom from another source forming a hybrid product (Tanimoto et al. 1992, Su et al. 2004). This has been shown to occur in fungi (Tanimoto et al. 1992) and bacteria (Garber and Hollocher 1982), but the reasons why and under which conditions the process occurs are unknown. In *P. stutzeri* and *Fusarium* spp. both conventional denitrification and codenitrification occur simultaneously (Shoun et al. 1992, Tanimoto et al. 1992). Laughlin and Stevens (2002) estimated that 92 percent of the measured N_2 production in grassland soils was due to codenitrification. The significance and regulation of codenitrification remains to be determined.

Denitrification in Soils

Assessment of denitrification by soils is hard to study because of the difficulties to quantify its gaseous end products (N_2O and N_2) and high spatial and temporal variability (Groffman et al. 2006, 2009, Philippot et al. 2008, Hallin et al. 2009, Bru et al. 2011, Keil et al. 2011). Although molecular methods (Philippot and Hallin 2006) have contributed to the understanding of denitrification, gene expression, denitrifier community composition and enzyme activities could not be easily related with the simultaneous production and emission of denitrification products (Wallenstein et al. 2006, Cuhel et al. 2010). Methodological problems continue hampering our understanding of denitrification at site, from landscape to continental scale, as well as the controls and magnitude of net N_2O losses (Butterbach-Bahl and Dannenmann 2011).

Depending on the methodology used to calculate denitrification, total N losses from arable soils are globally estimated to be in the range of 22 to 87 Tg N/year (Hofstra and Bowman 2005). This represents a considerable loss of N for agriculture as denitrification rates in agricultural soils are supposed to be approximately one order of magnitude larger than in natural soils (Barton et al. 1999). Agricultural soils are also the main anthropogenic source for atmospheric N_2O, contributing approximately 5.3–6.8 Tg N_2O-N/year, which represents 40–60 percent of the total global atmospheric N_2O source strength (Butterbach-Bahl and Dannenmann 2011, Shyakila and Kroeze 2011). Out of the 6.8 Tg N_2O-N/year, about 4.2 Tg is due to direct emission of N added to the soil as fertilizer; indirectly, 2.1 Tg N_2O-N/year derives from dung and manure management, and 0.5 Tg N_2O-N/year is formed during biomass burning (Baggs and Philippot 2011).

N_2O emissions from natural grasslands, wetlands and forest soils are low compared to agricultural soils. There are, however, some uncertainties over this contribution due to paucity in data compared to agricultural soils, and to the high spatial and temporal variability associated with these emissions (Bagss and Philippot 2011). Fungal denitrification appears to be of importance in forest systems due to the ectomycorrhizal denitrifying activity (Prendergast-Miller et al. 2011). Differences in N_2O emission have been observed in different agricultural systems, more likely reflecting differences in crop specific water demand, vegetation type, crop management and climate (Skiba and Smith 2000). In soils amended with plant residues, N_2O emissions vary according to the type of residue, chemical composition (quality) of the residue and the amount of biomass incorporated into the soil (García-Ruiz and Bagss 2007, Bagss and Philippot 2011). The importance of Arctic peat lands as sources of N_2O is just emerging (Marushchak et al. 2011, Palmer et al. 2012).

Contribution of livestock agriculture has been estimated to account for up to 65 percent of N_2O emission coming from denitrification of N excreted as dung and urine, either directly returned to soil, or applied to land as manure or effluent, from animal housing (Steinfeld et al. 2006).

Control of Denitrification

The general requirements for biological denitrification are: a) the presence of bacteria possessing the metabolic capacity; b) suitable electron donors such as organic carbon compounds; 3) anaerobic conditions or restricted O_2 availability; and 4) the presence of a nitrogen oxide (NO_3^-, NO_2^-, NO, or N_2O) as terminal electron acceptor. The process of denitrification is therefore generally promoted under anaerobic conditions, high levels of soil NO_3^-, and a readily available source of carbon. Nevertheless, denitrification events in the field occur irregularly in time and space because of weather conditions, heterogeneity of soil conditions and management practices.

Denitrification is suggested to be limited under field conditions by nitrate availability (Mahmood et al. 2005). Thus, application of N-fertilizer promotes denitrification activity in agricultural soils and, consequently, in short-term increased N_2O emissions (van Cleemput et al. 1994). Fertilization has also been shown to affect the N_2O to N_2 ratio (Skiba and Smith 2000). Since denitrification is usually higher at neutral rather than acidic conditions (Šimek and Cooper 2002), changes in pH due to fertilization affect denitrification activity. The amount of N_2O emitted varies with the type of fertilizer applied and its physical state (liquid or solid), but emissions are generally greater after application of organic rather than mineral sources (Clayton et al. 1997). The presence, however, of heavy metals, such as those found in sewage sludge, can decrease denitrification. Fertilizer application rates and timing have also been shown to influence denitrification in arable soils.

Temperature and water content of a soil are also factors that affect both total N_2O emission and the N_2O/N_2 ratio. Under controlled laboratory conditions, this ratio is heightened with increasing temperature, more likely due to proliferation of microorganisms in the soil and increased denitrifying activity. At elevated temperature, however, such correlation may be not observed since decreased activity, loss of water content and even cell death may occur. Linked to soil water is oxygen availability. Soils produced most N_2O at 60–80 percent water-filled pore space capacity, with denitrification being the predominant N_2O-producing process (Bateman and Baggs 2005). At lower water-filled pore space, N_2O production was mainly due to ammonia oxidation (Bateman and Baggs 2005). It is also possible that aerobic soils support anaerobic denitrification because nitrate produced by nitrification can be further denitrified in anaerobic niches within the soil aggregates (Morley and Baggs 2010). Quantity and quality of C is thought to be important in regulation of the N_2O/N_2 ratio (Henry et al. 2008), and shows an interactive effect with O_2 availability (Morley and Baggs 2010). Several reports show that the enzyme nitrous oxide reductase is more sensitive to low pH that the other denitrification enzymes (Thomsen et al. 1994), which results in higher N_2O/N_2 ratios as pH declines (Šimek and Cooper 2002, Dannenmann et al. 2008, Cuhel et al. 2010). Moreover, differences in net N_2O emission and in the microbial source of this N_2O

between short- and long-term pH changed soils have been reported (Baggs et al. 2010).

Freeze-thaw cycles and wet-dry cycles are other factors causing variations in denitrification activity. Because the growth and activity of the root system induce significant modifications in the physicochemical and biological properties of the soil surrounding the roots, the crop species also influences denitrification. Pesticides and heavy metals also affect the type and amount of carbon available for denitrification. A comprehensive review dealing with the ecology of denitrifying prokariotes in agricultural soils has been published elsewhere (Philippot et al. 2007).

Nitrogen Oxides in Root Nodules

Diazotrophic microorganisms can fix N_2 either under free-living conditions or in related and symbiotic associations with plants, mainly legumes and actinorhizal plants. Bacteria of the order Rhizobiales, collectively referred to as rhizobia, are best characterized by their ability to establish a N_2-fixing symbiosis on legume roots and on the stems of some aquatic leguminous plants. Within the order Actinomicetales, bacteria of the genus *Frankia* also form nodules on roots of a diverse woody species of the genera *Alnus, Casuarina, Myrica, Comptonia, Dryas, Chamaebatia, Coriaria,* etc. Within the nodules, the vegetative rhizobial cells transform into bacteroids which synthesize nitrogenase, the enzyme responsible for N_2 reduction.

The significance of symbiotic denitrification has been best studied in rhizobia. Within the nodules formed during the plant-bacteria interaction, simultaneous N_2 fixation by nitrogenase and nitrate reduction by denitrification can take place. Most rhizobia do not perform complete denitrification as, so far, only *Bradyrhizobium japonicum* and *Azorhizobium caulinodans* have been shown to denitrify (Delgado et al. 2007). Out of 250 strains of *Bradyrhizobium* sp. isolated from nodules of soybeans cultivated in Argentina, only 41 were considered true denitrifiers (Fernández et al. 2008). N_2O emissions from nodules of soybeans were observed during the plant growing season, and nitrogen fertilization increased N_2O production (Ciampitti et al. 2008). Emission of N_2O by root nodules was also detected after inoculation of lentils and peas with *Rhizobium leguminosarum* (Zhong et al. 2009). Consequently, if one considers the vast areas of the world cultivated with leguminous plants, N_2O emission from root nodules into the atmosphere is of great environmental importance. Denitrification in nodules could also have a role in detoxification of the cytotoxic compounds nitrite and NO produced as intermediates during denitrification reactions or emerging from the host plant. In fact, nitrite and NO have been reported as inhibitors of nitrogenase activity (Trinchant 1980, 1982). Denitrifying activity could work as a mechanism to generate ATP for survival of rhizobia in the rhizosphere, and also to maintain nodule functioning under conditions of environmental stress such as flooding or drought.

In the soybean microsimbiont *B. japonicum* USDA110, denitrification depends on the *napEDABC*, *nirK*, *norCBQD*, and *nosRZDFYLX* genes encoding nitrate-, nitrite-, nitric oxide, and nitrous oxide-reductase, respectively (reviewed in Bedmar et al. 2005, Delgado et al. 2007, Sánchez et al. 2011). Expression of the *nirK*, *norC* and *nosZ* genes has been reported both in soybean root nodules and bacteroids (Mesa et al. 2004), but its physiological significance is still unclear. In plant cells, assimilatory nitrate reductase reduces nitrate to nitrite, but also catalyzes a 1-electron transfer from NAD(P)H to nitrite resulting in NO formation (reviewed in Meyer et al. 2005, Planchet and Kaiser 2006). NO may also be produced from arginine via an unidentified nitric oxide synthase (NOS)-like enzyme, as the NOS inhibitor *N*-methyl-L-arginine impaired NO detection in alfalfa nodules (Baudouin et al. 2006). An as yet uncharacterized process that uses polyamines as substrates has also been involved in NO production (reviewed in Besson-Bard et al. 2008).

Since nitrite and NO are produced as intermediates during reduction of nitrate to N_2, bacterial denitrification is another likely source of NO in nodules (Mesa et al. 2004, Sameshima-Saito et al. 2006). Production of NO in nodules of soybean was demonstrated after detection of nitrosyl-leghaemoglobin complexes (Meakin et al. 2007), and that free NO formation is significantly reduced in nodules formed by a *B. japonicum* mutant lacking respiratory nitrate reductase (Sánchez et al. 2010). In addition, formation of nitrosyl-leghemoglobin was caused by nitrite and NO generated by respiratory nitrate- and nitrite-reductases denitrification enzymes, respectively (Sánchez et al. 2010). In contrast, inoculation of *Medicago truncatula* with a *Sinorhizobium meliloti* mutant strain impaired in nitrite reductase activity lead to the formation of nodules with an unaffected NO level, which suggests that denitrification is not required for NO production in alfalfa nodules (Pauly et al. 2011). Basal levels of nitrosyl-leghaemoglobin and free NO have been detected in soybean nodules produced by a *B. japonicum* strain lacking nitrate reductase (Sánchez et al. 2010). This suggests the presence of other denitrification-independent sources of NO, more likely those arising from the action of plant enzymes such as nitrate reductase and/or nitric oxide synthase. In addition to NO synthethized during enzymatic pathways, NO can also be produced in plant cells from nitrite via non-enzymatic reactions (Bethke et al. 2004).

NO has been involved in the establishment of the symbiosis between *Lotus japonicus* and *M. loti* (Nagata et al. 2008) and in the auxin-signaling pathway (Pii et al. 2007). Also, a transcriptomic study on NO-responsive genes in *M. truncatula* has revealed that NO regulates nodule development and functioning (Ferrarini et al. 2008).

Inhibition of nitrogenase activity in legume nodules by nitrate is well established. This might be due to the toxic effects of nitrite or NO produced as intermediates during denitrification reactions or emerging from the host plant. In nodules of *L. japonicus* the inhibitory effect of nitrate was suppressed by scavenging NO, showing that nitrate is likely involved in inhibition of nitrogenase activity (Kato et al. 2009). Other authors have also observed that a decrease in NO production in nodules results in increased rates of N_2 fixation activity (Shimoda et al. 2009, Sánchez et al. 2010, Tominaga et al. 2010). Nitrogenase activity and expression of the *nifH* and *nifD* genes, which encode the Fe protein and the α-chain of the FeMo protein, were affected in nodules produced by a *B. japonicum* lacking nitrous oxide reductase, which accumulates NO produced through denitrification, but not in those produced by a mutant deficient in respiratory nitrite reductase, where NO production from denitrification is blocked (Sánchez et al. 2010).

By binding to leghemoglobin, which would impair its ability to compete with O_2 for binding sites, NO could also interfere with N_2 fixation. This would diminish O_2 supply to bacteroids and, consequently, N_2 fixation would decrease. However, only a small proportion of leghemoglobin is bound to O_2. Since affinity of leghemoglobin for NO is higher than that for O_2, then leghemoglobin could act as NO scavenger modulating its bioactivity (Herold and Puppo 2005). Leghemoglobin has been shown to have a major role in detoxification of the nitrite and NO produced by during denitrification by soybean nodules in response to hypoxia (Sánchez et al. 2010). It is possible that NO could react with any of the proteins involved in regulation of N_2 fixation. Among them, NifA is the direct transcriptional activator of the *nif* and *fix* genes responsible for N_2 fixation, whose downregulation by NO would result in a negative effect on nitrogenase synthesis and/or activity.

Acknowledgments

This study was supported by ERDF-cofinanced grants CVI-3177 from Consejeria de Economia, Innovacion y Ciencia (Junta de Andalucía, Spain) and AGL2010-18607, and AGL2011-23383 from Ministerio de Economía y Competitividad (Spain). Support from Junta de Andalucía to Group BIO-275 is also acknowledged.

References

Arnoux, P., M. Sabaty, J. Alric, B. Frangioni, B. Guigliarelli, J.M. Adriano and D. Pignol. 2003. Structural and redox plasticity in the heterodimeric periplasmic nitrate reductase. Nat. Struct. Biol. 10: 928–934.

Barton, L., C.D.A. McLay, L.A. Schipper and C.T. Smith. 1999. Annual denitrification rates in agricultural and forest soils: a review. Aust. J. Soil Res. 37: 1073–1093.

Baggs, E.M. 2011. Soil microbial sources of nitrous oxide: recent advances in knowledge, emerging challenges and future direction. Curr. Op. Environ. Sust. 3: 321–327.

Bagss, M.E. and L. Philippot. 2011. Nitrous oxide production in the terrestrial environment. *In*: J.W.B. Moir [ed.]. Nitrogen cycling in bacteria: molecular analysis. Caister Academic Press, Norfolk, UK. pp. 211–232.

Baggs, E.M., C.M. Smales and E.J. Bateman. 2010. Changing pH shifts the microbial source as well as the magnitude of N_2O emission from soil. Biol. Fert. Soils. 46: 793–805.

Bateman, E.J. and E.M. Baggs. 2005. Contribution of nitrification to N_2O emission from soils at different water-filled pore space. Biol. Fert. Soils. 41: 379–388.

Bedmar, E.J., E.F. Robles and M.J. Delgado. 2005. The complete denitrification pathway of the symbiotic, nitrogen-fixing bacterium *Bradyrhizobium japonicum*. Biochem. Soc. Trans. 33: 141–144.

Baudouin, E., L. Pleuchot, G. Engler, N. Pauly and A. Puppo. 2006. Nitric oxide is formed in *Medicago truncatula-Sinorhizobium meliloti* functional nodules. Mol. Plant-Microbe Interact. 19: 970–975.

Bertero, M.G., R.A. Rothery, M. Palak, C. Hou, D. Lim, F. Blasco, J.H. Weiner and N.C. Strynadka. 2003. Insights into the respiratory electron transfer pathway from the structure of nitrate reductase. A. Nat. Struct. Biol. 10: 681–687.

Berks, B.C., D. Baratta, J. Richardson and S.J. Ferguson. 1993. Purification and characterization of a nitrous oxide reductase from *Thiosphaera pantotropha*. Implication for the mechanism of aerobic nitrous oxide reduction. Eur. J. Biochem. 212: 467–476.

Besson-Bard, A., A. Pugin and D. Wendehenne. 2008. New insights into nitric oxide signaling in plants. Annu. Rev. Plant Biol. 59: 21–39.

Bethke, P.C., M.R. Badger and R.L. Jones. 2004. Apoplastic synthesis of nitric oxide by plant tissues. Plant Cell. 16: 332–341.

Blasco, F., C. Iobbi, J. Ratouchniak, V. Bonnefoy and M. Chippaux. 1990. Nitrate reductases of *Escherichia coli*: sequence of the second nitrate reductase and comparison with that encoded by the *narGHJI* operon. Mol. Gen. Genet. 222: 104–111.

Blasco, F., J. Pommier, V. Augier, M. Chippaux and G. Giordano. 1992. Involvement of the *narJ* or *narW* gene product in the formation of active nitrate reductase in *Escherichia coli*. Mol. Microbiol. 6: 221–230.

Bremner, J.M. 1997. Sources of nitrous oxide in soils. Nutr. Cycling Agroecosyst. 49: 7–16.

Brondijk, T.H., D. Fiegen, D.J. Richardson and J.A. Cole. 2002. Roles of NapF, NapG and NapH, subunits of the *Escherichia coli* periplasmic nitrate reductase in ubiquinol oxidation. Mol. Microbiol. 44: 245–255.

Brondijk, T.H., A. Nilavongse, N. Filenko, D.J. Richardson and J.A. Cole. 2004. NapGH components of the periplasmic nitrate reductase of *Escherichia coli* K-12: location, topology and physiological roles in quinol oxidation and redox balancing. Biochem. J. 379: 47–55.

Bru, D., A. Ramette, N.P.A. Saby, S. Dequiedt, L. Ranjard, C. Joliver D. Arrouays and L. Philippot. 2011. Determinants of the distribution of nitrogen cycling microbial communities at the landscape scale. Int. Soc. Mol. Ecol. J. 5: 532–542.

Butterbach-Bahl, K. and M. Dannenmann. 2011. Denitrification and associated soil N_2O emissions due to agricultural activities in a changing climate. Curr. Op. Environ. Sust. 3: 389–395.

Cartron, M.L., M.D. Roldán, S.J. Ferguson, B.C. Berks and D.J. Richardson. 2002. Identification of two domains and distal histidine ligands to the four haems in the bacterial *c*-type cytochrome NapC; the prototype connector between quinol/quinone and periplasmic oxido-reductases. Biochem. J. 368: 425–432.

Cebron, A. and J. Garnier. 2005. *Nitrobacter* and *Nitrospira* genera as representatives of nitrite-oxidizing bacteria: detection, quantification and growth along the lower Seine River (France). Water Res. 39: 4979–4992.

Ciampitti, I.A., E.A. Ciarlo and M.E. Conti. 2008. Nitrous oxide emissions from soil during soybean (*Glycine max*) crop phenological stages and stubbles decomposition period. Biol. Fert. Soils. 44: 581–588.

Clayton, H., I.P. McTaggart, J. Parker, L. Swan and K.A. Smith. 1997. Nitrous oxide emission from fertilised grassland: A 2-year study of the effects of N fertiliser form and environmental conditions. Biol. Fert. Soils. 25: 252–260.

Cuhel, J., M. Ŝimek, R.J. Laughlin, D. Bru, D. Chéneby, C.J. Watson and L. Philippot. 2010. Insights into the effect of soil pH on N_2O and N_2 emissions and denitrifier community size and activity. Appl. Environ. Microbiol. 76: 1870–1878.

Dannenmann, M., K. Butterbach-Bahl, R. Gasche, G. Willibald and H. Papen. 2008. Dinitrogen emissions and the N_2:N_2O emission ratio of a Rendzic Leptosol as influenced by pH and forest thinning. Soil Biol. Biochem. 40: 2317–2323.

Delgado, M.J., S. Casella and E.J. Bedmar. 2007. Denitrification in Rhizobia-Legume Symbiosis. *In*: H. Bothe, S.J. Ferguson and W.E. Newton [eds.]. Biology of the Nitrogen Cycle. Elsevier, The Netherlands. pp. 83–92.

de Vries, S., M.J. Strampraad, S. Lu, P. Moenne-Loccoz and I. Schroder. 2003. Purification and characterization of the MQH2:NO oxidoreductase from the hyperthermophilic archaeon *Pyrobaculum aerophilum*. J. Biol. Chem. 278: 35861–35868.

de Vries, R., S. Suharti and L.A.M. Pouvreau. 2007. Nitric oxide reductase: structural variations and catalytic mechanism. *In*: H. Bothe, S.J. Ferguson and W.E. Newton [eds.]. Biology of the Nitrogen Cycle. Elsevier, The Netherlands. pp. 57–66.

Dias, J.M., M.E. Than, A. Humm, R. Huber, G.P. Bourenkov, H.D. Bartunik, S. Bursakov, J. Calvete, J. Caldeira, C. Carneiro, J.J. Moura, I. Moura and M.J. Romao. 1999. Crystal structure of the first dissimilatory nitrate reductase at 1.9 A solved by MAD methods. Structure. 7: 65–79.

Diaz, R.J. and R. Rosenberg. 2008. Spreading dead zones and consequences for marine ecosystems. Science. 321: 926–929.

Erisman, J.W., M.A. Sutton, J. Galloway, Z. Klimont and W. Winiwarter. 2009. How a century of ammonia synthesis changed the world. Nature Geoscience. 1: 636–639.

Fernández., A., E.B. Perotti, M.A. Sagardoy and M.A. Gómez. 2008. Denitrification activity of *Bradyrhizobium* sp. isolated from argentine soybean cultivated soils. World J. Microbiol. Biotechnol. 24: 2577–2585.

Ferrarini, A., M. De Stefano, E. Baudouin, C. Pucciariello, A. Polverari, A. Puppo and M. Delledonne. 2008. Expression of *Medicago truncatula* genes responsive to nitric oxide in pathogenic and symbiotic conditions. Mol. Plant-Microbe Interact. 21: 781–790.

Galloway, J.N., A.R. Townsend, J.W. Erisman, Z.C. Bekinda, J.R. Freney, L.A. Martinelli, S.P. Seitzinger and M.A. Sutton. 2008. Transformation of the nitrogen cycle: recent trends, questions and potential solutions. Science. 320: 889–892.

Garber, E.A. and T.C. Hollocher. 1982. Positional isotopic equivalence of nitrogen in nitrous oxide produced by the denitrifying bacterium *Pseudomonas stutzeri*: indirect evidence for a nitroxyl pathway. J. Biol. Chem. 257: 4705–4708.

García-Ruíz, R. and E.M. Baggs. 2007. N_2O emission from soil following combined application of fertiliser-N and ground weed residues. Plant Soil. 299: 263–274.

Gonzalez, P.J., C. Correia, I. Moura, C.D. Brondino and J.J. Moura. 2006. Bacterial nitrate reductases: molecular and biological aspects of nitrate reduction. J. Inorg. Biochem. 100: 1015–1023.

Groffman, P.M., M.A. Altabet, J.K. Böhlke, K. Butterbach-Bahl, M.B. David, M.K. Firestone, A.E. Giblin, T.M. Kana, L.P. Nielsen and M.A. Voytek. 2006. Methods for measuring denitrification: diverse approaches to a difficult problem. Ecol. Applications. 16: 2091–2122.

Groffman, P.M., K. Butterbach-Bahl, R.W. Fulweiler, A.J. Gold, J.L. Morse, E.K. Stander, C. Tague, C. Tonitto and P. Vidon. 2009. Challenges to incorporating spatially and temporally explicit phenomena (hotspots and hot moments). Biogeochemistry. 93: 49–77.

Hallin, S., C.M. Jones, M. Schloter and L. Philippot. 2009. Relationship between N-cycling communities an ecosystem functioning in a 50 year-old fertilization experiment. Int. Soc. Microbial Ecol. J. 3: 597–605.

Hartsock, A. and J.P. Shapleigh. 2010. Identification, functional studies, and genomic comparisons of new members of the NnrR regulon in *Rhodobacter sphaeroides*. J. Bacteriol. 192: 903–911.

Henry, S., S. Texier, S. Hallet, D. Bru, C. Dambreville, D. Cheneby, F. Bizouard, J.C. Germon and L. Philippot. 2007. Disentangling the rizosphere effect on nitrate reducers and denitrifiers: Insight into the role of root exudates. Environ. Microbiol. 10: 3082–3092.

Herold, S. and A. Puppo. 2005. Oxyleghemoglobin scavenges nitrogen monoxide and peroxynitrite: a possible role in functioning nodules? J. Biol. Inorg. Chem. 10: 935–945.

Hosftra, N. and A.F. Bowman. 2005. Denitrification in agricultural soils: summarizing published data and estimating global annual rates. Nut. Cycl. Agroecosyst. 72: 267–278.

IPCC (Intergovernmental Panel on Climate Change). 2007. Fourth Assessment Report. Climate Change (AR4.) Cambridge University Press. Cambridge.

Jepson, B.J., S. Mohan, T.A. Clarke, A.J. Gates, J.A. Cole, C.S. Butler, J.N. Butt, A.M. Hemmings and D.J. Richardson. 2007. Spectropotentiometric and structural analysis of the periplasmic nitrate reductase from *Escherichia coli*. J. Biol. Chem. 282: 6425–6437.

Jetten, M.S.M. 2008. The microbial nitrogen cycle. Environ. Microbiol. 10: 2903–2909.

Jormakka, M., D.J. Richardson, B. Byrne and S. Iwata. 2004. Architecture of NarGH reveals a structural classification of MobisMGD enzymes. Structure. 12: 95–104.

Kato, K., K. Kanahama and Y. Kanayama. 2009. Involvement of nitric oxide in the inhibition of nitrogenase activity by nitrate in *Lotus* root nodules. J. Plant. Physiol. 167: 38–241.

Keil, D., A. Meyer, D. Berner, C. Poll, A. Schützenmeister, H.P. Piepho, A. Vlasenko, L. Philippot, M. Schloter, E. Kandeler and S. Marhan. 2011. Influence of land-use intensity on the spatial distribution of N-cycling microorganism in grassland soils. FEMS Microbiol. Ecol. 77: 95–106.

Laughlin, R.J. and R.J. Stevens. 2002. Evidence for fungal dominance of denitrification and codenitrification in a grassland soil. Soil Sci. Soc. Am. J. 66: 1540–548.

Lassey, D. and M. Harvey. 2007. Nitrous oxide: the serious side of laughing gas. Science. 15: 10–11.

Mahmood, T., R. Ali, K.A. Malik, Z. Aslam and S. Ali. 2005. Seasonal pattern of denitrification under an irrigated wheat-maize cropping system fertilized with urea and farmyard manure in different combinations. Biol. Fert. Soils. 42: 1–9.

Martínez-Espinosa, R.M., E.J. Dridge, M.J. Bonete, J.N Butt, C.S. Butler, F. Sargent and D.J. Richardson. 2007. Look on the positive side! The orientation, identification and bioenergetics of 'Archaeal' membrane-bound nitrate reductases. FEMS Microbiol. Lett. 276: 129–139.

Marushchak, M.E., A. Pitkämäki, H. Koponen, C. Biasi, M. Seppälä and P.J. Martikainen. 2011. Hot spots for nitrous oxide emissions found in different types of permafrost peatlands. Glob. Change Biol. 17: 2601–2614.

Mathieu, C., S. Moreau, P. Frendo, A. Puppo and M.J. Davies. 1998. Direct detection of radicals in intact soybean nodules: presence of nitric oxide-leghemoglobin complexes. Free. Radic. Biol. Med. 24: 1242–1249.

Matsuda, Y., K. Inamori, T. Osaki, A. Eguchi, A. Watanabe, S. Kawabata, K. Iba and H. Arata. 2002. Nitric oxide-reductase homologue that contains a copper atom and has cytochrome c-oxidase activity from an aerobic phototrophic bacterium *Roseobacter denitrificans*. J. Biochem. 131: 791–800.

Meakin, G.E., E. Bueno, B. Jepson, E.J. Bedmar, D.J. Richardson and M.J. Delgado. 2007. The contribution of bacteroidal nitrate and nitrite reduction to the formation of nitrosylleghaemoglobin complexes in soybean root nodules. Microbiology. 153: 411–419.

Mehta, M.P. and J.A. Baross. 2006. Nitrogen fixation at 92 °C by a hydrothermal vent archaeon. Science. 314: 1783–1786.

Mesa, S., J.D. Alché, E.J. Bedmar and M.J. Delgado. 2004. Expression of the *nir, nor* and *nos* denitrification genes from *Bradyrhizobium japonicum* in soybean root nodules. Physiol. Plant. 120: 205–211.

Meyer, C., U.S. Lea, F. Provan, W.M. Kaiser and C. Lillo. 2005. Is nitrate reductase a major player in the plant NO (nitric oxide) game? Photosyn. Res. 83: 181–189.

Morley, N. and E.M. Baggs. 2010. Carbon and oxygen controls on N_2O and N_2 production during nitrate reduction. Soil Biol. Biochem. 42: 1864–1871.

Nagata, M., E. Murakami, Y. Shimoda, F. Shimoda-Sasakura, K. Kucho, A. Suzuki, M. Abe, S. Higashi and T. Uchiumi. 2008. Expression of a class 1 hemoglobin gene and production of nitric oxide in response to symbiotic and pathogenic bacteria in *Lotus japonicus*. Mol. Plant-Microbe Interact. 21: 1175–1183.

Nilavongse, A., T.H. Brondijk, T.W. Overton, D.J. Richardson, E.R. Leach and J.A. Cole. 2006. The NapF protein of the *Escherichia coli* periplasmic nitrate reductase system: demonstration of a cytoplasmic location and interaction with the catalytic subunit, NapA. Microbiology. 152: 3227–32376.

Olmo-Mira, M.F., M. Gavira, D.J. Richardson, F. Castillo, C. Moreno-Vivian and M.D. Roldán. 2004. NapF is a cytoplasmic iron-sulfur protein required for Fe-S cluster assembly in the periplasmic nitrate reductase. J. Biol. Chem. 279: 49727–49735.

Palmer, K., C. Biasi and M.A. Horn. 2012. Contrasting denitrifier communities relate to contrasting N_2O emission patterns from acidic peat soil in arctic tundra. Int. Soc. Microb. Ecol. J. 6: 1058–77.

Pauly, N., C. Ferrari, E. Andrio, D. Marino, S. Piardi, R. Brouquisse, E. Baudouin and A. Puppo. 2011. MtNOA1/RIF1 modulates *Medicago truncatula-Sinorhizobium meliloti* nodule development without affecting its nitric oxide content. J. Exp. Bot. 62: 939–948.

Pearson, I.V., M.D. Page, R.J. van Spanning and S.J. Ferguson. 2003. A mutant of *Paracoccus denitrificans* with disrupted genes coding for cytochrome c_{550} and pseudoazurin establishes these two proteins as the *in vivo* electron donors to cytochrome cd_1 nitrite reductase. J. Bacteriol. 185: 6308–6315.

Philippot, L. and S. Hallin. 2006. Molecular analysis of soil denitrifying bacteria. *In*: J.E. Cooper and J.R. Rao [eds.]. Molecular techniques for soil, rizosphere and plant microorganisms. CABI Publishing. Wallingford, UK. pp. 146–165.

Philippot, L., S. Hallin and M. Schloter. 2007. Ecology of denitrifying prokaryotes in agricultural soils. Ad. Agron. 96: 249–305.

Philippot, L., J. Cuhel, N.P.A. Saby, D. Chèneby, A. Chronakova, D. Bru, D. Arrouays, F. Martin-Laurent and M. Ŝimek. 2008. Mapping field-scale spatial patterns of size and activity of the denitrifier community. Environ. Microbiol. 11: 1518–1526.

Pii, Y., M. Crimi, G. Cremonese, A. Spena and T. Pandolfini. 2007. Auxin and nitric oxide control indeterminate nodule formation. BMC Plant Biol. 8: 7–21.

Planchet, E. and W.M. Kaiser. 2006. Nitric oxide detection by DAF fluorescence and chemiluminescence: A comparison using abiotic and biotic NO sources. J. Exp. Bot. 57: 3043–3055.

Potter, L., H. Angrove, D.J. Richardson and J. Cole. 2001. Nitrate reduction in the periplasm of gram-negative bacteria. Adv. Microb. Physiol. 45: 51–112.

Prendergast-Miller, M.T., E.M. Baggs, and D.J. Johnson. 2011. Nitrous oxide production by the ectomycorrhizal fungi *Paxillus involutus* and *Tylospora fibrillose*. FEMS Microbiol. Lett. 316: 31–35.

Raghoebarsin, A.A., A. Pol, K.T. van de Pas-Schoonen, A.J. Smolders, K.F. Ettwig, W.I. Rijpstra et al. 2006. A microbial consortium couples anaerobic methane oxidation to denitrification. Nature. 440: 918–921.

Ravishankara, A.R., J.S. Daniel and R.W. Portmann. 2009. Nitrous oxide (N_2O): the dominant ozone-depleting substance emitted in the 21st century. Science. 326: 123–125.

Richardson, D.J. 2011. Redox complexes of the nitrogen cycle. *In*: J.W.B. Moir [ed.]. Nitrogen Cycling in Bacteria: Molecular analysis. Caister Academic Press. Norkfolk, UK. pp. 23–37.

Richardson, D.J. 2011. Redox complexes of the nitrogen cycle. *In*: J.W.B. Moir [ed.]. Nitrogen Cycling in Bacteria: Molecular analysis. Caister Academic Press. Norkfolk, UK. pp. 23–37.

Richardson, D.J., B.C. Berks, D.A. Russell, S. Spiro and C.J. Taylor. 2001. Functional, biochemical and genetic diversity of prokaryotic nitrate reductases. Cell Mol. Life Sci. 58: 165–178.

Richardson, D.J., R.J. van Spanning and S.J. Ferguson. 2007. The prokaryotic nitrate reductases. *In*: H. Bothe, S.J. Ferguson and W.E. Newton [eds.]. Biology of the Nitrogen Cycle. Elsevier. The Netherlands. pp. 21–35.

Rinaldo, S. and F. Cutruzzola. 2007. Nitrite reductases in denitrification. *In*: H. Bothe, S.J. Ferguson and W.E. Newton [eds.]. Biology of the Nitrogen Cycle. Elsevier. The Netherlands. pp. 37–56.

Rinaldo, S., A. Arcovito, G. Giardina, N. Castiglione, M. Brunori and F. Cutruzzola. 2008. New insights into the activity of *Pseudomonas aeruginosa* cd_1 nitrite reductase. Biochem. Soc. Trans. 36: 1155–1159.

Roldán, M.D., H.J. Sears, M.R. Cheesman, S.J. Ferguson, A.J. Thomson, B.C. Berks and D.J. Richardson. 1998. Spectroscopic characterization of a novel multiheme *c*-type cytochrome widely implicated in bacterial electron transport. J. Biol. Chem. 273: 28785–28790.

Sameshima-Saito, R., K. Chiba, J. Hirayama, M. Itakura, H. Mitsui, S. Eda and K. Minamisawa. 2006. Symbiotic *Bradyrhizobium japonicum* reduces N_2O surrounding the soybean root system via nitrous oxide reductase. Appl. Environ. Microbiol. 72: 2526–2532.

Sánchez, C., A.J. Gates, G.E. Meakin, T. Uchiumi, L. Girard, D.J. Richardson, E.J. Bedmar and M.J. Delgado. 2010. Production of nitric oxide and nitrosyl-leghaemoglobin complexes in soybean nodules in response to flooding. Mol. Plant-Microbe Interact. 23: 702–711.

Sánchez, C., E.J. Bedmar and M.J. Delgado. 2011. Denitrification in legume-associated endosymbiotic bacteria. *In*: J.W.M. Moir [ed.]. Nitrogen Cycling in Bacteria: Molecular Analysis. Caister Academic Press. Norfolk, UK. pp. 195–210.

Shaw, L.J., G.W. Nicol, Z. Smith, J. Fear, J.I. Proser and E.M. Baggs. 2006. *Nitrosospira* spp. can produce nitrous oxide via a nitrifier denitrification pathway. Environ. Microbiol. 8: 214–222.

Shimoda, Y., F. Shimoda-Sasakura, K. Kucho, N. Kanamori, M. Nagata, A. Suzuki, M. Abe, S. Higashi and T. Uchiumi. 2009. Overexpression of class 1 plant hemoglobin genes enhances symbiotic nitrogen fixation activity between *Mesorhizobium loti* and *Lotus japonicus*. Plant J. 57: 254–263.

Shoun, H., D. Kim, H. Uchiyama and J. Sugiyama. 1992. Denitrification by fungi. FEMS Microbiol. Lett. 73: 277–281.

Shyakila, A. and K. Kroeze. 2011. The global nitrous oxide budget revisited. Greenhouse Gas Measurement and Management. 1: 17–26.

Šimek, M. and J.E. Cooper. 2002. The influence of soil pH on denitrification: Progress toward the understanding of this interaction over the last 50 years. Eur. J. Soil Sci. 53: 345–354.

Simon, J., R.J. van Spanning and D.J. Richardson. 2008. The organisation of proton motive and non-proton motive redox loops in prokaryotic respiratory systems. Biochim. Biophys. Acta. 1777: 1480–1490.

Skiba, U. and K.A. Smith. 2000. The control of nitrous oxide emissions from agricultural and natural soils. Chemosphere-Global Change Sci. 2: 379–3986.

Spector, M.P., F. Garcia del Portillo, S.M. Bearson, A. Mahmud, M. Magut, B.B. Finlay, G. Dougan, J.W. Foster and M.J. Pallen. 1999. The *rpoS*-dependent starvation-stress response locus *stiA* encodes a nitrate reductase (*narZYWV*) required for carbon starvation-inducible thermotolerance and acid tolerance in *Salmonella typhimurium*. Microbiology. 145: 3035–3045.

Steinfeld, H., P.T.W. Greber, V. Castel, M. Rosales and C. de Haan. 2006. Livestock's Long Shadow-Environmental Issues and Options. Food and Agriculture Organization of the United Nations, Rome.

Su, F., N. Takaya and H. Shoun. 2004. Nitrous oxide forming codenitrification catalysed by cytochrome P_{450nor}. Biosci. Biotechnol. Biochem. 68: 473–475.

Suharti, S., H.A. Heering and S. de Vries. 2004. NO reductase from *Bacillus azotoformans* is a bifunctional enzyme accepting electrons from menaquinol and a specific endogenous membrane-bound cytochrome c_{551}. Biochemistry. 43: 13487–13495.

Sutton, M.A., C.M. Howard, J.W. Erisman, G. Billen, A. Bleeker, P. Grenfelt, H. van Grinsven and B. Grizeti. 2011. The European Nitrogen Assesment. Cambridge University Press, Cambridge, UK.

Takaya, N. 2002. Dissimilatory nitrate reduction metabolisms and their control in fungi. J. Biosci. Bioeng. 94: 506–510.

Tanimoto, T., K. Katano and D.H. Kim. 1992. Codenitrification by the denitrifying system of the fungus *Fusarium oxysporum*. FEMS Microbiol. Lett. 93: 177–180.

Thomsen, J.K., T. Geest and R.P. Cox. 1994. Mass spectrometric studies of the effect of pH on the accumulation of intermediates in denitrification by *Paracoccus denitrificans*. Appl. Environ. Microbiol. 60: 536–541.

Tominaga, A., M. Nagata, K. Futsuki, H. Abe, T. Uchiumi, M. Abe, K. Kucho, M. Hashiguchi, R. Akashi, A.M. Hirsch, S. Arima and A. Suzuki. 2010. Enhanced nodulation and nitrogen fixation in the abscisic acid low-sensitive mutant *enhanced nitrogen fixation1* of *Lotus japonicus*. Plant Physiol. 151: 1965–1976.

Treusch, A.H., S. Leininger, A. Kletzin, S.C. Schuster, H.P. Klenk and C. Schleper. 2005. Novel genes for nitrate reductase and Amo-related proteins indicate a role of uncultivated mesophilic crenarchaeota in nitrogen cycling, Environ. Microbiol. 7: 1985–1995.

Trinchant, J.C. and J. Rigaud. 1980. Nitrite inhibition of nitrogenase from soybean bacteroids. Appl. Environ. Microbiol. 124: 49–54.

Trinchant, J.C. and J. Rigaud. 1982. Nitrite and nitric oxide as inhibitors of nitrogenase from soybean bacteroids. Appl. Environ. Microbiol. 44: 1385–1388.

van Cleenput, O., A. Vermoese, C.J. De Groot and K. van Ryckhegem. 1994. Nitrous oxide emission out of grassland. Environ. Monitoring Assess. 31: 145–152.

van Spannig, R.J. 2011. Structure, function, regulation and evolution of the nitrite and nitrous oxide reductases: denitrification enzymes with a β-propeller fold. *In*: J.W.B. Moir [ed.]. Nitrogen Cycling in Bacteria. Caister Academic Press. Norkfolk, UK. pp. 135–161.

van Spanning, R.J., M.J. Delgado and D.J. Richardson. 2005. The nitrogen cycle: denitrification and its relationship to N_2 fixation. *In*: D. Werner and W.E. Newton [eds.]. Fixation in Agriculture, Forestry, Ecology and the Environment. Elsevier. The Netherlands. pp. 277–342.

van Spannig, R.J., D.J. Richardson and S.J. Ferguson. 2007. Introduction to the biochemistry and molecular biology of denitrification. *In*: H. Bothe, S.J. Ferguson and W.E. Newton [eds.]. Biology of the Nitrogen Cycle. Elsevier. The Netherlands. pp. 83–93.

Venterea, R.T. and D.E. Ralston. 2000. Mechanisms and kinetics of nitric and nitrous oxide production during nitrification in agricultural soil. Global Change Biol. 6: 303–316.

Wallenstein, M.D., D.D. Myrold, M. Firestone and M. Voytek. 2006. Environmental controls on denitrifying communities and denitrification rates: insights from molecular methods. Ecological Applications. 16: 2143–2152.

Zhong, Z., R.L. Lemke and L.M. Nelson. 2009. Nitrous oxide emissions associated with nitrogen fixation by grain legumes. Soil Biol. Biochem. 41: 2283–2291.

Zumft, W.G. 1997. Cell biology and molecular basis of denitrification. Microbiol. Mol. Biol. Rev. 61: 533–616.

Zumft, W.G. 2005. Nitric oxide reductases of prokaryotes with emphasis on the respiratory, heme-copper oxidase type. J. Inorg. Biochem. 99: 194–215.

Zumft, W.G. and P.M. Kroneck. 2007. Respiratory transformation of nitrous oxide (N_2O) to dinitrogen by bacteria and archaea. Adv. Microb. Physiol. 52: 107–227.

9

Bacterial Protein Secretion Systems: Implications in Beneficial Associations with Plants

I. Jiménez-Guerrero, M.T. Cubo, F. Pérez-Montaño, F.J. López-Baena, B. Guash-Vidal, F.J. Ollero, R. Bellogín and *M.R. Espuny**

Introduction

Beneficial bacteria that live in the plant rhizosphere must cope with different environmental and biotic stresses in order to establish a successful relationship with their host plants. These bacteria have evolved different protein secretion systems that enable them to survive in hostile environments and facilitate colonization of their eukaryotic hosts.

Collectively, Gram-negative bacteria produce eight protein secretion pathways, Sec, Tat and Types 1 to 6, to deliver chemical signal molecules from the bacterial cytoplasm to the outside of the cell or directly into the cytoplasm of the host cells, playing a central role in modulating all these interactions. In this case, secretion requires translocation across the outer as well as the inner membrane. These protein effectors and toxins modify the

Department of Microbiology, Faculty of Sciences, U. of Seville. Avda. Reina Mercedes, 6. 41012-Sevilla.
*Corresponding author: espuny@us.es

host physiology to promote colonization. Bacterial protein secretion systems can be broadly categorized as either one- or two-step. Gram-positive bacteria share some of the same secretion systems as Gram-negative bacteria and also display one system specific to that group, the Type 7 system.

All the bacterial secretion systems, in general terms, are described in this chapter. We particularly focus on the machinery structure, proteins that form part of the secretor channel and other proteins implicated in the regulation of the secretion. When possible the proteins or molecules that are secreted by each system and their implication on the bacteria-hosts interactions, especially in the beneficial interactions are described. A general schematic view of the systems is shown in the Fig. 9.1. Table 9.1 compares the main properties of the protein secretion systems.

1. A

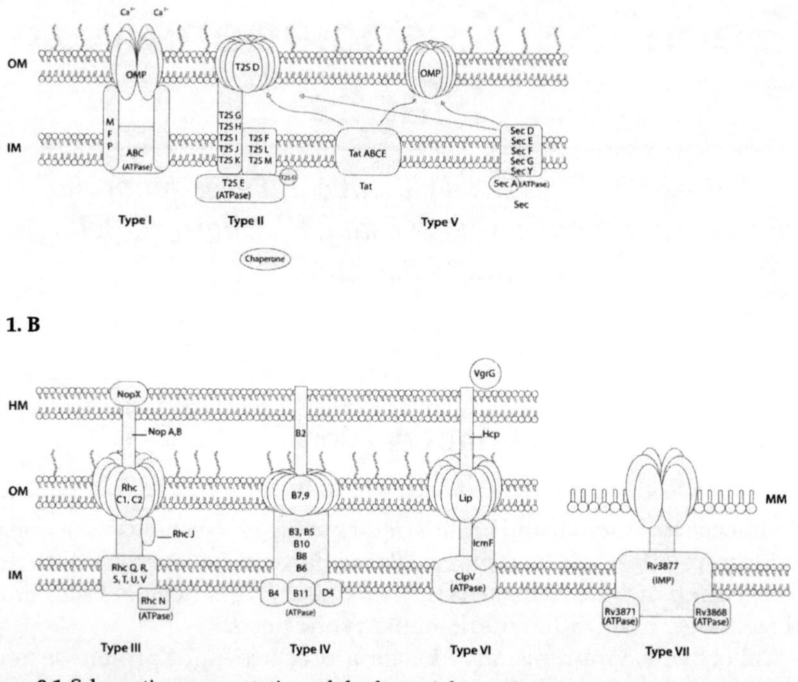

Figure 9.1 Schematic representation of the bacterial protein secretion systems. 1A) Types I, II and V. 1B) Types III, IV, VI and VII. IM: inner membrane; HM: host membrane; IMP: inner membrane protein; MM: mycobacterial membrane; OM: outer membrane; OMP: outer membrane protein. Type I: OMP; MFP: membrane fusion protein; ABC: ATP binding cassette transporter. Type II: T2S D: OMP; T2S F, L, M: IMP; T2S C: IMP (substrate recognition); T2S G, H, I, J, K: pseudopilins; T2S E: ATPase. Type V: Tat ABCE: IMP; SecDEFGY: IMP; SecA: ATPase. Type III: NopX: translocase NopA, B: pilus; RhcC1,C2: OMP; RhcJ:OMP associated protein; RhcQ, R, S, T, U, V: IMP; RhcN: ATPase; Type IV: B2: pilus; B3,B5: pilus associated; B7, B9: OMP; B10, B8, B6,: IMP, B4, B11, D4: ATPase; Type VI: VgrG: secreted substrate; Hcp: pilus protein; Lip: OMP; IcmF: IMP; ClpV: ATPase. Type VII: Rv3877: IMP; Rv3871, Rv3868: ATPases.

Table 9.1 Comparison of the protein secretory systems described in bacteria.

Characteristics	Sec	Tat	Type I	Type II	Type III	Type IV	Type V	Type VI	Type VII
Number of constituents	6	3	3	4-10	10-12	12	1	12-25	14
Spans the envelope (one step transport)	-	-	+	- (exported to periplasm via Sec/Tat)	+	+	- (exported to periplasm via Sec/Tat)	+	+
Energy	ATP (SecA)	PMF	ATP (ABC)	ATP (Sec) PMF(Tat)	ATP	ATP	ATP (Sec) PMF (Tat)	ATP	ATP
Leader sequence	Hydrophobic N-terminal	Basic N-terminal (twin-arginine motif: S-R-R-x-F-L-K)	Tandem nonapeptide glycine rich repeats at the C-terminal	Hydrophobic N-terminal (Sec) Basic N-terminal (Tat)	N-terminal	C-terminal	N-terminal (Sec) C-terminal domain (T5SS)	No signal	C-terminal?
Protein substrate conformation	Unfolded	Fully assembled enzyme complexes	Unfolded	Unfolded (Sec) Folded (MTB)	Partially folded subunits	Folded subunits	α-helical (secondary) and β-barrel (tertiary)	ND	Folded pairwise proteins
Secretory substrate	Proteins	Redox proteins	Pore-forming, glycanase, adhesin, bacteriocin, agglutinin, and nodulation factor proteins, β-glucans, polysaccharides	Hydrolytic enzymes, toxins proteins	Effector proteins	Single proteins, protein-protein and protein-DNA	Auto-aggregation, adherence, invasion, cytotoxicity, serum resistance, proteolysis proteins	Virulence, colonization proteins	Unknown, possibly pathogenic proteins
Distribution	Eubacteria, archaea and eukaryotes	Eubacteria, archaea and some eukaryotic organelles	Eubacteria,	Gram-negative bacteria	Eubacteria,	Eubacteria,	Eubacteria,	Eubacteria,	Gram-positive bacteria

ND: No data

General Secretion Pathways

Sec Pathway

The majority of secretory proteins cross the inner membrane via the Sec pathway. This pathway comprises a group of membrane and cytosolic proteins called translocon, that operate together to facilitate protein translocation (Papanikou et al. 2007). Proteins are targeted to their final location, the inner membrane or the periplasm, by their respective hydrophobic transmembrane segments (TMS) or signal sequences that will be finally removed. Thus, secretory proteins are usually referred to as pre-proteins. While Sec signal peptides are structurally similar, they do not show sequence conservation (von Heijne 1990). Sec pathway signal peptides in Gram-negative bacteria comprise three distinct regions: i) an N-terminal positively charged region (n-region), ii) a central hydrophobic α-helical region (h-region) and a iii) c-hydrophilic domain that contains the site of cleavage (such as Ala-X-Ala) by signal peptidase, an enzyme that cleaves off the signal peptide after translocation. The N-terminus of the signal peptide usually remains on the cytosolic side, and the hydrophobic region inserts the membrane in a loop-like configuration (Izard and Kendall 1994, Cristobal et al. 1999).

The translocase consists of the membrane SecYEG pre-protein-conducting channel and the ATPase motor SecA and a subset of partner proteins (SecDFYaj and YidC inserted in the membrane and the cytosolic chaperone SecB). A fundamental property of the Sec translocon is that it cannot accept tightly folded domains of a preprotein for translocation (Ito and Mori 2009).

The main model of the Sec pathway is established in *Escherichia coli*. Two major pathways are currently known, the SRP and the SecB pathways. At an early stage during translation, when the N-terminal signal sequence emerges from the ribosome, the ribonucleoprotein signal recognition particle (SRP) and the trigger factor (TF) compete for binding to the nascent chain (Beck et al. 2000, Ullers et al. 2006). If the N-terminal region of the nascent chain is strongly hydrophobic, SRP displaces the trigger factor and binds to the hydrophobic segment. SRP then guides the nascent translation complex to the membrane through its interaction with FtsY (SRP-membrane receptor), followed by its presentation to the SecYEG translocon. The bacterial SRP pathway is thought to be used primarily for integration of membrane proteins into the lipid phase of the membrane. If the nascent chain contains an export signal sequence, it is handed over from trigger factor to SecB and targeted to the Sec translocon for secretion to the periplasm.

The chaperone SecB keeps the pre-proteins, in an unfolded conformation, and directs them to the motor protein SecA (Ito and Mori 2009). SecA uses ATP as energy source and threads the unfolded polypeptide through the channel driving the stepwise translocation of pre-proteins across the SecYEG channel. The heterotrimeric protein complex SecYEG is the central player in protein translocation and functions as the membrane channel where cytosolic binding proteins provide the energy to translocate unfolded polypeptides through its aqueous interior.

The SecDF complex is involved in later stages of protein translocation and possibly pulls translocating proteins from the channel at the periplasmic side of the membrane (Lycklama et al. 2012). After signal peptidase cleaves the signal peptides, correct folding of the polypeptide chain is initiated at the *trans* side of the membrane or integrated into the lipid bilayer.

Most rhizobia, symbiotic nitrogen-fixing bacteria, secrete many proteins via the general export pathway and it is likely that some of them will be expressed during symbiosis although the symbiotic significance of most of them is not known. In *Rhizobium leguminosarum* bv. *trifolii* one protein of particular symbiotic significance is a cellulase (CelC2) that is predicted to be exported via the Sec pathway. This protein can erode the cellulose in the root-hair cell wall which can allow rhizobial penetration during the initial phase of infection in root-hair curls. Mutations in this gene lead to an inability of the bacteria to initiate infections in root hairs and those nodules that did form were uninfected (Robledo et al. 2008).

Tat Pathway

The bacterial twin-arginine translocation (Tat) pathway also translocates proteins across the cytoplasmic membrane (Bogsch et al. 1998). The differences between the Tat and the Sec pathway are that i) Tat substrates are secreted in a folded conformation (Hynds et al. 1998), ii) Tat signal peptides contain a highly conserved twin-arginine motif (Berks 1996), iii) the energy driving translocation is provided simply by the proton motive force (Cline et al. 1992), and iv) the Tat pathway is not a universally conserved secretion mechanism (Wu et al. 2000).

The major role of this pathway in bacteria is to translocate redox proteins that integrate their cofactors in the cytoplasm and therefore possess some degree of tertiary structure prior to secretion (Berks 1996, Rodrigue et al. 1999). Such proteins function predominantly in respiratory and photosynthetic electron transport chains and are vital for many types of bacterial energy metabolism (Berks et al. 2000).

Proteins exported by the Tat pathway are synthesized as precursors with N-terminal signal peptides with conserved twin-arginine amino acid sequence motifs (Berks et al. 1996). These precursors are transported across the membrane by a membrane "translocase" that catalyzes proteins transport. Three types of integral membrane proteins have been identified as components of the Tat translocase: TatA, TatB, and TatC. The TatC protein is the largest, most hydrophobic and highly sequence-conserved Tat component.

To date, two distinct high-molecular-mass complexes have been identified: a transport channel module containing multiple TatA protomers (Gohlke et al. 2005, Dabney-Smith et al. 2006) and a signal-recognition module comprising a complex of TatB and TatC (Cline and Mori 2001, Bolhuis et al. 2001).

Early in the transport process, twin-arginine signal peptides are recognized by the TatBC signal-recognition complex. The twin-arginine dipeptide is close to the TatC protein, while TatB cross-links more readily to the hydrophobic h-region of the signal peptide (Cline et al. 2001, Alami et al. 2003, Gérard and Cline 2006).

In relation to the implication of this system in symbiotic bacteria, it has been reported that mutation of *tatABC* genes in *R. leguminosarum* bv. *viciae* resulted in loss of the ability to fix nitrogen, probably because the mutants are defective for an essential exported iron-sulphur protein required for bacteroid respiration (Krehenbrink and Downie 2008).

Type 1 Secretion Systems

Type 1 secretion systems (T1SS) are widely distributed among Gram-negative bacteria. This secretion system allows the delivery of proteins with different sizes and functions from the bacterial cytoplasm to the extracellular medium in only one step, without a stable periplasmic intermediate. The structural feature for many secreted proteins is usually the presence of tandem nonapeptide glycine-rich repeats known as RTX motifs that form a β-roll structure stabilized by coordinated calcium ions (Jenewein et al. 2009). The secretion signal is usually located at the C-terminal end of the secreted protein and is not cleaved after translocation, which might be important for subsequent folding of the secreted protein. Moreover, the T1SS is also involved in secretion of non-proteinaceous substrates like cyclic β-glucans or polysaccharides (Feng et al. 2004).

The best characterized of these Type I secretion systems is that which exports the haemolysin A of *E. coli* and several model systems have also contributed to the understanding of this system (Delepelaire 2004). The Type 1 secretion machinery consists of three proteins localized in the cell envelope: i) an ATPase of the ATP-binding cassette (ABC) family of transporters able to hydrolyze ATP and use the energy obtained to transport various substrates across the bacterial cell membranes. It consists of one cytosolic nucleotide-binding domain (NBD) fused to a transmembrane domain (TMD), and it is believed to work as a dimer. The ABC protein recognizes the specific secretion signal of the substrate (Delepelaire 2004), ii) a membrane fusion protein (MFP) that connects the components of the secretion machinery located in the inner and the outer membrane (Jenewein et al. 2009), and iii) the outer membrane protein (OMP), which is the outer membrane component required for protein export, although it is absent in the T1SS of some bacteria (Delepelaire 2004).

The T1SS has been identified in some rhizobial species. These species may contain several distinct Type 1 exporters, and others have only one T1SS, although harbouring several potential substrates. Thus, *R. leguminosarum* bv. *viciae* 3841 has four T1SSs: PrsD-PrsE, BltD-BltE, ToaD-ToaE, TobD-TobE. Each pair corresponds to homologues of ABC transporters and MFP, respectively, and no outer membrane protein (OMP) in any case (Krehenbrink and Downie 2008). *Sinorhizobium meliloti* has two different T1SS (ExpD1-ExpD2, PrsD-PrsE and TolC as OMP in both cases) that are involved in symbiosis (York and Walker 1997, Moreira et al. 2000).

With respect to the protein substrates secreted through the T1SS in rhizobia, 13 proteins out of 14 have been identified as secreted by the *R. leguminosarum* bv. *viciae* PrsDE system (Krehenbrink and Downie 2008). This is a much broader specificity than typically observed with T1SSs in other bacteria, where the secretion system is usually limited to the secretion of only one or a narrow range of closely-related substances.

The first and most studied rhizobial protein identified as secreted by the T1SS was the *R. leguminosarum* bv. *viciae* NodO (Economou et al. 1990). This protein is secreted via the PrsDE system (Krehenbrink and Downie 2008). NodO is a calcium-binding protein that forms cation-selective channels in membranes (Economou et al. 1990, Sutton et al. 1994) and can extend the ability to nodulate different legumes (van Rhijn et al. 1996, Vlassak et al. 1998). It has been proposed that NodO may complement Nod factor function by promoting the movement of cations across the root-hair membrane (Miwa et al. 2006) because in the absence of host-specific Nod factors *nodO* is required to initiate infection in root hairs (Walker and Downie 2000).

Sixteen potential Type 1 substrates have been identified in *S. meliloti* and some of them are proteins involved in the synthesis or secretion of EPS (Delepelaire 2004). The protein ExsH is a succinoglycan depolymerase secreted by PrsDE proteins and is involved, in conjunction with the protein ExoK, in the production of low-molecular-weight succinoglycan (York and Walker 1997). The protein ExpE1 is secreted by ExpD1D2 proteins (Becker et al. 1997). The *expE1* gene is located immediately downstream of the *expD1-expD2* operon, as usually happens that genes encoding T1SS are situated adjacent to the genes encoding the secreted proteins (Delepelaire 2004).

Proteins secreted by the T1SS are also found in *R. leguminosarum* bv. *trifolii* TA1 (Downie 2010), *Mesorhizobium loti* MAFF303099, and *Bradyrhizobium japonicum* USDA110 (Delepelaire 2004).

In *Pseudomonas fluorescens* WCS365, a soil bacteria that is used as a biological control agent against plant pathogenic fungi, the LapA protein (large adhesion protein) is secreted by a T1SS composed of the LapE, LapB and LapC proteins (Delepelaire 2004, Hinsa et al. 2003). The four *lap* genes are located closely in the same region of the chromosome. LapA protein is a large adhesion protein involved in biofilm formation, which is thought to be important for its survival in a variety of environments. The *lap* genes are also present in other soil pseudomonads such as *P. putida* KT2440 and *P. fluorescens* PfO1, but are absent from pathogenic pseudomonads such as *P. aeruginosa* and *P. syringae* (Hinsa et al. 2003).

Type 2 Secretion Systems

A wide spectrum of Gram-negative bacteria that interact with plants and animals use the Type 2 secretion system (T2SS), also known as the secretion-dependent pathway, for secretion of extracellular proteins, such as hydrolytic enzymes and toxins.

T2SS is a two-step process that can be separated both genetically and biochemically. Proteins are synthesized with N-terminal signal peptides, which allows for Sec- or Tat-dependent secretion across the cytoplasmic membrane (Pugsley 1993, Voulhoux et al. 2001). This is followed by removal of the signal peptide and folding, and release of the mature proteins into the periplasm. In this compartment they may undergo chemical modifications before they are translocated across the outer membrane via the Type 2 secretion apparatus (Desvaux et al. 2004b, Filloux 2004, Rossier and Cianciotto 2005, Voulhoux et al. 2001). This apparatus is highly specific and is able to distinguish proteins to be secreted from resident periplasmic proteins and discriminate between its own secreted proteins and those introduced from other species (Filloux et al. 1990, Wong and Buckley 1993, Michel et al. 1995, Lindeberg et al. 1996).

The T2SS consists of 12 core components: the outer membrane secretin (T2S D), a cytoplasmic ATPase (T2S E), an inner (trans)membrane protein (T2S F), the major (T2S G) and minor (T2S H, I, J, K) pseudopilins facilitators of the ATPase attachment to the inner membrane that form together with T2S F an inner membrane platform (T2S L, M), the pre-pseudopilin peptidase/methyltransferase (T2S O), and a protein that might be involved in substrate recognition and/or secretin interactions (T2S C) (Filloux 2004, Peabody et al. 2003). The components of the Type 2 secreton were first discovered in *Klebsiella oxytoca* (d'Enfert et al. 1987), and their subsequent identification in *P. aeruginosa* demonstrated that they are conserved among most Gram-negative bacteria of the phylum Proteobacteria (Filloux et al. 1990).

Type 2 secretion in bacterial pathogenesis is important because: i) genes encoding the core components of T2SS are present in mammalian, fish and plant pathogens, ii) the generally degradative nature of the Type 2-dependent enzymes suggests that the system promotes the damage of host cells and tissue, iii) in some instances, individual Type 2 exoenzymes contribute to virulence [e.g., ADP-ribosylating toxins of *E. coli*, the cholera toxin of *Vibrio cholerae* and exotoxin A from *P. aeruginosa* (Sandkvist 2001, Tauschek et al. 2002)], and iv) mutations in T2S-specific genes attenuate virulence in animal and plant disease-models. The T2SS promotes pathogenesis by tissue destruction, cytotoxicity, adherence, spread and transmission (Cianciotto 2005). Furthermore, the T2SS can work along with other secretion systems to achieve full virulence (Cazalet et al. 2004).

T2S genes are common in pathogens, but they are also found in non-pathogenic, free-living environmental bacteria as well as symbionts of animals and plants (Cianciotto 2005). Recent studies have shown that Type 2 secretion can also significantly promote growth of bacteria in some environmental niches. These bacteria range from free-living inhabitants of water and soil (e.g., *Acinetobacter calcoaceticus*, *Shewanella oneidensis*) to commensals of plants (e.g., *P. fluorescens*) to symbionts of plants (e.g., *B. japonicum*) and animals (e.g., *V. fischeri*) to parasites of other bacteria (e.g., *Bdellovibrio bacteriovorus*). As a group, these bacteria are metabolically diverse and include, among others, anaerobic, aromatic-degrading denitrifiers such as *Azoarcus* spp. (Rabus et al. 2005), nitrogen-fixing bacteria (e.g., *B. japonicum*) (Kaneko et al. 2002) and methanotrophic bacteria (e.g., *Methylococcus capsulatus*) (Ward et al. 2004). Many nonpathogenic organisms secrete proteins, such as cellulases and other degradative enzymes, which help them adapt to their lifestyle.

Gluconacetobacter diazotrophicus is a nonpathogenic, nitrogen-fixing endophyte of sugarcane and other predominantly sucrose-rich crops. In addition, some strains of *Gluconacetobacter* produce considerable amounts of indol-acetic acid (IAA). Several varieties of sugar cane in Brazil are capable of obtaining over 60 percent of their nitrogen from biological nitrogen fixation with the association with this bacterium (Cavalcante and Döbereiner 1988, Jimenez-Salgado et al. 1997). This Gram-negative bacterium lacks a sucrose transport system (Alvarez and Martínez-Dretz 1995) and depends on the secretion of a constitutively expressed levansucrase (LsdA) secreted across the T2SS to obtain plant sucrose. This process plays a key role in the beneficial interaction between *G. diazotrophicus* and sugar cane or other host plants (Arrieta et al. 2004).

Type 3 Secretion Systems

The Type 3 secretion system (T3SS) has been intensively studied in the last few years due to its importance in the induction of diseases in plants and animals. The secretion machinery, which comprises about 20 highly conserved proteins, extends from the inner membrane to the outer membrane crossing the periplasmic space and extending to the exterior through a cylindrical structure called *pilus*, which forms a narrow conduit necessary for protein secretion and injection of effectors directly into host cells. The term effector groups a set of proteins related to the T3SS which exert their function mainly into the host cell. Thus, this protein complex forms a channel which is used by effectors to move from the bacterial cytoplasm to the cytoplasm of the host cell (Pallen et al. 2003).

The T3SS is present in animal pathogens such as *Yersinia* spp., *Salmonella* spp., *E. coli* or *P. aeruginosa*, and in phytopathogens such as *P. syringae* or *Xanthomonas* spp. This secretion system is also involved in the activation of the defence responses that block the infection usually accompanied by an induction of a hypersensitive response (HR), a programmed cell death in the site of infection that restricts colonization of the pathogen (Galan and Collmer 1999). Genes that code for the T3SS in bacteria are called *hrp* (hypersensitive response and pathogenicity) or *hrc* (hypersensitive response and conserved) because mutants affected in the T3SS can not trigger a hypersensitive response, a programmed cell death or pathogenicity. While proteins that form the secretion machinery are well conserved among bacteria, each microorganism secretes a specific cocktail of effectors (Lindgren 1997, Nümberger et al. 2004).

Effectors secreted through the T3SS share some structural characteristics: i) the N-terminal region harbours the information for secretion, ii) the central region constitutes a domain for binding to chaperones in those effectors that need these proteins for secretion and, iii) the rest of the protein defines the activity of the effector (Guttman et al. 2002). Some of the best studied effectors are called Avr (Avirulence function) because, if functional, the pathogen will be avirulent in those plants that have a specific resistance gene (R) (Keen 1990, Staskawicz et al. 1984). There are many effectors from plant or animal pathogens that share functions or targets in the host so it is probable that they have developed similar functions to avoid defence mechanisms of the host. These effectors modify signal transduction pathways, specifically alter proteins from the host or their patterns of phosphorylation, suppress defence responses or facilitate translocation of other effectors (Galan 2007).

The complete sequence of the genes involved in the biosynthesis of the rhizobial T3SS, grouped in the *tts* region, was obtained after sequencing the symbiotic plasmid of *Ensifer fredii* NGR234 (Viprey et al. 1998). The T3SS was later found in other rhizobial strains such as *E. fredii* strains HH103 and USDA257, *M. loti* MAFF303099, *B. japonicum* USDA110 and *B. elkanii* USDA61 (Deakin and Broughton 2009).

To date, many rhizobial proteins, called Nops (Nodulation outer proteins), have been identified as secreted through the T3SS: NopA, NopB, NopC, NopD, NopE1/E2, NopF, NopH, NopJ, NopL, NopM, NopP, NopT, NopX, Mlr6361, GunA2, and Pgl (Deakin and Broughton 2009).

NopA shares characteristics with proteins that form the Hrp pilus from phytopathogens. *E. fredii* NGR234 forms structures, called T3SS-pili, which would help bacteria to aggregate during attachment to plant cells. NopA, together with NopB and NopX, would be the major components of these appendages (Deakin et al. 2005, Saad et al. 2005, 2007).

On the other hand, NopC, NopD, NopE1/E2, NopJ, NopL, NopM, NopP, NopT, and Mlr6361 seem to be real effectors. While NopD, NopJ, NopM, NopT, and Mlr6361 are similar to effectors found in different animal and plant pathogens, NopL, NopP, and NopE1/E2 are specific for rhizobia. Other secreted proteins, such as GunA2 and Pgl, could locally degrade pectin from the plant cell wall but their poligalacturonidase activity has not been proved yet (Deakin and Broughton 2009).

NopL is a substrate for plant quinases. It has been proposed that NopL, once delivered to the interior of the host cell, would modulate signal transduction pathways regulated by MAP kinases that culminate in the expression of PR proteins. NopP is also phosphorylated by plant proteins with kinase activity and it is possible that both proteins could interfere with host defence pathways mediated by MAP kinases (Bartsev et al. 2003, 2004, Skorpil et al. 2005, Zhang et al. 2011).

Little is known about the exact role of NopM in symbiosis. This protein belongs to the IpaH-SspH-YopM family of effectors found in animal pathogens that travel to the nucleus of the host cell (Benabdillah et al. 2004). YopM from *Yersinia* spp. modulates the activity of two eukaryotic kinases (McDonald et al. 2003) and both IpaH from *Shigella flexneri* and SlrP from *Salmonella* spp. are E3 ubiquitin ligases (Rohde et al. 2007, Bernal-Bayard and Ramos-Morales 2009).

NopD is homologous to XopD from *X. campestris*, to the hypothetic protein Blr1693 from *B. japonicum* and to Msi059 from *M. loti* R7A secreted through the type IV secretion system. The C-terminal region of these proteins show homology to the catalytic domain of members of the family of C48 cysteine proteases involved in the de-ubiquitinisation of eukaryotic proteins. XopD posseses peptidase/isopeptidase activity *in planta* and it travels to the interior of the plant nucleus. The targets for this effector are SUMO (Small Ubiquitin-like Modifiers) proteins involved in the regulation of the activity of diverse transcriptional factors that coordinate the expression of genes necessary for plant development and for plant responses to the environment, such as abiotic stress, adaptation to changes in the environment, plant responses mediated by ABA (abscisic acid) or defence responses against pathogens (Hotson et al. 2003, Lee et al. 2007).

NopJ and NopT are homologous to Avr proteins that are determinants of virulence in phytopathogens. NopJ belongs to the family of YopJ effectors found in most of the plant pathogens that possess a T3SS. These proteins have a C55 domain with cysteine protease activity. This family includes many Avr proteins from phytopathogens able to induce the hypersensitive response (HR). The acetyl-transferase activity of these effectors blocks phosphorylation of MAP kinases by acetylation of the sites of phosphorylation. The rhizobial homologue NopJ seems to be involved in the differentiation of the symbiosome. NopT is homologous to AvrPphB of *P. syringae* and to YopT of *Yersinia* spp. This family of proteins has C58 domains with cysteine protease activity (Jenner et al. 1991). The *E. fredii* NGR234 NopT is responsible for the negative effect of Type 3 secretion in the symbiosis with *Crotalaria juncea* (Dai et al. 2008).

Mlr6361 of *M. loti* MAFF303099 possesses a C-terminal domain similar to that found in enzymes with shikimato-kinase activity. This domain is present in regions of proteins related with the hypersensitive response and pathogenicity. In addition, Mlr6316 is the major determinant of the nodulation restriction observed for *Lotus halophilus* (Okazaki et al. 2010).

Finally, NopE1 is an effector secreted by *B. japonicum* USDA110 that is expressed in nodules. It exhibits self-cleavage in a duplicated domain of unknown function in the presence of Ca^{2+} (Wenzel et al. 2009).

The role of Nops in symbiosis depends on the host plant. Thus, in *E. fredii* NGR234 the absence of Nops can be highly beneficial, have no effect or be detrimental for the symbiotic interaction. It seems that the negative effects are due to the action or the recognition of only one effector. However, when the presence of the T3SS is beneficial for nodulation, the final symbiotic phenotype would be the result of the effects of different Nops (Deakin and Broughton 2009).

Leucaena leucocephala is a plant in which the presence or absence of the Nops secreted by *E. fredii* NGR234 has no effect on nodulation. However, the inoculation with a mutant of *M. loti* MAFF303099 that does not secrete Nops induces the formation of some nitrogen-fixing nodules in this plant, in which the wild-type strain only forms pseudonodules. In addition, the presence of the T3SS in *M. loti* MAFF303099 is beneficial for nodulation in *L. corniculatus* subsp. *frondosus* and *L. filicaulis* and detrimental in *L. halophilus*, *L. peregrinus* var. *carmeli* and *L. subbiflorus* (Okazaki et al. 2010).

In *E. fredii* USDA257, mutants in any of the genes involved in the synthesis of the T3SS machinery gain the capacity to induce nitrogen-fixing nodules in American soybean cultivars in which the parental strain can not nodulate (Meindhart et al. 1993). In contrast, the inactivation of the same genes in *E. fredii* HH103 is detrimental for nodulation with American and Asiatic soybeans (López-Baena et al. 2008). However, the absence of Nops in both strains is beneficial for nodulation with the tropical legume *Erythrina variegata*.

In spite the absence of the whole set of Nops is detrimental for nodulation with soybeans and beneficial for nodulation with *E. variegata*, the inactivation of the *nopP* gene from *E. fredii* HH103, which codes for an effector protein, exerts the opposite effect on nodulation: it is highly beneficial for nodulation with soybeans and detrimental for nodulation with *E. variegata* (López-Baena et al. 2009). In the case of the *B. japonicum* USDA110, the complete absence of Nops only causes a delay in nodulation in soybean and *Macroptilium atropurpureum* (Krause et al. 2002). However, the inactivation of the *nopE1/nopE2* genes causes a reduction in the number of nodules formed in both plants but an increase in *Vigna radiata*. (Wenzel et al. 2009).

Plant defence responses associated with the role of the rhizobial T3SS in symbiosis are currently being studied. Thus, the absence of Nops secreted by *E. fredii* HH103 increases the expression of PR1 in soybean roots and leaves. This increase in the expression of proteins related to defence responses correlates with a decrease in nodule number. Therefore, Nops seem to suppress plant defence responses triggered by the inoculation with rhizobia in the early stages of symbiosis (López-Baena et al. 2009). Recently, Yang et al. (2010) described the positional cloning of two soybean genes, *Rj2* and *Rfg1*, involved in the restriction of nodulation with specific strains of *B. japonicum* and *E. fredii*, respectively. Both genes encode a member of the Toll-interleukin receptor/nucleotide-binding site/leucine-rich repeat (TIR-NBS-LRR) class of plant resistance (R) proteins. The presence of host R genes in the control of genotype-specific infection and nodulation shows that there are similarities between mechanisms of infection used by plant pathogenic and symbiotic bacteria and suggests the existence of their cognate avirulence genes derived from rhizobia. Thus, the establishment of specific symbiotic interactions between rhizobial strains that secrete Nops and some of their host plants requires the evasion of plant immune responses triggered by these effectors.

In *M. loti* MAFF303099, inoculation of *L. japonicus* with a mutant strain that does not secrete Nops causes a reduction in the expression of the plant gene *NIN* (nodule inception) 12 d after inoculation. The expression of this gene is necessary for the development of the infection threads and the beginning of the cellular divisions that initiate the formation of nodule primordia (Sanchez et al. 2009).

The T3SS is also present in other rhizobacteria that promote plant growth such as *P. fluorescens* SBW25 and KD (Preston et al. 2001, Rezzonico et al. 2005). The inactivation of the T3SS in the strain KD reduces the biocontrol capacity of this strain against the oomycete *Pythium ultimum*, which induces different diseases in cucumber plants. In the case of the bacteria-fungi interactions, the T3SS would secrete different effectors that only affect the phytopathogen and not the plant, contributing to its protection. Some effectors can induce cell death or modify fungal development when expressed in the model fungus *Neurospora crassa* and in yeast (Salomon et al. 2011).

Type 4 Secretion Systems

Type 4 secretion systems (T4SS) translocate DNA and protein substrates across the cell envelope by a mechanism that requires direct contact with the target cell. Three types of T4SS have been described (Cascales and Christie 2003): i) conjugation systems, machines that translocate DNA substrates to recipient cells by a contact-dependent process, ii) effector translocator systems, which deliver proteins or other effector molecules to eukaryotic target cells, and iii) systems that translocate DNA to or from the extracellular milieu.

The T4SS delivers nucleic acids and proteins into plant and animal cells, yeast and even other bacteria. The T4SS machinery spans both inner and outer membranes of Gram-negative bacteria or the membrane and the cell envelope of Gram-positive bacteria (Christie and Cascales 2005). Many organisms have T4SS, including the pathogens *Agrobacterium tumefaciens* C58, *Helicobacter pylori, P. aeruginosa, Bordetella pertussis, E. coli, Legionella pneumophila* and the rhizobacterium *M. loti* R7A (Christie and Cascales 2005, Hubber et al. 2007). While these systems may share functional similarities, not all systems contain the same sets of genes. The only common protein is VirB10 (TrbI) among all characterized Type 4 secretion systems.

The best-characterized T4SS is the VirB/D4 transfer system of the plant pathogen *A. tumefaciens* that causes crown gall disease in a wide range of plants. In general, T4SS include an extracellular pilus that is composed of a major (VirB2) and a minor (VirB5) subunit. Three ATPases, VirB4, VirB11 and VirD4, power substrate secretion and possibly assist in the assembly of the system. The inner membrane channel is composed of the polytopic membrane protein VirB6 and the bitopic membrane proteins VirB8 and VirB10. At the outer membrane, the composition of the pore is unknown; VirB9 in complex with the short lipoprotein VirB7 could be part of this structure (Fronzes et al. 2009). *Agrobacterium* translocates the VirD2/T-DNA complex (Zhu et al. 2000) into plant cells and at least three effector proteins, VirE2, VirE3, and VirF (Vergunst et al. 2000, 2003, Schrammeijer et al. 2003) that play a role in the infection process.

The mating-pair formation (Mpf) proteins belongs to T4SS and functions as secretion machinery for intracellular DNA transfer during bacterial conjugation. It comprises a minimal set of 10 conserved proteins and forms a membrane-spanning protein complex and a surface-exposed sex pilus, which both serve to establish intimate physical contacts with a recipient bacterium. The Mpf complex additionally requires the coupling protein which interacts with the DNA substrate and couples its secretion (Christie 2004, Gomis-Ruth et al. 2004, Lawley et al. 2003, Schroder and Lanka 2005). In Gram-negative bacteria, the Mpf proteins form the secretion channel as well as a pilus or other surface filament to promote attachment to target cells (Lawley et al. 2003, Christie and Cascales 2005). In Gram-positive bacteria, surface adhesins rather than conjugative pili mediate attachment (Grohmann et al. 2003).

Another large subfamily of T4SS, described so far only for Gram-negative bacteria, the effector translocators, has gained considerable attention because of its prominent roles in the infection processes of many bacterial pathogens. These systems deliver effector proteins or other macromolecules directly to the cytosols of eukaryotic target cells to aid bacterial colonization and survival within host cells or tissues (Backert and Meyer 2006, Cascales and Christie 2003, Fischer et al. 2002, Nagai et al. 2005). Effector translocator systems usually deliver their substrates to eukaryotic target cells through direct cell-to-cell contact (Burns 2003).

The T4SS of *M. loti* R7A is uniquely characterized in plant beneficial bacteria, although this secretion system has been detected in several rhizobia like other *M. loti* strains, *E. fredii* NGR234, *S. meliloti* strains AK83 and BL225C and *R. leguminosarum* bv. *viciae* strain 3841 (Hubber et al. 2004, Schmeisser et al. 2009, Galardini et al. 2011, Krehenbrink and Downie 2008). *M. loti* is the model microsymbiont of *Lotus* species. A few strains of *M. loti* are also able to form effective nodules on a wider range of hosts including *L. leucocephala*. However, most strains including the strains R7A and MAFF303099 form ineffective nodules that lack infected plant cells on this host (Pankhurst et al. 1987). The symbiosis island of *M. loti* R7A encodes a T4SS that is highly similar to the VirB/D4 T4SS of *A. tumefaciens* (Sullivan et al. 2002). In contrast, in *M. loti* strain MAFF303099 the T4SS is absent from the 610 kb symbiosis island, which contains a T3SS similar to that found in *E. fredii* NGR234 (Kaneko et al. 2000). Strain R7A T4SS mutants form large nitrogen-fixing nodules on *L. leucocephala* in contrast to the wild-type strain that forms only uninfected tumour-like structures. Strain MAFF303099 T3SS mutants also nodulate *L. leucocephala* and exhibit a similar phenotype to the strain R7A T4SS mutants on *L. leucocephala*, suggesting that Type 3 and Type 4 secretion systems play interchangeable roles in different strains of *M. loti* (Hubber et al. 2004). In *L. leucocephala*, the effector proteins may induce plant defence responses resulting in abortion of infections; whereas, in *L. corniculatus*, they may facilitate infection, perhaps by suppressing plant defence reactions. The host-dependent symbiotic phenotypes of the mutants may reflect differences in the reactions of the hosts to secreted effector proteins (Hubber et al. 2004).

Type 5 Secretion Systems

Type 5 secretion systems (T5SS) are the simplest protein secretion mechanisms with regard to the number of proteins required for effector delivery. Proteins secreted through T5SS are called autotransporters because they drive their own secretion. These proteins have three domains: the signal sequence, passenger domain and the translocation unit. The signal sequence (also called the signal peptide or leader sequence) is present at the N-terminal end of the protein and allows targeting of the protein to the inner membrane for its further export into the periplasm. The next domain is the passenger domain (also called the β-domain, N-passenger domain, or N-domain), which confers the diverse effector functions of the various autotransporters. The last main domain, located at the C-terminal end of the protein, is the translocation unit (also called the β-domain, helper domain, C-domain, transporter domain, or autotransporter domain), consisting of a short linker region with an α-helical secondary structure and a β-core that adopts a β-barrel tertiary structure when embedded in the outer membrane, facilitating translocation of the passenger domain through the outer membrane (Henderson et al. 2004).

A very large number of proteins involved in aggregation, biofilm formation or host cell invasion (Wells et al. 2007) are secreted via the T5SS and most of them contribute to the virulence of animal or human pathogens (Henderson et al. 2004, Jacob-Dubuisson et al. 2004, Bernstein 2007, Dautin and Bernstein 2007).

This family of secreted proteins includes those secreted via the autotransporter system (Type 5a), the two partner secretion pathway (Type 5b), and the Type 5c system (Desvaux et al. 2004a). The Type 5a autotransporter pathway is a widely distributed, two step pathway used in the secretion of exoenzymes. In this secretion pathway, a single polypeptide with an effector-like passenger domain and a C-terminal transporter domain are synthesized. Following delivery into the periplasm by the Sec translocon, substrates mediate their own transport outside the cell by forming pores in the outer membrane, through which they autotransport their passenger domains (Henderson et al. 2004). The secretion mechanism used by Type 5b secretion systems also called two-partner secretion (TPS) is similar to that used by the autotransporter secretion pathway. However, in contrast to the autotransporter pathway, where the protein is produced as a single polypeptide, the passenger domain or exoprotein and the pore-forming β-domain (also called the transporter domain) are translated as two separate proteins (Jacob-Dubuisson et al. 2001).

Recently, the members of the Oca (Oligomeric Coiled-coil Adhesins) family have been described as a subfamily of surface-attached oligomeric autotransporters, Type 5c, with the *Y. pestis* YadA, an adhesin protein, as the prototypical example (Roggenkamp et al. 2003, Nummelin et al. 2004). The proteins secreted via the Type 5c are trimeric proteins in which a single β-barrel is formed by contributions from all three polypeptides (Jacob-Dubuisson et al. 2004).

Although T5SSs have some functions in the interaction between bacteria and their hosts, only a few studies investigating the role of these secretion systems in the symbiosis between rhizobia and legumes have been undertaken.

The sequence analysis of the *Azorhizobium caulinodans* ORS571 a putative high-molecular-weight outer membrane protein, called azorhizobial outer membrane autotransporter A (AoaA), revealed that this protein was an autotransporter protein belonging to the T5SS. *A. caulinodans* forms nitrogen-fixing nodules on the stems and roots of *Sesbania rostrata* (Suzuki et al. 2007). An *aoaA* mutant formed nitrogen-fixing nodules in this plant with similar sizes as those formed by the wild-type strain, but with lower nitrogen-fixing activity. This mutant was defective in exopolysaccharide (EPS) production, suggesting that AoaA could have an important role in sustaining the symbiosis with this plant by probably suppressing defence responses by a mechanism in which the exopolysaccharides (EPS) might be involved (Suzuki et al. 2008).

Three autotransporters have been identified in *R. leguminosarum* bv. *viciae* 3841 but mutation of all three autotransporter genes had no observed effect on the symbiotic phenotype (Krehenbrink and Downie 2008).

Type 6 Secretion Systems

Type 6 secretion systems (T6SS) use a very large cell surface protein that forms a filament that likely interacts with target bacterial cells. T6SSs appears to constitute a phage-tail-spike-like injectisome that has the potential to introduce effector proteins directly into the cytoplasm of host cells, analogous to the T3SS and T4SS machineries (Bingle et al. 2008, Cascales 2008, Filloux et al. 2008, Shrivastava and Mande 2008), and share protein sequence and structural features with phages, indicating an evolutionary relationship (Leiman et al. 2009).

More than a quarter of sequenced bacterial genomes contain genes for T6SS components, mostly within the proteobacteria, but also within the planctomycetes and acidobacteria. The T6SS is required for virulence in human and animal pathogens such as *V. cholerae*, *Edwardsiella tarda*, *P. aeruginosa*, *Francisella tularensis* and *Burkholderia mallei*, and also in plant pathogens such as *A. tumefaciens*, *Pectobacterium atrosepticum* and *X. oryzae* (Bingle et al. 2008, Cascales 2008, Filloux et al. 2008, Liu et al. 2008, Shrivastava and Mande 2008, Wu et al. 2008). T6SS loci are highly conserved in bacterial genomes, including those of non-pathogenic species. This suggests that bacteria may use the T6SS in situations other than pathogenic interactions with a host. Furthermore it is required for efficient root colonization by *M. loti* and *R. leguminosarum*. Intriguingly, genes encoding the T6SS are also found in some non-symbionts such as *Myxococcus xanthus*,

Dechloromonas aromatica and *Rhodopirellula baltica,* where it may contribute to environmental adaptation such as biofilm formation.

Based on the available experimental evidence, as well as sequence similarities with some components of the T4SS and of the tail-spike complex of T4 phage, a model of the T6SS injectisome was proposed that includes a cytoplasmic chaperone with ATPase activity, a channel bridging from the inner to the outer membrane, and a needle tipped with a pore-forming protein (Shrivastava and Mande 2008). Some components of the machinery may also act as effectors, translocated into host cells. However, the identities and functions of T6SS effectors are still poorly understood.

T6SS genes are mostly not induced in laboratory growth conditions but are induced *in vivo* during infection. Several two-component systems have been identified that may play a crucial role in inducing/repressing expression of the T6SS genes while the bacteria are colonizing the host and encountering different environmental conditions. The T6SS seems to be involved in the intracellular behaviour of bacteria and allows chronic and persistent infection (Yahr 2006). Therefore, expression of T6SS genes is highly regulated. In *R. leguminosarum,* T6SS secretion is temperature-dependent, since putative T6SS effectors were optimally secreted at 24°C and only weakly at 20°C (Bladergroen et al. 2003).

Mutation of the locus involved in the synthesis of the T6SS of *R. leguminosarum* bv. *trifolii* allows this bacterium to fix nitrogen on pea-plants (Roest et al. 1997), which is normally a non-host plant. Because the locus is important for nodulation or subsequent stages of symbiosis, it was named *imp* (impaired in nodulation). Later on, another study reported the sequence of a 33-kb region revealing a putative operon of 14 genes named *impA-impN.* These genes mostly encode proteins with unknown function. However, ImpK and ImpL have similarities with DotU and IcmF, respectively, which are known as T4SS-like components. ImpK is similar to DotU but has a C-terminal extension with similarity to OmpA, an *E. coli* outer-membrane protein, and to the flagellar torque-generating protein MotB. This conserved C-terminal domain is described as a peptidoglycan-anchoring domain. Four proteins, including a homologue of the signal-peptide-containing protein RbsB (ribose-binding protein), were missing in the culture supernatant of the *imp* mutants suggesting that they were defective in protein secretion. In conclusion, the Type 6 (Imp) secretion system could block the colonization/infection process in pea-plants by *R. leguminosarum* bv. *trifolii* (Filloux et al. 2008).

Transcriptome profiling led to the discovery of a T6SS locus in the genome of *A. tumefaciens* C58, the causal agent of crown gall disease. Expression of genes in the *A. tumefaciens imp* locus was induced when cells were grown under mildly acidic conditions mimicking the rhizosphere (Yuan et al. 2008). Functionality of the *A. tumefaciens* T6SS was demonstrated through a study of the *A. tumefaciens* secretome, which led to the identification of an Hcp-family protein that is involved in virulence and is secreted via T6SS. The Hrp function is still unknown, and probably has a dual role: besides being secreted by T6SS, it may also be used to assemble a conduit at the bacterial cell surface. The protein Hrp was expressed and secreted under various growth conditions and, interestingly, deletion of *hcp* resulted in reduced tumourigenesis efficiency on potato tuber discs (Wu et al. 2008).

Transcriptome profiling also led to the identification of a *P. atrosepticum* T6SS locus, which appears to be regulated by quorum sensing. T6SS gene expression was reduced in a derivative with a deletion of *expI*, a gene responsible for *N*-(3-oxohexanoyl)-L-homoserine lactone synthesis (Liu et al. 2008). In the same study, mutant derivatives with deletions in either ECA3438 (*impJ*) or ECA3444 (*vipB*), both T6SS genes, exhibited slightly reduced virulence in potato stems and tubers. Mattinen et al. (2007) demonstrated that expression of the *P. atrosepticum* T6SS was induced upon exposure to potato tuber extracts. In their study, however, mutation of the *icmF* gene of *P. atrosepticum* ECA3432 (a structural T6SS gene) resulted in increased potato tuber maceration, which suggests that the *P. atrosepticum* T6SS may participate in antipathogenesis activity.

QS in pectobacteria regulates plant cell wall degrading enzymes (Jones et al. 1993), the Type III secreted harpin HrpN (Smadja et al. 2004) and other virulence factors, a very small number of virulence regulators (*expR, rsmA* and *virR*, Chatterjee et al. 1995, Cui et al. 2005, Burr et al. 2006), and the antibiotic carbapenem (McGowan et al. 2005). All these factors are controlled by the AHL, *N*-(3-oxohexanoyl)-L-homoserine lactone (OHHL), synthesized by ExpI. *P. atrosepticum* must secrete a multitude of plant cell wall degrading enzymes and other proteins, many of which are under QS control, to successfully cause disease. Liu et al. (2008) observed that both T1SS and T2SS, which can be considered as "accessory virulence factors", are modulated by QS. Either positive or negative QS regulation of the T3SS has been observed in other pathogens, e.g., *P. aeruginosa* (Bleves et al. 2005), *V. harveyi* (Henke and Bassler 2004), enteropathogenic *E. coli* (Sircili et al. 2004), *Ralstonia solanacearum* (Genin et al. 2005), and QS regulation of *hrpN* has been shown in *P. atrosepticum* (Pemberton et al. 2005). However, Liu et al. (2008) published evidence that QS plays a role in regulating the entire T3SS and its effectors in the enterobacterial plant pathogens and in the regulation of the T6SS and its putative substrates.

An *hcp* mutation in the genome of *Lysobacter enzymogenes* reduced the ability of the bacterium to colonize fungal hosts (Patel et al. 2009). *L. enzymogenes* is a pathogen of lower eukaryotes, including nematodes, fungi and lower plants. The bacterium is used as a biocontrol agent to limit a wide range of fungal plant diseases and has the potential to reduce bacterial disease as well (Patel et al. 2009, Qian et al. 2010).

Some plant growth-promoting rhizobacteria (PGPR) also express T6SS genes. *P. fluorescens* strains are widely used as biocontrol agents capable of suppressing various soil borne fungal and nematode diseases. *P. fluorescens* strain Pf29Arp is an efficient suppressor of take-all disease of wheat (Chapon et al. 2002). Analysis of the Pf29Arp transcriptome revealed that the expression of the protein VgrG (as protein Hcp, secreted via T6SS and it may also be used as component of T6SS) is induced when the bacteria are exposed to necrotic conditions associated with the colonization of wheat roots by the take-all fungus *Gaeumannomyces graminis* var. *tritici* (Barret et al. 2009). Microarray studies of the *P. fluorescens* Pf-5 genome revealed that, like that of *P. syringae* B728a, the Pf-5 T6SS locus is regulated by global regulator GacA (Hassan et al. 2010, Records and Gross 2010). Mutation of *gacA* resulted in significantly reduced expression of 14 genes in the Pf-5 T6SS locus. Functional studies are needed to augment the *P. fluorescens* transcriptome data so that we may begin to understand T6SS influence on the ecology and social activities of this important biocontrol agent.

The function and characteristics of the T6SS are far from being understood but the system appears as a novel key player in bacterial pathogenesis and bacteria–host interaction. In the future it may become another target in the hunt for novel antimicrobials, which will help fight human morbidity and mortality due to persistent bacterial infections.

Type 7 Secretion Systems

The Type 7 Secretion Sysems (T7SS) are the most recently categorized secretion systems and are specifically present in Gram-positive bacteria, mostly belonging to the Actinomycetales order as mycobacteria (Abdallah et al. 2007). Also a few components related to this secretion system have been identified in some species belonging to distant phyla like Firmicutes (Pallen 2002, Burst et al. 2005).

The T7SS secretes proteins lacking classical signal peptides and most of the proteins secreted by this system follow a pairwise dependency, both for secretion and function (Fortune et al. 2005).

The T7SS was first identified in the pathogenic organism *Mycobacterium tuberculosis* H37Rv, which requires this protein secretion system for growth during infection and virulence. *M. tuberculosis* is the etiological agent of the disease tuberculosis. Clearly, this system is a major determinant of mycobacterial pathogenesis as it mediates the early contact with the host cell, and functions to modulate the host cell immune response (DiGiuseppe and Cox 2007), but the way in which this system affects the biology of the host cell remains unknown.

One of the first well characterized substrates of the T7SS was the CFP-10 (Culture filtrate protein)/ESAT-6 (Early secreted antigen target) pair (Pym et al. 2003). Both substrates adopt helix-turn-helix folds and interact with each other forming a tight four-helix-bundle. The proteins constituting these pairs have been known to belong to the WXG100 superfamily because of having certain conserved structural properties such as length of approximately 100 residues forming a probable helix-turn-helix structure and containing a conserved WXG motif in the region linking the two helices.

The presence of the T7SS was initially predicted *in silico* based on clustering of genes encoding secreted proteins that lacked signal sequences with those encoding membrane proteins, ATPases and/or chaperones. The first T7SS region was discovered during the comparative genomic analysis of the attenuated strain *M. bovis* Bacille Calmette-Guerin (BCG) and other pathogenic mycobacterial species (Gordon et al. 1999). Sequencing of the *M. bovis* BCG vaccine strain and mutational analysis of this cluster in *M. tuberculosis* confirmed the hypothesis. The T7SS is also required for virulence and haemolysis in the fish pathogen *M. marinum* (Volkman et al. 2004), and for conjugation in the non-pathogenic species *M. smegmatis* (Converse and Cox 2005).

Mycobacterial genomes contain up to five T7SS gene clusters that do not functionally complement one another. T7SS gene clusters are also found in the closely related pathogens *Corynebacterium diphtheriae* and *Nocardia*. More distantly related gene clusters are also found in the genomes of pathogenic and non-pathogenic Gram-positive species that lack mycomembranes such as *Streptomyces* species and firmicutes such as *Bacillus* and *Clostridium* spp., *Staphylococcus aureus*, *Streptococcus agalactiae* and *Listeria monocytogenes*. The T7SS is required for virulence in *S. aureus* but not in *L. monocytogenes* (Abdallah et al. 2007). The complete structure and operation of the T7SS are still being pieced together.

Although it appears that this secretion system evolved independently from the Gram-negative systems, there are significant parallels between the mechanisms of secretion of the T7SS and the T4SS at the molecular level. First, Type 4 substrates are also targeted for secretion using unstructured C-terminal transport signals. Second, Type 4 translocators use the coupling protein (CP) to facilitate the interaction between the substrate and secretion machine, and to link cytosolic and membrane components. And finally, Type 4 secretion systems have examples of chaperone-substrate pairs that are targeted and secreted from the bacterial cell (DiGiuseppe and Cox 2007).

Conclusions

The significance of the role that secretory proteins play in adaptation of beneficial bacteria to environmental conditions, especially in the interior of the host plants is becoming clearer. The major source of information comes from bacterial genome-sequence analysis. The studies of the interaction between pathogenic bacteria and their eukaryotic hosts suggest that pathogens and plant-beneficial bacteria (rhizobia and PGPR) share not only secretion pathways, but some effector proteins. Understanding the role of the bacterial protein secretion systems on the initiation and maintenance of the beneficial interactions with plants will produce better understanding of these relationships and will contribute to a better use of these bacteria in sustainable agriculture.

Acknowledgements

The authors are indebted to Lourdes Bellogín Espuny for elaborating Fig. 9.1. Due to space limitations, we apologize to the researchers whose work could not be cited . This work was supported partially by Ministerio de Ciencia e Innovación (España): Proyecto AGL2009-13487-C04-03/AGR

References

Abdallah, A.M., N.C. Gey van Pittius, P.A. Champion, J. Cox, J. Luirink, C.M. Vandenbroucke-Grauls, B.J. Appelmelk and W. Bitter. 2007. Type VII secretion-mycobacteria show the way. Nat. Rev. Microbiol. 5: 883–891.

Alami, M., I. Lüke, S. Deitermann, G. Eisner, H.G. Koch, J. Brunner and M. Müller. 2003. Differential interactions between a twin-arginine signal peptide and its translocase in *Escherichia coli*. Mol. Cell. 12: 937–946.

Álvarez, B. and G. Martinez-Dretz. 1995. Metabolic characterization of *Acetobacter diazotrophicus*. Can. J. Microbiol. 41: 918–924.

Arrieta, J.G., M. Sotolongo, C. Menéndez, D. Alfonso, L.E. Trujillo, M. Soto, R. Ramírez and L. Hernández. 2004. A type II protein secretory pathway required for levansucrase secretion *Gluconacetobacter diazotrophicus*. J. Bacteriol. 186: 5031–5039.

Backert, S. and T.F. Meyer. 2006. Type IV secretion systems and their effectors in bacterial pathogenesis. Curr. Opin. Microbiol. 9: 207–217.

Barret, M., P. Frey-Klett, A.Y. Guillerm-Erckelboudt, M. Boutin, G. Guernec and A. Sarniguet. 2009. Effect of wheat roots infected with the pathogenic fungus *Gaeumannomyces graminis* var. *tritici* on gene expression of the biocontrol bacterium *Pseudomonas fluorescens* Pf29Arp. Mol. Plant Microbe Interact. 22: 1611–1623.

Bartsev, A.V., N.M. Boukli, W.J. Deakin, C. Staehelin and W.J. Broughton. 2003. Purification and phosphorylation of the effector protein NopL from *Rhizobium* sp. NGR234. FEBS Lett. 554: 271–274.

Bartsev, A.V., W.J. Deakin, M. Boukli, C.B. McAlvin, G. Stacey, P. Malnoë, W.J. Broughton and C. Staehelin. 2004. NopL, an effector protein of *Rhizobium* sp. NGR234, thwarts activation of plant defense reactions. Plant Physiol. 134: 871–879.

Beck, K., L.F. Wu, J. Brunner and M. Muller. 2000. Discrimination between SRP- and SecA/SecB dependent substrates involves selective recognition of nascent chains by SRP and trigger factor. EMBO J. 19: 134–143.

Becker, A., S. Rüberg, H. Küster, A.A. Roxlau, M. Keller, T. Ivashina, H.P. Cheng, G.C. Walker and A. Pühler. 1997. The 32-kilobase *exp* gene cluster of *Rhizobium meliloti* directing the biosynthesis of galactoglucan: genetic organization and properties of the encoded gene products. J. Bacteriol. 179: 1375–1384.

Benabdillah, R., L.J. Mota, S. Lützelschwab, E. Demoinet and G.R. Cornelis. 2004. Identification of a nuclear targeting signal in YopM from *Yersinia* spp. Microb. Pathog. 36: 247–261.

Berks, B.C. 1996. A common export pathway for proteins binding complex redox cofactors? Mol. Microbiol. 22: 393–404.

Berks, B.C., F. Sargent and T. Palmer. 2000. The Tat protein export pathway. Mol. Microbiol. 35: 260–274.

Bernal-Bayard, J. and F. Ramos-Morales. 2009. Salmonella type III secretion effector SlrP is an E3 ubiquitin ligase for mammalian thioredoxin. J. Biol. Chem. 284: 27587–27595.

Bernstein, H.D. 2007. Are bacterial 'autotransporters' really transporters? Trends Microbiol. 15: 441–447.

Bingle, L.E., C.M. Bailey and M.J. Pallen. 2008. Type VI secretion: a beginner's guide. Curr. Opin. Microbiol. 11: 3–8.

Bladergroen, M.R., K. Badelt and H.P. Spaink. 2003. Infection-blocking genes of a symbiotic *Rhizobium leguminosarum* strain that are involved in temperature-dependent protein secretion. Mol. Plant Microbe Interact. 16: 53–64.

Bleves, S., C. Soscia, P. Nogueira-Orlandi, A. Lazdunski and A. Filloux. 2005. Quorum sensing negatively controls type III secretion regulon expression in *Pseudomonas aeruginosa* PAO1. J. Bacteriol. 187: 3898–3902.

Bogsch, E.G., F. Sargent, N.R. Stanley, B.C. Berks, C. Robinson and T. Palmer. 1998. An essential component of a novel bacterial protein export system with homologues in plastids and mitochondria. J. Biol. Chem. 273: 18003–18006.

Bolhuis, A., J.E. Mathers, J.D. Thomas, C.M. Barrett and C. Robinson. 2001. TatB and TatC form a functional and structural unit of the twin-arginine translocase from *Escherichia coli*. J. Biol. Chem. 276: 20213–20219.

Burns, D.L. 2003. Type IV transporters of pathogenic bacteria. Curr. Opin. Microbiol. 6: 29–34.

Burr, T., A.M. Barnard, M.J. Corbett, C.L. Pemberton, N.J. Simpson and G.P. Salmond. 2006. Identification of the central quorum sensing regulator of virulence in the enteric phytopathogen, *Erwinia carotovora*: the VirR repressor. Mol. Microbiol. 59: 113–125.

Burst, M.L., W.A. Williams, K. DeBord and D.M. Missiakas. 2005. EsxA and EsxB are secreted by an ESAT-6-like system that is required for the pathogenesis of *Staphylococcus aureus* infections. Proc. Natl. Acad. Sci. U.S.A. 102: 1169–1174.

Cascales, E. 2008. The type VI secretion toolkit. EMBO rep. 9: 735–741.

Cascales, E. and E.P. Christie. 2003. The versatile bacterial type IV secretion systems. Nat. Rev. Microbiol. 1: 137–150.

Cavalcante, V.A. and J. Döbereiner. 1988. A new acid-tolerant nitrogenfixation bacterium associated with sugarcane. Plant Soil. 108: 23–31.

Cazalet, C., C. Rusniok, H. Brüggemann, N. Zidane, A. Magnier, L. Ma, M. Tichit, S. Jarraud, C. Bouchier, F. Vandenesch, F. Kunst, J. Etienne, P. Glaser and C. Buchrieser. 2004. Evidence in the *Legionella pneumophila* genome for exploitation of host cell functions and high genome plasticity. Nat. Genet. 36: 1165–1173.

Chapon, A., A.Y. Guillerm, L. Delalande, L. Lebreton and A. Sarniguet. 2002. Dominant colonisation of wheat roots by *Pseudomonas fluorescens* Pf29A and selection of indigenous microflora in the presence of the take-all fungus. Eur. J. Plant Pathol. 108: 449–459.

Chatterjee, A., Y. Cui, Y. Liu, C.K. Dumenyo and A.K. Chatterjee. 1995. Inactivation of *rsmA* leads to overproduction of extracellular pectinases, cellulases, and proteases in *Erwinia carotovora* subsp. *carotovora* in the absence of the starvation/cell density-sensing signal, N-(3-oxohexanoyl)-L-homoserine lactone. Appl. Environ. Microbiol. 61: 1959–1967.

Christie, P.J. 2004. Bacterial type IV secretion: the *Agrobacterium* VirB/D4 and related conjugation systems. Biochim. Biophys. Acta. 1694: 219–234.

Christie, P.J. and E. Cascales. 2005. Structural and dynamic properties of bacterial type IV secretion systems (review). Mol. Membr. Biol. 22: 51–61.

Cianciotto N.P. 2005. Type II secretion: a protein secretion system for all seasons. Trends Microbiol. 13: 581–588.

Cline, K. and H. Mori. 2001. Thylakoid Delta pH-dependent precursor proteins bind to a cpTatC-Hcf106 complex before Tha4-dependent transport. J. Cell Biol. 154: 719–729.

Cline, K., W.F. Ettinger and S.M. Theg. 1992. Protein-specific energy requirements for protein transport across or into thylakoid membranes. Two lumenal proteins are transported in the absence of ATP. J. Biol. Chem. 267: 2688–2696.

Converse, S.E. and J.S. Cox. 2005. A protein secretion pathway critical for *Mycobacterium tuberculosis* virulence is conserved and functional in *Mycobacterium smegmatis*. J. Bacteriol. 187: 1238–1245.

Cristobal, S., J.-W. de Gier, H. Nielsen and G. von Heijne. 1999. Competition between Sec- and Tat-dependent protein translocation in *Escherichia coli*. EMBO J. 18: 2982–2990.

Cui, Y., A. Chatterjee, H. Hasegawa, V. Dixit and N. Leigh. 2005. ExpR, a LuxR homolog of *Erwinia carotovora* subsp. *carotovora*, activates transcription of *rsmA*, which specifies a global regulatory RNA-binding protein. J. Bacteriol. 187: 4792–4803.

Dabney-Smith, C., H. Mori and K. Cline. 2006. Oligomers of Tha4 organize at the thylakoid Tat translocase during protein transport. J. Biol. Chem. 281: 5476–5483.

Dai, W.J., Y. Zeng, Z.P. Xie and C. Stehelin. 2008. Symbiosis-promoting and deleterious effects of NopT, a novel type 3 effector of *Rhizobum* sp. NGR234. J. Bacteriol. 190: 5101–5110.

Dautin, N. and H.D. Bernstein. 2007. Protein secretion in Gram-negative bacteria via the autotransporter pathway. Ann. Rev. Microbiol. 61: 89–112.

Deakin, W.J. and W.J. Broughton. 2009. Symbiotic use of pathogenic strategies: rhizobial protein secretion systems. Nat. Rev. Microbiol. 7: 312–320.

Deakin, W.J., C. Marie, M. Saad, H.B. Krishnan and W.J. Broughton. 2005. NopA is associated with cell surface appendages produced by the type III secretion system of *Rhizobium* sp. NGR234. Mol. Plant Microbe Interact. 18: 499–507.

Delepelaire, P. 2004. Type I secretion in Gram-negative bacteria. Biochim. Biophys. Acta. 1694: 149–161.

d'Enfert, C., A. Ryter and A.P. Pugsley. 1987. Cloning and expression in *Escherichia coli* of the *Klebsiella pneumoniae* genes for production, surface localization and secretion of the lipoprotein pullulanase. EMBO J. 6: 3531–3538.

Desvaux, M., N.J. Parham and I.R. Henderson. 2004a. Type V protein secretion: simplicity gone awry? Curr. Iss. Mol. Biol. 6: 111–124.

Desvaux, M., N.J. Parham, A. Scott-Tucker and I.R. Henderson. 2004b. The general secretory pathway: a general misnomer? Trends Microbiol. 12: 306–309.

DiGiuseppe, P.A. and J.S. Cox. 2007. Microreview: Protein secretion systems in Mycobacteria. Cell. Biol. 9: 1376–1384.

Downie, J.A. 2010. The roles of extracellular proteins, polysaccharides and signals in the interactions of rhizobia with legume roots. FEMS Microbiol. Rev. 34: 150–170.

Economou, A., W.D.O. Hamilton, A.W.B. Johnston and J.A. Downie. 1990. The *Rhizobium* nodulation gene *nodO* encodes a Ca^{2+} binding protein that is exported without N-terminal cleavage and is homologous to hemolysin and related proteins. EMBO J. 9: 349–354.

Feng, L., S.N. Senchenkova, J. Yang, A.S. Shashkov, J. Tao, H. Guo, J. Cheng, Y. Ren, Y.A. Knirel, P.R. Reeves and L. Wang. 2004. Synthesis of the heteropolysaccharide O antigen of *Escherichia coli* O52 requires an ABC transporter: structural and genetic evidence. J. Bacteriol. 186: 4510–4519.

Filloux, A. 2004. The underlying mechanisms of type II protein secretion. Biochim. Biophys. Acta. 1694: 163–179.

Filloux, A., M. Bally, G. Ball, M. Akrim, J. Tommassen and A. Lazdunski. 1990. Protein secretion in gram-negative bacteria: transport across the outer membrane involves common mechanisms in different bacteria. EMBO J. 9: 4323–4329.

Filloux, A., A. Hachani and S. Bleves. 2008. The bacterial type VI secretion machine: yet another player for protein transport across membranes. Microbiology. 154: 1570–1583.

Fischer, W., R. Haas and S. Odenbreit. 2002. Type IV secretion systems in pathogenic bacteria. Int. J. Med. Microbiol. 292: 159–168.

Fortune, S.M., A. Jaeger, D.A. Sarracino, M.R. Chase, C.M. Sassetti, D.R. Sherman, B.R. Bloom and E.J. Rubin. 2005. Mutually dependent secretion of proteins required for mycobacterial virulence. Proc. Natl. Acad. Sci. U.S.A. 102: 10676–10681.

Fronzes, R., P.J. Christie and G. Waksman. 2009. The structural biology of type IV secretion systems. Nat. Rev. Microbiol. 7: 703–714.
Galan, J.E. 2007. Snapshot: effector proteins of type III secretion systems. Cell. 130.
Galan, J.E. and A. Collmer. 1999. Type III secretion machines: bacterial devices for protein delivery into host cells. Science. 284: 1322–1328.
Galardini, M., A. Mengoni, M. Brilli, F. Pini, A. Fioravanti, S. Lucas, A. Lapidus, J.F. Cheng, L. Goodwin, S. Pitluck, M. Land, L. Hauser, T. Woyke, N. Mikhailova, N. Ivanova, H.M. Daligault, D. Bruce, C. Detter, R. Tapia, C. Han, H. Teshima, S. Mocali, M. Bazzicalupo and E.G. Biondi. 2011. Exploring the symbiotic pangenome of the nitrogen-fixing bacterium *Sinorhizobium meliloti*. BMC Genomics. 12: 235–250.
Genin, S., B. Brito, T.P. Denny and C. Boucher. 2005. Control of the *Ralstonia solanacearum* Type III secretion system (Hrp) genes by the global virulence regulator PhcA. FEBS Lett. 579: 2077–2081.
Gérard, F. and K. Cline. 2006. Efficient twin arginine translocation (Tat) pathway transport of a precursor protein covalently anchored to its initial cpTatC binding site. J. Biol. Chem. 281: 6130–6135.
Gohlke, U., L. Pullan, C.A. McDevitt, I. Porcelli, E. de Leeuw, T. Palmer, H.R. Saibil and B.C. Berks. 2005. The TatA component of the twin-arginine protein transport system forms channel complexes of variable diameter. Proc. Natl. Acad. Sci. U.S.A. 102: 10482–10486.
Gomis-Ruth, F.X., M. Sola, F. de la Cruz and M. Coll. 2004. Coupling factors in macromolecular type-IV secretion machineries. Curr. Pharm. Des. 10: 1551–1565.
Gordon, S.V., R. Brosch, A. Billault, T. Garnier, K. Eiglmeier and S.T. Cole. 1999. Identification of variable regions of the genomes of the tubercle bacilli using bacterial artificial chromosome arrays. Mol. Microbiol. 32: 643–655.
Grohmann, E., G. Muth and M. Espinosa. 2003. Conjugative plasmid transfer in Gram-positive bacteria. Microbiol. Mol. Biol. Rev. 67: 277–301.
Guttman, D.S., B.A. Vinatzer, S. Sarkar, M.V. Ranall, G. Kettler and J.T. Greenberg. 2002. A functional screen for the type III (Hrp) secretome of the plant pathogen *Pseudomonas syringae*. Science. 295: 1722–1726.
Hassan, K.A., A. Johnson, B.T. Shaffer, Q. Ren, T.A. Kidarsa, L.D.H. Elbourne, S. Hartney, R. Duboy, N.C. Goebel, T.M. Zabriskie, I.T. Paulsen and J.E. Loper. 2010. Inactivation of the GacA response regulator in *Pseudomonas fluorescens* Pf-5 has far-reaching transcriptomic consequences. Environ. Microbiol. 12: 899–915.
Henderson, I.R., F. Navarro-García, M. Desvaux, R.C. Fernandez and D. Ala'Aldeen. 2004. Type V protein secretion pathway: the autotransporter story. Microbiol. Mol. Biol. Rev. 68: 692–744.
Henke, J.M. and B.L. Bassler. 2004. Quorum sensing regulates type III secretion in *Vibrio harveyi* and *Vibrio parahaemolyticus*. J. Bacteriol. 186: 3794–3805.
Hinsa, S.M., M. Espinosa-Urgel, J.L. Ramos and G.A. O'Toole. 2003. Transition from reversible to irreversible attachment during biofilm formation by *Pseudomonas fluorescens* WCS365 requires an ABC transporter and a large secreted protein. Mol. Microbiol. 49: 905–918.
Hotson, A., R. Chosed, H. Shu, K. Orth and M.B. Mudgett. 2003. *Xanthomonas* type III effector XopD targets SUMO-conjugated proteins *in planta*. Mol. Microbiol. 50: 377–389.
Hubber, A., A.C. Vergunst, J.T. Sullivan, P.J.J. Hooykaas and C.W. Ronson. 2004. Symbiotic phenotypes and translocated effector proteins of the *Mesorhizobium loti* strain R7A VirB/D4 type IV secretion system. Mol. Microbiol. 54: 561–574.
Hubber, A.M., J.T. Sullivan and C.W. Ronson. 2007. Symbiosis-Induced Cascade Regulation of the *Mesorhizobium loti* R7A VirB/D4 Type IV Secretion System. Mol. Plant Microbe Interact. 20: 255–261.
Hynds, P.J., D. Robinson and C. Robinson. 1998. The sec-independent twin-arginine translocation system can transport both tightly folded and malfolded proteins across the thylakoid membrane. J. Biol. Chem. 273: 34868–34874.

Ito, K. and H. Mori. 2009. The Sec Protein Secretion System. *In*: K. Wooldridge [ed.]. Bacterial secreted proteins: secretory mechanisms and role in pathogenesis. Caster Academic Press. Norfolk, UK. pp. 3–22.

Izard, J.W. and D.A. Kendall. 1994. Signal peptides: exquisitely designed transport promoters. Mol. Microbiol. 13: 765–773.

Jacob-Dubuisson, F., C. Locht and R. Antoine. 2001. Two-partner secretion in Gram negative bacteria: a thrifty, specific pathway for large virulence proteins. Mol. Microbiol. 40: 306–313.

Jacob-Dubuisson, F., R. Fernandez and L. Coutte. 2004. Protein secretion through autotransporter and two partner pathways. Biochim. Biophys. Acta. 1694: 235–257.

Jenewein, S., I.B. Holland and L. Schmitt. 2009. Type I bacterial secretion systems. *In*: K. Wooldridge [ed.]. Bacterial secreted proteins: secretory mechanisms and role in pathogenesis. Caster Academic Press. Norfolk, UK. pp. 45–65.

Jenner, C., E. Hitchin, J. Mansfield, K. Walters, P. Betteridge, D. Teverson and J. Taylor. 1991. Gene-for-gene interactions between *Pseudomonas syringae* pv. *phaseolicola* and *Phaseolus*. Mol. Plant Microbe Interact. 4: 553–562.

Jiménez-Salgado, T., L.E. Fuentes-Ramírez, A. Tapia-Hernández, M.A. Mascarua-Esparza, E. Martínez-Romero and J. Caballero-Mellado. 1997. *Coffea arabica* L., a new host plant for *Acetobacter iazotrophicus*, and isolation of other nitrogen-fixing acetobacteria. Appl. Environ. Microbiol. 63: 3676–3683.

Jones, S., B. Yu, N.J. Bainton, M. Birdsall, B.W. Bycroft, S.R. Chhabra, A.J. Cox, P. Golby, P.J. Reeves and S. Stephens. 1993. The *lux* autoinducer regulates the production of exoenzyme virulence determinants in *Erwinia carotovora* and *Pseudomonas aeruginosa*. EMBO J. 12: 2477–2482.

Kaneko, T., Y. Nakamura, S. Sato, E. Asamizu, T. Kato and S. Sasamoto. 2000. Complete genome structure of the nitrogen-fixing symbiotic bacterium *Mesorhizobium loti*. DNA Res. 7: 331–338.

Kaneko, T., Y. Nakamura, S. Sato, K. Minamisawa, T. Uchiumi, S. Sasamoto, A. Watanabe, K. Idesawa, M. Iriguchi, K. Kawashima, M. Kohara, M. Matsumoto, S. Shimpo, H. Tsuruoka, T. Wada, M. Yamada and S. Tabata. 2002. Complete genomic sequence of nitrogen-fixing symbiotic bacterium *Bradyrhizobium japonicum* USDA110. DNA Res. 9: 189–197.

Keen, N.T. 1990. Gene-for-gene complementarity in plant-pathogen interactions. Annu. Rev. Genet. 24: 447–463.

Krause, A., A. Doerfel and M. Göttfert. 2002. Mutational and transcriptional analysis of the type III secretion system of *Bradyrhizobium japonicum*. Mol. Plant Microbe Interact. 15: 1228–1235.

Krehenbrink, M. and J.A. Downie. 2008. Identification of protein secretion systems and novel secreted proteins in *Rhizobium leguminosarum* bv. *viciae*. BMC Genomics. 9: 55–73.

Lawley, T.D., W.A. Klimke, M.J. Gubbins and L.S. Frost. 2003. F factor conjugation is a true type IV secretion system. FEMS Microbiol. Lett. 224: 1–15.

Lee, J., J. Nam, H.C. Park, G. Na, K. Miura, J.B. Jin, C.Y. Yoo, D. Baek, D.H. Kim, J.C. Jeong, D. Kim, S.Y. Lee, D.E. Salt, T. Mengiste, Q. Gong, S. Ma, H.J. Bohnert, S.S. Kwak, R.A. Bressan, P.M. Hasegawa and D.J. Yun. 2007. Salicylic acid-mediated innate immunity in *Arabidopsis* is regulated by SIZ1 SUMO E3 ligase. Plant J. 49: 79–90.

Leiman, P.G., M. Basler, U.A. Ramagopal, J.B. Bonanno, J.M. Sauder, S. Pukatzki, S.K. Burley, S.C. Almo and J.J. Mekalanos. 2009. Type VI secretion apparatus and phage tail-associated protein complexes share a common evolutionary origin. Proc. Natl. Acad. Sci. U.S.A. 106: 4154–4159.

Lindeberg, M., G.P. Salmond and A. Collmer. 1996. Complementation of deletion mutations in a cloned functional cluster of *Erwinia chrysanthemi* out genes with *Erwinia carotovora* out homologues reveals OutC and OutD as candidate gatekeepers of species-specific secretion of proteins via type II pathway. Mol. Microbiol. 20: 175–190.

Lindgren, P.B. 1997. The role of *hrp* genes during plant-bacterial interactions. Annu. Rev. Phytopathol. 35: 129–152.

Liu, H., S.J. Coulthurst, L. Pritchard, P.E. Hedley, M. Ravensdale, S. Humphris, T. Burr, G. Takle, M.B. Brurberg, P.R.J. Birch, G.P.C. Salmond and I.K. Toth. 2008. Quorum sensing coordinates brute force and stealth modes of infection in the plant pathogen *Pectobacterium atrosepticum*. PLoS Pathog. 4: e1000093 (published online).

López-Baena, F.J., J.M. Vinardell, F. Pérez-Montaño, J.C. Crespo-Rivas, R.A. Bellogín, M.R. Espuny and F.J. Ollero. 2008. Regulation and symbiotic significance of nodulation outer proteins secretion in *Sinorhizobium fredii* HH103. Microbiology. 154: 1825–1836.

López-Baena, F.J., J.A. Monreal, F. Pérez-Montaño, B. Guasch-Vidal, R.A. Bellogín, J.M. Vinardell and F.J. Ollero. 2009. The absence of Nops secretion in *Sinorhizobium fredii* HH103 increases *GmPR1* expression in Williams soybean. Mol. Plant Microbe Interact. 22: 1445–1454.Mattinen, L., R. Nissinen, T. Riipi, N. Kalkkinen and M. Pirhonen. 2007. Host-extract induced changes in the secretome of the plant pathogenic bacterium *Pectobacterium atrosepticum*. Proteomics. 7: 3527–3537.

Lycklama, A., J.A. Nijeholt and A.J. Driessen. 2012. The bacterial Sec-translocase: structure and mechanism. Philos. Trans. R. Soc. Lond. B. Biol. Sci. 367: 1016–1028.

McDonald, C., P.O. Vacratsis, J.B. Bliska and J.E. Dixon. 2003. The *Yersinia* virulence factor YopM forms a novel protein complex with two cellular kinases. J. Biol. Chem. 278: 18514–18523.

McGowan, S.J., A.M. Barnard, G. Bosgelmez, M. Sebaihia, N.J. Simpson, N.R. Thomson, D.E. Todd, M. Welch, N.A. Whitehead and G.P. Salmond. 2005. Carbapenem antibiotic biosynthesis in *Erwinia carotovora* is regulated by physiological and genetic factors modulating the quorum sensing-dependent control pathway. Mol. Microbiol. 55: 526–545.

Meinhardt, L.W., H.B. Krishnan, P.A. Balatti and S.G. Pueppke. 1993. Molecular cloning and characterization of a sym plasmid locus that regulates cultivar-specific nodulation of soybean by *Rhizobium fredii* USDA257. Mol. Microbiol. 9: 17–29.

Michel, L.O., M. Sandkvist and M. Bagdasarian. 1995. Specifity of the protein secretory apparatus: secretion of the heat-labile enterotoxin B subunit pentamers by different species of Gram-bacteria. Gene. 152: 41–45.

Miwa, H., J. Sun, G.E.D. Oldroyd and J.A. Downie. 2006. Analysis of Nod-factor-induced calcium signalling in root hairs of symbiotically defective mutants of *Lotus japonicus*. Mol. Plant Microbe Interact. 19: 914–923.

Moreira, L.M., J.D. Becker, A. Pühler and A. Becker. 2000. The *Sinorhizobium meliloti* ExpE1 protein secreted by a type I secretion system involving ExpD1 and ExpD2 is required for biosynthesis or secretion of the exopolysaccharide galactoglucan. Microbiology. 146: 2237–2248.

Nagai, H., E.D. Cambronne, J.C. Kagan, J.C. Amor, R.A, Kahn and C.R. Roy. 2005. A C-terminal translocation signal required for Dot/Icm-dependent delivery of the *Legionella* RalF protein to host cells. Proc. Natl. Acad. Sci. U.S.A. 102: 826–831.

Nümberger, T., F. Brunner, B. Kemmerling and L. Piater. 2004. Innate immunity in plants and animals: striking similarities and obvious differences. Immunol. Rev. 198: 249–266.

Nummelin, H., M.C. Merckel, J.C. Leo, H. Lankinen, M. Skurnik and A. Goldman. 2004. The *Yersinia* adhesin YadA collagen-binding domain structure is a novel left-handed parallel beta-roll. EMBO J. 23: 701–711.

Okazaki, S., S. Okabe, M. Higashi, Y. Shimoda, S. Sato, S. Tabata, M. Hashigushi, R. Akashi, M. Göttfert and K. Saeki. 2010. Identification and functional analysis of type III effector protein in *Mesorhizobium loti*. Mol. Plant Microbe Interact. 23: 223–234.

Pallen, M.J. 2002. The ESAT-6/WXG100 superfamily - and a new Gram-positive secretion system? Trends Microbiol. 10: 209–212.

Pallen, M.J., R.R. Chaudhuri and I.R. Henderson. 2003. Genomic analysis of secretion systems. Curr. Opin. Microbiol. 6: 519–527.

Pankhurst, C.E., D.H. Hopcroft and W.T. Jones. 1987. Comparative morphology and flavolan content of *Rhizobium loti* induced effective and ineffective root nodules on *Lotus* species, *Leuceana leucocephala*, *Carmichaelia flagelliformis*, *Ornithopus sativus*, and *Clianthus*.
Papanikou, E., S. Karamanou and A. Economou. 2007. Bacterial protein secretion through the trasnlocase nanomachine. Nat. Rev. Microbiol. 5: 839–851.
Patel, N., M. Blackmoore, B. Hillman and D. Kobayashi. 2009. Evidence for the role of Type VI secretion during *Lysobacter enzymogenes* pathogenesis of fungal hosts. Phytopathol. 99: 100–101.
Peabody, C.R., Y.J. Chung, M.R. Yen, D. Vidal-Ingigliardi, A.P. Pugsley and M.H. Saier, Jr. 2003. Type II protein secretion and its relationship to bacterial type IV pili and archaeal flagella. Microbiology. 149: 3051–3072.
Pemberton, C.L., N.A. Whitehead, M. Sebaihia, K.S. Bell, L.J. Hyman, S.J. Harris, A.J. Matlin, N.D. Robson, P.R. Birch, J.P. Carr, I.K. Toth and G.P. Salmond. 2005. Novel quorum-sensing-controlled genes in *Erwinia carotovora* subsp. *carotovora*: identification of a fungal elicitor homologue in a soft-rotting bacterium. Mol. Plant Microbe Interact. 18: 343–353.
Preston, G.M., N. Bertrand and P.B. Rainey. 2001. Type III secretion in plant growth-promoting *Pseudomonas fluorescens* SBW25. Mol Microbiol. 41: 999–1014.
Pugsley, A.P. 1993. The complete general secretory pathway in gram-negative bacteria. Microbiol. Rev. 57: 50–108.
Pym, A.S., P. Brodin, L. Majlessi, R. Brosch, C. Demangel, A. Williams, K.E. Griffiths, G. Marchal, C. Leclerc and S.T. Cole. 2003. Recombinant BCG exporting ESAT-6 confers enhanced protection against tuberculosis. Nat. Med. 9: 533–539.
Qian, G.L., J.Q. Fan, D.F. Chen, Y.J. Kang, B. Han, B.S. Hu and F.Q. Liu. 2010. Reducing *Pectobacterium* virulence by expression of an *N*-acyl homoserine lactonase gene *Plpp-aiiA* in *Lysobacter enzymogenes* strain OH11. Biol. Control. 52: 17–23.
Rabus, R., M. Kube, J. Heider, A. Beck, K. Heitmann, F. Widdel and R. Reinhardt. 2005. The genome sequence of an anaerobic aromatic-degrading denitrifying bacterium, strain EbN1. Arch. Microbiol. 183: 27–36.
Records, A.R. and D.C. Gross. 2010. Sensor kinases RetS and LadS regulate *Pseudomonas syringae* type VI secretion and virulence factors. J. Bacteriol. 192: 3584–3596.
Rezzonico, F., C. Binder, G. Défago and Y. Moënne-Loccoz. 2005. The type III secretion system of biocontrol *Pseudomonas fluorescens* KD targets the phytopathogenic Chromista *Pythium ultimum* and promotes cucumber protection. Mol. Plant Microbe Interact. 18: 991–1001.
Robledo, M., J.I. Jiménez-Zurdo, E. Velázquez, M.E. Trujillo, J.L. Zurdo-Piñero, M.H. Ramírez-Bahena, B. Ramos, J.M. Díaz-Mínguez, F. Dazzo, E. Martínez-Molina and P.F. Mateos. 2008. *Rhizobium* cellulose CelC2 is essential for primary symbiotic infection of legume host roots. Proc. Nat. Acad. Sci. (U.S.A.). 105: 7064–7069.
Rodrigue, A., A. Chanal, K. Beck, M. Muller and L.F. Wu. 1999. Cotranslocation of a periplasmic enzyme complex by a hitchhiker mechanism through the bacterial tat pathway. J. Biol. Chem. 274: 13223–13228.
Roest, H.P., I.H. Mulders, H.P. Spaink, C.A. Wijffelman and B.J. Lugtenberg. 1997. A *Rhizobium leguminosarum* biovar *trifolii* locus not localized on the sym plasmid hinders effective nodulation on plants of the pea cross-inoculation group. Mol. Plant Microbe Interact. 10: 938–941.
Roggenkamp, A., N. Ackermann, C.A. Jacobi, K. Truelzsch, H. Hoffmann and J. Heesemann. 2003. Molecular analysis of transport and oligomerization of the *Yersinia enterocolitica* adhesin YadA. J. Bacteriol. 185: 3735–3744.
Rohde, J.R., A. Breitkreutz, A. Chenal, P.J. Sansonetti and C. Parsot. 2007. Type III secretion effectors of the IpaH family are E3 ubiquitin ligases. Cell Host Microbe. 1: 77–83.
Rossier, O. and N.P. Cianciotto. 2005. The *Legionella pneumophila tatB* gene facilitates secretion of phospholipase C, growth under iron-limiting conditions, and intracellular infection. Infect. Immun. 73: 2020–2032.

Saad, M.M., H. Kobayashi, C. Marie, I.R. Brown, J.W. Mansfield, W.J. Broughton and W.J. Deakin. 2005. NopB, a type III secreted protein of *Rhizobium* sp. strain NGR234, is associated with pilus-like surface appendages. J. Bacteriol. 187: 1173–1181.

Saad, M.M., C. Staehelin, W.J. Broughton and W.J. Deakin. 2007. Protein-protein interactions within type three secretion system-dependent pili of *Rhizobium* sp. strain NGR234. J. Bacteriol. 190: 750–754.

Salomon, D., D. Dar, S. Sreeramulu and G. Sessa. 2011. Expression of *Xanthomonas campestris* pv. *vesicatoria* type III effectors in yeast affects cell growth and viability. Mol. Plant Microbe Interact. 24: 305–314.

Sánchez, C., F. Iannino, W.J. Deakin, R.A. Ugalde and V.C. Lepek. 2009. Characterization of the *Mesorhizobium loti* MAFF303099 type-three protein secretion system. Mol. Plant Microbe Interact. 22: 519–528.

Sandkvist, M. 2001. Type II secretion and pathogenesis. Infect. Immun. 69: 3523–3535.

Schmeisser, C., H. Liesegang, D. Krysciak, N. Bakkou, A. Le Quéré, A. Wollherr, I. Heinemeyer, B. Morgenstern, A. Pommerening-Röser, M. Flores, R. Palacios, S. Brenner, G. Gottschalk, R.A. Schmitz, W.J. Broughton, X. Perret, A.W. Strittmatter and W.R. Streit. 2009. *Rhizobium* sp. strain NGR234 possesses a remarkable number of secretion systems. Appl. Environ. Microbiol. 75: 4035–4045.

Schrammeijer, B., A. den Dulk-Ras, A.C. Vergunst, E.J. Jacome and P.J.J. Hooykaas. 2003. Analysis of Vir protein translocation from *Agrobacterium tumefaciens* using *Saccharomyces cerevisiae* as a model: evidence for transport of a novel effector protein VirE3. Nucleic Acids Res. 31: 860–868.

Schroder, G. and E. Lanka. 2005. The mating pair formation system of conjugative plasmids-a versatile secretion machinery for transfer of proteins and DNA. Plasmid. 54: 1–25.

Shrivastava, S. and S.S. Mande. 2008. Identification and functional characterization of gene components of Type VI secretion system in bacterial genomes. PLoS ONE. 3: e2955(published online).

Sircili, M.P., M. Walters, L.R. Trabulsi and V. Sperandio. 2004. Modulation of enteropathogenic *Escherichia coli* virulence by quorum sensing. Infect. Immun. 72: 2329–2337.

Skorpil, P., M.M. Saad, N.M. Boukli, H. Kobayashi, F. Ares-Orpel, W.J. Brougthon and W.J. Deakin. 2005. NopP, a phosphorylated effector of *Rhizobium* sp. strain NGR234, is a major determinant of nodulation of the tropical legumes *Flemingia congesta* and *Tephrosia vogelii*. Mol. Microbiol. 57: 1304–1317.

Smadja, B., X. Latour, D. Faure, S. Chevalier, Y. Dessaux and N. Orange. 2004. Involvement of N-acylhomoserine lactones throughout plant infection by *Erwinia carotovora* subsp. *atroseptica* (*Pectobacterium atrosepticum*). Mol. Plant Microbe Interact. 17: 1269–1278.

Staskawicz, B.J., D. Dahlbeck and N.T. Keen. 1984. Cloned avirulence gene of *Pseudomonas syringae* pv. *glycinea* determines race specific incompatibility on *Glycine max* (L.) Merr. Proc. Natl. Acad. Sci. U.S.A. 81: 6024–6028.

Sullivan, J.T., J.R. Trzebiatowski, R.W. Cruickshank, J. Gouzy, S.D. Brown and R.M. Elliot. 2002. Comparative sequence analysis of the symbiosis island of *Mesorhizobium loti* strain R7A. J. Bacteriol. 184: 3086–3095.

Sutton, J.M., E.J.A. Lea and J.A. Downie. 1994. The nodulation signalling protein NodO from *Rhizobium leguminosarum* biovar. *viciae* forms ion channels in membranes. Proc. Natl. Acad. Sci. U.S.A. 91: 9990–9994.

Suzuki, S., T. Aono, K.B. Lee, T. Suzuki, C.T. Liu, H. Miwa, S. Wakao, T. Iki and H. Oyaizu. 2007. Rhizobial factors required for stem nodule maturation and maintenance in *Sesbania rostrata Azorhizobium caulinodans* ORS571 symbiosis. Appl. Environ. Microbiol. 73: 6650–6659.

Suzuki, T., T. Aono, C.T. Liu, S. Suzuki, T. Iki, K. Yokota and H. Oyaizu. 2008. An outer membrane autotransporter, AoaA, of *Azorhizobium caulinodans* is required for sustaining high N_2-fixing activity of stem nodules. FEMS Microbiol. Lett. 285: 16–24.

Tauschek, M., R.J. Gorrell, R.A. Strugnell and R.M. Robins-Browne. 2002. Identification of a protein secretory pathway for the secretion of heat-labile enterotoxin by an enterotoxigenic strain of *Escherichia coli*. Proc. Natl. Acad. Sci. U.S.A. 99: 7066–7071.

Ullers, R.S., E.N. Houben, J. Brunner, B. Oudega, N. Harms and J. Luirink. 2006. Sequence-specific interactions of nascent *Escherichia coli* polypeptides with trigger factor and signal recognition particle. J. Biol. Chem. 281: 13999–14005.

van Rhijn, P., E. Luyten, K. Vlassak and J. Vanderleyden. 1996. Isolation and characterization of a pSym locus of *Rhizobium* sp. BR816 that extends nodulation ability of narrow host range *Phaseolus vulgaris* symbionts to *Leucaena leucocephala*. Mol. Plant Microbe Interact. 9: 74–77.

Vergunst, A.C., B. Schrammeijer, A. den Dulk-Ras, C.M.T. de Vlaam, T.J.G. Regensburg-Tuink and P.J.J. Hooykaas. 2000. VirB/D4-dependent protein translocation from *Agrobacterium* into plant cells. Science. 290: 979–982.

Vergunst, A.C., M.C. van Lier, A. den Dulk-Ras and P.J. Hooykaas. 2003. Recognition of the *Agrobacterium tumefaciens* VirE2 translocation signal by the VirB/D4 transport system does not require VirE1. Plant Physiol. 133: 978–88.

Viprey, V., A. del Greco, W. Golinowski, W.J. Broughton and X. Perret. 1998. Symbiotic implications of type III protein secretion machinery in *Rhizobium*. Mol. Microbiol. 28: 1381–1389.

Vlassak, K.M., E. Luyten, C. Verreth, P. van Rhijn, T. Bisseling and J. Vanderleyden. 1998. *Rhizobium* sp. BR816 *nodO* gene can function as a determinant for nodulation of *Leucaena leucocephala, Phaseolus vulgaris*, and *Trifolium repens* by a diversity of *Rhizobium* spp. Mol. Plant Microbe Interact. 11: 383–392.

Volkman, H.E., H. Clay, D. Beery, J.C. Chang, D.R. Sherman and L. Ramakrishnan. 2004. Tuberculous granuloma formation is enhanced by a mycobacterium virulence determinant. PLoS Biol. 2: 1946–1956.

von Heijne, G. 1990. The signal peptide. J. Membr. Biol. 115: 195–201.

Voulhoux, R., G. Ball, B. Ize, M.L. Vasil, A. Lazdunski, L.F. Wu and A. Filloux. 2001. Involvement of the twin-arginine translocation system in protein secretion via the type II pathway. EMBO J. 20: 6735–6741.

Walker, S.A. and J.A. Downie. 2000. Entry of *Rhizobium leguminosarum* biovar. *viciae* into root hairs requires minimal Nod factor specificity, but subsequent infection thread growth requires *nodO* or *nodE*. Mol. Plant Microbe Interact. 13: 754–762.

Ward, N., Ø. Larsen, J. Sakwa, L. Bruseth, H. Khouri, A.S. Durkin, G. Dimitrov, L. Jiang, D. Scanlan, K.H. Kang, M. Lewis, K.E. Nelson, B. Methé, M. Wu, J.F. Heidelberg, I.T. Paulsen, D. Fouts, J. Ravel, H. Tettelin, Q. Ren, T. Read, R.T. DeBoy, R. Seshadri, S.L. Salzberg, H.B. Jensen, N.K. Birkeland, W.C. Nelson, R.J. Dodson, S.H. Grindhaug, I. Holt, I. Eidhammer, I. Jonasen, S. Vanaken, T. Utterback, T.V. Feldblyum, C.M. Fraser, J.R. Lillehaug and J.A. Eisen. 2004. Genomic insights into methanotrophy: the complete genome sequence of *Methylococcus capsulatus* (Bath). PLoS Biol. 2: e303 (published online).

Wells, T.J., J.J. Tree, G.C. Ulett and M.A. Schembri. 2007. Autotransporter proteins: novel targets at the bacterial cell surface. FEMS Microbiol. Lett. 274: 163–172.

Wenzel, M., L. Friedrich, M. Göttfert and S. Zehner. 2009. The type-III secreted protein NopE1 affects symbiosis and exhibits a calcium-dependent autocleavage activity. Mol. Plant Microbe Interact. 23: 124–129.

Wong, K.R. and J.T. Buckley. 1993. *Aeromonas* spp. can secrete *Escherichia coli* phosphatase into the culture supernatant, and its release requires a functional general secretion pathway. Mol. Microbiol. 9: 955–963.

Wu, H.Y., P.C. Chung, H.W. Shih, S.R. Wen and E.M. Lai. 2008. Secretome analysis uncovers an Hcp-family protein secreted via a type VI secretion system in *Agrobacterium tumefaciens*. J. Bacteriol. 190: 2841–2850.

Wu, L.F., B. Ize, A. Chanal, Y. Quentin and G. Fichant. 2000. Bacterial twin-arginine signal peptide-dependent protein translocation pathway: evolution and mechanism. J. Mol. Microbiol. Biotechnol. 2: 179–189.

Yahr, T.L. 2006. A critical new pathway for toxin secretion? New Engl. J. Med. 355: 1171–1172.

Yang, S., F. Tang, M. Gao, H.B. Krishnan and H. Zhu. 2010. R gene-controlled host specificity in the legume-rhizobia symbiosis. Proc. Natl. Acad. Sci. U.S.A. 107: 18735–18740.

York, G.M. and G.C. Walker. 1997. The *Rhizobium meliloti exoK* gene and *prsD/prsE/exsH* genes are components of independent degradative pathways which contribute to production of low-molecular-weight succinoglycan. Mol. Microbiol. 25: 117–134.

Yuan, Z.C., P. Liu, P. Saenkham, K. Kerr and E.W. Nester. 2008. Transcriptome profiling and functional analysis of *Agrobacterium tumefaciens* reveals a general conserved response to acidic conditions (pH 5.5) and a complex acid-mediated signalling involved in *Agrobacterium* plant interactions. J. Bacteriol. 190: 494–507.

Zhang, L., X.J. Chen, H.B. Lu, Z.P. Xie and C. Staehelin. 2011. Functional analysis of the type 3 effector nodulation outer protein L (NopL) from *Rhizobium* sp. NGR234: symbiotic effects, phosphorylation, and interference with mitogen-activated protein kinase signaling. J. Biol. Chem. 286: 32178–32187.

Zhu, J., P.M. Oger, B. Schrammeijer, P.J.J. Hooykaas, S.K. Farrand and S.C. Winans. 2000. The bases of crown gall tumorigenesis. J. Bacteriol. 182: 3885–3895.

10

Nodular Endophytes: An Untapped Diversity

E. Velázquez,[1] *P. Martínez-Hidalgo,*[1] *L. Carro,*[1] *P. Alonso,*[1] *Á. Peix,*[2] *M. Trujillo*[1] and *E. Martínez-Molina*[1,*]

Introduction

The natural condition of plants closely interacts with endophytic microorganisms. Endophytes help in increasing crop yields, removing contaminants, inhibiting pathogens, and producing fixed nitrogen or novel substances. The repertoire of their effects and functions in plants has not been comprehensively defined. The challenge and goal is to be able to manage microbial communities to favour plant colonization by beneficial bacteria. This would be possible when more knowledge on endophyte ecology and their molecular interaction is attained. The contributions of this research field may have economic and environmental impacts.

The most extensively studied mutualistic bacteria interacting with plants are Gram-negative because they are readily isolated from plant tissues, easily handled, amenable to genetic approaches and because they include the symbiotic nitrogen fixing microorganisms, namely, rhizobia. However, the impact of Gram-positive bacteria on plants is more important

[1]Universidad de Salamanca, C/Doctores de la Reina s/n 37007, Salamanca, Spain.
[2]Instituto de Recursos Naturales y Agrobiología. IRNASA-CSIC. C/ Cordel de Merinas, 52. 37008 Salamanca, Spain.
*Corresponding author: emm@usal.es

than was initially thought and though it is far less documented than Gram-negatives, it should not be underestimated (Francis et al. 2010). An overview with some specific examples will be given of the beneficial nodule-inhabiting microorganisms that interact with plants belonging to Gram-negative phylum *Proteobacteria* and the two Gram-positive phyla: *Actinobacteria* and *Firmicutes*. Studies regarding this niche for new microorganisms are still underdeveloped because not all of them can be isolated with standard procedures (Vincent 1970). Improving protocols in isolation and detection, new species with very interesting characteristics have been described that make these isolates potentially useful as plant growth promoting rhizobacteria (PGPR). Hence, in the last few decades research on the diversity of legume nodule endophytes has been increasing and currently it is an attractive field of research complementary to that of rhizobial diversity.

Proteobacteria Endophytes

The phylum *Proteobacteria* comprises Gram-negative aerobic or facultatively anaerobic bacteria and is divided into several classes from which alpha, beta and gamma contain legume nodule endophytes as well as the rhizobial endosymbionts (Velázquez et al. 2011, Gyaneshwar et al. 2011, Shiraishi et al. 2010).

The classical rhizobia, which belong to the class *alpha-Proteobacteria*, are able to induce legume nodules but they are also frequently found as endophytes which, despite their inability to reinfect the legume from which they were isolated, often promote its growth. For example *R. leguminosarum* bv. *phaseoli* and *Mesorhizobium loti* were isolated from nodules of *Trifolium pratense* and combinations of endophytic *R. leguminosarum* bv. phaseoli strains and *R. leguminosarum* bv. *trifolii* resulted in the promotion of clover growth (Sturtz et al. 1997). Other endophytic rhizobia were isolated from *Vicia* nodules which after sequencing in different genes were classified into the genera *Ensifer* and *Shinella* as well as into the species *Rhizobium tropici* (Lei et al. 2008).

The abundance of endophytic *alpha-Proteobacteria* in legume nodules has been revealed in several studies such as those from Zakhia et al. (2006), Deng et al. (2011a), Hoque et al. (2011) and Li et al. (2011) and are good examples of the genetic diversity of legume nodule endophytes. Strains belonging to very diverse families, genera and species were identified on the basis of their 16S rRNA gene sequences in nodules of different legumes in Tunisia, China and Australia (Table 10.1). Although these strains did not form nodules, some of them have *nifH* genes close to those of the endosymbiont *Sinorhizobium* (currently *Ensifer*) *meliloti* (Zakhia et al. 2006).

Table 10.1 Endophytes found in legume nodules.

Phylum *Proteobacteria*:		
Class *alpha-Proteobacteria*		
Agrobacterium rhizogenes‡	*Trifolium pratense*	Sturtz et al. (1997)
Agrobacterium rubi†	*Caragana sinica*	Hou et al. (2009)
Agrobacterium tumefaciens¥	*Acacia abyssinica, A. seyal, Acacia* sp., *Amorpha fruticosa, Amphicarpaea trisperma, Argyrolobium uniflorum, Campylotropis delavayi, C. prainii, Caragana microphylla, C. pruinosa, C. sinica, Cassia leschenaultiana,Colytea nepalensis, Erythrina brucei, Glycine max, Gueldenstaedtia niultiyora, Lespedeza* sp., *Medicago sativa, Melilotus albus, M. dentatus, M. officinalis, Mimosa pudica, Ononis viciaefolia, Phaseolus vulgaris, Sesbania cannabina, Sophora alopecuroides, Thermopsis lanceolata, V. angustifolia, V. bungei, V. faba, V. sativa*	Chen et al. (2000), Tan et al. (1999), Han et al. (2005), Mhamdi et al. (2005), Wolde-Meskel et al. (2005), Li et al. (2008), Wang et al. (2006), Diouf et al. (2007), Kan et al. (2007), Liu et al. (2007), Mahdhi et al. (2008), Cummings et al. (2009), Hou et al. (2009), Yu et al. (2009), Palaniappan et al. (2010), Zhao et al. (2010), Djedidi et al. (2011), Hoque et al. (2011)
Agrobacterium vitis§	*Acacia salina*	Wolde-Meskel et al. (2005)
Agrobacterium sp.∫	*Caragana sinica, Colytea nepalensis, Glycyrrhiza eurycarpa, G. glabra, G. squamulosa, G. uralensis, Melilotus officinalis, Thermopsis lanceolata, Vicia charca, V. sepium, V. unijuga, Wisteria sinensis*	Liu et al. (2005), Lei et al. (2008), Hou et al. (2009)
Bosea thiooxidans, B. mexicana	*Astragalus gombiformis, Ononis vaginalis, Retama raetam*	Zakhia et al. (2006)
Caulobacter sp.	*Acacia* sp.	Hoque et al. (2011)
Ensifer sp.	*Vicia amoena, V. charca*	Lei et al. (2008)
Kaitsia sp.	*Acacia* sp.	Hoque et al. (2011)
Inquilinus limosus	*Astragalus algerianus, Calycotome villosa, Sphaerophysa salsula*	Zakhia et al. (2006), Deng et al. (2011a)
Mesorhizobium loti	*Trifolium pratense*	Sturtz et al. (1997)
Methylobacterium fujisawaense	*Lespedeza* sp.	Palianappan et al. (2010)

Ochrobactrum ciceri	*Cicer arietinum*	Imran et al. (2010)
Ochrobactrum grignonense	*Retama raetam*	Zakhia et al. (2006)
Paracraurococcus ruber	*Retama raetam*	Zakhia et al. (2006)
Paracoccus halophilus, *P. sphaerophisae*	*Sphaerophysa salsula*	Deng et al. (2011a, b)
Phyllobacterium myrsinacearum	*A. algerianus*, *A. uniflorus*, *Calycotome villosa*, *Lathyrus numidicus*	Zakhia et al. (2006)
Phyllobacterium sp.	*Glycyrrhiza squamulosa*, *Vicia charca*, *V. multicaules*, *V. pseudoorobus*	Lei et al. (2008), Li et al. (2011)
Rhodopseudomonas rhenobacensis	*Ononis natrix*	Zakhia et al. (2006)
Rhizobium leguminosarum bv. *phaseoli*	*Trifolium pratense*	Sturtz et al. (1997)
Rhizobium tropici	*Vicia sativa*, *V. unijuga*, *V. venosa*	Lei et al. (2008)
Rhodobacter changlensis	*Glycyrrhiza* sp.	Li et al. (2011)
Shinella sp.	*Vicia charca*	Lei et al. (2008)
Sphingomonas asaccharolytica	*Astragalus gombiformis*, *Calycotome villosa*, *Lotus argenteus*	Zakhia et al. (2006)
Sphingomonas pruni	*Sphaerophysa salsula*	Deng et al. (2011a)
Class *beta-Proteobacteria*		
Bordetella avium	*Trifolium pratense*	Sturtz et al. (1997)
Burkholderia cepacia	*Acacia seyal*	Diouf et al. (2007)
Burkholderia phenazinium, *B. phytofirmans*, *B. caledonica*, *B. glathei*, *B. sediminicola*, *B. sordidicola*	*Lespedeza* sp.	Palianappan et al. (2010)
Burkholderia sp.	*Glycine max*, *Mimosa pudica*	Pandey et al. (2005), Li et al. (2008)

Table 10. contd....

Table 10.1 contd.

Class *beta-Proteobacteria*		
Comamonas sp.	*Acacia* sp.	Hoque et al. (2011)
Herbaspirillum lusitanum	*Phaseolus vulgaris*	Valverde et al. (2003)
Herbaspirillum sp.	*Acacia salina, A. stenophylla*	Hoque et al. (2011)
Roseateles sp.	*Acacia* sp.	Hoque et al. (2011)
Variovorax sp.	*Acacia* sp.	Hoque et al. (2011)
Acinetobacter sp.	*Glycine max, Acacia* sp.	Li et al. (2008), Hoque et al. (2011)
Ancylobacter	*Acacia* sp.	Hoque et al. (2011)
Dyella koreensis, D. japonica, D. marensis	*Lespedeza* sp.	Palianappan et al. (2010)
Enterobacter sp.	*Acacia* sp., *Arachis hypogaea, Glycyrrhiza uralensis*	Ibáñez et al. (2009), Hoque et al. (2011), Li et al. (2011)
Erwinia sp.	*Glycyrrhiza uralensis*	Li et al. (2011)
Klebsiella sp.	*Arachis hypogaea*	Ibáñez et al. (2009)
Labrys neptuniae	*Neptunia oleracea*	Chou et al. (2007)
Pantoea agglomerans	*Glycyrrhiza uralensis*	Li et al. (2011)
Pantoea agglomerans, Enterobacter kobei, E. cloacae, Escherichia vulneris, Leclercia adecarboxylata,	*Hedysarum carnosum, H. pallidum, H. spinosissimum subsp. capitatum*	Benhizia et al. (2004)
Pantoea agglomerans, Leclercia decarboxilata	*Oxitropis ochrocephala, O. glabra, Vicia angustifolia, V. faba*	Kan et al. (2007)
Pantoea sp., *Serratia* sp.	*Glycine max*	Li et al. (2008)
Pseudomonas fluorescens	*Sphaerophysa salsula*	Deng et al. (2011a)
Pseudomonas syringae, Pseudomonas brassicacearum	*Hedysarum carnosum, Medicago truncatula*	Zakhia et al. (2006)

Pseudomonas sp.	*Acacia* sp., *Arachis hypogaea*, *Hedysarum carnosum*, *H. spinosissimum subsp. capitatum*, *H. pallidum*	Benhizia et al. (2004), Ibáñez et al. (2009), Hoque et al. (2011)
Serratia plymuthica	*Sphaerophysa salsula*	Deng et al. (2011a)
Starkeya novella	*Retama raetam*	Zakhia et al. (2006)
Stenotrophomonas sp.	*Acacia* sp., *Vicia angustifolia*	Kan et al. (2007), Hoque et al. (2011)
Phylum *Firmicutes*		
Bacillus cereus	*Sophora alopecuroides*	Zhao et al. (2010)
Bacillus circulans	*Calycotome villosa*	Zakhia et al. (2006)
Bacillus megaterium	*Cajanus cajan*, *Lespedeza* sp., *Medicago sativa*	Rajendran et al. (2008), Stajković et al. (200[illegible]), Palianappan et al. (2010)
Bacillus megaterium, *B. insolitus*, *B. brevis*	*Trifolium pratense*	Sturz et al. (1997)
Bacillus simplex, *B.licheniformis*, *B. cereus*, *B. pumilus*	*Sphaerophysa salsula*	Deng et al. (2011a)
Bacillus subtilis, *B. thuringiensis*	*Glycine max*	Bai et al. (2007)
Bacillus subtilis	*Phaseolus vulgaris*	Lee et al. (2005)
Bacillus thuringiensis	*Pueraria thunbergiana*	Selvakumar et al. (2008)
Bacillus sp.	*Acacia sp*, *Glycine max*, *Robinia pseudoacacia*,	Li et al. (2008), Shiraishi et al. (2010), Hoque et al. (2011)
Brevibacillus borstelensis	*Sphaerophysa salsula*	Deng et al. (2011a)
Brevibacillus brevis	*Acacia* sp.	Hoque et al. (2011)
Cohnella phaseoli	*Phaseolus vulgaris*	García-Fraile et al. (2008)
Exiguobacterium sp.	*Trigonella foenum-graecum*	Rajendran et al. (2012)
Lysinibacillus fusiformis	*Sphaerophysa salsula*	Deng et al. (2011a)
Paenibacillus amylolyticus	*Sphaerophysa salsula*, *Glycyrrhiza* sp.	Deng et al. (2011a), Li et al. (2011)
Paenibacillus lautus	*Lotus argenteus*	Zakhia et al. (2006)

Table 10.[illegible] contd....

Table 10.1 contd.

Phylum *Firmicutes*		
Paenibacillus pasadenensis, P. polymyxa, Paenibacillus sp.	*Glycyrrhiza eurycarpa, G. uralensis, Glycyrrhiza* sp.	Li et al. (2011)
Paenibacillus prosopidis	*Prosopis farcta*	Valverde et al. (2010)
Paenibacillus sp.	*Acacia* sp.	Hoque et al. (2011)
Staphylococcus saprophyticus	*Sphaerophysa salsula*	Deng et al. (2011a)
Staphylococcus warneri	*Lespedeza* sp.	Palianappan et al. (2010)
Phylum *Actinobacteria*		
Agromyces cerinum	*Argyrolobium uniflorum*	Zakhia et al. (2006)
Arthrobacter nitroguaiacolicus	*Lespedeza* sp.	Palianappan et al. (2010)
Arthrobacter sp.		
Brevibacterium chosinensis	*Medicago sativa*	Stajković et al. (2009)
Curtobacterium luteum	*Trifolium pratense*	Sturz et al. (1997)
Curtobacterium sp.	*Ornithopus pinnatus*	Muresu et al. (2008)
Kribbella lupini	*Lupinus angustifolius*	Trujillo et al. 2006
Microbacterium barkeri	*Retama raetam*	Zakhia et al. (2006)
Microbacterium flavescens	*Ononis natrix, Retama raetam*	Zakhia et al. (2006)
Microbacterium ginsengisoli	*Lespedeza* sp.	Palianappan et al. (2010)
Microbacterium trichothecenolyticum	*Medicago sativa*	Stajković et al. (2009)
Microbacterium sp.	*Acacia* sp., *Glycyrrhiza* sp., *Ornithopus pinnatus,*	Muresu et al. (2008), Hoque et al. (2011), Li et al. (2011)
Micromonospora lupini, M. saelisecensis	*Lupinus angustifolius*	Trujillo et al. 2007
Micromonospora spp.	*Lupins angustifolius, L. gredensis*	Trujillo et al. 2010
Micromonospora pisi	*Pisum sativum*	García et al. 2011

Micromonospora spp.	*Arachis hypogaea, Phaseolus vulgaris, Pisum sativum*	Carro et al. (2012), Cerda (2008)
Mycobacterium frederiksbergensis	*Astragalus armatus*	Zakhia et al. (2006)
Mycobacterium sacrum	*Sphaerophysa salsula*	Deng et al. (2011a)
Mycobacterium sp.	*Acacia* sp.	Hoque et al. (2011)
Nocardia uniformis	*Sphaerophysa salsula*	Deng et al. (2011a)
Ornithicoccus hortensis	*Medicago truncatula*	Zakhia et al. (2006)
Streptomyces bottropensis	*Sphaerophysa salsula*	Deng et al. (2011a)

‡Currently *Rhizobium rhizogenes*.
¥Currently *Rhizobium tumefaciens*.
†Currently *Rhizobium rubi*.
§Currently *Rhizobium vitis*.
ʃCurrently *Rhizobium* sp.

A common legume nodule endophyte of *alpha-Proteobacteria* is the genus *Phyllobacterium* that has been found as an endophyte in nodules of *Agyrolobium, Astragalus, Calycotome* and *Lathyrus* in Tunisia (Zakhia et al. 2006) and *Vicia, Oxytropis* and *Glycyrrhiza* in China (Kan et al. 2007, Lei et al. 2008, Li et al. 2011). Recently, our group found a new species of this genus, named *P. endophyticum*, in nodules of *Phaseolus vulgaris* in Spain (Flores-Felix et al. unpublished).

Other *alpha-Proteobacteria* endophytes such as *Caulobacter* and *Kaitsia* have been found in nodules of *Acacia* in Australia (Hoque et al. 2011) and *Rhodobacter, Erwinia* and *Pantoea* were found in those from *Glycyrrhiza* in China (Li et al. 2011). Also the new species *Ochrobactrum cicer* was found in nodules of *Cicer arietinum* in Pakistan (Imran et al. 2010).

The class *beta-Proteobacteria* contains many new species of the genus *Burkholderia* able to induce nitrogen fixing nodules in several legumes (revised by Velázquez et al. 2011) and since the current research on nodular *Burkholderia* strains is focused on their ability for nodulation there are still very few reports about endophytic *Burkholderia.* Nevertheless, they have been reported in nodules of *Mimosa pudica* in India (Pandey et al. 2005), *Glycine max* in China (Li et al. 2008) and *Acacia seyal* in Senegal (Diouf et al. 2007). Although some of these strains are able to induce ineffective nodules (Diouf et al. 2007), most of them are endophytes presenting *in vitro* plant growth mechanisms such as phytohormone secretion, ACC deaminase, phosphate solubilization and biocontrol activity (Pandey et al. 2005).

In addition to *Burkholderia*, within *beta-Proteobacteria* the species *Herbaspirillum lusitanum* has been isolated in Portugal from nodules of *Phaseolus vulgaris* by our research group (Valverde et al. 2003). *Herbaspirillum* sp., *Comamonas* and *Variovorax* were found in *Acacia stenophylla* nodules and *Roseateles* in nodules of *Acacia* sp. in Australia (Hoque et al. 2011). Also some animal pathogenic *beta-Proteobacteria* such as *Bordetella avium* have been identified in nodules of *Trifolium pratense* (Sturz et al. 1997). Interestingly, this species was able to promote clover growth either individually and in combination with *R. leguminosarum* bv. *trifolii* (Sturz et al. 1997).

Gamma-Proteobacteria only contain a species able to induce legume nodules, belonging to the genus *Pseudomonas* (Shiraishi et al. 2010), but comprises several genera and species that have been widely isolated from nodules of different legumes. Specifically, members of this genus were one of the first described legume nodule endophytes in several species of *Hedysarum* (Benhizia et al. 2004). Later, *P. syringae* and *P. brassicacearum* identified by 16S rRNA gene sequencing were isolated from nodules of *Medicago truncatula* and *Hedysarum carnosum*, respectively, in Tunisia (Zakhia et al. 2006). The species *P. fluorescens* was found in nodules of *Sphaerophysa salsula* in China (Deng et al. 2011a) and several *Pseudomonas* strains able to promote plant growth and to colonize the pre-nodules were found in *Arachis hypogaea* nodules (Ibáñez et al. 2009).

Several species from *Enterobacteriaceae*, such as *Pantoea agglomerans*, *Enterobacter kobei*, *Enterobacter cloacae*, *Leclercia adecarboxylata* and *Escherichia vulneris*, were identified after 16S rRNA gene sequencing in nodules of several species of *Hedysarum* in Algeria (Benhizia et al. 2004). *Enterobacter* sp., *Pantoea* sp. and *Klebsiella* sp. were identified in nodules of *Arachis hypogaea* in Argentina (Ibáñez et al. 2009) and *Enterobacter* sp. has been also found in nodules of *Acacia* sp. in Australia (Hoque et al. 2011). Besides, there are several studies in China reporting the occurrence of *Enterobacteriaceae* strains in nodules of *Vicia faba*, *Vicia angustifolia*, *Oxytropis glabra* and *Oxytropis ochrocephala* (Kan et al. 2007). *Serratia plymuthica* was found in nodules of *Sphaerophysa salsula* (Deng et al. 2011a) and *Pantoea* sp. and *Serratia* sp. were found in *Glycine max* (Li et al. 2008). Although some of these strains present *in vitro* plant growth promotion mechanisms such as the production of indole acetic acid, phosphate solubilization and free-living nitrogen fixation ability, no effect on growth and nodulation of soybean was reported.

Other *gamma-Proteobacteria* were found in several legume nodules, such as *Stenotrophomonas* isolated from nodules of *Vicia angustifolia* and *Acinetobacter* isolated from *Glycine max* in China (Kan et al. 2007, Li et al. 2008). These two genera together with *Ancylobacter* were found in *Acacia* sp. nodules in Australia (Hoque et al. 2011), *Starkeya novella* was found in nodules of *Retama raetam* in Tunisia (Zakhia et al. 2006), and *Labrys neptuniae* was isolated from *Neptunia oleracea* in Taiwan (Chou et al. 2007).

Nevertheless, within the phylum *Proteobacteria*, the genus *Agrobacterium* belonging to the class *alpha-Proteobacteria* and currently classified into the genus *Rhizobium* (Young et al. 2001), is the endophyte more widely found in legume nodules up to date. This genus was one of the first reported endophytes of legume nodules because it is routinely isolated from them together with nodulating rhizobia. Since the colony morphologies of "*Agrobacterium*" strains are similar to those of rhizobia, this bacterium was commonly not discarded in the initial steps of the recollection of nodule isolates as occurred with other endophytes. Although some strains have been reported as endosymbionts able to form effective nodules in soybean (Chen et al. 2000) and *Sophora* (Zhao et al. 2010) and ineffective nodules in *Acacia* (Diouf et al. 2007) and *Trifolium*, *Crotalaria* and *Mimosa* (Liu et al. 2007), most of strains from this old genus are endophytes unable to produce legume nodules.

One of the first reports on the presence of "*Agrobacterium*" in nodules as an endophyte was published by Sturtz et al. in 1997. In this work the species "*Agrobacterium rhizogenes*" (currently *Rhizobium rhizogenes*) was isolated from *Trifolium* nodules being able to promote the nodulation of this legume by *R. leguminosarum* bv. trifoli. Strains from other species such as "*Agrobacterium vitis*" (currently *R. vitis*) have been later isolated from *Acacia saligna* (Wolde-Meskel et al. 2005) and strains from "*Agrobacterium rubi*" (currently *Rhizobium rubi*) have been isolated from *Caragana sinica* (Hou et al. 2009). Strains of genus "*Agrobacterium*" probably belonging to undescribed species have also been found in these nodules as well as in nodules of *Wisteria sinensis* (Liu et al. 2005), several *Vicia* species (Lei et al. 2008), and in medicinal species from genus *Glycyrrhiza* (Li et al. 2011).

Nevertheless, up to date the species "*Agrobacterium tumefaciens*" (currently *Rhizobium radiobacter*) is the most frequent in nodules of herbaceous and woody legumes worldwide. In Tunisia this species has been isolated from nodules of herbaceous legumes such as *Phaseolus vulgaris* (Mhamdi et al. 2005) and shrubs such as *Argyrolobium uniflorum* (Mahdhi et al. 2008). In China it is particularly frequent in nodules of herbaceous legumes such as *Gueldenstaedtia* and *Amphicarpaea* (Tan et al. 1999), *Phaseolus vulgaris* (Han et al. 2005), *Melilotus dentatus* (Wang et al. 2006), *Crotalaria pallida* and *Trifolium fragiferum* (Liu et al. 2007), *Lespedeza cuneata* (Yu et al. 2009), in different species of *Vicia* (Kan et al. 2007, Lei et al. 2008), in *Melilotus albus, Melilotus dentatus* and *Ononis viciaefolia* (Kan et al. 2007) and in *Caragana sinica, Colytea nepalensis, Melilotus officinalis* and *Thermopsis lanceolata* (Hou et al. 2009). In India two strains of this species isolated from *Vicia faba* nodules have been proposed as a new biovar of this species (Tiwary et al. 2007).

"*A. tumefaciens*" (currently *Rhizobium radiobacter*) has been also found in nodules of woody legumes in China such as *Caragana* (Tan et al. 1999), *Campylotropis* and *Cassia* (Han et al. 2005), *Sophora alopecuroides* (Zhao et al. 2010), *Amorpha fruticosa* (Liu et al. 2005) and *Mimosa pudica* (Kan et al. 2007). In some African countries such as Ethiopia *A. tumefaciens* has been isolated from nodules of *Erythrina brucei* and *Acacia abyssinica* (Wolde-Meskel et al. 2005) and in Senegal it has been isolated from *Acacia seyal* nodules (Diouf et al. 2007). In Australia several strains of *Rhizobium radiobacter* have been found in nodules of *Acacia* (Hoque et al. 2011).

Usually the nodule endophytic strains of genus "*Agrobacterium*" (currently *Rhizobium*) are identified by 16S rRNA gene sequencing, but additional core genes such as *rpoB* gene have been used to identify strains from this genus in nodules of *Sesbania cannabina* (Cummings et al. 2009). The *recA* and *atpD* genes were analyzed by Saïdi et al. (2011) in isolates from legume nodules in Tunisia who concluded that most of them belong to the species "*Agrobacterium radiobacter*". The *recA* gene was also analyzed in strains isolated from *Sophora alopecuroides* nodules (Zhao et al. 2010) and *atpD* and *glnII* genes were analyzed in isolates from several wild species from *Vicia* (Lei et al. 2008).

In several studies the presence of auxiliary genes encoding for virulence (*vir*) or symbiosis (*nod* genes) in strains of *A. tumefaciens* has also been analyzed. Genes encoding nod factors have been found in some strains of this species isolated from nodules of *Medicago sativa* that carry a *nodA* gene similar to that from *Ensifer meliloti* and were able to nodulate this host (Djedidi et al. 2011). Also, strains of this species carrying *nodA* genes close to those of *Mesorhizobium* species were found in nodules of *Sophora alopecuroides* by Zhao et al. (2010). Finally, a strain classified into the genus "*Agrobacterium*", but probably belonging to a new species, does not carry t *vir* genes but nodulation genes that allow the nodulation in *Sesbania cannabina* (Cummings et al. 2009).

Some authors have analyzed the effect of "*Agrobacterium tumefaciens*" strains on legume nodulation and it has been reported that some strains cause a reduction of nodulation of *P. vulgaris* by *R. gallicum* (Mrabet et al. 2006). However, no effect on nodulation was found in the case of soybean when a strain of this species isolated from *Glycine* nodules was inoculated with *Bradyrhizobium japonicum* in *Glycine max* (Li et al. 2008). Other authors have found positive effects on the nodulation since strains of "*Agrobacterium*" can induce unspecific nodulation of *Sinorhizobium meliloti* in the woody legume *Wisteria sinensis* (Liu et al. 2010).

In recent works the ability of "*Agrobacterium*" to promote the plant growth is also being investigated and despite some strains such as those from soybean nodules reported by Li et al. (2008) did not show any *in vitro* plant growth promotion mechanisms, other strains presented several of them. For example, strains close to "*Agrobacterium tumefaciens*" (currently *Rhizobium radiobacter*) recently isolated from nodules of the medicinal legume *Glycyrrhiza* in China were able to solubilize phosphate and to produce indole acetic acid, ACC deaminase and siderophores (Li et al. 2011). A strain of this species isolated from nodules of *Lespedeza* presenting different *in vitro* plant growth promotion mechanisms such as phosphate solubilization and indole acetic acid, ACC deaminase and siderophore production was also able to promote plant growth *in vivo* (Palaniappan et al. 2010).

All these results showed that strains from the former genus *Agrobacterium* are common inhabitants of legume nodules together with rhizobia, but few strains carry symbiotic information and the nodules formed by these strains are usually ineffective. The ability to promote the growth of *Agrobacterium* strains has not been widely studied although in some strains *in vitro* mechanisms to promote legumes and non-legumes have been detected and there are even reports of the legume plant growth promotion *in vivo*.

Firmicutes Endophytes

As stated in the introduction, the best studied mutualistic interactions between plants and microorganisms almost always involve Gram-negative bacteria. However, the impact of Gram-positive bacteria in plants is more important than was initially thought.

Gram-positive bacteria (*Firmicutes* and *Actinobacteria*) form spores, are usually pigmented and produce a huge amount of bioactive metabolites.

The phylum *Firmicutes* is defined by a low G+C content and includes the classes *Clostridia, Erysipelotrichi, Thermolithobacteria, Mollicutes* and *Bacilli,* that comprises rod-shaped sporulated bacteria (Francis et al. 2010). These microorganisms are often isolated from legume nodules and other plant tissues. Some studies relate these spore forming taxa with lack of geographical region-specificity and genetic homogeneity (Li et al. 2011).

In the last decades, legume nodules have been a source of a great variety of microorganisms that belong to the phylum *Firmicutes*. Recently, strains from the genera *Staphylococcus* and *Exiguobacterium* have been isolated (Rajendran et al. 2008, Palaniappan et al. 2010, Deng et al. 2011a) from *Lespedeza* sp. and *Trigonella foenum-graecum* nodules and are the only isolates that belong to these genera found up to date.

However, the occurrence of *Bacillus* species inside root nodules has been reported repeatedly from a range of plants such as *Acacia, Cajanus, Glycine, Calycotome, Medicago, Lespedeza, Phaseolus, Pueraria, Robinia, Sphaerophysa* and *Trifolium* (Table 10.1). They have also been shown to act as symbiotic organisms to their hosts by promoting nodulation and growth.

For example, *Bacillus megaterium* isolated from *Trifolium pratense* nodules was able to promote its growth in absence of *Rhizobium,* and *Bacillus brevis* and *Bacillus insolitus* isolated from the same nodules were able to promote nodulation by *R. leguminosarum* bv. trifolii (Sturz et al. 1997). *Bacillus subtilis* and *Bacillus thuringiensis* were isolated from *Glycine max* (soybean) in China and were capable of promoting plant growth when coinoculated with *Bradyrhizobium* nodulating this same host (Bai et al. 2002). It has been found that another strain of the species *B. thuringiensis,* isolated from the root nodules of the wild legume *Pueraria thunbergiana* (Kudzu), was responsible for an enhancement of plant growth and nodulation in pea and lentil. The same results were found with *B. pumilus* when it was coinoculated with *Bradyrhizobium* in soybean (Li et al. 2008). *B. thuringiensis* strains able to promote plant growth of wheat were also isolated in China from nodules of *Pueraria thunbergiana.* An interesting discovery of one of the *B. thuringiensis* isolates from *P. thunbergiana* is its ability to produce IAA and HCN in appreciable quantities. IAA is a growth-promoting hormone and HCN production by microorganisms from the rhizosphere could be used as a possible defence mechanism against plant pathogens (Selvakumar et al. 2007).

Bacillus and *Paenibacillus* strains were also found in nodules of wild legumes in Tunisia (Zakhia et al. 2006), *Robinia pseudoacacia* in Japan (Shiraishi et al. 2010), and medicinal species of *Glycyrrhiza* in China (Li et al. 2011). After conducting isolations from nodules of *Glycyrrhiza* plants in China, it was found that more than 20 percent of all the strains were Gram-positive bacteria, and 30 *Paenibacillus* strains were found, representing several species, such as *P. amylolyticus, P. graminis, P. pasadenensis* and *P. polymyxa*. When studying the *Paenibacillus* isolates, it was found that a high percentage of the strains had protease activity (61 percent), cellulase (57 percent), siderophores production (38 percent) and also small amounts of IAA (44 percent) (Li et al. 2011). *Paenibacillus* is the second most abundant genera found in legume nodule isolates. It has been isolated from *Hedysarum, Glycyrrhiza, Lotus, Phaseolus, Prosopis* and *Sphaerophysa* (Table 10.1) and is also a very good candidate for PGPR studies and biotechnological applications.

This is not the only example of large numbers of Gram-positive bacteria isolation: in studies with the legume *Trigonella foenum-graecum*, about 29.5 percent of the strains isolated from nodules were Gram-positive bacilli, also characterized for their PGPR potential (Rajendran et al. 2012).

PGPR potential in Gram-positive bacteria has been studied also with *Phaseolus vulgaris* under field conditions, with single, dual and triple inoculations with *Rhizobium*, N_2-fixing *Bacillus subtilis* and P-solubilizing *Bacillus megaterium*. Results showed an increase in growth, nodulation, seed yield, chlorophyll and protein content and PNE uptake (Elkoca et al. 2010).

Also, in *Medicago sativa*, inoculations done only with all non-rhizobial strains studied caused significant increase in shoot and root parameters compared to a non inoculated control (Stajkovi et al. 2009).

Strains from several *Firmicutes* genera were identified after sequencing of the 16S rRNA gene in nodules of *Sphaerophysa salsula* in China. These bacteria were studied and found unable to induce nodules or fix nitrogen even though they live inside nodules and do not visibly harm host plants. These strains belong to *Staphylococcus saprophyticus* from family *Staphylococcaceae, Bacillus cereus, Bacillus licheniformis, Bacillus pumilus, Bacillus simplex* and *Lysinibacillus fusiformis* from family *Bacillaceae*, and *Brevibacillus borstelensis* and *P. amylolyticus* from family *Paenibacillaceae* (Deng et al. 2011a).

It is interesting how new species have been described from root nodule isolates. For example, *Cohnella phaseoli* is a new species isolated from root nodules of *Phaseolus coccineus* in Spain (García-Fraile et al. 2008). This species promotes the growth of *Phaseolus vulgaris* increasing 10 percent of dry matter with respect to the uninoculated plants (García-Fraile 2008). Also, there are some recently described new species of *Paenibacillus* that were isolated from legume nodules such as *Paenibacillus prosopidis*, isolated from *Prosopis farcta* in Tunisia (Valverde et al. 2010). This genus comprises many described species isolated from plant rhizospheres. Moreover, several species of this genus, including the type species, *Paenibacillus polymyxa*, have been reported as plant growth-promoting rhizobacteria (Rivas et al. 2005).

Actinobacteria Endophytes

Actinobacteria are a diverse group of Gram-positive bacteria that range in shape from unicellular organisms to branching filaments stretching to form a complex network called a mycelium. These bacteria contain a high amount of guanine plus cytosine (450 mol% G+C) in their DNA. Many actinobacteria produce specialized reproductive structures called spores which are resistant to ultraviolet light and survive well under dry conditions. The great versatility of the secondary metabolism of many Gram-positive bacteria makes them suitable as biocontrol organisms against insects, nematodes, fungi and other bacteria, and their broad catabolic capacities provide a basis for their use as bioremediation agents.

Few reports are available regarding the presence of non-rhizobial microorganisms (associated or endophytic) in the nodules in spite of the fact that nodules are much richer in nutrients as compared to roots. In the case of actinobacteria, these microorganisms have been isolated from nodule tissues and include the genera *Agromyces* (Zakhia et al. 2006), *Curtobacterium* (Sturz et al. 1997), *Microbacterium* (Zakhia et al. 2006, Muresu et al. 2008, Stajković et al. 2009, Hoque et al. 2011, Li et al. 2011), *Micromonospora* (Trujillo et al. 2007, García et al. 2010, Trujillo et al. 2010, Carro et al. 2012) and *Streptomyces* (Deng et al. 2011a).

Deng and colleagues (2011a) carried out a sampling of the legume *Sphaerophysa salsula* at two locations of the Loess Plateau in China. These authors reported the isolation of 115 endophytic bacteria which included six actinobacterial isolates. These strains were identified by 16S rRNA gene sequencing and showed the closest similarity to *Mycobacterium sacrum* (2 strains), detected in both sampling sites, *Nocardia* and *Streptomyces*. Coinoculation tests carried out with three of these strains had no negative effects in the host plants, in comparison with the non inoculated controls (Deng et al. 2011a) However, the interactions between symbiotic and nonsymbiotic endophytes are still unclear.

Agromyces sp. and *Microbacterium* spp. were reported from nodules collected from wild legumes sampled in geographically distant sites in the infra-arid zone of Tunisia (Zakhia et al. 2006). In this study seven actinobacterial strains were isolated from the diverse legumes sampled. Three strains representing a different genus each were recovered from these nodules. Specifically one strain isolated from *Argyrolobium uniflorum* was identified as *Agromyces cerinum* based on partial 16S rRNA sequencing (550 bp), a second strain recovered from *Astragalus armatus* was identified as *Mycobacterium frederiksbergense* (100 percent similarity based on 552 bp), and a third strain, isolated from *Medicago truncatula*, showed 97 percent sequence similarity (based on 573 bp) to *Orthinicoccus hortensis*. The remaining four strains reported in this study belonged to the genus *Microbacterium* and were isolated from *Retama raetam* and *Ononis natrix* subsp. *falcata*.

Zakhia et al. (2006) carried out nodulation tests with the six actinobacterial strains using *Macroptilium atropurpureum* and *Retama raetam* plants. In all cases, nodulation tests were negative, however, these researchers reported that *Microbacterium flavescens* strain ORS 1472 induced a black pigment, which covered the whole root system of both plants.

In addition, these strains were tested for the presence of *nifH*-like gene sequences by PCR amplification. Positive results were reported for three *Microbacterium* strains ORS1417, ORS1418 and ORS1472, and *Agromyces*-like strain ORS1437. The sequence contigs obtained (314–318 bp) showed a similarity between 93–96 percent to *Sinorhizobium meliloti* suggesting a lateral gene transfer from this microorganism. Nitrogen fixing activity within the actinobacteria is well demonstrated for *Frankia* but it has also been reported for the genera *Arthrobacter* (Cacciari et al. 1971), *Streptomyces* (Ding et al. 1981, Knapp and Jurtshuk 1988) and *Propionibacterium* (Baranova and Gogotov 1974). Furthermore, Ruppel (1989) reported that *Mycobacterium* and *Microbacterium* strains had some low nitrogen-fixing activity, as measured by acetylene-reducing capacity.

The genera *Microbacterium* (four strains) and *Curtobacterium* (one strain) were also reported from nodules of *Ornithopus pinnatus* (Muresu et al. 2008) but no further studies were carried out with these strains. *Microbacterium trichothecenolyticum* was also found in alfalfa nodules (Stajković et al. 2009). *Microbacterium* strains have also been isolated from *Lezpedeza* sp. (Palaniappan et al. 2010), *Acacia salicina* and *A. stenophylla* (Hoque et al. 2011). These studies suggest that *Microbacterium* may be a common inhabitant of legumes.

The genus *Micromonospora* which is widely distributed in diverse habitats has recently been reported as a major component of legume nitrogen fixing root nodules. This microorganism has been isolated from the wild legume *Lupinus angustifolius* (Trujillo et al. 2007, Trujillo et al. 2010) and also from cultivated plants such as *Pisum sativum* (Carro et al. 2012). Successful isolation of *Micromonospora* from internal nodule tissues has been reported from several legume species including *Arachis hypogaea*, *Cicer arietinum*, *Glycine max*, *Lens culinaris*, *Lupinus gredensis*, *Medicago sativa*, *Ononis* sp., *Ornithopus* sp., *Phaseolus* sp., *Trifolium* sp., and *Vicia* sp. (Cerda 2008, Rodriguez 2008, Alonso-Vega 2010, Trujillo et al. unpublished results).

These studies have shown that *Micromonospora* is widespread in legumes and that it can be found in both wild and cultivated legumes. Unlike the previous actinobacterial genera mentioned in this chapter, *Micromonospora* strains have been isolated in significant numbers from at least 20 different legume species suggesting a wide distribution among this important group of plants. Our laboratory currently contains a collection of approximately 1800 *Micromonospora* strains isolated from this niche.

In lupine plants (*Lupinus* sp.), the distribution pattern is rather complex ranging from no isolates for some nodules to as many as 29 colonies arising from a single nodule (Alonso de la Vega 2010, Trujillo et al. 2010). Variation has also been observed from plant to plant or even from different nodules of the same plant (Trujillo et al. 2010). In the case of *P. sativum*, the distribution of the microorganism was similar to that reported for *L. angustifolius*, however for *P. sativum*, at least one *Micromonospora* colony was isolated from every nodule sampled (Carro et al. 2012). The presence of *Micromonospora* within the nodule tissues of lupin is strongly supported by *in situ* localization using FISH and TEM studies, suggesting a close interaction between the host plant and the bacterium (Trujillo et al. 2010).

The genetic diversity of *Micromonospora* strains isolated from legume root nodules has been assessed by BOX–PCR and ARDRA fingerprinting studies (Trujillo et al. 2010, Carro et al. 2012). In both cases, a high degree of genetic variation was observed indicating that the strains isolated from the nodules are not clones; the diversity found was unexpectedly high considering such a specific niche. Phylogenetic analyses based on 16S rRNA gene sequencing demonstrated the high diversity at the species level of many of these *Micromonospora* nodule isolates. Many of these strains probably represent novel species, indeed, *M. lupini*, *M. saelicesensis* (Trujillo et al. 2007), and *M. pisi* (García et al. 2010) isolated from *L. angustifolius* and *P. sativum* respectively, have been formally described as new species. Furthermore, a comprehensive multilocus sequence analysis based on five genes and more than 90 *Micromonospora* strains isolated from *P. sativum* has shown excellent correlation with the 16S rRNA gene sequencing data and confirm the high diversity at the species level (Carro et al. 2012).

Given the recent discovery of the presence of *Micromonospora* in legume root nodules, very little is known about its ecological role and interaction with the plant and/or the rhizobial strains inhabiting these nodules. Coinoculation studies suggest that *Micromonospora* has a positive effect on the plant, although nodulation tests using *Lupinus albus* plants showed that *Micromonospora* is not able to induce nodules (Cerda 2008). Current studies are aimed at understanding the potential role of this actinobacterium in legumes.

The great versatility of the secondary metabolism of many Gram-positive bacteria is well known. Studies of bacterial secondary metabolism have largely targeted the discovery of new compounds and the mechanisms for their biosynthesis. However, little is known about the ecological functions of secondary metabolites. These molecules may act as allelochemicals, signalling molecules and siderophores. A recent analysis of the biosynthetic potential of natural products by *Frankia* resulted in the identification of an important number of clusters which appeared to be unique and for which products have not been observed or characterized. The discovery of the biosynthetic potential of *Frankia* could have a significant impact on the study of host-microbe interactions at the molecular level (Udwary et al. 2011) as some of these molecules may function as cell signalling molecules during the establishment and development of the symbiotic association between *Frankia* and its host plant. Pujic and colleagues have recently reported on the potential role of lectins in the establishment of symbiosis with its host plant (Pujic et al. 2012). Furthermore, it was also recently reported that a polyketide synthetase/non-ribosomal peptide synthetase hybrid enzyme is involved in *Trichoderma*-plant interactions resulting in induction of defence responses (Mukherjee et al. 2012). These examples highlight the relevant role that secondary metabolites may play in plant-microbe communications.

Actinobacterial genomes are larger than those of most other bacteria and have a high capacity for the synthesis of secondary metabolites. The sequenced actinobacterial genomes (currently ~165) are a valuable resource, which have revealed that actinobacteria have the biosynthetic potential to make far more natural products than was realized before genome sequences were available. In this sense, *Micromonospora lupini* Lupac 08 is no exception and our research group has reported a family of new anthraquinone molecules with antitumoural activity from this strain (Igarashi et al. 2007, Igarashi et al. 2011). Preliminary data obtained from the genome sequence of *M. lupini* Lupac 08 indicates that it contains a significant number of genes that code for polyketides, non-ribosomal peptides, polyketide/nonribosomal and peptide hybrids in addition to pigments (e.g., melanin) and lectins. Some of these molecules may act as communication molecules between the microorganism and the plant to allow colonization by the bacterium.

Actinobacteria are a unique group of microorganisms which are usually pigmented, produce spores, produce a plethora of bioactive secondary metabolites and have a complex lifestyle. Understanding these new plant-microbe interactions as well as the ecological roles played by these microorganims is likely to have a positive impact in agricultural, biological control and bioremediation processes.

In summary, all the work reviewed in this chapter showed that the diversity of endophytes in legume nodules is higher than that suspected until now and some work revealed the presence of a high number of non-culturable rhizobia together with many non-rhizobia endophytes (Muresu et al. 2008, Zakhia et al. 2006). Taking these results into account, it is possible that some endophytes may affect some groups of rhizobia found in the same nodule in a limiting way under some circumstances; instead of being part of a fully mutualistic relationship, they could be competing for a niche. Nevertheless, many endophytic strains are isolated from nodules together with rhizobia and their ability for plant growth promotion opens a new way for the inoculants design applicable to legumes and non-legumes that will allow the reduction of the chemicals applied to crops.

Acknowledgments

The authors would like to thank our numerous collaborators and students involved in this research over the years. Funding was provided by Ministerio de Ciencia e Innovación (MICINN) and Junta de Castilla y León from Spain.

References

Alonso de la Vega, P. 2010. Distribución, caracterización e importancia ecológica de *Micromonospora* en nódulos fijadores de nitrógeno de *Lupinus*. Ph.D. Thesis, Universidad de Salamanca, Salamanca, Spain.

Bai, Y., F. D'Aoust, D.L. Smith and B.T. Driscoll. 2002. Isolation of plant-growth-promoting *Bacillus* strains from soybean root. Can. J. Microbiol. 48: 230–238.

Baranova, N.A. and I.N. Gogotov. 1974. Fixation of molecular nitrogen by propionic bacteria. Mikrobiologiya. 43: 791–794 (in Russian).

Benhizia, Y., H. Benhizia, A. Benguedouar, R. Muresu, A. Giacomini and A. Squartini. 2004. Gamma proteobacteria can nodulate legumes of the genus *Hedysarum*. System. Appl. Microbiol. 27: 462–468.

Cacciari, I., G. Giovannozi-Sermanni, A. Grapelli and D. Lippi. 1971. Nitrogen fixation by *Arthrobacter* species. Ann. Microbiol. Enzimol. 21: 97–105.

Carro, L., C. Spröer, P. Alonso and M.E. Trujillo. 2012. Diversity of *Micromonospora* strains isolated from nitrogen fixing nodules and rhizosphere of *Pisum sativum* analyzed by multilocus sequence analysis. System. Appl. Microbiol. 35: 73–80.

Cerda, E. 2008. Aislamiento de *Micromonospora* de nódulos de leguminosas tropicales y análisis de su interés como promotor del crecimiento vegetal. Ph.D. Thesis, Universidad de Salamanca, Salamanca, Spain.

Chen, L.S., A. Figueredo, F.O. Pedrosa and M. Hungria. 2000. Genetic characterization of soybean rhizobia in Paraguay. Appl. Environ. Microbiol. 66: 5099–5103.

Chou, Y.J., G.N. Elliott, E.K. James, K.Y. Lin, J.H. Chou, S.Y. Sheu, D.S. Sheu, J.I. Sprent and W.M. Chen. 2007. *Labrys neptuniae* sp. nov., isolated from root nodules of the aquatic legume *Neptunia oleracea*. Int. J. Syst. Evol. Microbiol. 57: 577–581.

Cummings, S.P., P. Gyaneshwar, P. Vinuesa, F.T. Farruggia, M. Andrews, D. Humphry, G.N. Elliott, A. Nelson, C. Orr, D. Pettitt, G.R. Shah, S.R. Santos, H.B. Krishnan, D. Odee, F.M. Moreira, J.I. Sprent, J.P. Young and E.K. James. 2009. Nodulation of *Sesbania* species by *Rhizobium* (*Agrobacterium*) strain IRBG74 and other rhizobia. Environ. Microbiol. 11: 2510–2525.

Deng, Z.S., L.F. Zhao, Z.Y. Kong, W.Q. Yang, K. Lindström, E.T. Wang and G.H. Wei. 2011a. Diversity of endophytic bacteria within nodules of the *Sphaerophysa salsula* in different regions of Loess Plateau in China. FEMS Microbiol. Ecol. 76: 463–475.

Deng, Z.S., L.F. Zhao, L. Xu, Z.Y. Kong, P. Zhao, W. Qin, J.L. Chang and G.H. Wei. 2011b. *Paracoccus sphaerophysae* sp. nov., a siderophore-producing, endophytic bacterium isolated from root nodules of *Sphaerophysa salsula*. Int. J. Syst. Evol. Microbiol. 61: 665–669.

Ding, J., H. Sun, F. Su, Q. Xu, Y. Huang and P. Ling. 1981. Studies on nitrogen fixation by actinomycetes. Acta Microbiol. Sin. 21: 424–427.

Diouf, D., R. Samba-Mbaye, D. Lesueur, A.T. Ba, B. Dreyfus, P. de Lajudie and M. Neyra. 2007. Genetic diversity of *Acacia seyal* Del. rhizobial populations indigenous to Senegalese soils in relation to salinity and pH of the sampling sites. Microbial Ecol. 54: 553–566.

Djedidi, S., T. Yokoyama, N. Ohkama-Ohtsu, C.P. Risal, C. Abdelly and H. Sekimoto, H. 2011. Stress tolerance and symbiotic and phylogenic features of root nodule bacteria associated with *Medicago* species in different bioclimatic regions of Tunisia. Microbes Environ. 26: 36–45.

Elkoca, E., M. Turan and M.F. Donmez. 2010. Effects of single, dual and triple inoculations with *Bacillus subtilis, Bacillus megaterium* and *Rhizobium leguminosarum* bv. phaseoli on nodulation, nutrient uptake, yield and yield parameters of common bean (*Phaseolus vulgaris* L). J. Plant Nutr. 33: 2104–2119.

Faghire, M., B. Mandri, K. Oufdou, A. Bargaz, C. Ghoulam, M.H. Ramírez-Bahena, E. Velázquez and A. Peix. 2012. Identification at the species and symbiovar levels of strains nodulating *Phaseolus vulgaris* in saline soils of the Marrakech region (Morocco) and analysis of the *otsA* gene putatively involved in osmotolerance. Syst. Appl. Microbiol. 35: 156–64.

Francis, I., M. Holsters and D. Vereecke. 2010. The Gram-positive side of plant-microbe interactions. Environ. Microbiol. 12: 1–12.

García, L.C., E. Martinez-Molina and M.E. Trujillo. 2010. *Micromonospora pisi* sp. nov., isolated from root nodules of *Pisum sativum*. Int. J. Syst. Evol. Microbiol. 60: 331–337.

García-Fraile, P. 2008. Identificación y caracterización de bacterias que nodulan *Phaseolus* en diferentes regiones de España. Ph.D. Thesis, Universidad de Salamanca, Salamanca, Spain.

García-Fraile, P., E. Velázquez, P.F. Mateos, E. Martínez Molina and R. Rivas. 2008. *Cohnella phaseoli* sp. nov., isolated from root nodules of *Phaseolus coccineus* in Spain, and emended description of the genus *Cohnella*. Int. J. Syst. Evol. Microbiol. 58: 1855–1859.

Gyaneshwar, P., A.M. Hirsch, L. Moulin, W.M. Chen, G.N. Elliott, C. Bontemps, P. Estrada-de Los Santos, E. Gross, F.B. Dos Reis, J. Sprent, J.P. Young and E.K. James. 2011. Legume-nodulating betaproteobacteria: diversity, host range, and future prospects. Mol. Plant Microbe Interact. 24: 1276–88.

Han, S.Z., E.T. Wang and W.X. Chen. 2005. Diverse bacteria isolated from root nodules of *Phaseolus vulgaris* and species within the genera *Campylotropis* and *Cassia* grown in China. System. Appl. Microbiol. 28: 265–276.

Hoque, M.S., L.M. Broadhurst and P.H. Thrall. 2011. Genetic characterization of root-nodule bacteria associated with *Acacia salicina* and *A. stenophylla* (Mimosaceae) across south-eastern Australia. Int. J. Syst. Evol. Microbiol. 61: 299–309.

Hou, B.C., E.T. Wang, Y. Li, R.Z. Jia, W.F. Chen, C.X. Man, X.H., Sui W.X. Chen. 2009. Rhizobial Resource Associated with Epidemic Legumes in Tibet. Microbial Ecol. 57: 69–81.

Ibáñez, F., J. Angelini, T. Taurian, M.L. Tonelli and A. Fabra. 2009. Endophytic occupation of peanut root nodules by opportunistic Gammaproteobacteria. System. Appl. Microbiol. 32: 49–55.

Igarashi, Y., M.E. Trujillo, E. Martínez-Molina, S. Yanase, S. Miyanaga, T. Obata, H. Sakurai, I. Saiki, T. Fujita and T. Furumai, 2007. Antitumor anthraquinones from an endophytic actinomycete *Micromonospora lupini* sp. nov. Bioorg. Med. Chem. Lett. 17: 3702–3705.

Igarashi, Y., S. Yanase, K. Sugimoto, M. Enomoto, S. Miyanaga, M.E. Trujillo, I. Saiki and S. Kuwahara. 2011. Lupinacidin C, an inhibitor of tumor cell invasion from *Micromonospora lupini*. J. Nat. Prod. 74: 862–865.

Imran, A., F.Y. Hafeez, A. Frühling, P. Schumann, K.A. Malik and E. Stackebrandt. 2010. *Ochrobactrum ciceri* sp. nov., isolated from nodules of *Cicer arietinum*. Int. J. Syst. Evol. Microbiol. 60: 1548–1553.

Kan, F.L., Z.Y. Chen, E.T. Wang, C.F. Tian, X.H. Sui and W.X. Chen. 2007. Characterization of symbiotic and endophytic bacteria isolated from root nodules of herbaceous legumes grown in Qinghai-Tibet plateau and in other zones of China. Arch. Microbiol. 188: 103–115.

Knapp, R. and P. Jurtshuk. 1988. Characterization of free-living nitrogen-fixing *Streptomyces* species and factors which affect their rates of acetylene reduction. Abstr. Annu. Meet. Am. Soc. Microbiol. 88 (219).

Kuykendall, L.D., J.M. Young, E. Martínez-Romero, A. Kerr and H. Sawada. 2005. Order *Rhizobiales* (new) Family *Rhizobiaceae* Genus *Rhizobium*. *In*: D.J. Brenner, N.R. Krieg, J.T. Staley and G.M. Garrity [eds.]. The Alpha-, Beta-, Delta- and Epsilon-proteobacteria, The Proteobacteria, part C, Bergey's Manual of Systematic Bacteriology, 2nd edn. vol. 2, Springer, New York. pp. 324–340.

Lei, X., E.T. Wang, W.F. Chen, X.H. Sui and W.X. Chen. 2008. Diverse bacteria isolated from root nodules of wild *Vicia* species grown in temperate region of China. Arch. Microbiol. 190: 657–671.

Li, J.H., E.T. Wang, W.F. Chea and W.X. Chen. 2008. Genetic diversity and potential for promotion of plant growth detected in nodule endophytic bacteria of soybean grown in Heilongjiang province of China. Soil Biol. Biochem. 40: 238–246.

Li, L., H. Sinkko, L. Montonen, G. Wei, K. Lindström and L.A. Räsänen. 2011. Biogeography of symbiotic and other endophytic bacteria isolated from medicinal *Glycyrrhiza* species in China. FEMS Microbiol. Ecol. 79: 46–68.

Liu, J., E.T. Wang, and W.X. Chen. 2005. Diverse rhizobia associated with woody legumes *Wisteria sinensis, Cercis racemosa* and *Amorpha fruticosa* grown in the temperate zone of China. System. Appl. Microbiol. 28: 465–477.

Liu, J., E.T. Wang, Y. Li and W.X. Chen. 2007. Diverse bacteria isolated from root nodules of *Trifolium, Crotalaria* and *Mimosa* grown in the subtropical regions of China. Arch. Microbiol. 188: 1–14.

Liu, J., E.T. Wang, D.W. Ren and W.X. Chen. 2010. Mixture of endophytic *Agrobacterium* and *Sinorhizobium meliloti* strains could induce nonspecific nodulation on some woody legumes. Arch. Microbiol. 192: 229–34.

Mahdhi, M., P. de Lajudie and M. Mars. 2008. Phylogenetic and symbiotic characterization of rhizobial bacteria nodulating *Argyrolobium uniflorum* in Tunisian arid soils. Can. J. Microbiol. 54: 209–217.

Mhamdi, R., M. Mrabet, G. Laguerre, R. Tiwari and M.E. Aouani. 2005. Colonization of *Phaseolus vulgaris* nodules by *Agrobacterium*-like strains. Can. J. Microbiol. 51: 105–111.

Mrabet, M., B. Mnasri, S.B. Romdhane, G. Laguerre, M.E. Aouani and R. Mhamdi. 2006. *Agrobacterium* strains isolated from root nodules of common bean specifically reduce nodulation by *Rhizobium gallicum*. FEMS Microbiol. Ecol. 56: 304–309.

Mukherjee, P.K., N. Buensanteai, M.E. Moran-Diez, I.S. Druzhinina and C.M. Kenerley. 2012. Functional analysis of non-ribosomal peptide synthetases (NRPSs) in *Trichoderma virens* reveals a polyketide synthase (PKS)/NRPS hybrid enzyme involved in the induced systemic resistance response in maize. Microbiol. 158: 155–165.

Muresu, R., E. Polone, L. Sulas, B. Baldan, A. Tondello, G. Delogu, P. Cappuccinelli, S. Alberghini, Y. Benhizia, H. Benhizia, A. Benguedouar, B. Mori, R. Calamassi, F.B. Dazzo and S. Squartini. 2008. Coexistence of predominantly nonculturable rhizobia with diverse, endophytic bacterial taxa within nodules of wild legumes. FEMS Microbiol. Ecol. 63: 383–400.

Palaniappan, P., P.S. Chauhan, V.S. Saravanan, R. Anandham and T. Sa. 2010. Isolation and characterization of plant growth promoting endophytic bacterial isolates from root nodule of *Lespedeza* sp. Biol. Fertil. Soils. 46: 807–816.

Pandey, A., S.C. Kang and D.K. Maheshwari. 2005. Isolation of endophytic plant growth promoting *Burkholderia* sp. MSSP from root nodules of *Mimosa pudica*. Curr. Sci. 89: 177–180.

Pujic, P., P. Fournier, N. Alloisio, A.E. Hay, J. Maréchal, S. Anchisi and P. Normand. 2012. Lectin genes in the *Frankia alni* genome. Arch. Microbiol. 194: 46–57.

Rajendran, G., F. Sing, A.J. Desai and G. Archana. 2008. Enhanced growth and nodulation of pigeon pea by co-inoculation of *Bacillus* strains with *Rhizobium* spp. Biores. Technol. 99: 4544–4550.

Rajendran, G., H.M. Patel and S.J. Joshi. 2012. Isolation and Characterization of Nodule-Associated *Exiguobacterium* sp. from the Root Nodules of Fenugreek (*Trigonella foenum-graecum*) and Their Possible Role in Plant Growth Promotion. Int. J. Microbiol. 2012: 1–8.

Rivas, R., C. Gutiérrez, A. Abril, P.F. Mateos, E. Martínez-Molina, A. Ventosa and E. Velázquez. 2005. *Paenibacillus rhizosphaerae* sp. nov., isolated from the rhizosphere of *Cicer arietinum*. Int. J. Syst. Evol. Microbiol. 55: 1305–1309.

Rodríguez, R. 2008. Análisis de la población bacteriana endofita presente en nódulos de *Lupinus*: interacción y localización *in situ*. Ph.D. Thesis, Universidad de Salamanca, Salamanca, Spain.

Ruppel, S. 1989. Isolation and characterization of dinitrogenfixing bacteria from the rhizosphere of *Triticum aestivum* and *Ammophila arenaria*. *In*: V. Vancûra and F. Kunc [eds.]. Interrelationships Between Microorganisms and Plants in Soil. Proceedings of an International Symposium. Liblice, Prague. pp. 253–262.

Saïdi, S., B. Mnasri and R. Mhamdi. 2011. Diversity of nodule-endophytic agrobacteria-like strains associated with different grain legumes in Tunisia. System. Appl. Microbiol. 34: 524–530.

Selvakumar, G., S. Kundu, A.D. Gupta, Y.S. Shouche and H.S. Gupta. 2007. Isolation and Characterization of Nonrhizobial Plant Growth Promoting Bacteria from Nodules of Kudzu (*Pueraria thunbergiana*) and Their Effect on Wheat Seedling Growth. Curr. Microbiol. 56: 134–139.

Shiraishi, A., N. Matsushita and T. Hougetsu. 2010. Nodulation in blacklocust by the Gammaproteobacteria *Pseudomonas* sp. and the Betaproteobacteria *Burkholderia* sp. Syst. Appl. Microbiol. 33: 269–274.

Stajković, O., S. de Meyer, B. Mili, A. Willems and D. Deli. 2009. Isolation and characterization of endophytic non-rhizobial bacteria from root nodules of alfalfa (*Medicago sativa* L.). Botanica Serbica. 33: 107–114.

Sturz, A.V., B.R. Christie, B.G. Matheson and J. Nowak. 1997. Biodiversity of endophytic bacteria which colonize red clover nodules, roots, stems and foliage and their influence on host growth. Biol. Fertil. Soils. 25: 13–19.

Tan, Z.Y., E.T. Wang, G.X. Peng, M.E. Zhu, E. Martínez-Romero and W.X. Chen. 1999. Characterization of bacteria isolated from wild legumes in the north-western regions of China. Int. J. Syst. Bacteriol. 49: 1457–1469.

Tiwary, B.N., B. Prasad, A. Ghosh, S. Kumar and R.K. Jain. 2007. Characterization of two novel biovar of *Agrobacterium tumefaciens* isolated from root nodules of *Vicia faba*. Curr. Microbiol. 55: 328–333.

Tokala, R.K., Strap, J.L., Jung, C.M., Crawford, D.L., Salove, M.H., Deobald, L.A., Bailey, J.F. and M.J. Morra. 2002. Novel plant-microbe rhizosphere interaction involving *Streptomyces lydicus* WYEC108 and the pea plant (*Pisum sativum*). Appl. Environ. Microbiol. 68: 2161–2171.

Trujillo, M.E., R.M. Kroppenstedt, P. Schumann and E. Martínez-Molina. 2006. *Kribbella lupini* sp. nov., isolated from the roots of Lupinus angustifolius. Int. J. Syst. Evol. Microbiol. 56: 407–411.

Trujillo, M.E., R.M. Kroppenstedt, C. Fernández-Molinero, P. Schumann and E. Martínez-Molina. 2007. *Micromonospora lupini* sp. nov. and *Micromonospora saelicesensis* sp. nov., isolated from root nodules of *Lupinus angustifolius*. Int. J. Syst. Evol. Microbiol. 57: 2799–2804.

Trujillo, M.E., P. Alonso-Vega, R. Rodríguez, L. Carro, E. Cerda, P. Alonso and E. Martínez-Molina. 2010. The genus *Micromonospora* is widespread in legume root nodules: the example of *Lupinus angustifolius*. ISME J. 4: 1265–1281.

Udwary, D.W., E.A. Gontang, A.C. Jones, C.S. Jones, A.W. Schultz, J.M. Winter, J.Y. Yang, N. Beauchemin, T.L. Capson, B.R. Clark, E. Esquenazi, A.S. Eustaquio, K. Freel, L. Gerwick, W.H. Gerwick, D. Gonzalez, W.-T. Liu, K.L. Malloy, K.N. Maloney, M. Nett, J.K. Nunnery, K. Penn, A. Prieto-Davo, T.L. Simmons, S. Weitz, M.C. Wilson, L.S. Tisa, P.C. Dorrestein and B.S. Moore. 2011. Significant natural product biosynthetic potential of actinorhizal symbionts of the genus *Frankia*, as revealed by comparative genomic and proteomic analyses. Appl. Environ. Microbiol. 77: 3617–3625.

Valverde, A., E. Velázquez, C. Gutiérrez, E. Cervantes, A. Ventosa and J.M. Igual. 2003. *Herbaspirillum lusitanum* sp. nov., a novel nitrogen-fixing bacterium associated with root nodules of *Phaseolus vulgaris*. Int. J. Syst. Evol. Microbiol. 53: 1979–1983.

Valverde, A., A. Fterich, M. Mahdhi, M.H. Ramírez-Bahena, M.A. Caviedes, M. Mars, E. Velázquez and I. Rodríguez-Llorente. 2010. *Paenibacillus prosopidis* sp. nov., isolated from the nodules of *Prosopis farcta*. Int. J. Syst. Evol. Microbiol. 60: 2182–2186.

Velázquez, E., P. García-Fraile, M.H. Ramírez-Bahena, A. Peix and R. Rivas. 2011. Proteobacteria forming nitrogen fixing symbiosis with higher plants. *In*: M.L. Sezenna [ed.]. Proteobacteria: Phylogeny, Metabolic Diversity and Ecological Effects, 1st edn. Nova Science Publishers Inc. New York, U.S.A. pp. 37–56.

Vincent, J.M. 1970. A manual for the practical study of the root nodule bacteria. IBP Handbook. Oxford, England.

Wang, L.L., E.T. Wang, J. Liu, Y. Li and W.X. Chen. 2006. Endophytic occupation of root nodules and roots of *Melilotus dentatus* by *Agrobacterium tumefaciens*. Microb. Ecol. 52: 436–443.

Wolde-Meskel, E., Z. Terefework, A. Frostegård and K. Lindström. 2005. Genetic diversity and phylogeny of rhizobia isolated from agroforestry legume species in southern Ethiopia. Int. J. Syst. Evol. Microbiol. 55: 1439–1452.

Young, J.M., L.D. Kuykendall, E. Martínez-Romero, A. Kerr and H. Sawada. 2001. A revision of *Rhizobium* Frank 1889, with an emended description of the genus, and the inclusion of all species of *Agrobacterium* Conn 1942 and *Allorhizobium undicola* de Lajudie et al. 1998 as new combinations: *Rhizobium radiobacter, R. rhizogenes, R. rubi, R. undicola* and *R. vitis*. Int. J. Syst. Evol. Microbiol. 51: 89–103.

Yu, J., L. Fan, S. Yang, M. Tang, W. Yang, H. Li and G. Wei. 2009. Characterization of copper-resistant *Agrobacterium* isolated from legume nodule in mining tailings. Bull. Environ. Contam. Toxicol. 82: 354–357.

Zakhia, F., H. Jeder, A. Willems, M. Gillis, B. Dreyfus and P. de Lajudie. 2006. Diverse bacteria associated with root nodules of spontaneous legumes in Tunisia and first report for nifH-like gene within the genera *Microbacterium* and *Starkeya*. Microb. Ecol. 51: 375–393.

Zhao, L., Z. Deng, W. Yang, Y. Cao, E. Wang and G. Wei. 2010. Diverse rhizobia associated with *Sophora alopecuroides* grown in different regions of Loess Plateau in China. System. Appl. Microbiol. 33: 468–477.

11

Azospirillum-Plant Interaction: from Root Colonization to Plant Growth Promotion

F. Wisniewski-Dyé, B. Drogue, S. Borland* and *C. Prigent-Combaret*

Introduction

Bacteria of the genus *Azospirillum* (alpha-proteobacteria) have been known for many years as plant growth-promoting rhizobacteria (PGPR) (Döbereiner and Day 1976). These free-living nitrogen-fixing highly motile bacteria can be isolated from the rhizosphere of many grasses and cereals under tropical and temperate climates (Döbereiner et al. 1976, Patriquin et al. 1983). Azospirilla are predominantly surface-colonizing bacteria, whereas other diazotrophs such as *Azotobacter diazotrophicus*, *Herbaspirillum seropedicae* and *Azoarcus* sp. are endophytic; only a few *Azospirillum* strains are able to penetrate roots (Baldani et al. 1986, Assmus et al. 1995), probably via localized enzymatic degradation of plant cell wall or via root cracks.

UMR-CNRS 5557, Ecologie Microbienne, Université de Lyon, Villeurbanne, France.
*Corresponding author: florence.wisniewski@univ-lyon1.fr

Azospirilla have beneficial effects on plant growth and yield of many agronomically important crops (Okon 1985, Dobbelaere et al. 2001, Helman et al. 2011). This phytostimulatory effect has been attributed to several mechanisms; the two main characteristics that defined the genus, production of plant growth-promoting substances (such as auxins) and biological nitrogen (N) fixation, are the most documented. Besides these two mechanisms, other properties are thought to be implicated in plant growth and it is likely that the *Azospirillum* effect results from a combination of different mechanims (Bashan and de-Bashan 2010). Moreover, *Azospirillum* spp. can to a lesser extent enhance plant health by inhibiting plant parasites and/or by stimulating plant defence.

Azospirillum inoculation results in an increased number of lateral roots and root hairs allowing an improved uptake of minerals and water (Okon and Labandera-Gonzalez 1994, Lin et al. 1983, Sarig et al. 1988). In several developing and developed countries, notably in Latin America, *Azospirillum* is used as the bacterial inoculant of choice, alone or together with other PGPR and arbuscular mycorrhizal fungi, for many crops (Fuentes-Ramirez and Caballero-Mellado 2005, Helman et al. 2011, Hungria et al. 2010).

Among the 16 *Azospirillum* species described so far, 12 have been isolated from the soil or rhizosphere: *A. amazonense* (Magalhães et al. 1983), *A. brasilense* (Tarrand et al. 1978), *A. canadense* (Mehnaz et al. 2007a), *A. dobereinerae* (Eckert et al. 2001), *A. formosense* (Lin et al. 2011), *A. halopraeferens* (Reinhold et al. 1987), *A. irakense* (Khammas et al. 1989), *A. lipoferum* (Tarrand et al. 1978), *A. melinis* (Peng et al. 2006), *A. oryzae* (Xie and Yokota 2005), *A. palatum* (Zhou et al. 2009), *A. zeae* (Mehnaz et al. 2007b). The first two described species, *A. brasilense* and *A. lipoferum*, remain the most studied (Baldani et al. 1986). Isolation usually takes place on various N-free semi-solid media (Bashan et al. 1993); an alternative technique using soil extract medium can facilitate isolation of azospirilla (Trân Vân et al. 1997).

This chapter is divided into six sections : (1) features contributing to adaptation to the soil and rhizosphere, (2) colonization of plant roots, (3) attachment to roots, (4) mechanims involved in plant-growth promotion and plant health, (5) effects on plants, and (6) agronomical applications.

Adaptation to the Soil and Rhizosphere

Pioneer studies have shown that genomes of *Azospirillum* are constituted of multiple replicons and their sizes vary among species from 4.8 Mbp to 9.7 Mpb (Caballero-Mellado et al. 1999, Martin-Didonet et al. 2000). Recently, the complete genomes of four strains belonging to different species, isolated from various host plants and locations, were sequenced and published (Kaneko et al. 2010, Wisniewski-Dyé et al. 2011, Wisniewski-

Dyé et al. 2012), as well as a draft genome (Sant'Anna et al. 2011). The size range of complete genomes is 6.5 to 7.6 Mpb and they all possess six (*A. brasilense* CBG497) or seven replicons (*A. brasilense* Sp245, *A. lipoferum* 4B, *Azospirillum* sp. B510). Whereas the biggest replicon (around 3 Mbp) has all the features of a bacterial chromosome, several replicons have been defined as chromids (Harrison et al. 2010); some *Azospirillum* strains appear to possess the largest number of chromids among all prokaryotic genomes sequenced to date (Wisniewski-Dyé et al. 2011, Wisniewski-Dyé et al. 2012). A clear lack of synteny was found between replicons of *Azospirillum* strains; compared to rhizobial genomes known for their genome plasticity, more genomic rearrangements have occurred in *Azospirillum* genomes and these may have been facilitated by a composite organization (Wisniewski-Dyé et al. 2011).

Compared to other members of the family *Rhodospirillaceae*, who live in aquatic habitats, members of the genus *Azospirillum* have the particularity of being terrestrial and plant-associated. By using a robust scheme for detecting ancestral and horizontally transferred genes in *Azospirillum*, it was shown that nearly half of the genes whose origins could be resolved, appear to have been acquired horizontally from soil and plant-associated bacteria, notably from Rhizobiales and Burkholderiales (Wisniewski-Dyé et al. 2011). The majority of genes encoding functions critical for survival in the rhizosphere and association with plants are among those acquired horizontally: genes involved in iron acquisition (siderophore uptake and biosynthesis), transport and metabolism of compounds abundant in plant exudates (aromatic compounds, carbohydrates and organic acids), multidrug efflux systems and osmoprotection. Elements of a type VI secretion system, also acquired laterally, may provide a means of interspecies competition to enhance environmental survival (MacIntyre et al. 2010).

Horizontal gene transfer was thus a major driving force in the transition of these bacteria from aquatic to terrestrial environments (Wisniewski-Dyé et al. 2011). These acquisitions were likely promoted by conjugation and transduction; prophage regions were highlighted in *Azospirillum* genomes (Kaneko et al. 2010; Wisniewski-Dyé et al. 2012) and the release of phage particles was reported for several strains (Boyer et al. 2008).

The extraordinary genome plasticity disclosed by whole genome alignments has also been witnessed experimentally. Spontaneaous rearrangements between replicons were observed, leading to variants of a same strain being used in different laboratories (Pothier et al. 2008). Deletion or cointegration of replicons can also occur at high frequency during the process of phase variation leading to intraspecific diversity (see below). *Azospirillum* genomes carry other elements that may contribute to genome plasticity such as direct repeats, palindromic repeats, CRISPR and insertion sequences (Wisniewski-Dyé et al. 2011).

One of the main features of Azospirilla is to synthetize polymers such as polyhydroxybutyrate (PHB), which can serve as a carbon and energy source under conditions of stress or starvation (Tal et al. 1985, Tal et al. 1990). The production of PHB granules is induced when growth occurs under suboptimal conditions, such as a high C:N ratio medium (Kadouri et al. 2002). The ability to survive upon starvation is impaired for an *A. brasilense phbC* (encoding PHB synthase) mutant compared to the wild-type (Kadouri et al. 2002). Tolerance and survival to various stresses, including heat, osmotic shock, dessication, UV irradiation, and oxidative stress are reduced in the *A. brasilense phbC* mutant, as well as chemotactic responses towards attractants (Kadouri et al. 2003). The wild-type strain exhibits better survival than the mutant strain in carrier materials used for soil inoculants, but no difference in the ability to colonize roots and to promote plant growth was detected between the strains (Kadouri et al. 2003). Consequently, biosynthesis of PHB is of critical importance for improving shelf life and efficiency of commercial inoculants (Dobbelaere et al. 2001, Kadouri et al. 2003).

Unfavorable conditions, such as nutrient limitation, induce flocculation in some *A. brasilense* and *A. lipoferum* strains; cells convert to nonmotile highly refractile cyst-like forms entangled within a fibrillar matrix. The cyst-like cells contain abundant PHB granules and are surrounded by a layer of exopolysaccharides (Sadasivan and Neyra 1985). Under this dormant state, *Azospirillum* can resist desiccation, displays long-term maintenance of viability and higher heat resistance as compared with vegetative cells (Sadasivan and Neyra 1985, 1987). Interestingly, such a property has also been reported in the phylogenetic aquatic relative *Rhodospirillum centenum* (Berleman and Bauer 2004). More importantly, *A. brasilense* mutants that are impaired in cyst formation are less efficient in root colonization and nitrogenase activity (Katupitiya et al. 1995, Pereg-Gerk et al. 1998). Cyst-like *A. brasilense* cells were observed in the rhizosphere of water-stressed plants (Bashan et al. 1991). All these observations suggest that aggregation and encystment are of great relevance for agricultural applications.

Under cultural conditions conducive for encystment, *A. brasilense* produces compounds that form a brown pigment similar to melanin (Sadasivan and Neyra 1987, Gowri and Srivastava 1996). Melanization was also observed under some circumstances in *A. lipoferum* 4B and was correlated to a laccase activity (Faure et al. 1994). Laccase-positive strains are less sensitive to the inhibitory action of quinone analogs due to rearrangements of their respiratory chain, a feature that might be a competitive advantage in the rhizosphere in the presence of quinone compounds (Alexandre et al. 1999a). A survey of bacterial laccases suggests they are an advantageous trait for a rhizosphere bacterium as they are involved in various functions such as copper resistance, manganese oxidation, pigmentation, oxidation of toxic compounds and destruction of reactive oxygen species (Sharma et al. 2007).

Many azospirilla also produce carotenoids that have been shown to protect nitrogenase against oxidative damage (Hartmann and Hurek 1988). In *A. brasilense* Sp7, the regulation of carotenoid biosynthesis involves the sigma factor RpoE; RpoE belongs to the extracytoplasmic function (ECF) sigma factors, implicated in responses to changes in the extra-cytoplasmic compartment of the cell and assumed to be critical for adaptation, survival, and growth of rhizosphere bacteria evolving in fluctuating environments. A *rpoE* deletion mutant is carotenoidless and slow-growing, and is more sensitive than the wild-type to various stresses, such as salt, ethanol and methylene blue (Mishra et al. 2011); moreover, a mutation in a gene encoding an anti-sigma factor causes overproduction of carotenoids in *A. brasilense* (Thirunavukkarasu et al. 2008).

Phase and antigenic variation is used by several bacterial species to generate intra-population diversity that increases bacterial fitness and is important in niche adaptation, or to escape host defences. By this adaptive process, bacteria undergo frequent and usually reversible phenotypic changes resulting from genetic or epigenetic alterations at specific genetic loci (Wisniewski-Dyé and Vial 2008). Several strains of *Azospirillum* can generate phenotypic variants *in vitro* and this was correlated with genome plasticity such as plasmid loss or reorganization (Vial et al. 2006). *A. lipoferum* 4B, a strain isolated from a rice rhizosphere, generates *in vitro* at high frequencies (10^{-4} to 10^{-3} per cell per generation) a stable phase variant named $4V_I$ exhibiting pleiotropic modifications: loss of ability to swim, to assimilate certain sugars, to reduce triphenyl tetrazolium chloride, to bind some dyes, to reduce nitrous oxide, and to deaminate 1-aminocyclopropane-1-carboxylic acid (ACC) (Alexandre and Bally 1999, Alexandre et al. 1999b, Prigent-Combaret et al. 2008). Those modifications are correlated with the loss of a 750-kb replicon (Vial et al. 2006). *A. lipoferum* 4T, a nonswimming strain displaying all the features of the $4V_I$ variant, and *A. lipoferum* 4B have been isolated simultaneously from the rice rhizosphere at the same frequency (Bally et al. 1983); as *A. lipoferum* 4T was found to be genetically very close to *A. lipoferum* 4B, it was suggested that *A. lipoferum* 4T could in fact be a $4V_I$ variant of strain 4B generated within the soil ecosystem. After exposure of $4V_I$ to low oxygen concentrations, a laccase-positive variant can be obtained (Alexandre and Bally 1999).

Variants of *A. brasilense* Sp7 appear after exposure to prolonged starvation; these variants differ from the parental strain in several features, such as pigmentation, aggregation ability, EPS amount and composition, LPS structure and profile of outer membrane proteins (Lerner et al. 2010). Interestingly, one of these variants displays enhanced resistance to various stresses (Lerner et al. 2010). Whether phenotypic variation occurs in the rhizosphere and how it impacts survival in bacterial inoculants and plant growth promotion ability remains to be investigated.

All these key physiological properties may contribute to rhizosphere adaptation and ultimately to plant growth promotion abilities of *Azospirillum*.

Colonization of Plant Roots by *Azospirillum*

Colonization and attachment are two key processes required for a successful bacteria-plant interaction. In the rhizosphere, plant roots exude significant amounts of organic acids, sugars, amino acids and aromatic compounds. Azospirilla are highly motile and several species exhibit a mixed pattern of flagellation: one polar flagellum is synthesized during growth in liquid medium and is primarily used for swimming, whereas lateral flagella are induced during growth on solid media and are responsible for swarming over solid surfaces (Tarrand et al. 1978, Khammas et al. 1989, Moens et al. 1995). Motility enables the bacterium to move towards favorable nutrient conditions. Azospirilla exhibit chemotaxis towards several root compounds, which constitutes the very early step leading to effective root colonization. Evidence supporting this affirmation was obtained by demonstrating that non-chemotactic and non-motile mutants are strongly impaired in root colonization (Vande Broek et al. 1998). First reports highlighting *Azospirillum* chemotactic behavior revealed that some strains exhibit positive chemotaxis *in vitro* towards several attractants such as sugar, amino acids, aromatic compounds (Reinhold et al. 1985, Zhulin and Armitage 1993, Lopez-De-Victoria and Lowell 1993, Alexandre et al. 2000) and towards root exudates (Zhulin et al. 1988). Migration of Azospirilla towards wheat seedlings grown in the soil was shown to be limited by soil moisture, indicating that free swimming through water films rather than swarming plays a major role in the chemotactic behavior in the rhizosphere (Bashan 1986). Interestingly, differential attraction of the bacterium, both at the species and strain levels, has been reported. As for organic acids, *Azospirillum brasilense* SpT60 (isolated from wheat rhizosphere) is strongly attracted by malate, oxalate and citrate, whereas *A. brasilense* JM6A2 (isolated from maize rhizosphere) and *A. lipoferum* ER15 (isolated from Kallar grass) do not react chemotactically to these substrates (Reinhold et al. 1985). These differential chemotactic responses to organic acids correlate with the exudation of these compounds by the host plant, the best attractants being generally the best growth substrates (Van Bastelaere et al. 1999, Alexandre et al. 2000). Consequently, much evidence supports the existence of preferential chemotactic behavior between certain bacterial species/strains and plant species/genotypes, likely reflecting bacterial adaptation to the environmental conditions provided by the plant (Drogue et al. 2012).

Azospirilla also display aerotaxis, i.e., the directed movement towards optimal oxygen concentrations; this behavioral response can be advantageous to guide bacteria to optimal niches for nitrogen fixation (Barak et al. 1982). In *A. brasilense*, energy taxis, that encompasses aerotaxis, redox taxis, taxis to alternative electron acceptors, and chemotaxis to carbon sources, is the most dominant behavior (Alexandre et al. 2000, Alexandre et al. 2004). Bacteria monitor their cellular energy levels and respond to a decrease in energy by swimming to a more suitable microenvironment that can re-energize the cells. This mechanism has the advantage of allowing migration towards environmental niches that are optimal for the bacterium metabolism (Alexandre et al. 2004).

Chemotaxis system integrates environment signals into an appropriate bacterial response by using a dedicated signal transduction pathway, allowing bacteria to detect changes in environmental conditions and respond by navigating towards niches optimal for growth. In *Escherichia coli*, this signal transduction system consists of a set of conserved proteins: chemotactic proteins (Che) CheA, CheW, CheY, CheB and CheR and a set of chemoreceptors known as methyl-accepting proteins (MCPs) that perceive environmental cues (Wadhams and Armitage 2004). Upon binding a chemotactic ligand, MCPs generate chemotactic signals that are communicated to the flagellar motor via Che proteins. Among the *Azospirillum* genus, genes involved in chemotaxis have been characterized mainly in the genetically amenable species *A. brasilense*. Genes encoding a central signal transduction pathway for chemotaxis were initially identified by genetic complementation of two non-chemotactic mutants (Hauwaerts et al. 2002). This study has revealed the presence of *cheR* (encoding a methylesterase) and *cheB* (encoding a methyltransferase) in *A. brasilense*, thus demonstrating that responses to certain stimuli could undergo methylation and demethylation of the chemotaxis transducers, contrary to what was previously assessed (Zhulin and Armitage 1993). Although mutants lacking CheB and/or CheR homologs from this pathway are defective in chemotaxis, a mutant in which the entire chemotaxis pathway has been mutated displays a chemotaxis phenotype very similar to that of the parent strain, suggesting that the primary function of this pathway is not the control of motility behavior; these results also put forward the presence of multiple chemotaxis systems in *A. brasilense* (Stephens et al. 2006). Further characterization of this pathway (named Che1) showed that it modulates motility, cell length and flocculation and contributes indirectly to attachment to plant roots (Bible et al. 2008, Siuti et al. 2011). Recently, with the availability of four complete *Azospirillum* genomes, the existence of multiple chemotaxis systems has been confirmed (Wisniewski-Dyé et al.

2011). Indeed, the four strains host three chemotaxis systems of ancestral origin (including Che1), but they also have horizontally acquired up to three more chemotaxis operons resulting in four, five and six chemotaxis systems in *A. brasilense* Sp245, *A. lipoferum* 4B and *Azospirillum* sp. 510, respectively (Wisniewski-Dyé et al. 2011). The biological role of these different chemotaxis operons is not clear, but may fine-tune behavioral responses to different environmental conditions, and provide another level of control for the motile chemosensory behavior (Alexandre et al. 2004). Several MCPs have been identified in *A. brasilense*: Tlp1 involved in energy taxis and in colonization of plant roots (Greer-Phillips et al. 2004) and AerC functioning as a redox sensor (Xie et al. 2010). ChsA, displaying a PAS sensory domain and an EAL transmitter domain, was recently shown to be involved in chemotaxis as a partially reduced chemotactic response could be observed in a *chsA* mutant (Carreño-Lopez et al. 2009). The search for plant-inducible genes from *A. brasilense* led to the identification of a sugar-binding protein SpbA, specifically induced by wheat root exudates. Further characterization revealed that SpbA is involved in *A. brasilense* chemotaxis response towards D-galactose, L-arabinose and D-fucose (Van Bastelaere et al. 1999). This protein is very similar to ChvE of *Agrobacterium tumefaciens*, a periplamic component of a sugar ABC transporter; this suggests that the first steps of rhizobacteria-plant interaction are conserved among different bacteria.

As well as for chemotactic signaling and molecular mechanisms governing chemotaxis, the ecological role of this complex behavior is not yet fully understood. Nonetheless it must provide significant advantage to rhizobacteria that must perpetually seek for optimal growth conditions and compete with other microorganisms in such a stochastic environment constituted by the rhizosphere. Unraveling the genetic determinants involved in chemotactic response may represent an initial step in selecting *Azospirillum* as inoculants of different crops.

Attachment to Roots

Attachment of bacteria to plant roots is an important step for the establishment of an effective bacteria-plant association. A study of inner-root colonization of wheat by *A. brasilense* Sp7 revealed discrepancies from one cultivar to another, preferential PGPR/plant pair associations were observed, even if other pairs were effective (Mostajeran et al. 2007). Similarly, bacterial strain/plant genotype affinities were observed, with *A. brasilense* cells adhering more efficiently to the cultivar from which they had been isolated (Egorenkova et al. 2000). Preferential colonization sites have been determined by using direct microscopy examination of the bacterium on the root surface. By monitoring the colonization pattern of *A. brasilense* Sp245 constitutively expressing the *gusA* gene, it was observed that the sites of primary root colonization are the root hair zones and the site of lateral root emergence, but not the older part of the root system (Vande Broek et al. 1993). Interestingly, colonization of the root surface by the bacterium involves a morphological change reminiscent of the morphology of cysts. Visualization of rice root colonization by two *Azospirillum* species reveals that colonization sites depend on the species; indeed, while *A. irakense* cells are mainly associated with rice root hairs, *A. brasilense* cells are mainly located on root surfaces (Zhu et al. 2002). These different colonization patterns were attributed to bacterial shapes (vibroid for *A. irakense*, cyst-like for *A. brasilense*), which probably are an adaptation to their natural environment (wetland for *A. irakense* and dry land for *A. brasilense*). Moreover, *Azospirillum* displays strain-specific differences in the way they colonize roots. *In situ* monitoring with fluorescently labeled rRNA-targeted oligonucleotide probes in combination with scanning confocal microscopy revealed that *A. brasilense* Sp245 forms microcolonies in the intercellular spaces of the inner root tissue of wheat, while *A. brasilense* Sp7 is restricted to the root hair zone (Assmus et al. 1995). These differences in spatial distribution suggest the existence of specificity between bacterial strain surface components and some plant receptors on the root surface (Drogue et al. 2012).

Two different modes of attachment of *Azospirillum* have been proposed: i) a rapid, reversible and weak adsorption mediated by proteins, and ii) a strong and irreversible anchoring mediated by bacterial extracellular polysaccharides (Michiels et al. 1991, Croes et al. 1993). Several bacterial components have previously been described as important features for root attachment but the exact mechanism involved has not been described in detail as yet. In *A. brasilense*, a major outer membrane protein was shown to act as an adhesin, and may be consequently involved in bacterial aggregation and adsorption to plant roots (Burdman et al. 2000a, 2001). The polar flagellum also mediates the initial adsorption step. Indeed, purified polar flagellum of *A. brasilense* binds to wheat root, whereas lateral flagella do not (Croes et al. 1993). Further characterization revealed that the polar flagellum Fla1 is a glycoprotein which can mediate adsorption to roots (Moens et al. 1995). The TAD (tight adherence) secretion system is another structural component essential for host-colonization and biofilm formation in numerous other bacterial genera (Tomich et al. 2007). Analysis of the *A. brasilense* Sp245 genome revealed the presence of genes encoding TAD pili. A mutant deficient in TAD pili had a severe defect in adhesion to abiotic surface (Wisniewski-Dyé et al. 2011). Thus, TAD pili could also play a role in the bacterium attachment process but its role in *Azospirillum*-plant colonization needs further investigation.

The second step, root anchoring, is mediated by extracellular components exposed to plant root surface, such as expolysaccharides (EPS), capsular polysaccharides (CPS), lipopolysaccharides (LPS). In *A. brasilense*, a mutant deficient in EPS was shown to be impaired in root anchoring, suggesting a strong role of EPS in root attachment (Michiels et al. 1991). Moreover, an *A. brasilense* pleiotropic mutant, displaying a thicker layer of EPS than the wild-type, is defective in aggregation process (Blaha and Schrank 2003). A gene encoding for a LuxR-FixJ transcriptional regulator, termed *flcA*, controls exopolysaccharides production, flocculation and wheat root colonization in *A. brasilense* Sp7 (Pereg-Gerk et al. 1998). EPS and CPS composition vary among different *Azospirillum* strains, and is correlated with the ability of these strains to aggregate *in vitro* (Burdman et al. 2000b). The aggregation ability of different *A. brasilense* strains depends on both the concentration and composition of expolysaccharides (Burdman et al. 1998, 2000b, Bahat-Samet et al. 2004). Among five of the main monosaccharides found in the extracellular bacterial cell surface, L-arabinose is able to strongly inhibit *in vitro* aggregation and none of this monosaccharide could be detected in *A. brasilense* strains defective in aggregation, providing evidence for the involvement of L-arabinose in aggregation (Jofré et al. 2004, Bahat-Samet et al. 2004).

As previously described in other plant-bacteria interactions, there is some evidence showing that lectins are implicated in root attachment (Umali-Garcia et al. 1980). Several strains of *A. lipoferum* and *A. brasilense* can bind the lectin wheat germ agglutinin (WGA), suggesting the presence of sugar-binding receptors for WGA on the bacterial cell surface (Del Gallo et al. 1989). In *A. lipoferum*, a 32 kDa capsular glycoprotein was found to be the WGA-ligand (Karpati et al. 1999). Surface attachment of *A. brasilense* can be increased by nitrogen limitation and can be promoted *in vitro* by lectins, suggesting that it depends on the interaction with surface-exposed residues within the extracellular matrix of cells (Siuti et al. 2011). In addition, the occurrence of cell-surface lectins by several *Azospirillum* species was demonstrated (Castellanos et al. 1998). More particularly, *A. brasilense* Sp7 produces a 67 kDa outer-membrane lectin that can bind surface-exposed extracellular polysaccharides of other bacterium (Mora et al. 2008). Thus, it was suggested that bacterial lectins could also play a role in cell-to-cell adhesion and subsequent aggregation to plant root, together leading to effective root colonization.

As occurs in several plant-bacteria associations, LPS are also involved in the attachment process to roots by *Azospirillum*. *A. brasilense* LPS are composed of glucose, galactose, xylose, rhamnose, fucose and glucosamine, whereas those of *A. lipoferum* contain mainly glucose (Jofré et al. 2004, Vanbleu et al. 2005). The O-antigen chains of LPS from several *Azospirillum* strains are composed of linear pentasaccharide repeats containing only

D-rhamnose residues (Konnova et al. 2008). In *A. brasilense* Cd, the disruption of the *rmlD* gene involved in rhamnose biosynthesis lead to a pleiotropic phenotype; the *rmlD* mutant displays altered LPS core structure, an increased EPS production and colonizes maize roots to a lesser extent than the wild-type (Bahat-Samet et al. 2004, Jofré et al. 2004). However, no LPS plant receptor has been characterized as yet.

Attachment of *Azospirillum* to plant roots undoubtedly acts as a key factor in determining bacterial competitiveness to colonize the root. Even if the precise mechanism has not been discovered so far, efficient adhesion to plant cells seems to be a complex interplay between *Azospirillum* cell surface components and plant root receptors. The well-adapted phenotype of azospirilla to the rhizosphere of their host plant is actually a prerequisite for its growth promoting effects.

Mechanisms Involved in Plant Growth Promotion and Plant Health

Direct Stimulation of Plant Growth

Most *Azospirillum* strains are able (i) to stimulate plant nutrition through nitrogen fixation, phosphate solubilization or through their impact on plant root system architecture, and (ii) to enhance plant growth through production of phytohormones (Table 11.1).

Azospirillum is a free-nitrogen fixing bacterium which has been isolated for a long time on the basis of its positive response to acetylene reduction assays. Several reports have shown that following *Azospirillum* inoculation, there is a significant increase in the total N in shoots but only part of fixed N by the bacterium appeared to be transferred to the inoculated plant (Kapulnik et al. 1981). Thus, many studies showed that the contribution of N_2 fixation by *Azospirillum* to the plant represents less than 20 percent of the total N increase in the plant. Mutants deficient in nitrogenase activity (i.e., Nif mutants) have been shown in several cases to retain their ability to promote plant growth of certain crops (Bashan et al. 1989). This questions the relative contribution of N_2 fixation to the growth promotion effect. In contrast to symbiotic N_2 fixation, where there is direct transfer of N across the symbiotic interface, *Azospirillum* like other root surface associated diazotrophs seems not able to readily release fixed N to the host plant and this process occurs mainly through microbial turnover (Lethbridge and Davidson 1983, Rao et al. 1998).

Azospirillum can solubilize phosphate through the release of gluconic acid such as in *A. brasilense* and *A. lipoferum* or through uncharacterized mechanisms (Bashan and de-Bashan 2010, Puente et al. 2004, Ramachandran et al. 2007, Seshadri et al. 2000). The ability of *Azospirillum* to enhance plant

Table 11.1 Proposed modes of action of *Azospirillum* on plants, besides nitrogen fixation.

Mode of action	Specific mechanism	References
Production of phytohormones	Abscisic acid	Cohen et al. 2008, 2009
	ACC deaminase activity	Glick et al. 2007a, 2007b Prigent-Combaret et al. 2008
	Cytokinins	de García Salamone et al. 2001 Perrig et al. 2007 Tien et al. 1979
	Ethylene	Perrig et al. 2007 Ribaudo et al. 2006 Thuler et al. 2003
	Gibberellins	Bottini et al. 1989, 2004 Cassán et al. 2001, 2009a Perrig et al. 2007 Piccoli et al. 1997, 1999
	IAA	Barbieri and Galli 1993 Carreño-Lopez et al. 2000 Dobbelaere et al. 1999 Fallik et al. 1989 Malhotra and Srivastava 2008 Prinsen et al. 1993 Spaepen et al. 2007b Vande Broek et al. 1999
	NO	Creus et al. 2005 Molina-Favero et al. 2008 Pothier et al. 2007
	Other auxins	Costacurta et al. 1994 Fallik et al. 1989 Hartmann et al. 1983 Somers et al. 2005
Production of polyamines		Cassán et al. 2009b Perrig et al. 2007
Solubilization of phosphate		Puente et al. 2004 Ramachandran et al. 2007 Seshadri et al. 2000
Biological control	ACC deaminase activity	Toklikishvili et al. 2010
	Competition with bacterial pathogens	Bashan and de-Bashan 2002
	Induced Systemic Resistance	Yasuda et al. 2009
	Inhibition of germination of parasitic weeds	Bouillant et al. 1997 Dadon et al. 2004 Miché et al. 2000
	Production of cyanidric acid	Gonçalves and de Oliveira 1998
	Production of bacteriocins	Oliveira and Drozdowicz 1987 Tapia-Hernández et al. 1990
	Production of siderophores	Shah et al. 1992 Tapia-Hernández et al. 1990

uptake of key minerals like nitrates, ammonium, potassium, iron and several metallic micro-nutrients has been reported (Barton et al. 1986, Lin et al. 1983, Murty and Ladha 1988, Ogut and Er 2006).

Azospirillum spp. was shown to modulate the plant hormonal balance, particularly by producing a wide diversity of plant hormones, i.e., auxins, gibberellins (GAs), cytokinins, and ethylene that have a different effect on plant root system architecture. The production of phytohormones by *Azospirillum* mostly enhances root branching, resulting in a greater surface of the root system. Thus, plant roots can explore a wider soil volume and thereby improve the mineral and aqueous nutrition of the plant. Phytohormone production rather than N_2 fixation is considered as the main way *Azospirillum* promotes plant growth (Bashan and de-Bashan 2010, Bottini et al. 2004, Spaepen et al. 2007a). But, even if transfer of hormones from bacteria to the plant may occur, it has been rarely observed *in vivo*, during the course of the associative symbiosis.

Indole-3-acetic acid (IAA) is the best characterized auxin produced by PGPR including *Azospirillum* strains. It controls a wide variety of processes in plant development and plant growth and plays a key role in shaping plant root architecture such as regulation of lateral root initiation, root vascular tissue differentiation, and polar root hair positioning (Aloni et al. 2006, Fukaki et al. 2007). IAA is usually synthesized from tryptophan, which can be found by PGPR in root exudates at different concentrations according to the plant genotype (Kamilova et al. 2006). However, in *Azospirillum*, a tryptophan-independent pathway has been reported (Carreño-Lopez et al. 2000; Prinsen et al. 1993). The known routes of tryptophan dependent IAA biosynthesis in *Azospirillum* includes: IAA formation via (i) indole-3-pyruvic acid (IPyA), (ii) indole-3-acetonitrile (IAN), (iii) tryptamine or (iii) indole-3-acetamide (IAM) formation (Baca and Elmerich 2003, Carreño-Lopez et al. 2000, Zakharova et al. 1999, Kaneko et al. 2010, Wisniewski-Dyé et al. 2012). The indole pyruvate decarboxylase, encoded by the *ipdC/ppdC* gene, is a key enzyme in the IPyA pathway that mediates conversion of IPyA into indole-3-acetaldehyde. Various *ipdC* mutants from *Azospirillum* spp. displayed altered phenotypes compared to the wild type strains in their ability to alter plant root morphology (Dobbelaere et al. 1999, Malhotra and Srivastava 2008). But, due to the redundancy in IAA biosynthesis pathways, most *ipdC* mutants were shown to still produce around 10 percent of the wild-type IAA production level, and to retain the property of modifying root system architecture (Barbieri and Galli 1993, Prinsen et al. 1993, Spaepen et al. 2007b, Vande Broek et al. 1999). The involvement of IAA production in plant growth promotion by *Azospirillum* was also highlighted by increased IAA levels observed in inoculated maize plants (Fallik et al. 1989). Besides IAA production, *Azospirillum* seems to produce several other key auxin-type molecules like indole butyric acid, phenyl acetate, indole acetaldehyde or indole acetamide (Costacurta et al. 1994, Fallik et al. 1989, Hartmann et al. 1983, Somers et al. 2005).

Production of bacterial gibberellins, which corresponds mostly to GA1, GA3 and GA4 has been reported in *Azospirillum lipoferum* (Perrig et al. 2007). Gibberellins enhance the development of plant tissues, particularly stem tissue, promote root elongation and extension of lateral roots and are involved in breaking dormancy during seed germination (Barlow et al. 1991, Yaxley et al. 2001). A combination of both gibberellin production and hydrolysis of glucosyl-conjugates of gibberellic acid, which correspond to storage or transport forms of gibberellic acid produced by plants, occurs in *Azospirillum* (Cassán et al. 2001, Piccoli et al. 1997, 1999). Production of GA by *Azospirillum* has been shown to be involved in growth promotion of maize (Lucangeli and Bottini 1997), and to improve maize and soybean seed germination (Cassán et al. 2009a). However, the bacterial genetic determinants involved in this mechanism remain to be identified, as does the precise role of gibberellins in plant growth promotion by PGPR.

Cytokinin production (especially zeatin) has been documented in *Azospirillum brasilense* (de García Salamone et al. 2001, Perrig et al. 2007, Tien et al. 1979). Cytokinins stimulate plant cell division, control root meristem differentiation, induce proliferation of root hairs, whereas they inhibit lateral root formation, and primary root elongation (Riefler et al. 2006, Silverman et al. 1998). However, the real contribution of cytokinin production by PGPR to plant growth promotion is rather speculative so far, as genetic determinants genes involved in putative cytokinin bacterial biosynthetic pathways were only identified by comparative genomics and their role has not been validated by functional studies (Frébort et al. 2011).

Bacterial production of abscisic acid (ABA), which is involved in plant stress alleviation, has been studied less. The ability of bacterial ABA to promote plant growth has been documented in *A. lipoferum* and *A. brasilense*, but a positive effect has been only observed in water-stressed plants (Cohen et al. 2008, 2009).

Ethylene is another key phytohormone produced in small amounts by some *A. brasilense* from methionine as a precursor (Perrig et al. 2007, Thuler et al. 2003). It inhibits root elongation and auxin transport, and promotes seed germination, senescence and abscission of various organs and fruit ripening (Bleecker and Kende 2000, Glick et al. 2007b). Ethylene is also involved in plant defense pathways (Glick et al. 2007b). The ability of *A. brasilense* to produce ethylene seems to promote root hair development in tomato plants (Ribaudo et al. 2006). However, to evaluate the contribution of ethylene in *Azospirillum* growth promoting effect, characterization of the biosynthetic pathway and genetic determinants is required. On the contrary, certain *Azospirillum* strains especially from the *lipoferum* species are able to lower plant ethylene levels through deamination of ACC. ACC deaminase (encoded by the *acdS* gene) catalyses the conversion of ACC, the immediate plant precursor for ethylene, into NH_3 and α-ketobutyrate. Since ethylene inhibits growth and elongation of root, this may lead to enhanced root system development (Glick et al. 2007a). In the case of *A. brasilense* strains, complementation of AcdS⁻ strains with an *acdS* gene from *P. putida* enhanced the plant-beneficial effects of these PGPR on both tomato and canola (Holguin and Glick 2001, 2003).

Polyamines are considered as plant growth regulating compounds; among them, cadaverine production was shown in some *Azospirillum* strains (Cassán et al. 2009b, Perrig et al. 2007) and its possible role in rice plant growth promotion and osmotic stress mitigation was reported (Cassán et al. 2009b). Production of vitamins was also reported for *Azospirillum* but whether this trait is involved in plant growth-promotion remains to be investigated (Rodelas et al. 1993).

Finally, some *Azospirillum* strains like *A. brasilense* Sp245 are able to produce nitric oxide (NO), during root colonization, due to the activity of nitrite reductases (Creus et al. 2005, Molina-Favero et al. 2008, Pothier et al. 2007). Increasing evidence indicates that NO is a key signaling molecule involved in a wide range of effects on plants, including control of the formation of lateral and adventitious roots (Creus et al. 2005, Molina-Favero et al. 2008).

Biocontrol Properties of *Azospirillum*

The ability of some bacterial strains to protect plant from parasites, mainly described for the genus *Pseudomonas*, has also been observed with some *Azospirillum* strains; this enhanced resistance against diseases caused by virulent fungus and bacteria relies on different mechanisms such as (i) inhibition or retardation of phytopathogens growth by competition or production of antimicrobial compounds, (ii) inhibition of parasitic weeds, like *Striga hermontica*, or (iii) enhancement of plant resistance to pathogen infection.

Besides its direct plant growth promoting effects, *A. brasilense* can inhibit *Agrobacterium*-induced crown gall, bacterial speck disease caused by *Pseudomonas syringae* pv. *tomato* or bacterial canker caused by *Clavibacter michiganensis* subsp. *michiganensis* on tomato (Bakanchikova et al. 1993, Bashan and de-Bashan 2002, Romero et al. 2003). The biocontrol effect towards *Agrobacterium* seems to be correlated with the ability to deaminate ACC (Toklikishvili et al. 2010). Prevention of bacterial speck disease development by *A. brasilense* may be the consequence of plant growth promotion leading to a more robust plant, or competition between *Azospirillum* and the pathogen (Bashan and de-Bashan 2002). *Azospirillum* also protects plants against fungus pathogen attacks, such as *Rhizoctonia* spp. and *Colletotrichum acutatum* (Russo et al. 2008, Tortora et al. 2011). Reduction of anthracnose symptoms, observed when *A. brasilense* is inoculated on strawberry plants 15 d before the infection with the fungus *Colletotrichum acutatum*, coincides with the ability of *Azospirillum* strains to produce siderophores displaying antifungal effects *in vitro* (Tortora et al. 2011). All these observed antimicrobial activities could be related to the ability of *Azospirillum* spp. to produce bacteriocins, siderophores, cyanidric acid and phenylacetic acid (Gonçalves and de Oliveira 1998, Oliveira and Drozdowicz 1987, Shah et al. 1992, Somers et al. 2005, Tapia-Hernández et al. 1990, Tortora et al. 2011).

Rice inoculated with the endophyte *Azospirillum* sp. B510, display significant resistance against rice blast disease caused by the fungus *Magnaporthe oryzae* and rice blight disease caused by the bacterium *Xanthomonas oryzae* pv. *oryzae* (Yasuda et al. 2009). Interestingly, pathogenesis-related genes and salicylic acid accumulation are not induced, suggesting that the priming effect of *Azospirillum* endophytes may involve a novel type of plant resistance mechanism, independent of salicylic acid signaling.

Several strains of the *Azospirillum* genus can inhibit the parasitic plant *Striga* by stopping radicle development of GR24-germinated *Striga* seeds (Bouillant et al. 1997, Miché et al. 2000). The ability of a low-molecular-weight alcohol-soluble compound of *A. brasilense* to inhibit the germination and radicle growth of GR24-conditioned seeds of *Orobanche aegyptiaca* was also reported (Dadon et al. 2004).

It must be noted that the production of both antimicrobial (biocontrol) and plant growth-promoting compounds, implicated in the functioning of the associative symbiosis, appears to be strain-specific (Tortora et al. 2011, Perrig et al. 2007). Many determinants involved in the plant growth-promoting effect, at least those characterized, are among those acquired horizontally (Wisniewski-Dyé et al. 2011). Such acquisitions may have occurred before or after speciation events.

Moreover, expression of the corresponding genes might be modulated by rhizosphere signals, some of them being produced by other members of the microbial community (Combes-Meynet et al. 2010), some being specifically produced by peculiar plant genotypes (Drogue et al. 2012, Drogue et al. (2013)). One of the best documented examples is the influence of root exudates composition on phytohormone production (Ona et al. 2006, Rothballer et al. 2005). Thus, regulation of phytohormone production by PGPR might constitute an important component of the specificity in the plant-*Azospirillum* associative symbiosis.

Effect on Plant Growth and Health Following Inoculation with *Azospirillum*

Initially described as a cereal growth promoter, it has become clear that the beneficial effects of *Azospirillum* are not limited to growth promotion and not even to cereal crop plants. More than a 100 plant species, including both monocots and dicots, were successfully inoculated with *Azospirillum* or consortia of microorganisms containing *Azospirillum* (Bashan et al. 2004). In most cases successful inoculation results in (i) roots and shoots morphological changes, (ii) yield enhancements, and/or (iii) nutrition improvements. Among the reported morphological changes, root and shoot length are currently being investigated. Other parameters such as root and leaf area, root diameter, root hair development and number of root tips were also reported to be modified. Yield enhancements result in increases in grain weight, number of grains per plant, germination rate and biomass. When considering plant nutrition, improvement in nitrogen, phosphorus, potassium, sugar and chlorophyll contents were noticed.

During the last 30 yr, plant growth promotion was mainly studied on wheat, maize and rice, probably because most of the *Azospirillum* strains were initially isolated from cereals (Bashan et al. 2004, Fuentes-Ramirez and Caballero-Mellado 2005, Veresoglou and Menexes 2010). However, an increasing number of strains are isolated from non-cereal plants, leading scientists to consider the beneficial effects of *Azospirillum* on a wider range of ornamental and agricultural plants. Thus, strawberry plants, sugar cane and blanket flower were reported to be natural hosts for *Azospirillum* (Pedraza et al. 2007, Gadagi et al. 2004, Tejera et al. 2005). Natural isolates of *A. brasilense* inoculated on strawberry plants induce significant changes in root morphology, resulting in enhancement or decreases in root length and dry weight (Pedraza et al. 2010). These changes are also observed on shoot dry weight, which emphasizes the importance of rhizospheric interactions in health and yield of the whole plant. Similar results were reported on the ornamental plant *Gaillardia pulchella*, as natural isolates promote plant height, number of leaves and branches as well as dry matter of the whole plant (Gadagi et al. 2004). Interestingly, these studies point out that both plant and *Azospirillum* genotypes are determining factors for the establishment of a successful interaction between PGPR and plants. Such observations were already made on cereals and especially on rice. In a study where 37 isolates of *Azospirillum* spp. were inoculated on lowland rice, five isolates enhanced plant height and root and shoot dry weights, 20 isolates promoted only root and shoot dry weights, nine isolates enhanced only shoot dry weight and three isolates showed no beneficial effect (Gunarto et al. 1999). Similarly, stem height, stem dry matter, root length and N accumulation of rice differ according to the isolate of *A. amazonense* that was inoculated (Rodrigues et al. 2008).

From the earliest field experiments with *Azospirillum* in the 1980s, the most striking effects on plant growth and yield were obtained when the growth conditions were suboptimal. Mitigation of salt stress was observed on maize and wheat (Bacilio et al. 2004, Creus et al. 1997, Hamdia and El-Komy 1997, Hamdia et al. 2004); in maize, bacteria seem to restrict Na^+ uptake, enhance the uptake of K^+ and Ca^{2+}, and stimulate nitrate reductase activity (Hamdia et al. 2004). Inoculation with *Azospirillum* also improves growth under drought conditions, as demonstrated for sorghum and wheat (Creus et al. 2004, El-Komy et al. 2003, Sarig et al. 1990); this could result from better water uptake as a response to inoculation. *Azospirillum* inoculation slightly enhances root length and biomass of barley seedling treated with cadmium and the amount of nutrients absorbed by the inoculated plants increases significantly, demonstrating that *Azospirillum* can also mitigate metal toxicity (Belimov and Dietz 2000). So far, ABA and cadaverine were reported to be involved in stress mitigation (Cassán et al. 2009b, Cohen et al. 2009) but alleviation of these stresses may operate via other mechanisms that remain to be characterized.

Recently, a new facet of the interaction between *Azospirillum* and cereals was explored, i.e. its impact on secondary metabolism of the host plant (Walker et al. 2011). In addition to shoot biomass enhancement, metabolites profiles of maize are modified at a qualitative and quantitative level, depending on the inoculated strain and the maize cultivar. Interestingly, the modification is strain-specific and differences at the strain level are more important than differences between bacterial species or geographic origins. Indeed, the inoculation of *A. brasilense* CFN-535 (from Mexico) and *A. lipoferum* CRT1 (from France) triggers similar profiles, whereas *A. brasilense* UAP-154 (from Mexico) inoculation triggers a different one. Among the compounds showing quantitative changes, individual benzoxazinoids may be important for plant interactions with bacteria, and they may serve as early markers of effective PGPR-maize interactions (Walker et al. 2011).

Among all the unresolved issues about associative symbiosis between *Azospirillum* and plants, the less addressed is that of interaction specificity. On one hand, some studies point out a lack of specificity (Isawa et al. 2010, Naiman et al. 2009, Puente and Bashan, 1993). On the other hand, studies described above underline the importance of plant and *Azospirillum* genotypes in the establishment of an effective interaction (Gadagi et al. 2004, Pedraza et al. 2009, Rodrigues et al. 2008, Walker et al. 2011). Then, genotypes must be taken into account when considering direct plant growth promotion of *Azospirillum* and other PGPR (Moutia et al. 2010). Interestingly, examples of both demonstrated specificity and known mechanisms of recognition that suggest specificity can be underlined for each step of the accomplishment of an effective associative symbiosis, i.e., chemotactic attraction of *Azospirillum*, root colonization, production of antimicrobial (biocontrol) and plant growth-promoting compounds (see above, Drogue et al. 2012). However, the lack of studies including a large range of host genotypes and bacterial strains prevents a clear appraisal of the level of specificity in *Azospirillum*-plant interactions.

Agronomic Applications and Key Determinants of Future Successes

The agronomic potential of bacteria has been studied for long, based on farmers' experience, and plant inoculation with *Rhizobium* sp. has been common since the end of the 19th century (Bashan 1998, Nobbe and Hiltner 1896). It was not until the 1930s that PGPR, mainly *Azotobacter* and *Bacillus*, were used on large scale field experiments in Europe, with rather unsatisfying results (Macdonald 1989). Plant growth-promoting abilities of *Azospirillum*, highlighted in the mid-1970s, revived the interest of PGPR for plant inoculation, mainly on economically important non-legume plants (Dobereiner and Day 1976).

During the last 40 yr, *Azospirillum* effects on plant growth were extensively studied, and the amount of field experiments is increasing, due to economic and environmental benefits of biofertilizers. Most of the studies were conducted in India and Latin America where several *Azospirillum* inoculants are now commercialized (Bashan et al. 2004, Fuentes-Ramirez and Caballero-Mellado 2005). The analysis of data accumulated during the 1980s and 1990s revealed that *Azospirillum* inoculation induces statistically significant yield improvement, ranging from 5 to 30 percent, in 60–70 percent of the experiments (Okon and Itzigsohn 1995). Evaluations made by farmers and agronomists in a large field experiment carried out in 1999 in Mexico, indicated that inoculated maize, sorghum and barley display mean grain yield enhancement of 8, 11.5 and 63.5 percent respectively (Fuentes-Ramirez and Caballero-Mellado 2005). A recent survey conducted on data published from 1981 to 2008, evaluated the impact of *Azospirillum* inoculation on wheat growth and yield (Veresoglou and Menexes 2010). It appears that grain yield is increased by 8.9 percent and aboveground dry weight by 17.8 percent on average, when *Azospirillum* is inoculated on wheat. Despite these numerous consistent field results, the commercialization of *Azospirillum* biofertilizers remains limited, due to the observed variability, and to the lack of consideration of key determinants for the plant growth-promoting effect.

Many field experiments have been carried out without considering the importance of climate, soil, host plant, mineral fertilization and *Azospirillum* strain combinations. Meta-analysis of experiments conducted on wheat, points out that (i) applied N fertilization, (ii) plant cultivar and (iii) bacterial strains are key determinants for the success of plant growth promotion (Veresoglou and Menexes 2010). Thus, inoculations of wheat are more efficient when no N fertilization is applied. Similar observations were made on barley as yield increases, associated with *Azospirillum* inoculation, diminished when fertilizers are applied at high levels (Ozturk et al. 2003). Whereas the inoculation of *Azospirillum* positively impacted grain yield of maize in 93 percent of the non fertilized sites, beneficial effects were observed in only 50 percent of the highly fertilized sites (Fuentes-Ramirez and Caballero-Mellado 2005). So, rational application of fertilizers and bacterial inoculants may generate savings of chemical fertilizers. Some authors suggested that these reductions could be around 20 to 50 percent (Bashan et al. 2004, Okon and Labandera-Gonzalez 1994). Moreover, taking into account the importance of plant and bacterial genotypes in the establishment of an effective interaction (see above) and exploring interaction specificity in the associative symbiosis, inoculants containing *Azospirillum* will be more effective in the future. However, it is clear that a particular strain cannot be universally successful, in any soil, with any host plant and agricultural practices (Fuentes-Ramirez and Caballero-Mellado 2005).

A major progress in the early 1990s was the coinoculation of *Azospirillum* with other microorganisms (Bashan et al. 1998). Inoculation of consortia conducted to more successful and more beneficial impacts on plant health and growth (Bashan et al. 2004). When *Azotobacter* and *Azospirillum* are coinoculated on wheat, beneficial effect on seed yield seems to be further increased than for single inoculations (Veresoglou and Menexes 2010). Similarly, nitrogen and phosphorus concentrations in tomato, red pepper and rice are higher when *Methylobacterium oryzae* and *A. brasilense* are coinoculated (Madhaiyan et al. 2010). The combination of the fungus *Trichoderma* and *Azospirillum* can improve bean seed yield if mineral phosphate is applied at one ton per ha (Ogut et al. 2005). *A. brasilense* also stimulates the development of vesicular arbuscular mycorrhiza, notably on maize (Barea et al. 1983, Ratti and Janardhanan 1996). Interestingly, *Azospirillum* is also described to be a *Rhizobium*-"helper", as several studies reported positive effects of *Azospirillum*/*Rhizobium* combinations (Delgallo and Fabbri 1991, Dardanelli et al. 2008, Itzigsohn et al. 1993, Sarig et al. 1986, Star et al. 2011). Coinoculation of *Azospirillum* and *Rhizobium* or *Bradyrhizobium* conduces earlier nodulation or enhancement of nodule number and nodule weight (Burdman et al. 1997, Groppa et al. 1998, Molla et al. 2001). *A. brasilense* also allows a more persistent exudation of flavonoids by bean roots under salt stress (Dardanelli et al. 2008). However, some combinations do not appear more effective than single inoculations, and other seem even less effective (Cassán et al. 2009a, Felici et al. 2008).

The formulation of *Azospirillum* inoculants and the choice of the carrier also determine the success of agronomic applications (Bashan 1998). Organic or not, defined or not, the optimal carrier should be economical and easy to use (Bashan et al. 2004); the carrier must ensure microorganisms survival and plant root colonization in field.

To conclude, in addition to technical issues, a few key determinants must be taken into account to optimize future uses of *Azospirillum* containing inoculants. Interaction specificity in *Azospirillum*-plant associative symbiosis is still an open issue, and the existence of preferential interactions between a particular bacterial strain and a particular plant genotype is overlooked. Because one strain cannot be successful worldwide, the consideration of microbial consortia containing both a large variety of *Azospirillum* strains and a wide range of beneficial microorganisms appears to be the most efficient strategy. The agronomical interest of *Azospirillum* is not limited to cereal growth promotion, as biocontrol and *Rhizobium*-"helping" are promising properties. However, a rational balance between the levels of chemical and biological fertilizers should enhance fertilization efficiency.

Concluding Remarks

There is no definite agreement on exactly how *Azospirillum* achieves plant growth promotion; it is very likely that *Azospirillum* effect results from a combination of different mechanisms (Bashan and de-Bashan 2010), and that this combination differs from one strain to another. Numerous studies have been dedicated to nitrogen fixation and production of phytohormones, the two hallmarks of this genus. Biocontrol abilities and induction of resistance by *Azospirillum* deserve to be further investigated; however, plant growth promotion is supposed to enhance the competitiveness of *Azospirillum* in the rhizosphere and the robustness of the plant, suggesting that biocontrol and stimulation of growth are not independent from each other.

The potential of *Azospirillum* for agronomic uses is undeniable, especially to replace chemical agents employed for growth promotion and control of pathogens. Thus, it is essential to better understand the mechanisms that constrain the accomplishment of an effective associative symbiosis between *Azospirillum* and the plants. The consideration of potential host specificity, underlined by the bacterial and plant genotype-specific effects observed, appears essential. For this purpose, studies of bacterial and plant genomes, transcriptomes and proteomes should provide important information on key genetic determinants involved in associative symbiosis, and could bring important clues on evolution dynamics and on events that led to specific adaptation.

Acknowledgements

We wish to thank ANR for funding the project Azoriz (ANR-08-BLAN-0098). B. Drogue and S. Borland received fellowships from Région Rhône-Alpes and French Ministère de l'Enseignement supérieur et de la Recherche, respectively.

References

Alexandre, G. and R. Bally. 1999. Emergence of a laccase-positive variant of *Azospirillum lipoferum* occurs via a two-step phenotypic switching process. FEMS Microbiol. Lett. 174: 371–378.

Alexandre, G., R. Bally, B.L. Taylor and I.B. Zhulin. 1999a. Loss of cytochrome c oxidase activity and acquisition of resistance to quinone analogs in a laccase-positive variant of *Azospirillum lipoferum*. J. Bacteriol. 181: 6730–6738.

Alexandre, G., R. Rohr and R. Bally. 1999b. A phase variant of *Azospirillum lipoferum* lacks a polar flagellum and constitutively expresses mechanosensing lateral flagella. Appl. Environ. Microbiol. 65: 4701–4704.

Alexandre, G., S.E. Greer and I.B. Zhulin. 2000. Energy taxis is the dominant behavior in *Azospirillum brasilense*. J. Bacteriol. 182: 6042–6048.

Alexandre, G., S. Greer-Phillips and I.B. Zhulin. 2004. Ecological role of energy taxis in microorganisms. FEMS Microbiol. Rev. 28: 113–126.

Aloni, R., E. Aloni, M. Langhans and C.I. Ullrich. 2006. Role of cytokinin and auxin in shaping root architecture: regulating vascular differentiation, lateral root initiation, root apical dominance and root gravitropism. Ann. Bot. 97: 883–893.

Assmus, B., P. Hutzler, G. Kirchhof, R. Amann, J.R. Lawrence and A. Hartmann. 1995. *In Situ* localization of *Azospirillum brasilense* in the rhizosphere of wheat with fluorescently labeled, ribosomal-RNA-targeted oligonucleotide probes and scanning confocal laser microscopy. Appl. Environ. Microbiol. 61: 1013–1019.

Baca, B.E. and C. Elmerich. 2003. Associative and endophytic nitrogen-fixing bacteria and cyanobacterial associations. *In:* E.C.A.N. We [ed.]. Microbial production of plant hormones. Kluwer Academic Publishers. The Netherlands.

Bacilio, M., H. Rodriguez, M. Moreno, J.P. Hernandez and Y. Bashan. 2004. Mitigation of salt stress in wheat seedlings by a *gfp*-tagged *Azospirillum lipoferum*. Biol. Fertil. Soils. 40: 188–193.

Bahat-Samet, E., S. Castro-Sowinski and Y. Okon. 2004. Arabinose content of extracellular polysaccharide plays a role in cell aggregation of *Azospirillum brasilense*. FEMS Microbiol. Lett. 237: 195–203.

Bakanchikova, T.I., E.V. Lobanok, L.K. Pavlova-Ivanova, T.V. Red'kina, J.A. Nagapetyan and A.N. Maisuryan. 1993. Inhibition of tumor formation in dicotyledonous plants by *Azospirillum brasilense* strains. Microbiology. 62: 319–323.

Baldani, V.L.D., M.A.D.B. Alvarez, J.I. Baldani and J. Döbereiner. 1986. Establishment of inoculated *Azospirillum* spp. in the rhizosphere and in roots of field grown wheat and sorghum. Plant Soil. 90: 35–46.

Bally, R., D. Thomas-Bauzon, T. Heulin, J. Balandreau, J. Richard and J. De Ley. 1983. Determination of the most frequent N_2-fixing bacteria from the rhizosphere of rice. Can. J. Microbiol. 29: 881–887.

Barak, R., I. Nur, Y. Okon and Y. Henis. 1982. Aerotactic response of *Azospirillum brasilense*. J. Bacteriol. 152: 643–649.

Barbieri, P. and E. Galli. 1993. Effect on wheat root development of inoculation with an *Azospirillum brasilense* mutant with altered indole-3-acetic acid production. Res. Microbiol. 144: 69–75.

Barea, J.M., A.F. Bonis and J. Olivares. 1983. Interactions between *Azospirillum* and VA mycorrhiza and their effects on growth and nutrition of maize and ryegrass. Soil Biol. Biochem. 15: 705–709.

Barlow, P.W., P. Brain and J.S. Parker. 1991. Cellular growth in roots of a gibberellin-deficient mutant of tomato (*Lycopersicon esculentum* Mill.) and its wild-type. J. Exp. Bot. 42: 339–351.

Barton, L.L., G.V. Johnson and S.O. Miller. 1986. The effect of *Azospirillum brasilense* on iron absorption and translocation by sorghum. J. Plant Nutr. 9: 557–565.

Bashan, Y. 1986. Migration of the rhizosphere bacteria *Azospirillum brasilense* and *Pseudomonas fluorescens* towards wheat roots in the soil. J. Gen. Microbiol. 132: 3407–3414.

Bashan, Y. 1998. Inoculants of plant growth-promoting bacteria for use in agriculture. Biotechnol. Adv. 16: 729–770.

Bashan, Y. and L.E. de-Bashan. 2002. Protection of tomato seedlings against infection by *Pseudomonas syringae* pv. tomato by using the plant growth-promoting bacterium *Azospirillum brasilense*. Appl. Environ. Microbiol. 68: 2637–2643.

Bashan, Y. and L.E. de-Bashan. 2010. How the plant growth-promoting bacterium *Azospirillum* promotes plant growth - A critical assessment. Adv. Agron. 108: 77–136.

Bashan, Y., M. Singh and H. Levanony. 1989. Contribution of *Azospirillum brasilense* Cd to growth of tomato seedlings is not through nitrogen fixation. Can. J. Bot. 67: 2429–2434.

Bashan, Y., H. Levanony and R.E. Whitmoyer. 1991. Root surface colonization of non-cereal crop plants by pleomorphic *Azospirillum brasilense* Cd. J. Gen. Microbiol. 137: 187–196.

Bashan, Y., G. Holguin and R. Lifshitz. 1993. Isolation and characterization of plant growth-promoting rhizobacteria. *In:* B.R. Glick and J.E. Thompson [eds.] Methods in plant molecular biology and biotechnology. CRC Press. Boca Raton. pp. 331–345.

Bashan, Y., G. Holguin and L.E. de-Bashan. 2004. *Azospirillum*-plant relationships: physiological, molecular, agricultural, and environmental advances (1997–2003). Can. J. Microbiol. 50: 521–577.

Belimov, A.A. and K.J. Dietz. 2000. Effect of associative bacteria on element composition of barley seedlings grown in solution culture at toxic cadmium concentrations. Microbiol. Res. 155: 113–121.

Berleman, J.E. and C.E. Bauer. 2004. Characterization of cyst cell formation in the purple photosynthetic bacterium *Rhodospirillum centenum*. Microbiology. 150: 383–390.

Bible, A.N., B.B. Stephens, D.R. Ortega, Z. Xie and G. Alexandre. 2008. Function of a chemotaxis-like signal transduction pathway in modulating motility, cell clumping, and cell length in the alphaproteobacterium *Azospirillum brasilense*. J. Bacteriol. 190: 6365–6375.

Blaha, C.A.G. and I.S. Schrank. 2003. An *Azospirillum brasilense* Tn5 mutant with modified stress response and impaired in flocculation. Antonie Leeuwenhoek. 83: 35–43.

Blaha, D., C. Prigent-Combaret, M.S. Mirza and Y. Moenne-Loccoz. 2006. Phylogeny of the 1-aminocyclopropane-1-carboxylic acid deaminase-encoding gene *acdS* in phytobeneficial and pathogenic Proteobacteria and relation with strain biogeography. FEMS Microbiol. Ecol. 56: 455–470.

Bleecker, A.B. and H. Kende. 2000. A gaseous signal molecule in plants. Annu. Rev. Cell Dev. Biol. 16: 1–18.

Bottini, R., M. Fulchieri, D. Pearce and R.P. Pharis. 1989. Identification of gibberellins A1, A_3, and iso-A_3 in cultures of *Azospirillum lipoferum*. Plant Physiol. 90: 45–47.

Bottini, R., F. Cassán and P. Piccoli. 2004. Gibberellin production by bacteria and its involvement in plant growth promotion and yield increase. Appl. Microbiol. Biotechnol. 65: 497–503.

Bouillant, M.L., L. Miche, O. Ouedraogo, G. Alexandre, C. Jacoud, G. Salle and R. Bally. 1997. Inhibition of *Striga* seed germination associated with sorghum growth promotion by soil bacteria. Comptes Rendus de l'Académie des Sciences. 320: 159–162.

Boyer, M., J. Haurat, S. Samain, B. Segurens, F. Gavory, V. Gonzalez, P. Mavingui, R. Rohr, R. Bally and F. Wisniewski-Dyé. 2008. Bacteriophage prevalence in the genus *Azospirillum* and analysis of the first genome sequence of an *Azospirillum brasilense* integrative phage. Appl. Environ. Microbiol. 74: 861–874.

Burdman, S., J. Kigel and Y. Okon. 1997. Effects of *Azospirillum brasilense* on nodulation and growth of common bean (*Phaseolus vulgaris* L.). Soil Biol. Biochem. 29: 923–929.

Burdman, S., E. Jurkevitch, B. Schwartsburd, M. Hampel and Y. Okon. 1998. Aggregation in *Azospirillum brasilense*: effects of chemical and physical factors and involvement of extracellular components. Microbiology. 144: 1989–1999.

Burdman, S., R. De Mot, J. Vanderleyden, Y. Okon and E. Jurkevitch. 2000a. Identification and characterization of the *omaA* gene encoding the major outer membrane protein of *Azospirillum brasilense*. DNA Seq. 11: 225–237.

Burdman, S., E. Jurkevitch, M.E. Soria-Diaz, A.M.G. Serrano and Y. Okon. 2000b. Extracellular polysaccharide composition of *Azospirillum brasilense* and its relation with cell aggregation. FEMS Microbiol. Lett. 189: 259–264.

Burdman, S., G. Dulguerova, Y. Okon and E. Jurkevitch. 2001. Purification of the major outer membrane protein of *Azospirillum brasilense*, its affinity to plant roots, and its involvement in cell aggregation. Mol. Plant-Microbe Interac. 14: 555–561.

Caballero-Mellado, J., L. Lopez-Reyes and R. Bustillos-Cristales. 1999. Presence of 16S rRNA genes in multiple replicons in *Azospirillum brasilense*. FEMS Microbiol. Lett. 178: 283–288.

Carreño-Lopez, R., N. Campos-Reales, C. Elmerich and B.E. Baca. 2000. Physiological evidence for differently regulated tryptophan-dependent pathways for indole-3-acetic acid synthesis in *Azospirillum brasilense*. Mol. Gen. Genet. 264: 521–530.

Carreño-Lopez, R., A. Sanchez, N. Camargo, C. Elmerich and B.E. Baca. 2009. Characterization of *chsA*, a new gene controlling the chemotactic response in *Azospirillum brasilense* Sp7. Arch. Microbiol. 191: 501–507.

Cassán, F., R. Bottini, G. Schneider and P. Piccoli. 2001. *Azospirillum brasilense* and *Azospirillum lipoferum* hydrolyze conjugates of GA(20) and metabolize the resultant aglycones to GA(1) in seedlings of rice dwarf mutants. Plant Physiol. 125: 2053–2058.

Cassán, F., D. Perrig, V. Sgroy, O. Masciarelli, C. Penna and V. Luna. 2009a. *Azospirillum brasilense* Az39 and *Bradyrhizobium japonicum* E109, inoculated singly or in combination, promote seed germination and early seedling growth in corn (*Zea mays* L.) and soybean (*Glycine max* L.). Eur. J. Soil Biol. 45: 28–35.

Cassán, F., S. Maiale, O. Masciarelli, A. Vidal, L. Luna and O. Ruiz. 2009b. Cadaverine production by *Azospirillum brasilense* and its possible role in plant growth promotion and osmotic stress mitigation. Eur. J. Soil Biol. 45: 12–19.

Castellanos, T., F. Ascencio and Y. Bashan. 1998. Cell-surface lectins of *Azospirillum* spp. Curr. Microbiol. 36: 241–244.

Cohen, A.C., R. Bottini and P.N. Piccoli. 2008. *Azospirillum brasilense* Sp245 produces ABA in chemically-defined culture medium and increases ABA content in arabidopsis plants. Plant Growth Reg. 54: 97–103.

Cohen, A.C., C.N. Travaglia, R. Bottini and P.N. Piccoli. 2009. Participation of abscisic acid and gibberellins produced by endophytic *Azospirillum* in the alleviation of drought effects in maize. Botany. 87: 455–462.

Combes-Meynet, E., J.F. Pothier, Y. Moënne-Loccoz and C. Prigent-Combaret. 2010. The *Pseudomonas* secondary metabolite 2,4-diacetylphloroglucinol is a signal inducing rhizoplane expression of *Azospirillum* genes involved in plant-growth promotion. Mol. Plant-Microbe Interact. 24: 271–284.

Costacurta, A., V. Keijers and J. Vanderleyden. 1994. Molecular cloning and sequence analysis of an *Azospirillum brasilense* indole-3-pyruvate decarboxylase gene. Mol. Gen. Genet. 243: 463–472.

Creus, C.M., R.J. Sueldo and C.A. Barassi. 1997. Shoot growth and water status in *Azospirillum*-inoculated wheat seedlings grown under osmotic and salt stresses. Plant Physiol. Biochem. 35: 939–944.

Creus, C.M., R.J. Sueldo and C.A. Barassi. 2004. Water relations and yield in *Azospirillum*-inoculated wheat exposed to drought in the field. Can. J. Bot. 82: 273–281.

Creus, C.M., M. Graziano, E.M. Casanovas, M.A. Pereyra, M. Simontacchi, S. Puntarulo, C.A. Barassi and L. Lamattina. 2005. Nitric oxide is involved in the *Azospirillum brasilense*-induced lateral root formation in tomato. Planta. 221: 297–303.

Croes, C.L., S. Moens, E. Van Bastelaere, J. Vanderleyden and K.W. Michiels. 1993. The polar flagellum mediates *Azospirillum brasilense* adsorption to wheat roots. J. Gen. Microbiol. 139: 2261–2269.

Dadon, T., N. Bar Nun and A.M. Mayer. 2004. A factor from *Azospirillum brasilense* inhibits germination and radicle growth of *Orobanche aegyptiaca*. Isr. J. Plant Sci. 52: 83–86.

Dardanelli, M.S., F.J. Fernandez de Cordoba, M.R. Espuny, M.A. Rodriguez Carvajal, M.E. Soria Diaz, A. Gil Serrano, Y. Okon and M. Megias. 2008. Effect of *Azospirillum brasilense* coinoculated with *Rhizobium* on *Phaseolus vulgaris* flavonoids and Nod factor production under salt stress. Soil Biol. Biochem. 40: 2713–2721.

de García Salamone, I.E., R.K. Hynes and L.M. Nelson. 2001. Cytokinin production by plant growth promoting rhizobacteria and selected mutants. Can. J. Microbiol. 47: 404–411.

Del Gallo, M. and P. Fabbri. 1991. Effect of soil organic matter on chickpea inoculated with *Azospirillum brasilense* and *Rhizobium leguminosarum* by ciceri. Plant Soil. 137: 171–175.

Del Gallo, M., M. Negi and C.A. Neyra. 1989. Calcofluor- and lectin-binding exocellular polysaccharides of *Azospirillum brasilense* and *Azospirillum lipoferum*. J. Bacteriol. 171: 3504–3510.

Dobbelaere, S., A. Croonenborghs, A. Thys, A. Vande Broek, and J. Vanderleyden. 1999. Phytostimulatory effect of *Azospirillum brasilense* wild type and mutant strains altered in IAA production on wheat. Plant Soil. 212: 155–164.

Dobbelaere, S., A. Croonenborghs, A. Thys, D. Ptacek, J. Vanderleyden, P. Dutto, C. Labandera-Gonzalez, J. Caballero-Mellado, A.J.F., Y. Kapulnik, S. Brener, S. Burdman, D. Kadouri, S. Sarig and Y. Okon. 2001. Responses of agronomically important crops to inoculation with *Azospirillum*. Aust. J. Plant Physiol. 28: 871–879.

Döbereiner, J. and L. Day. 1976. Associative symbiosis in tropical grasses: Characterization of microorganisms and dinitrogen fixing sites. *In:* W.E. Newton and C.J. Nyman [eds.]. Proceedings of the First International Symposium on Nitrogen Fixation. Washington State University Press, Pullman. pp. 518–538.

Döbereiner, J., I.E. Marriel and M. Nery. 1976. Ecological distribution of *Spirillum lipoferum* Beijerinck. Can. J. Microbiol. 22: 1464–1472.

Drogue, B., H. Doré, S. Borland, F. Wisniewski-Dyé and C. Prigent-Combaret. 2012. Which specificity in cooperation between phytostimulating rhizobacteria and plants? Res. Microbiol. 163: 500–510.

Drogue, B., E. Combes-Meynet, Y. Moënne-Loccoz, F. Wisniewski-Dyé and C. Prigent-Combaret. 2013. Control of the cooperation between plant growth-promoting rhizobacteria and crops by rhizosphere signals. *In:* F. De Bruin [ed.]. Molecular Microbial Ecology of the Rhizosphere. John Wiley & Sons, Inc., Hoboken, NJ, USA. pp. 2807–2812.

Eckert, B., O.B. Weber, G. Kirchhof, A. Halbritter, M. Stoffels and A. Hartmann. 2001. *Azospirillum doebereinerae* sp. nov., a nitrogen-fixing bacterium associated with the C_4-grass *Miscanthus*. Int. J. Syst. Evol. Microbiol. 51: 17–26.

Egorenkova, I.V., S.A. Konnova, I.M. Skvortsov and V.V. Ignatov. 2000. Investigation of the initial stages of interaction of the bacterium *Azospirillum brasilense* with wheat seedling roots: Adsorption and root hair deformation. Microbiology. 69: 103–108.

El-Komy, H.M., M.A. Hamdia and G.K.A. El-Baki. 2003. Nitrate reductase in wheat plants grown under water stress and inoculated with *Azospirillum* spp. Biol. Plantarum. 46: 281–287.

Fallik, E., Y. Okon, E. Epstein, A. Goldman and M. Fischer. 1989. Identification and quantification of IAA and IBA in *Azospirillum brasilense* inoculated maize roots. Soil Biol. Biochem. 21: 147–153.

Faure, D., M.L. Bouillant and R. Bally. 1994. Isolation of *Azospirillum lipoferum* 4T Tn5 mutants affected in melanization and laccase activity. Appl. Environ. Microbiol. 60: 3413–3415.

Felici, C., L. Vettori, E. Giraldi, L.M.C. Forino, A. Toffanin, A.M. Tagliasacchi and M. Nuti. 2008. Single and co-inoculation of *Bacillus subtilis* and *Azospirillum brasilense* on *Lycopersicon esculentum*: effects on plant growth and rhizosphere microbial community. Appl. Soil Ecol. 40: 260–270.

Frébort, I., M. Kowalska, T. Hluska, J. Frébortová and P. Galuszka. 2011. Evolution of cytokinin biosynthesis and degradation. J. Exp. Bot. 62: 2431–2452.

Fuentes-Ramirez, L.E. and J. Caballero-Mellado. 2005. Bacterial biofertlizers. *In:* S. ZA [ed.]. PGPR: Biological control and biofertilization. Springer-Verlab, Dordrecht. pp. 143–172.

Fukaki, H., Y. Okushima and M. Tasaka. 2007. Auxin-mediated lateral root formation in higher plants. Int. Rev. Cytol. 256: 111–137.

Gadagi, R.S., P.U. Krishnaraj, J.H. Kulkarni and T. Sa. 2004. The effect of combined *Azospirillum* inoculation and nitrogen fertilizer on plant growth promotion and yield response of the blanket flower *Gaillardia pulchella*. Sci. Hortic. 100: 323–332.

Glick, B.R., Z. Cheng, J. Czarny and J. Duan. 2007a. Promotion of plant growth by ACC deaminase-producing soil bacteria. Eur. J. Plant Pathol. 119: 329–339.

Glick, B.R., B. Todorovic, J. Czarny, Z. Cheng, J. Duan and B. McConkey. 2007b. Promotion of plant growth by bacterial ACC deaminase. Crit. Rev. Plant Sci. 26: 227–242.

Gonçalves, A.F.S. and R.D.B. de Oliveira. 1998. Cyanide production by Brazilian strains of *Azospirillum*. Rev. Microbiol. 29: 36–39.

Gowri, P.M. and S. Srivastava. 1996. Encapsulation as a response of *Azospirillum brasilense* Sp7 to zinc stress. World J. Microbiol. Biotechnol. 12: 319–322.

Greer-Phillips, S.E., B.B. Stephens and G. Alexandre. 2004. An energy taxis transducer promotes root colonization by *Azospirillum brasilense*. J. Bacteriol. 186: 6595–6604.

Groppa, M.D., M.S. Zawoznik and M.L. Tomaro. 1998. Effect of co-inoculation with *Bradyrhizobium japonicum* and *Azospirillum brasilense* on soybean plants. Eur. J. Soil Biol. 34: 75–80.

Gunarto, L., K. Adachi and T. Senboku. 1999. Isolation and selection of indigenous *Azospirillum* spp. from a subtropical island, and effect of inoculation on growth of lowland rice under several levels of N application. Biol. Fertil. Soils. 28: 129–135.

Hamdia, A.B.E., M.A.K. Shaddad and M.M. Doaa. 2004. Mechanisms of salt tolerance and interactive effects of *Azospirillum brasilense* inoculation on maize cultivars grown under salt stress conditions. Plant Growth Reg. 44: 165–174.

Hamdia, M.A. and H.M. El-Komy. 1997. Effect of salinity, gibberellic acid and *Azospirillum* inoculation on growth and nitrogen uptake of *Zea mays*. Biol. Plant. 40: 109–120.

Harrison, P.W., R.P. Lower, N.K. Kim and J.P. Young. 2010. Introducing the bacterial 'chromid': not a chromosome, not a plasmid. Trends Microbiol. 18: 141–148.

Hartmann, A. and T. Hurek. 1988. Effect of carotenoid overproduction on oxygen tolerance of nitrogen fixation in *Azospirillum brasilense* Sp7. J. Gen. Microbiol. 134: 449–455.

Hartmann, A., M. Singh and W. Klingmuller. 1983. Isolation and characterization of *Azospirillum* mutants excreting high amounts of indoleacetic acid. Can. J. Microbiol. 29: 916–923.

Hauwaerts, D., G. Alexandre, S.K. Das, J. Vanderleyden and I.B. Zhulin. 2002. A major chemotaxis gene cluster in *Azospirillum brasilense* and relationships between chemotaxis operons in alpha-proteobacteria. FEMS Microbiol. Lett. 208: 61–67.

Helman, Y., S. Burdman and Y. Okon. 2011. Plant growth promotion by rhizosphere bacteria through direct effects. *In:* E. Rosenberg and U. Gophna [eds.]. Beneficial microorganisms in multicellular life forms. Springer-Verlab, Berlin. pp. 89–103.

Holguin, G. and B.R. Glick. 2001. Expression of the ACC deaminase gene from *Enterobacter cloacae* UW4 in *Azospirillum brasilense*. Microb. Ecol. 41: 281–288.

Holguin, G. and B.R. Glick. 2003. Transformation of *Azospirillum brasilense* Cd with an ACC deaminase gene from *Enterobacter cloacae* UW4 fused to the Tet(r) gene promoter improves its fitness and plant growth promoting ability. Microb. Ecol. 46: 122–133.

Hungria, M., R.J. Campo, E.M. Souza and F.A. Pedroza. 2010. Inoculation with selected strains of *Azospirillum brasilense* and *A. lipoferum* improves yields of maize and wheat in Brazil. Plant Soil. 331: 413–425.

Isawa, T., M. Yasuda, H. Awazaki, K. Minamisawa, S. Shinozaki and H. Nakashita. 2010. *Azospirillum* sp. strain B510 enhances rice growth and yield. Microbes Environ. 25: 58–61.

Itzigsohn, R., Y. Kapulnik, Y. Okon and A. Dovrat. 1993. Physiological and morphological aspects of interactions between *Rhizobium meliloti* and Alfalfa (*Medicago sativa*) in association with *Azospirillum brasilense*. Can. J. Microbiol. 39: 610–615.

Jofré, E., A. Lagares and G. Mori. 2004. Disruption of dTDP-rhamnose biosynthesis modifies lipopolysaccharide core, exopolysaccharide production, and root colonization in *Azospirillum brasilense*. FEMS Microbiol. Lett. 231: 267–275.

Kadouri, D., S. Burdman, E. Jurkevitch and Y. Okon. 2002. Identification and isolation of genes involved in poly(beta-hydroxybutyrate) biosynthesis in *Azospirillum brasilense* and characterization of a *phbC* mutant. Appl. Environ. Microbiol. 68: 2943–2949.

Kadouri, D., E. Jurkevitch and Y. Okon. 2003. Involvement of the reserve material poly-beta-hydroxybutyrate in *Azospirillum brasilense* stress endurance and root colonization. Appl. Environ. Microbiol. 69: 3244–3250.

Kamilova, F., L.V. Kravchenko, A.I. Shaposhnikov, T. Azarova, N. Makarova and B. Lugtenberg. 2006. Organic acids, sugars and L-tryptophan in exudates of vegetables growing on stonewool and their effects on activities of rhizosphere bacteria. Mol. Plant-Microbe Interact. 9: 250–256.

Kaneko, T., K. Minamisawa, T. Isawa, H. Nakatsukasa, H. Mitsui, Y. Kawaharada, Y. Nakamura, A. Watanabe, K. Kawashima, A. Ono, Y. Shimizu, C. Takahashi, C. Minami, T. Fujishiro, M. Kohara, M. Katoh, N. Nakazaki, S. Nakayama, M. Yamada, S. Tabata and S. Sato. 2010. Complete genomic structure of the cultivated rice endophyte *Azospirillum* sp. B510. DNA Res. 17: 37–50.

Kapulnik, Y., J. Kigel, Y. Okon, I. Nur and Y. Henis. 1981. Effect of *Azospirillum* inoculation on some growth parameters and N-content of wheat, sorghum and panicum. Plant Soil. 61: 65–70.

Karpati, E., P. Kiss, T. Ponyi, I. Fendrik, M. de Zamaroczy and L. Orosz. 1999. Interaction of *Azospirillum lipoferum* with wheat germ agglutinin stimulates nitrogen fixation. J. Bacteriol. 181: 3949–3955.

Katupitiya, S., J. Millet, M. Vesk, L. Viccars, A. Zeman, Z. Lidong, C. Elmerich and I.R. Kennedy. 1995. A mutant of *Azospirillum brasilense* Sp7 impaired in flocculation with a modified colonization pattern and superior nitrogen fixation in association with wheat. Appl. Environ. Microbiol. 61: 1987–1995.

Khammas, K.M., E. Ageron, P.A. Grimont and P. Kaiser. 1989. *Azospirillum irakense* sp. nov., a nitrogen-fixing bacterium associated with rice roots and rhizosphere soil. Res. Microbiol. 140: 679–93.

Konnova, O.N., A.S. Boiko, G.L. Burygin, P. Fedorenko Iu, L. Matora, S.A. Konnova and V.V. Ignatov. 2008. Chemical and serological studies of liposaccharides of bacteria of the genus *Azospirillum*. Mikrobiologiia. 77: 350–357.

Lerner, A., A. Valverde, S. Castro-Sowinski, H. Lerner, Y. Okon and S. Burdman. 2010. Phenotypic variation in *Azospirillum brasilense* exposed to starvation. Environ. Microbiol. Rep. 2: 577–586.

Lethbridge, G. and M.S. Davidson. 1983. Root-associated nitrogen-fixing bacteria and their role in the nitrogen nutrition of wheat estimated by 15N isotope dilution. Soil Biol. Biochem. 15: 365–374.

Lin, S.Y., F.T. Shen, L.S. Young, Z.L. Zhu, W.M. Chen and C.C. Young. 2011. *Azospirillum formosense* sp. nov., a diazotroph from agricultural soil. Int. J. Syst. Evol. Microbiol. 62: 1185–1190.

Lin, W., Y. Okon and R.W.F. Hardy. 1983. Enhanced mineral uptake by *Zea mays* and *Sorghum bicolor* roots inoculated with *Azospirillum brasilense*. Appl. Environ. Microbiol. 45: 1775–1779.

Lopez-de-Victoria, G. and C.R. Lovell. 1993. Chemotaxis of *Azospirillum* species to aromatic compounds. Appl. Environ. Microbiol. 59: 2951–2955.

Lucangeli, C. and R. Bottini. 1997. Effects of *Azospirillum* spp. on endogenous gibberellin content and growth of maize (*Zea mays* L.) treated with uniconazole. Symbiosis. 23: 63–71.

Macdonald, R.M. 1989. An overview of crop inoculation. *In:* R. Campbell and R. M. Macdonald [eds.]. Microbial inoculation of crop plants, vol. 25. IRL Press, Oxford, New York. pp. 1–9.

MacIntyre, D.L., S.T. Miyata, M. Kitaoka and S. Pukatzki. 2010. The *Vibrio cholerae* type VI secretion system displays antimicrobial properties. Proc. Natl. Acad. Sci. U.S.A. 107: 19520–19524.

Madhaiyan, M., S. Poonguzhali, B.G. Kang, Y.J. Lee, J.B. Chung and T.M. Sa. 2010. Effect of co-inoculation of methylotrophic *Methylobacterium oryzae* with *Azospirillum brasilense* and *Burkholderia pyrrocinia* on the growth and nutrient uptake of tomato, red pepper and rice. Plant Soil. 328: 71–82.

Magalhães, F.M., J.I. Baldani, S.M. Souto, J.R. Kuykendall and J. Dobereiner. 1983. A new acid-tolerant *Azospirillum* species. An. Acad. Brasil Cienc. 55: 417–430.

Malhotra, M. and S. Srivastava. 2008. An *ipdC* gene knock-out of *Azospirillum brasilense* strain SM and its implications on indole-3-acetic acid biosynthesis and plant growth promotion. Antonie Leeuwenhoek. 93: 425–433.

Martin-Didonet, C.C. G., L.S. Chubatsu, E.M. Souza, M. Kleina, F.G. M. Rego, L.U. Rigo, M.G. Yates and F.O. Pedrosa. 2000. Genome structure of the genus *Azospirillum*. J. Bacteriol. 182: 4113–4116.

Mehnaz, S., B. Weselowski and G. Lazarovits. 2007a. *Azospirillum canadense* sp. nov., a nitrogen-fixing bacterium isolated from corn rhizosphere. Int. J. Syst. Evol. Microbiol. 57: 620–624.

Mehnaz, S., B. Weselowski and G. Lazarovits. 2007b. *Azospirillum zeae* sp. nov., a diazotrophic bacterium isolated from rhizosphere soil of *Zea mays*. Int. J. Syst. Evol. Microbiol. 57: 2805–2809.

Miché, L., M.L. Bouillant, R. Rohr, G. Salle and R. Bally. 2000. Physiological and cytological studies on the inhibition of *Striga* seed germination by the plant growth-promoting bacterium *Azospirillum brasilense*. Eur. J. Plant Pathol. 106: 347–351.

Michiels, K.W., C.L. Croes and J. Vanderleyden. 1991. Two different modes of attachment of *Azospirillum brasilense* Sp7 to wheat roots. J. Gen. Microbiol. 137: 2241–2246.

Mishra, M., S. Kumar, N. Gupta, S. Kaur, A. Gupta and A.K. Tripathi. 2011. An extracytoplasmic function sigma factor cotranscribed with its cognate anti-sigma factor confers tolerance to NaCl, ethanol and methylene blue in *Azospirillum brasilense* Sp7. Microbiology. 157: 988–999.

Moens, S., K. Michiels, V. Keijers, F. Vanleuven and J. Vanderleyden. 1995. Cloning, sequencing, and phenotypic analysis of Laf1, encoding the flagellin of the lateral flagella of *Azospirillum brasilense* Sp7. J. Bacteriol. 177: 5419–5426.

Molina-Favero, C., C.M. Creus, M. Simontacchi, S. Puntarulo and L. Lamattina. 2008. Aerobic nitric oxide production by *Azospirillum brasilense* Sp245 and its influence on root architecture in tomato. Mol. Plant-Microbe Interact. 21: 1001–1009.

Molla, A.H., Z.H. Shamsuddin, M.S. Halimi, M. Morziah and A.B. Puteh. 2001. Potential for enhancement of root growth and nodulation of soybean coinoculated with *Azospirillum* and *Bradyrhizobium* in laboratory systems. Soil Biol. Biochem. 33: 457–463.

Mora, P., F. Rosconi, L. Franco Fraguas and S. Castro-Sowinski. 2008. *Azospirillum brasilense* Sp7 prodcues an outer-membrane lectin that specifically binds to surface-exposed extracellular polysaccharide produced by the bacterium. Arch. Microbiol. 189: 519–524.

Mostajeran, A., R. Amooaghaie and G. Emtiazi. 2007. The participation of the cell wall hydrolytic enzymes in the initial colonization of *Azospirillum brasilense* on wheat roots. Plant Soil. 291: 239–248.

Moutia, J.F., S. Saumtally, S. Spaepen and J. Vanderleyden. 2010. Plant growth promotion by *Azospirillum* sp. in sugarcane is influenced by genotype and drought stress. Plant Soil. 337: 233–242.

Murty, M.G. and J.K. Ladha. 1988. Influence of *Azospirillum* inoculation on the mineral uptake and growth of rice under hydroponic conditions. Plant Soil. 108: 281–285.

Naiman, A.D., A. Latronico and I.E. Garcia de Salamone. 2009. Inoculation of wheat with *Azospirillum brasilense* and *Pseudomonas fluorescens*: impact on the production and culturable rhizosphere microflora. Eur. J. Soil Biol. 45: 44–51.

Nobbe, F. and L. Hiltner. 1896. Inoculation of the soil for cultivating leguminous plants. Patent 570813. U.S.A.

Ogut, M. and F. Er. 2006. Micronutrient composition of field-grown dry bean and wheat inoculated with *Azospirillum* and *Trichoderma*. J. Plant Nutr. Soil Sci. 169: 699–703.

Ogut, M., C. Akdag, O. Düzdemir and M.A. Sakin. 2005. Single and double inoculation with *Azospirillum/Trichoderma*: the effects on dry bean and wheat. Biol. Fertil. Soils. 41: 262–272.

Okon, Y. 1985. *Azospirillum* as a potential inoculant for agriculture. Trends Biotechnol. 3: 223–228.

Okon, Y. and C.A. Labandera-Gonzalez. 1994. Agronomic Applications of *Azospirillum*—an Evaluation of 20 Years Worldwide Field Inoculation. Soil Biol. Biochem. 26: 1591–1601.

Okon, Y. and R. Itzigsohn. 1995. The Development of *Azospirillum* as a Commercial Inoculant for Improving Crop Yields. Biotechnol. Adv. 13: 415–424.

Oliveira, R.G.B. and A. Drozdowicz. 1987. Inhibition of producing strains of *Azospirillum lipoferum* by their own bacteriocin. Zentralblatt Fur Mikrobiologie. 142: 387–391.

Ona, O., J. Impe, E. Prinsen and J. Vanderleyden. 2006. Growth and indole-3-acetic acid biosynthesis of *Azospirillum brasilense* Sp245 is environmentally controlled. FEMS Microbiol. Lett. 246: 125–132.

Ozturk, A., O. Caglar and F. Sahin. 2003. Yield response of wheat and barley to inoculation of plant growth promoting rhizobacteria at various levels of nitrogen fertilization. J. Plant Nutr. Soil Sci. 166: 262–266.

Patriquin, D.G., J. Döbereiner and D.K. Jain. 1983. Sites and processes of association between diazotrophs and grasses. Can. J. Microbiol. 29: 900–915.

Pedraza, R.O., J. Motok, M.L. Tortora, S.M. Salazar and J.C. Diaz-Ricci. 2007. Natural occurrence of *Azospirillum brasilense* in strawberry plants. Plant Soil. 295: 169–178.

Pedraza, R.O., J. Motok, S.M. Salazar, A.L. Ragout, M.I. Mentel, M.L. Tortora, M.F. Guerrero-Molina, B.C. Winik and J.C. Diaz-Ricci. 2010. Growth-promotion of strawberry plants inoculated with *Azospirillum brasilense*. World J. Microbiol. Biotechnol. 26: 265–272.

Peng, G.X., H.R. Wang, G.X. Zhang, W. Hou, Y. Liu, E.T. Wang and Z.Y. Tan. 2006. *Azospirillum melinis* sp. nov., a group of diazotrophs isolated from tropical molasses grass. Int. J. Syst. Evol. Microbiol. 56: 1263–1271.

Pereg-Gerk, L., A. Paquelin, P. Gounon, I.R. Kennedy and C. Elmerich. 1998. A transcriptional regulator of the LuxR-UhpA family, FlcA, controls flocculation and wheat root surface colonization by *Azospirillum brasilense* Sp7. Mol. Plant-Microbe Interact. 11: 177–187.

Perrig, D., M.L. Boiero, O.A. Masciarelli, C. Penna, O.A. Ruiz, F.D. Cassán and M.V. Luna. 2007. Plant-growth-promoting compounds produced by two agronomically important strains of *Azospirillum brasilense*, and implications for inoculant formulation. Appl. Microbiol. Biotechnol. 75: 1143–1150.

Piccoli, P., C.D. Lucangeli, G. Schneider and R. Bottini. 1997. Hydrolysis of [17,17-H-2(2)] gibberellin A(20) glucoside and [17,17-H-2(2)]gibberellin A(20) glucosyl ester by *Azospirillum lipoferum* cultured in a nitrogen-free biotin-based chemically-defined medium. Plant Growth Reg. 23: 179–182.

Piccoli, P., O. Masciarelli and R. Bottini. 1999. Gibberellin production by *Azospirillum lipoferum* cultured in chemically-defined medium as affected by oxygen availability and water status. Symbiosis. 27: 135–145.

Pothier, J.F., F. Wisniewski-Dyé, M. Weiss-Gayet, Y. Moenne-Loccoz and C. Prigent-Combaret. 2007. Promoter-trap identification of wheat seed extract-induced genes in the plant-growth-promoting rhizobacterium *Azospirillum brasilense* Sp245. Microbiology. 153: 3608–3622.

Pothier, J.F., C. Prigent-Combaret, J. Haurat, Y. Moenne-Loccoz and F. Wisniewski-Dyé. 2008. Duplication of plasmid-borne nitrite reductase gene *nirK* in the wheat-associated plant growth-promoting rhizobacterium *Azospirillum brasilense* Sp245. Mol. Plant-Microbe Interact. 21: 831–842.

Prigent-Combaret, C., D. Blaha, J.F. Pothier, L. Vial, M.A. Poirier, F. Wisniewski-Dyé and Y. Moenne-Loccoz. 2008. Physical organization and phylogenetic analysis of *acdR* as leucine-responsive regulator of the 1-aminocyclopropane-1-carboxylate deaminase gene *acdS* in phytobeneficial *Azospirillum lipoferum* 4B and other Proteobacteria. FEMS Microbiol. Ecol. 65: 202–219.

Prinsen, E., A. Costacurta, K. Michiels, J. Vanderleyden and H. Van Onckelen. 1993. *Azospirillum brasilense* indole-3-acetic acid biosynthesis: evidence for a non-tryptophan dependent pathway. Mol. Plant-Microbe Interact. 6: 609–615.

Puente, M.E. and Y. Bashan. 1993. Effect of inoculation with *Azospirillum brasilense* strains on the germination and seedlings growth of the giant columnar cardon cactus (*Pachycereus pringlei*). Symbiosis. 15: 49–60.

Puente, M.E., C.Y. Li and Y. Bashan. 2004. Microbial populations and activities in the rhizoplane of rock-weathering desert plants. II. Growth promotion of cactus seedlings. Plant Biol. 6: 643–649.

Ramachandran, K., V. Srinivasan, S. Hamza and M. Anandaraj. 2007. Phosphate solubilizing bacteria isolated from the rhizosphere soil and its growth promotion on black pepper (*Piper nigrum* L.) cuttings. Dev. Plant Soil Sci. 102: 324–331.

Rao, V.R., B. Ramakrishnan, T.K. Adhya, P.K. Kanungo and D.N. Nayak. 1998. Current status and future prospects of associative nitrogen fixation in rice. World J. Microbiol. Biotechnol. 14: 621–633.

Ratti, N. and K.K. Janardhanan. 1996. Response of dual inoculation with VAM and *Azospirilllum* on the yield and oil content of Plamarosa (*Cymbopogon martinii* var. *motia*). Microbial. Res. 151: 325–328.

Reinhold, B., T. Hurek and I. Fendrik. 1985. Strain-specific chemotaxis of *Azospirillum* spp. J. Bacteriol. 162: 190–195.

Reinhold, B., T. Hurek, I. Fendrik, B. Pot, M. Gillis, K. Kersters, S. Thielemans and J. De Ley. 1987. *Azospirillum halopraeferens* sp. nov., a nitrogen-fixing organism associated with roots of Kallar grass (*Leptochloa fusca* (L.) Kunth). Int. J. Syst. Bacteriol. 37: 43–51.

Ribaudo, C.M., E.M. Krumpholz, F.D. Cassán, R. Bottini, M.L. Cantore and J.A. Cura. 2006. *Azospirillum* sp. promotes root hair development in tomato plants through a mechanism that involves ethylene. J. Plant Growth Reg. 25: 175–185.

Riefler, M., O. Novak, M. Strnad and T. Schmülling. 2006. *Arabidopsis* cytokinin receptor mutants reveal functions in shoot growth, leaf senescence, seed size, germination, root development, and cytokinin metabolism. Plant Cell. 18: 40–54.

Rodelas, B., V. Salmeron, M.V. Martineztoledo and J. Gonzalezlopez. 1993. Production of vitamins by *Azospirillum brasilense* in chemically-defined media. Plant Soil. 153: 97–101.

Rodrigues, E.P., L.S. Rodrigues, A.L.M. de Oliveira, V.L.D. Baldani, K.R.D.S. Teixeira, S. Urquiaga and V.M. Reis. 2008. *Azospirillum amazonense* inoculation: effects on growth, yield, and N_2 fixation of rice (*Oryza sativa* L.). Plant Soil. 302: 249–261.

Romero, A.M., O.S. Correa, S. Moccia and J.G. Rivas. 2003. Effect of *Azospirillum*-mediated plant growth promotion on the development of bacterial diseases on fresh-market and cherry tomato. J. Appl. Microbiol. 95: 832–838.

Rothballer, M., M. Schmid, A. Fekete and A. Hartmann. 2005. Comparative *in situ* analysis of *ipdC-gfpmut3* promoter fusions of *Azospirillum brasilense* strains Sp7 and Sp245. Environ. Microbiol. 7: 1839–1846.

Russo, A., L. Vettori, C. Felici, G. Fiaschi, S. Morini and A. Toffanin. 2008. Enhanced micropropagation response and biocontrol effect of *Azospirillum brasilense* Sp245 on *Prunus cerasifera* L. clone Mr. S 2/5 plants. J. Biotechnol. 134: 312–319.

Sadasivan, L. and C.A. Neyra. 1987. Cyst production and brown pigment formation in aging cultures of *Azospirillum brasilense* ATCC 29145. J. Bacteriol. 169: 1670–1677.

Sadasivan, L. and C.A. Neyra. 1985. Flocculation in *Azospirillum brasilense* and *Azospirillum lipoferum*: exopolysaccharides and cyst formation. J. Bacteriol. 163: 716–723.

Sant'Anna, F.H., L.G. Almeida, R. Cecagno, L.A. Reolon, F.M. Siqueira, M.R. Machado, A.T. Vasconcelos and I.S. Schrank. 2011. Genomic insights into the versatility of the plant growth-promoting bacterium *Azospirillum amazonense*. BMC Genomics. 12: 409.

Sarig, S., Y. Kapulnik and Y. Okon. 1986. Effect of *Azospirillum* inoculation on nitrogen fixation and growth of several winter legumes. Plant Soil. 90: 335–342.

Sarig, S., A. Blum and Y. Okon. 1988. Improvement of the water status and yield of field-grown grain sorghum (*Sorghum bicolor*) by inoculation with *Azospirillum brasilense*. J. Agric. Sci. 110: 271–277.

Sarig, S., Y. Okon and A. Blum. 1990. Promotion of leaf area development and yield in *Sorghum bicolor* inoculated with *Azospirillum brasilense*. Symbiosis. 9: 235–245.

Seshadri, S., R. Muthukumarasamy, C. Lakshminarasimhan and S. Ignacimuthu. 2000. Solubilization of inorganic phosphates by *Azospirillum halopraeferans*. Curr. Sci. 79: 565–567.

Shah, S., V. Karkhanis and A. Desai. 1992. Isolation and characterization of siderophore, with antimicrobial activity, from *Azospirillum lipoferum* M. Curr. Microbiol. 25: 347–351.

Sharma, P., R. Goel and N. Capalash. 2007. Bacterial laccases. World J. Microbiol. Biotechnol. 23: 823–832.

Silverman, F.P., A.A. Assiamah and D.S. Bush. 1998. Membrane transport and cytokinin action in root hairs of *Medicago sativa*. Planta. 205: 23–31.

Siuti, P., C. Green, A.N. Edwards, M.J. Doktycz and G. Alexandre. 2011. The chemotaxis-like Che1 pathway has an indirect role in adhesive cell properties of *Azospirillum brasilense*. FEMS Microbiol. Lett. 323: 105–112.

Somers, E., D. Ptacek, P. Gysegom, M. Srinivasan and J. Vanderleyden. 2005. *Azospirillum brasilense* produces the auxin-like phenylacetic acid by using the key enzyme for indole-3-acetic acid biosynthesis. Appl. Environ. Microbiol. 71: 1803–1810.

Spaepen, S., J. Vanderleyden and R. Remans. 2007a. Indole-3-acetic acid in microbial and microorganism-plant signaling. FEMS Microbiol. Rev. 31: 425–448.

Spaepen, S., W. Versees, D. Gocke, M. Pohl, J. Steyaert and J. Vanderleyden. 2007b. Characterization of phenylpyruvate decarboxylase, involved in auxin production of *Azospirillum brasilense*. J. Bacteriol. 189: 7626–7633.

Star, L., O. Matan, M.S. Dardanelli, Y. Kapulnik, S. Burdman and Y. Okon. 2011. The *Vicia sativa* spp. *nigra–Rhizobium leguminosarum* bv. *viciae* symbiotic interaction is improved by *Azospirillum brasilense*. Plant Soil Doi: 10.1007/s11104-010-0713-7.

Stephens, B.B., S.N. Loar and G. Alexandre. 2006. Role of CheB and CheR in the complex chemotactic and aerotactic pathway of *Azospirillum brasilense*. J. Bacteriol. 188: 4759–4768.

Tal, S. and Y. Okon. 1985. Production of the reserve material polybeta-hydroxybutyrate and its function in *Azospirillum brasilense* Cd. Can. J. Microbiol. 31: 608–613.

Tal, S., P. Smirnoff and Y. Okon. 1990. The regulation of poly-beta-hydroxybutyrate metabolism in *Azospirillum brasilense* during balanced growth and starvation. J. Gen. Microbiol. 136: 1191–1196.

Tapia-Hernandez, A., M.A. Mascarua-Esparza and J. Caballero Mellado. 1990. Production of bacteriocins and siderophore-like activity by *Azospirillum brasilense*. Microbios. 64: 73–83.

Tarrand, J.J., N.R. Krieg and J. Dobereiner. 1978. A taxonomic study of the *Spirillum lipoferum* group, with descriptions of a new genus, *Azospirillum* gen. nov. and two species, *Azospirillum lipoferum* (Beijerinck) comb. nov. and *Azospirillum brasilense* sp. nov. Can. J. Microbiol. 24: 967–980.

Tejera, N., C. Lluch, M.V. Martinez-Toledo and J. Gonzalez-Lopez. 2005. Isolation and characterization of *Azotobacter* and *Azospirillum* strains from the sugarcane rhizosphere. Plant Soil. 270: 223–232.

Thirunavukkarasu, N., M.N. Mishra, S. Spaepen, J. Vanderleyden, C.A. Gross and A.K. Tripathi. 2008. An extra-cytoplasmic function sigma factor and anti-sigma factor control carotenoid biosynthesis in *Azospirillum brasilense*. Microbiology. 154: 2096–2105.

Thuler, D.S., E.I.S. Floh, W. Handro and H.R. Barbosa. 2003. Plant growth regulators and amino acids released by *Azospirillum* sp. in chemically defined media. Lett. Appl. Microbiol. 37: 174–178.

Tien, T.M., M.H. Gaskins and D.H. Hubbell. 1979. Plant growth substances produced by *Azospirillum brasilense* and their effect on the growth of Pearl millet (*Pennisetum americanum* L.). Appl. Environ. Microbiol. 37: 1016–1024.

Toklikishvili, N., N. Dandurishvili, A. Vainstein, M. Tediashvili, N. Giorgobiani, S. Lurie, E. Szegedi, B.R. Glick and L. Chernin. 2010. Inhibitory effect of ACC deaminase-producing bacteria on crown gall formation in tomato plants infected by *Agrobacterium tumefaciens* or *A. vitis*. Plant Pathol. 59: 1023–1030.

Tomich, M., P.J. Planet and D.H. Figurski. 2007. The tad locus: postcards from the widespread colonization island. Nat. Rev. Microbiol. 5: 363–375.

Tortora, M.L., J.C. Diaz-Ricci and R.O. Pedraza. 2011. *Azospirillum brasilense* siderophores with antifungal activity against *Colletotrichum acutatum*. Arch. Microbiol. 193: 275–286.

Trân Vân, V., S. Ngoke, O. Berge, D. Faure, R. Bally, P. Hebbar and T. Heulin. 1997. Isolation of *Azospirillum lipoferum* from the rhizosphere of rice by a new, simple method. Can. J. Microbiol. 43: 486–490.

Umali-garcia, M., D.H. Hubbell, M.H. Gaskins and F.B. Dazzo. 1980. Association of *Azospirillum* with grass roots. Appl. Environ. Microbiol. 39: 219–226.

Van Bastelaere, E., M. Lambrecht, H. Vermeiren, A. Van Dommelen, V. Keijers, P. Proost and J. Vanderleyden. 1999. Characterization of a sugar-binding protein from *Azospirillum brasilense* mediating chemotaxis to and uptake of sugars. Mol. Microbiol. 32: 703–714.

Vanbleu, E., B.P. Choudhury, R.W. Carlson and J. Vanderleyden. 2005. The *nodPQ* genes in *Azospirillum brasilense* Sp7 are involved in sulfation of lipopolysaccharides. Environ. Microbiol. 7: 1769–1774.

Vande Broek, A., J. Michiels, A. Van Gool and J. Vanderleyden. 1993. Spatial-temporal colonization patterns of *Azospirillum brasilense* on the wheat root surface and expression of the bacterial *nifH* gene during association. Mol. Plant-Microbe Interact. 6: 592–600.

Vande Broek, A., M. Lambrecht and J. Vanderleyden. 1998. Bacterial chemotactic motility is important for the initiation of wheat root colonization by *Azospirillum brasilense*. Microbiology. 144: 2599–606.

Vande Broek, A., M. Lambrecht, K. Eggermont and J. Vanderleyden. 1999. Auxins upregulate expression of the indole-3-pyruvate decarboxylase gene in *Azospirillum brasilense*. J. Bacteriol. 181: 1338–1342.

Veresoglou, S.D. and G. Menexes. 2010. Impact of inoculation with *Azospirillum* spp. on growth properties and seed yield of wheat: a meta-analysis of studies in the ISI Web of Science from 1981 to 2008. Plant Soil. 337: 469–480.

Vial, L., C. Lavire, P. Mavingui, D. Blaha, J. Haurat, Y. Moenne-Loccoz, R. Bally and F. Wisniewski-Dyé. 2006. Phase variation and genomic architecture changes in *Azospirillum*. J. Bacteriol. 188: 5364–5373.

Wadhams, G.H. and J.P. Armitage. 2004. Making sense of it all: bacterial chemotaxis. Nat. Rev. Mol. Cell Bio. 5: 1024–1037.

Walker, V., C. Bertrand, F. Bellvert, Y. Moënne-Loccoz, R. Bally and G. Comte. 2011. Host plant secondary metabolite profiling shows a complex, strain-dependent response of maize to plant growth-promoting rhizobacteria of the genus *Azospirillum*. New Phytol. 189: 494–506.

Wisniewski-Dyé, F. and L. Vial. 2008. Phase and antigenic variation mediated by genome modifications. Antonie Leeuwenhoek. 94: 493–515.

Wisniewski-Dyé, F., K. Borziak, G. Khalsa-Moyers, G. Alexandre, L.O. Sukharnikov, K. Wuichet, G.B. Hurst, W.H. McDonald, J.S. Robertson, V. Barbe, A. Calteau, Z. Rouy, S. Mangenot, C. Prigent-Combaret, P. Normand, M. Boyer, P. Siguier, Y. Dessaux, C. Elmerich, G. Condemine, G. Krishnen, I. Kennedy, A.H. Paterson, V. Gonzalez, P. Mavingui and I.B. Zhulin. 2011. *Azospirillum* genomes reveal transition of bacteria from aquatic to terrestrial environments. PLoS Genet. 7: e1002430.

Wisniewski-Dyé, F., L. Lozano, E. Acosta-Cruz, S. Borland, B. Drogue, C. Prigent-Combaret, Z. Rouy, V. Brabe, A. Mendoza Herrera, V. Gonzalez and P. Mavingui. 2012. Comparative analyses of *Azospirillum* core and accesory genomes provide insight into niche adaptation. Genes. 3: 576–602.

Xie, C.H. and A. Yokota. 2005. *Azospirillum oryzae* sp. nov., a nitrogen-fixing bacterium isolated from the roots of the rice plant *Oryza sativa*. Int. J. Syst. Bacteriol. 55: 1435–1438.

Xie, Z., L.E. Ulrich, I.B. Zhulin and G. Alexandre. 2010. PAS domain containing chemoreceptor couples dynamic changes in metabolism with chemotaxis. Proc. Natl. Acad. Sci. U.S.A. 107: 2235–2240.

Yasuda, M., T. Isawa, S. Shinozaki, K. Minamisawa and H. Nakashita. 2009. Effects of colonization of a bacterial endophyte, *Azospirillum* sp. B510, on disease resistance in rice. Biosci. Biotechnol. Biochem. 73: 2595–9259.

Yaxley, J.R., J.J. Ross, L.J. Sherriff and J.B. Reid. 2001. Gibberellin biosynthesis mutations and root development in pea. Plant Physiol. 125: 627–633.

Zakharova, E.A., A.A. Shcherbakov, V.V. Brudnik, N.G. Skripko, N.S. Bulkhin and V.V. Ignatov. 1999. Biosynthesis of indole-3-acetic acid in *Azospirillum brasilense*—Insights from quantum chemistry. Eur. J. Biochem. 259: 572–576.

Zhou, Y., W. Wei, X. Wang, L. Xu and R. Lai. 2009. *Azospirillum palatum* sp. nov., isolated from forest soil in Zhejiang province, China. J. Gen. Appl. Microbiol. 55: 1–7.

Zhu, G.Y., S. Dobbelaere and J. Vanderleyden. 2002. Use of green fluorescent protein to visualize rice root colonization by *Azospirillum irakense* and *A. brasilense*. Funct. Plant Biol. 29: 1279–1285.

Zhulin, I.B. and J.P. Armitage. 1993. Motility, chemokinesis, and methylation-independent chemotaxis in *Azospirillum brasilense*. J. Bacteriol. 175: 952–958.

Zhulin, I.B., S.E. Tretyakova and V.V. Ignatov. 1988. Chemotaxis of *Azospirillum brasilense* towards compounds typical of plant-root exudates. Folia Microbiol. 33: 277–280.

12

Biocontrol of Fungal Root Pathogens by Fluorescent Pseudomonas

M. Sánchez-Contreras, F. Martínez-Granero, M. Redondo-Nieto, R. Rivilla* and *M. Martín*

Introduction

This chapter provides an overview, illustrated with relevant examples, of the main mechanisms used by the fluorescent pseudomonas group to protect plants from pathogenic fungi. The biological control of fungal root pathogens can be achieved by a variety of mechanisms such as secretion of antibiotics and other antagonistic metabolites, niche occupation or exclusion and the induction of systemic resistance and other responses in the plant. It has been estimated that 10 percent of rhizobacteria possess biocontrol ability (Rezzonico et al. 2007). The genus *Pseudomonas*, together with bacteria belonging to the genera *Bacillus* and *Streptomyces*, constitute an important fraction of the total rhizospheric bacteria (Chin-A-Woeng et al. 2003) and their presence is associated with the suppression of diseases in soil (Haas and Defago 2005). *Pseudomonas* spp. can suppress a wide range of symptoms and diseases, for instance the damping-off caused by the oomycete *Pythium*

Department of Biology, Autonomous University of Madrid.
*Corresponding author: maria.contreras@uam.es

ultimum and the basidiomycete *Rhizoctonia solani* affecting sugar beet and radish; selected examples of activity against other pathogenic fungi are presented in Table 12.1.

Antagonism Mediated by Metabolites

Antagonism through the production of antibiotics and other secreted products is a direct and effective mechanism to neutralize the pathogen. As a result of their secondary metabolism, pseudomonads produce a broad range of compounds that alter the growth and development of phytopathogenic organisms. Antibiotics and antifungal metabolites constitute a heterogeneous group of low molecular weight chemical compounds that in low concentrations can produce a deleterious effect on other organisms (Raaijmakers et al. 2002). A range of antibiotic compounds have been isolated and their chemical structure analyzed from biocontrol strains. The diversity of compounds suggests that there are antibiotics specific for the suppression of certain diseases and others more general that are antagonistic towards several pathogens (Bender et al. 1999, Keel et al. 1996). The most common antibiotics directly responsible of the inhibition of fungal diseases in plant by *Pseudomonas* are phloroglucinols (DAPG), pyoluteorin (Plt), pyrrolnitrin (Prn), cyclic dipeptides, phenazines and cyanide. Secreted degrading enzymes and iron-scavenging siderophores are also produced by pseudomonads and confer them biocontrol properties. These groups of antagonistic macromolecules will be described below providing examples of their action *in vitro* and *in vivo* against root pathogenic fungi.

Antifungal Metabolites

2,4-diacetilphloroglucinol (DAPG or Phl) and its precursor monoacetylphloroglucinol (MAPG) are phenolic compounds of poliketyde origin isolated from culture extracts of *Pseudomonas fluorescens* strains Q2-87 and F113 (Bangera and Thomashow 1999, Shanahan et al. 1993), among others. The locus involved in DAPG biosynthesis is ancestral and conserved in fluorescent pseudomonads (Frapolli et al. 2012) and consists of structural genes (*phlABC* operon and *phlD*), a permease (*phlE*) and regulators (*phlF, phlG* and *phlH*) (Hothersall and Thomas 2004). The mode of action of DAPG has not been completely elucidated but it has been shown to damage the membrane of *Pythium* spp. and to cause the spores' lysis (de Souza et al. 2003). DAPG is a broad spectrum antibiotic effective in the protection of several crops against a range of pathogens (Rezzonico et al. 2007, Weller 2007). For example, DAPG produced by *Pseudomonas fluorescens* F113 is important in the suppression of beetroot damping-off caused by *Pythium*

Table 12.1 Examples of pseudomonads-produced metabolites involved in antagonism.

Metabolite	Species/strain	Host plant	Target fungal pathogen	Origin and Reference
DAPG (Phl)	*P. fluorescens*			
	Q2-87	Wheat	*Gaeumannomyces graminis* var *tritici*	WA, US (Vincent et al. 1991)
	F113	Sugar beet	*Pythium ultimum*	Ireland (Shanahan et al. 1992)
	CHA0	Tobacco	*Gaeumannomyces graminis* var *tritici*, *Pythium ultimum*, *Thielaviopsis basicola*	Switzerland (Keel et al. 1992, Laville et al. 1992)
	Pf-5	Cotton	*Pythium ultimum*, *Rhizoctonia solani*	Texas, US (Howell and Stipanovic 1979, 1980)
	Q8r1-96	Wheat	*Gaeumannomyces graminis* var *tritici*	WA, US (Raaijmakers and Weller 2001)
	PFM2	Wheat	*Septoria tritici*	OK, US (Levy et al. 1992)
Pyoluteorin (Plt)	*P. fluorescens*			
	Pf-5	Cotton	*Pythium ultimum*, *Rhizoctonia solani*	Texas, US (Howell and Stipanovic 1979, 1980)
	CHA0	Tobacco	*Pythium ultimum*, *Thielaviopsis basicola*	Switzerland (Keel et al. 1992, Laville et al. 1992)
Pyrrolnitrin (Prn)	*P. fluorescens*			
	BL915	Cotton	*Rhizoctonia solani*	US (Ligon et al. 2000, van Pee and Ligon 2000)
Phenazines (Phz)	*P. fluorescens*			
	2–79	Wheat	*Gaeumannomyces graminis* var *tritici*	WA, US (Gurusiddaiah et al. 1986, Weller 1983)

	P. aureofaciens			
	30–84	Wheat	*Gaeumannomyces graminis* var *tritici*	Kansas, US (Pierson and Thomashow 1992)
	PGS12	Corn	*Rhizoctonia solani, Gaeumannomyces graminis*	Belgium (Rosales et al. 1995)
Phenazines (Phz)	*P. chlororaphis*			
	PCL1391	Tomato	*Fusarium oxysporum*	Spain (Chin-A-Woeng et al. 1998)
	P. aeruginosa			
	PNA1	Chickpea	*Fusarium oxysporum, Pythium splendens*	India (Anjaiah et al. 1998)
	Pseudomonas spp.			
	In-b-109	Rice	*Fusarium oxysporum*	Philippines (Georgakopoulos et al. 1994)
HCN	*P. fluorescens*			
	CHA0	Tobacco	*Thielaviopsis basicola*	Switzerland (Voisard et al. 1989)
Viscosinamide	*P. fluorescens*			
	DR54	Sugar beet	*Rhizoctonia solani, Pythium ultimum*	Denmark (Nielsen et al. 1998, Nielsen et al. 1999)
Massetolide	*P. fluorescens*			
	SS101	Wheat	*Phytophthora infestans*	Netherlands (Tran et al. 2007)

Adapted from Raaijmakers et al. (2002).

ultimum (Shanahan et al. 1992). In *P. fluorescens* CHA0, suppression of tobacco black root rot, wheat take-all and *Fusarium* crown and root rot of tomato is achieved by DAPG (Keel et al. 1992, Stutz et al. 1986). The organization of the biosynthetic genes in a cluster allows the transfer of extra-copies of the *phl* locus to DAPG-producing strains to increase their biocontrol ability (Delany et al. 2001, Maurhofer et al. 1995) and to non-producer strains to give them this activity. For instance *Pseudomonas fluorescens* M114, which lacks DAPG production, showed biocontrol activity against *P. ultimum* when provided with the *phl* locus in a plasmid (Fenton et al. 1992).

Pyoluteorin is a chlorinated aromatic polyketide (Cuppels et al. 1986, Nowak-Thompson et al. 1997) which has been shown to inhibit *P. ultimum* (Howell and Stipanovic 1980). The biosynthesis is mediated by a polyketide synthase (PKS) and a non-ribosomal peptide synthase (NRPS) complex encoded by the *plt* genes (Kraus and Loper 1995). The degree of contribution of pyoluteorin to biocontrol seems to depend on the host plant. In *P. fluorescens* strains Pf-5 and CHA0, pyoluteorin was required to suppress *P. ultimum* damping-off on cotton and cress respectively, while its contribution was minimal for fungal suppression in cucumber by the same strains (Howell and Stipanovic 1980, Kraus and Loper 1992, Maurhofer et al. 1994). Production of pyoluterin is known to be affected by carbon sources (Kraus and Loper 1995). Given that plants secrete a variety of carbon-rich exudates, this could account for biocontrol differences observed in response to diverse plant hosts.

Pyrrolnitrin, a phenylpyrrol derived from tryptophan, has been described as an inhibitor of the respiratory chain in fungi (Tripathi and Gottlieb 1969). The genes responsible for its production in *P. fluorescens* have been identified and consist of a cluster of four genes, *prnABCD,* which are conserved in other *Pseudomonas* and *Burkholderia* strains (Hammer et al. 1999, Hammer et al. 1997). *Pseudomonas fluorescens* BL915 produces pyrrolnitrin against the cotton pathogen *Rhizoctonia solani* and other soil-borne pathogens (van Pee and Ligon 2000). Synthetic analogues of this compound have been used as fungicides in agriculture (Ligon et al. 2000).

Cyclic lipopeptides (CLP) are small molecules produced by non-ribosomal peptide synthases consisting of a fatty acid chain, variable in length and modifications, linked to a short cycled oligopeptide that often contains unusual aminoacids (Raaijmakers et al. 2006). CLPs can insert into membranes altering their function. This mode of action explains the fact that phytotoxins as well as biocontrol substances with broad antibacterial and antifungal properties are found among them. Viscosinamide and amphisin-group CLPs showed antagonist effect *in vitro* against *P. ultimum* and *R. solani,* reducing the hyphal growth (Nielsen et al. 2002). These effects have been confirmed during *in vivo* assays with the viscosinamide-producer strain *Pseudomonas fluorescens* DR54 which proved to reduce damping-off symptoms and to protect sugar beet root from *P. ultimum* and *R. solani* (Thrane et al. 2001, Thrane et al. 2000). Production of viscosinamide, amphisin and tensin CLPs also occurs in the rhizosphere of young sugar beet plants, and even when these compounds are introduced in bulk soil without the producing strain, they can persist for a few weeks contributing to root protection (Nielsen and Sorensen 2003). Massetolide A was identified from *Pseudomonas fluorescens* SS101. This CLP plays a role in swarming motility and biofilm formation (de Bruijn et al. 2008) and it is important in tomato root colonization since a mutant affected in one of the synthesis genes does not colonize as efficiently as the wild type (Tran et al. 2007). Massetolide A was shown to protect tomato leaves against infection by *Phytophthora infestans* both in assays with the isolated compound and in experiments comparing the non-producing mutant with the parental strain (Tran et al. 2007). In addition to direct fungal inhibition, the surfactant properties of CLPs can facilitate bacterial motility on surfaces (Nybroe and Sorensen 2004) and this may contribute to improve competitive rhizosphere colonization which is an important factor for overall biocontrol activity.

Phenazines are a large family of heterocyclic nitrogen-containing compounds that differ in the derivatives of their core and are synthesized by enzymes encoded by the conserved *phz* gene cluster (Mavrodi et al. 1998). Although their mode of action is not completely understood, it is thought that they diffuse through membranes and undergo redox cycling resulting in the accumulation of reactive oxygen species (ROS) such as superoxide and hydrogen peroxide that damage the cell, ultimately causing death. Superoxide dismutases in the *Pseudomonas* cells are probably responsible for the tolerance to the phenazines produced (Hassan and Fridovich 1980, Hassett et al. 1992). Some of the best studied among these compounds are phenazine-1-carboxilic acid (PCA), phenazine-1-carboxamide (PCN) and pyocyanin. Phenazines possess a broad spectrum of activity against fungi and bacteria contributing to the soil competence of fluorescent pseudomonads (Mazzola et al. 1992). In *Pseudomonas chlororaphis* PCL1391 phenazines are responsible for the protection of tomato against *Fusarium oxysporum* (Chin-A-Woeng et al. 1998). Phenazine secretion by *Pseudomonas aureofaciens* 30–84 and *Pseudomonas fluorescens* 2–79 suppresses take-all of wheat caused by *Gaeumannomyces graminis* var. *tritici* (Gurusiddaiah et al. 1986, Pierson and Thomashow 1992). Transfer of the phenazine biosynthetic operon into the non-producing strains *Pseudomonas putida* WCS358 and *Pseudomonas fluorescens* SBW25 enhanced their existing biocontrol abilities (Bakker et al. 2002, Glandorf et al. 2001, Timms-Wilson et al. 2000).

Hydrogen cyanide (HCN) or cyanide (CN^-) is a volatile compound produced by many fluorescent pseudomonads. The HCN synthase complex consists of three subunits encoded by the operon *hcnABC* and is anchored to the cell inner membrane by HcnC (Laville et al. 1998, Pessi and Haas

2000). The production of significant amounts, from 0.1 to 1 mM, of HCN has been reported in fluorescent pseudomonads (*P. fluorescens, P. chlororaphis*) and in the soil bacterium *Chromobacterium violaceum* (Castric 1981, Knowles and Bunch 1986). This volatile compound is effective against fungi and other organisms and interferes with the respiratory chain. One protection mechanism harboured by the HCN-producing pseudomonads is a cyanide-insensitive cytochrome oxidase encoded by the *cioAB* genes, as described for *Pseudomonas aeruginosa* (Comolli and Donohue 2002, Matsushita et al. 1983, Trutko et al. 1979) in addition to other cyanide detoxification mechanisms.

The effect of HCN-producing strains on plant disease suppression has been investigated. The correlation of biocontrol ability and HCN-producing strains was positive in a study with 29 isolates of fluorescent pseudomonads (Ellis et al. 2000), but no correlation was found with disease suppression and the number of cyanogenic fluorescent pseudomonas present in soils (Ramette et al. 2003). This indicates that the effect of HCN is not the only mechanism responsible for biocontrol and can be masked by the production of other compounds resulting in different outcomes for the host plant. However, experiments with specific strains showed the role of cyanide in disease suppression. The HCN-producing strain *P. fluorescens* CHA0 showed that a mutant unable to produce cyanide had reduced biocontrol ability against tobacco root rot by *Thielaviopsis basicola* compared to the wild-type (Laville et al. 1998, Voisard et al. 1989). In addition, the transfer of the HCN-production function by recombination to a *P. putida* strain, increased the biocontrol ability with respect to the parental strain against two different fungi infecting wheat seedling leaves (Flaishman et al. 1996).

Secreted Degrading Enzymes

Soil microorganisms produce a variety of degrading enzymes such as chitinases, beta-(1,3)-glucanases, lipases, cellulases and exoproteases that allow them to attack and parasitize fungi. Chitinases can degrade fungal cell wall, leading to cell damage and leaking (Lugtenberg et al. 1991, Thomashow and Weller 1996). The role of lytic enzymes of the biocontrol fungus *Tricoderma* spp. has been investigated for the suppression of fungal diseases (Markovich and Kononova 2003). Fewer examples are available in *Pseudomonas* spp. although many of the strains produce lytic enzymes (Sharifi-Tehrani et al. 1998). It is known that *Pseudomonas stutzeri* secretes beta 1,3-glucanase, chitinase and laminarinase that lyse the mycelia of *Fusarium solanii* (Lim and Kim 1995, Lim et al. 1991). The action of surfactants, such as CLPs mentioned above, can improve the access of degrading enzymes to more hydrophobic areas in the fungal mycelium and have an enhancing effect towards fungal parasitism resulting in more effective biocontrol.

Siderophores

Siderophores are extracellular diffusible pigments with high affinity for iron. Pyoverdin (Pvd, also named pseudobactin) is the main pigment conferring the green-yellowish colour and fluorescence to this *Pseudomonas* group. Pyoverdin is a strong iron chelator with high affinity for Fe^{3+}, this form of iron is poorly soluble in soil. The pyoverdin-iron complex binds to specific receptors (that can be found in Pvd-producing and non-producing strains) and is then transported to the cytoplasm where Fe^{3+} is reduced and used by the cell (Visca 2004). The production of siderophores has been also identified as a mechanism of biological control (Kloepper et al. 1980a). Production of pyoverdin can cause growth inhibition of other microorganisms by sequestering the scarce iron available and depriving the pathogen of the local iron (Kloepper et al. 1980b, Schroth and Hancock 1982). However, this mechanism of biocontrol is only effective in iron-depleted conditions, as the production of siderophores is repressed when the concentration of iron is high in the environment. For example, *P. putida* strain B10 suppressed *Fusarium* causing wilt and take-all, but this effect was reversed by adding iron to the soil (Kloepper et al. 1980b). It has been demonstrated that soil pH is an important factor influencing iron availability in the rhizosphere (Loper and Henkels 1997). Experiments in the cotton rhizosphere showed that *P. fluorescens* Pf-5 is not iron-limited after 2 days (Loper and Henkels 1997). Generally, as pH increases the solubility of Fe^{3+} species decreases (Elad and Baker 1985, Misaghi et al. 1988). Therefore the biocontrol effect due to siderophores may only be transient in soil conditions, where pH and iron concentrations may fluctuate.

Secondary siderophores have been identified in fluorescent pseudomonads as having an antifungal role. One of them is pyochelin (Phoebe et al. 2001); the mechanism involved is not certain, although it could be related with fungi deprivation of copper and zinc, since it is known that pyochelin is a weaker iron chelator than pyoverdin but a strong chelator of Cu^{2+} and Zn^{2+} (Cuppels et al. 1987, Visca et al. 1992). Another secondary siderophore with antifungal activity is thioquinolobactin, which in *Pseudomonas fluorescens* ATCC 17400 *in vitro* assays showed to have a greater effect in the antagonism towards the oomycete *Pythium* than pyoverdin (Matthijs et al. 2007).

Regulation

The synthesis and production of enzymes and secondary metabolites is regulated by the *Pseudomonas* sensing of environmental cues and neighbouring cells, as well as by the intrinsic state of the bacterium. One of the factors influencing the production of secondary metabolites is the availability and the specific type of carbon source. It has been shown that the presence of saccharose and mannitol in the media induce DAPG production in *P. fluorescens* F113, while succinate and glucose lower DAPG levels (Shanahan et al. 1992). Conversely, *P. fluorescens* Pf-5 increases DAPG production when glucose is added to the medium but this carbon source triggers a decrease in pyoluteorin production (Kraus and Loper 1995). Furthermore, the synthesis of HCN is dependent on the balance of carbon and nitrogen sources: *Pseudomonas fluorescens* In5 grown on nitrogen-rich medium produced HCN and inhibited growth of *R. solani* and *Pythium aphanidermatum* hyphae. In contrast, in a carbohydrate-rich medium the strain did not produce HCN and had no biocontrol activity (Michelsen and Stougaard 2012). The dramatic effect exerted on antibiotic production by the carbon sources suggests that the chemical composition of the rhizospheric habitat (including the production of root exudates) is critical for biocontrol mediated by antifungal metabolites (Nowak-Thompson et al. 1994). Plant-related parameters were found to exert the strongest influence on the genotypic and phenotypic variations in the diversity of fluorescent *Pseudomonas* spp., over other factors such as climate, soil type and management regime, geographic location and interactions with other organisms (Picard and Bosco 2008).

Environmental signals are perceived and integrated with the internal state of the cell through the two-component regulatory system GacA-GacS. This system positively regulates antibiotic production (Aarons et al. 2000, Kang et al. 2006, Laville et al. 1992), lipopeptides (Koch et al. 2002), extracellular enzymes and siderophores (Liao et al. 1996, Sacherer et al. 1994) in *Pseudomonas* at post-translational level (Fig. 12.1). Mutations in the *gacA* and *gacS* genes are common in certain conditions (Duffy and Defago 2000) and in the rhizosphere (Sánchez-Contreras et al. 2002), which results in less stability in the production of antibiotics by the population. The pseudomonas strains affected in this two-component system have been described as having less biocontrol ability respective of their parental strains (Heeb and Haas 2001, Ligon et al. 2000). The Gac system controls the production of a range of secondary metabolites through the transcription of small regulatory RNAs that titrate out RNA-binding proteins that repress the transcription of biosynthetic genes (Heeb and Haas 2001, Valverde et al. 2003). The transcription factors encoded by *rpoS*, *rpoD* and *algU* also play a role in the transcriptional regulation of metabolites such as DAPG (Mascher et al. 2002, Schnider et al. 1995, Whistler et al. 1998).

Cell to cell signalling networks, or quorum sensing, can also influence production of secondary metabolites (Chancey et al. 1999, Loh et al. 2002, Pierson et al. 1998). In *P. aureofaciens* 30–84 and *Pseudomonas chlororaphis* PCL1393, the production of phenazine is under the positive control of the

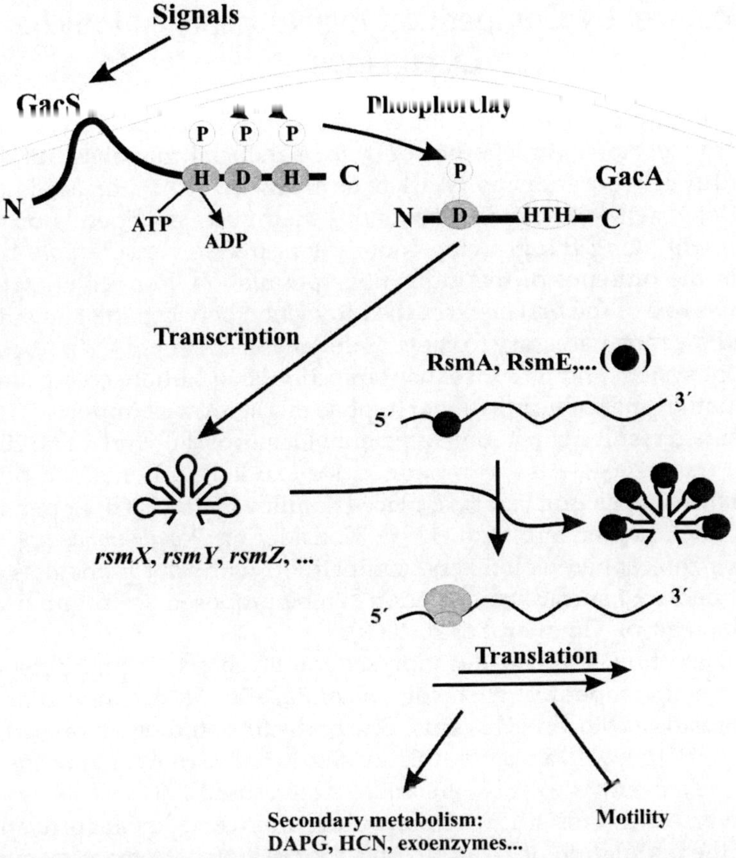

Figure 12.1 Scheme of the regulation of secondary metabolites and motility through the Gac system and small RNAs in *Pseudomonas fluorescens*.

quorum sensing system PhzI-PhzR (Chin-A-Woeng et al. 2001, Wood et al. 1997). A *phzR* mutant in *Pseudomonas chlororaphis* PA23, showed reduced production of the metabolites pyrrolnitrin, phenazine and protease and no longer inhibited the fungal pathogen *Sclerotinia sclerotium* (Selin et al. 2012). Furthermore, in this strain there was cross-regulation by the sigma factor RpoS and the quorum sensing system PhzI-PhzR (Selin et al. 2012). The mutual influence of regulators and transcription factors in the cell make the regulation of antagonistic metabolites a complex network integrating environmental and physiological signals. However, this complexity of control at transcriptional and post-transcriptional levels allows fine tuning of the production of these costly metabolites under the appropriate conditions.

Biocontrol by Competition for Nutrients and Niche Occupation

As a result of the release of root exudates, the rhizosphere becomes an attractive ecological niche not only for beneficial microbes but also for harmful microorganisms (Walker et al. 2003). Plants or seeds can be inoculated with biocontrol agents but their success depends on the establishment along the growing root system over a considerable time period, in the presence of the indigenous population. Root colonization seems to be one of the first steps in the interaction between the biocontrol agent and the root, necessary to prevent the establishment of the pathogen on the root system and its subsequent infection. In addition, colonization of phytopathogens including fungal hyphae might form an important trait for parasitism resulting in a more efficient biocontrol (Bolwerk et al. 2003, de Weert et al. 2004). Also, competition for exuded nutrients and other components such as iron can take place (Kamilova et al. 2005, Loper and Henkels 1999, Lugtenberg et al. 1999). Considering *Pseudomonas* spp. as aggressive rhizosphere colonizers, competition with root pathogens for nutrients and root surface colonization can be proposed as an important trait for biocontrol (Couillerot et al. 2009).

In this section, we will describe several studies that highlight the significance of competitive rhizosphere colonization in the utilization of pseudomonads as biocontrol agents. One of the first studies conducted by Bull et al. (1991) showed a positive relationship between root colonization by *Pseudomonas fluorescens* 2-79RN_{10} and root lesions caused by *Gaeumannomyces graminis* var. *tritici*. This study demonstrated an inverse linear correlation between the population of 2-79RN_{10} on a root section 4–6 cm far from the inoculation site and the number of lesions on the root system, indicating that colonization of growing roots is required to maintain the biocontrol capacities.

Pseudomonas chlororaphis PCL1391 was selected for the suppression of tomato root and foot rot caused by *Fusarium oxysporum* f. sp. *radicis-lycopersici* (Chin-A-Woeng et al. 1998). It has been shown that colonization is also essential for the biocontrol ability of *P. chlororaphis* PCL1391 since mutants impaired in known colonization traits such as motility (Capdevila et al. 2004, de Weger et al. 1987), auxotrophy for amino acids (Simons et al. 1997) and site-specific recombinases (Dekkers et al. 1998, Martínez-Granero et al. 2005, Mavrodi et al. 2006) cannot control tomato foot and root rot, despite the fact that they showed normal antifungal activity (Chin-A-Woeng et al. 2000).

Most biocontrol agents, including strains of *Pseudomonas* spp., are still too variable in their plant protection to be successfully used as inoculants.

This inconsistency has been attributed to different factors, including the differential expression of genes involved in disease suppression and poor root colonization by the applied biocontrol agent. Raaijmakers and Weller (2001) isolated three DAPG-producing *Pseudomonas* strains from roots of wheat grown in a soil naturally suppressive to take-all to demonstrate that the competence of their biocontrol abilities comprises an efficient root colonization combined with the ability to survive and proliferate over a considerable time period. The ability of strains Q8r1-96 (the dominant genotype), Q2-87 and 1M1-96 to colonize the rhizosphere of wheat over both a short and an extended period of time was tested. The rhizosphere competence of strain Q8r1-96 was only substantially greater than that of the other two strains in the case of long-term experiments where Q8r1-96 reached higher rhizosphere population densities. These results were correlated with those of the suppression of take-all, being Q8r1-96 a better biocontrol agent in long-term experiments but not in short-term studies. This experiment highlights the key role of root colonization in the suppression of take-all. DAPG production levels can be discarded to be responsible for the superior rhizosphere competence of Q8r1-96 since no significant differences were observed between the three strains.

Damping-off disease in sugar beet seedlings is caused by a number of soil-borne plant pathogens, including the basidiomycete *Rhizoctonia solani* and the oomycete *Pythium ultimum*. The fluorescent *Pseudomonas* sp. DSS73, which was isolated from the rhizoplane of sugar beet seedlings as a biosurfactant-producing strain capable of inhibiting the growth of these fungi (Nielsen et al. 2002, Sorensen et al. 2001), produces a battery of antagonistic agents such as proteases, chitinases and HCN controlled by the GacA/GacS system (Koch et al. 2002). Nevertheless, Andersen et al. (2003) have also demonstrated that surface/swarming motility is crucial to ensure an effective block of the spreading mycelium of these two plant-pathogenic fungi. In this study, the two surfactant-negative mutants DSS73-15C2 ($AmsY^-$, mutant affected in the putative amphisin synthase) and DSS73-12H8 ($GacS^-$) were tested for their motility and biocontrol phenotypes. Although both mutants showed functional flagella and a fully operational chemotaxis apparatus, they failed to carry out swarming motility unless the medium was supplemented with amphisin or biosurfactants such as tensin or viscosinamide. This demonstrated that amphisin is acting as a compound that lowers surface tension and enables *Pseudomonas* sp. DSS73 to move over surfaces. When the same motility assays were carried out in the presence of fungi, the non-motile strains DSS73-15C2 ($AmsY^-$) and DSS73-12H8 ($GacS^-$) did not inhibit the spreading mycelium of both *R. solani* and *P. ultimum*. However, when amphisin was added and motility was recovered, only the ability of the *amsY* mutant to block mycelium spreading was restored. These results clearly demonstrate that the efficient elimination of pathogenic fungi is due to the synergistic effect of surface motility and the production of antifungal compounds.

More recently, Pliego et al. (2007, 2008) have reported two rhizosphere-colonizing pseudomonads, named *P. alcaligenes* AVO73 and *P. pseudoalcaligenes* AVO110, which had different biocontrol abilities over *Rosellinia necatrix* on avocado roots. Although both AVO73 and AVO110 showed *in vitro* antagonism against *R. necatrix*, only AVO110 demonstrated significant protection against this fungus *in vivo*. As both strains were selected because of their efficient avocado root tip colonization and they did not produce antifungal compounds, differences in their biocontrol abilities were attributed to their different colonization patterns and their different abilities to utilize exudates from roots and hyphae. Despite the fact that both strains were able to establish and survive in the root system of avocado, the sites occupied by them were different. Abundant colonization of fungal hyphae was only observed for the biocontrol strain AVO110, while AVO110 was more abundant at junctions between epidermal root cells and root wounds (previously described as preferential penetration sites for *R. necatrix* infection, Mantell and Wheeler 1973), strain AVO73 was observed preferentially attached to lateral roots.

Valuable information about the importance of root colonization in biocontrol can be also obtained by *in situ* localization of microorganisms in the rhizosphere as it is the case in a study by Prieto et al. (2009) who used the confocal laser scanning microscopy (CLSM) technique to monitor the infection and colonization of olive plants by the fungal phytopathogen *Verticillium dahlia* Kleb and the beneficial endophytic *Pseudomonas fluorescens* PICF7. To address co-localization assays, an enhanced yellow fluorescent protein (eYFP)-tagged transformant of *V. dahlia* and an enhanced green fluorescent protein (eGFP)-tagged *P. fluorescens* PICF7 derivatives were generated. Results from colonization assays showed that wounds in root hairs or dead epidermal cells are preferred penetration sites by *V. dahlia* within olive root tissues. Olive root hairs were also confirmed as a preferred site for colonization and penetration into olive roots by *P. fluorescens* PICF7, indicating that both microorganisms share the same ecological niche. In fact, colonization of pathogen hyphae by PICF7 was also observed. A correlation between PICF7 colonization and reduction of infection was found; they specifically observed a decrease of pathogen biomass within root regions where PICF7 had already endophytically been established. Therefore, effective root colonization by *P. fluorescens* PICF7, including endophytic establishment, seems to be necessary to hamper *V. dahlia* colonization and to impair infection.

Given the importance of root colonization in biological control, the selection of strains that are rhizosphere-competent will significantly contribute to improve the efficacy of biocontrol agents. This desirable trait of good root colonization can be selected by isolating hypercompetitive bacteria or much easier by using some methods to enrich for excellent root colonizers.

One of these methods described by Kuiper and colleagues (2001) involves three enrichment cycles where a mixture of rhizobacteria is applied on a sterile seedling and after plant growth, those bacteria that have reached the root tip are isolated. In this case they selected a *Pseudomonas putida* strain that colonizes the root tip up to 100-fold better than the well-known efficient competitive root tip colonizer and biocontrol strain *P. fluorescens* WCS365. The effectiveness of this enrichment method to select for a novel class of biocontrol bacteria was subsequently tested by Kamilova et al. (2005). In this work, several strains which showed better competitive tomato root tip colonization than *P. fluorescens* WCS365, were isolated to test their biocontrol abilities. Surprisingly, one of the isolated strains called *Pseudomonas fluorescens* PCL1751, that did not show either antifungal activity *in vitro* or induction of systemic resistance, significantly controlled tomato foot and root rot and did so to a similar extent as the strain WCS365. A spontaneous mutant of PCL1751 impaired in motility was outcompeted by the wild-type strain both in the middle part of the root and on the root tip and showed no significant biocontrol activity against tomato foot and root rot. This result suggests that competition for niches and/or nutrients is required for biocontrol activity of *P. fluorescens* PCL1751.

A similar procedure based on the selection of competitive avocado root tip colonizers was carried out by Pliego et al. (2007). By using this technique they obtained a high frequency (24.4 percent) of strains that displayed antagonism towards the fungus *Rosellinia necatrix*, responsible for the white root rot in avocado. The antagonistic activity against *R. necatrix* shown by these isolates is probably due to a mode of action different from antibiosis, as most of them did not produce detectable amounts of known antifungal compounds. Indeed, it has been postulated that the biological control of *R. necatrix* might occur through competition for space and nutrients.

A second accelerated evolution method described next is based on previous results that highlight the importance of the site-specific recombinases Sss and XerD in competitive rhizosphere colonization: mutants of different *Pseudomonas* strains affected in either *sss* or *xerD* genes are severely impaired for competitive rhizosphere colonization (Dekkers et al. 1998, Martínez-Granero et al. 2005) and conversely, ectopic expression of the *sss* gene increases the competitive colonization ability of several pseudomonads (Dekkers et al. 2000). Taking these results as their starting point, Martínez-Granero et al. (2006) demonstrated a clear relationship between these site-specific recombinases and the appearance of hypermotile phenotypic variants that were more competitive, and how this enhanced motility phenotype was being selected by the rhizospheric environment.

So they proposed a new method to generate more competitive strains for rhizosphere colonization by overexpression of the site-specific recombinases Sss or XerD followed by a single rhizosphere passage.

Strains isolated by these two previous methods probably present several unknown mutations that cause their enhanced competitive colonization. But more knowledge of the motility regulation of a biocontrol strain as is the case of *P. fluorescens* F113 (Fenton et al. 1992, Naseby et al. 2001) can be also useful to generate hypermotile and more competitive strains by mutating specific regulatory genes (Barahona et al. 2010, Navazo et al. 2009). In F113, motility is repressed by at least three independent pathways including the two-component system GacA/S, the Wsp chemotaxis-like system and the sensor proteins SadB and KinB (Barahona et al. 2011, Navazo et al. 2009). Based on this information, Barahona and colleagues (2011) constructed a *kinB sadB wspR* triple mutant (KSW), which presents a hypermotile phenotype and is more competitive than the wild-type strain, to test whether a higher competitiveness can result in a better performance in biocontrol compared to the wild-type strain. To this end, F113 and its isogenic KSW mutant were compared in two pathogenicity systems (*Fusarium osysporum* f. sp. *radicis-lycopersici*/tomato and *Phytophthora cactorum*/strawberry). KSW mutant showed improved biocontrol compared to the wild-type strain in both pathosystems, although these effects were especially clear in the case of the *P. cactorum*/strawberry system. These results prove that a rational design of biocontrol agents through site-directed mutagenesis that enhances rhizosphere competitive colonization may result in an improvement in biocontrol.

Biocontrol through the Activation of Plant Defences

When plants are properly stimulated, they can reach a state of enhanced defensive capacity. As well as other PGPR (Plant-Growth Promoting Rhizobacteria) in the rhizosphere, several *Pseudomonas fluorescens* strains can trigger the defence mechanisms in plants. Two forms of resistance have been described: Systemic acquired resistance (SAR) and induced systemic resistance (ISR). SAR is a long term and broad spectrum induced disease resistance (Ross 1961) associated to increasing levels of salicylic acid (SA) and with the coordinate expression of a set of *PATHOGENESIS-RELATED* (*PR*) genes that encode PR proteins with antimicrobial activity (Sels et al. 2008). SA signal is perceived by the regulatory protein NPR1 (NONEXPRESSOR OF PR GENES1) that is the transcriptional co-activator of PR gene expression (Dong 2004). Besides SAR mechanisms against pathogens, certain non-pathogenic rhizospheric bacteria are responsible for triggering an enhanced disease resistance in plants (VanPeer et al. 1991, Wei et al. 1991) by ISR. Like SAR, ISR has been found in many plants and has a broad-spectrum of effectiveness. Although the signalling pathway is initiated in roots, the effects are present in the aerial part. This is possible due to the action of the jasmonic acid (JA) and ethylene (ET) signalling routes, which play an important role in this pathway (Knoester et al. 1999, Pieterse et al. 1990), though accumulation of both hormones is not necessarily required (van Wees et al. 1999).

As mentioned above *P. fluorescens* is one of the PGPR species able to promote ISR, although not all *P. fluorescens* strains have the same effectiveness against different pathogens and plants (Table 12.2). Therefore, not all the strains stimulate ISR with the same kind of molecules and not all the plants respond to the same signals (van Loon 2007). Two types of studies regarding ISR stimulation have been performed: specific strains against specific plants and pathogens, and isolates from the rhizosphere of plants affected by a specific pathogen. One of the main differences can be found in the elicitors that stimulate ISR. Little is known about how or whether plants can induce cross signalling with *P. fluorescens*. Some studies point out that *P. fluorescens* CHA0 in soil dedicates all its resources to growth but the alteration of barley exudates upon infection by *Pythium ultimum* is enough to modify bacterial metabolism in order to produce ISR-triggering molecules (Jousset et al. 2011). So far, several families of elicitors have been demonstrated to be involved in the process: flagella, lipopolysaccharides (LPS), iron-chelating siderophores, bacteria-synthesized antibiotics, quorum sensing *N*-acylhomoserine lactones, and volatile compounds such as 2,3-butanediol (van Loon 2007). These bacterial components, present in non-pathogenic strains and able to activate the defence responses of plant cells, have been termed microbe-associated molecular patterns (MAMPs). MAMPs can differentially activate ISR depending on: pathogen, *Pseudomonas* strain and plant. That is, PGPR that elicit ISR in one plant species may not do so in another, indicating a specific interaction between rhizobacteria and plant. Van Loon et al. (2008) prepared a battery of tests to measure the intensity of ISR in both suspension-cultured tobacco cells and whole plants against different *Pseudomonas* strains using bacterial whole cultures and fractions. The results indicated that flagella are not so important to trigger ISR for *Pseudomonas fluorescens* WCS417r and WCS374r strains. In fact, WCS374r protects rice from *Magnaporthe oryzae* through ISR induction by the siderophore pseudobactin (De Vleesschauwer et al. 2008). Bacterial cell walls are also important to elicit ISR in *Arabidopsis thaliana*, i.e., LPS fractions of WCS417r are enough to promote *A. thaliana* resistance against *Fusarium oxysporum* (Van Wees et al. 1997). Additionally, DAPG produced by *P. fluorescens* CHA0 can also be detected by *A. thaliana* roots stimulating ISR against *Peronospora parasitica* infection in leaves (Iavicoli et al. 2003).

Plant receptors recognizing fluorescent pseudomonas MAMPs are not fully characterized. There is a family of leucin rich receptors (LRR-RJ), FLS2 in *A. thaliana*, responsible for flagellin perception triggering a mitogen-

Table 12.2 Examples of biocontrol *Pseudomonas fluorescens* strains tested as ISR-promoting agents in different plants against fungal pathogens.

Strain	Plant	Pathogen	Reference
WCS417r	*A. thaliana*	*Fusarium oxysporum* f. sp. Raphani	(Van Wees et al. 1997)
		Peronospora parasitica	(Ton et al. 2002b)
		Hyaloperonospora arabidopsis	(Van der Ent et al. 2009)
		Alternaria brassicola	(Van Wees et al. 1997)
		Hyaloperonospora parasitica strain WACO9	(Pozo et al. 2008)
		Botritis cinerea strain B0510	(Van der Ent et al. 2008)
Pfl, EP1, VPT4 and VPT10	Sugarcane	*Colletotrichum falcatum*	(Viswanathan and Samiyappan 2002)
		Glomerella tucumanensis	
Mixture of Pfl, TDK1, PY15	Rice	*Sarocladium orizae*	(Saravanakumar et al. 2009)
2-79	Wheat	*Gaeumannimyces graminis* var tritici	(Thomashow and Weller 1988)
		Phytium ultimum	(Timms-Wilson et al. 2000)
89B61	Tomato	*Phythophthora infestans*	(Yan et al. 2002)
CHA0	*A. thaliana*	*Peronospora parasitica*	(Iavicoli et al. 2003)
	Grapevine	*Botytis cinerea*	(Verhagen et al. 2010)
	Barley	*Phytium ultimum*	(Jousset et al. 2011)
		Fussarium graminearum	(Henkes et al. 2011)
Pfl, FP7	Rice	*Rhizoctonia solani*	(Nandakumar et al. 2001)
WCS374	Radish	*Fusarium oxysporum* f. sp. Raphani	(Leeman et al. 1996)
	Rice	*Magnaporthe oryzae*	(De Vleesschauwer et al. 2008)

activated kinase pathway and the ethylene signalling route (Boutrot et al. 2010). So far, no other MAMPs receptors have been found.

Independent of the eliciting mechanism, triggering of ISR requires both a minimum number of bacterial cells (10^5 CFU per gram of plant root) and pre-incubation with roots for hours or days, depending on the system (Raaijmakers et al. 1995). This condition, termed as priming, has been extensively studied. Split root studies in barley showed that *Fusarium graminearum* infection induces carbon redistribution from the infected root to the non-infected one, as a plant defence system to control the pathogen spreading in the plant. Priming of barley with *P. fluorescens* CHA0 hours before *F. graminearum* treatment prevented carbon redistribution and promoted the growth of infected roots. However, a reduction in pathogen progression and an increase in the overall plant biomass production was observed (Henkes et al. 2011). These results raise the question as to whether the initial plant defence response for containing the pathogen progression can be compromised by the action of the biocontrol strain through the promotion of root growth to reach optimal priming conditions.

A. thaliana and *P. fluorescens* have been used as a main model system to establish the molecular signalling cascade preceding ISR due to its broad range of responses against different pathogens, either bacteria or fungi (Pieterse et al. 2002). The *A. thaliana* mutants *eir1-1* (insensitive to ethylene in roots) and *jar1-1* (insensitive to jasmonic acid and methyl-jasmonic acid) were unable to undergo ISR in the presence of *P. fluorescens* CHA0 and the oomycete *Peronospora parasitica* (Iavicoli et al. 2003). On the other hand, ethylene insensitive *eir1* plants showed induction of ISR against leaf pathogens when WCS417r was infiltrated in leaves but not when it was applied on roots (Knoester et al. 1999), demonstrating that ISR signal transduction from root to leaves in *A. thaliana* required an ethylene-dependent signalling pathway.

Transcriptome analyses have been crucial to identify genes implicated in ISR signalling in order to build a model that explains how MAMPs perception in roots can trigger a defence process in leaves (Verhagen et al. 2004). Genes with altered expression were predicted to be JA and/or ET responsive genes. Among the genes responding to strain WCS417r, the Myb family transcription factor MYB72 plays a critical role in ISR onset (Segarra et al. 2009, Van der Ent et al. 2008). *A. thaliana* carrying the recessive form of *MYB72* is defective in the activation of ISR (Verhagen et al. 2004) though is not impaired in SAR or resistance induced by methyl JA or the ethylene precursor 1-aminocyclopropane-1-carboxylate (ACC). *A. thaliana* transgenic plants that constitutively expressed *MYB72* could not initiate ISR without exposition to PGPR. MYB72 cannot launch ISR alone and there must be other elements participating. Yeast two-hybrid experiments showed an interaction between MYB72 and ethylene-insensitive like 3 protein (EIL3), in the regulation of the ethylene-dependent signalling pathway (Van der Ent et al. 2008). As ACC can initiate ISR, these results indicate that MYB72 acts in a synergistic way to the ethylene route preparing the plant for the JA/ET pathways.

WCS417r ISR-primed *A. thaliana* expression pattern also showed an upregulation of 442 putative methyl jasmonate (MeJA)-responsive genes. Most of these genes possess a G-box-like motif (5′-CACATG-3′), which is recognized by the transcription factor MYC2 (Dombrecht et al. 2007). Mutants in the *JASMONATE-INSENSITIVE1/MYC2* gene are blocked in their ability to express WCS417r-dependent ISR, indicating that the MYC2 transcription factor is an essential regulator of ISR induced by WCS417r (Anderson et al. 2004). Therefore, ISR induced by WCS417r leads to a systemic activation of the expression of the *MYC2* gene. In the absence of JA-inducing pathogens, MYC2 is repressed by JAZ proteins (Chini et al. 2007). When ISR-induced plants face a JA-inducing pathogen, JAZ proteins are ubiquitinated and degraded by the proteasome. Plants with unblocked MYC2 proteins trigger defence responses faster than non-induced plants (Pozo et al. 2008).

ET and JA routes do not work independently as they converge in the activation of *NPR1* gene expression. NPR1 is also an activator of SAR, so it is proposed that depending on the plant stimulation, either by a PGPR or a pathogen, ISR or SAR would be triggered (Pieterse et al. 1998). Enhanced disease susceptibility *A. thaliana* mutants revealed a regulation network reliant on NPR1, switching between ISR or SAR depending on the type of biocontrol and infection (Ton et al. 2002a). NPR1 activity is connected to gene regulation in the cell nucleus during SAR (Dong 2004), while NPR1 is linked to a cytosolic function through ISR (Leon-Reyes et al. 2009). This modulation relies on ethylene levels that favour JA response against SA. Hence, SAR will be effective against pathogens resisted through SA-dependent basal defences, while ISR will be the main defence response against pathogens resisted through the JA/ET pathway (Ton et al. 2002b).

As a result of these signalling cascades plants are primed. Induced plants acquire an enhanced defensive capacity that does not require immediate *de novo* defences, but allow a faster and stronger expression of basal defence responses upon pathogen attack at a lower cost than a constitutive activation. So, because of priming, activation of defence genes similar to those enhanced in SAR can be found (Ward et al. 1991). *P. fluorescens* strains PF1 and FP7 inhibit mycelial growth of sheath blight fungus *Rhizoctonia solani* in rice by activating and maintaining in time rice chitinase and peroxidase activities (Nandakumar et al. 2001). Plant peroxidase activity leads to a burst of reactive oxygen species (ROS) considered as a significant marker for ISR activation by PGPR (van Loon et al. 2008). ROS also promote the polymerization of phenolic compounds that together with callose deposition strengthens the plant cell wall limiting pathogen entrance (De Vleesschauwer et al. 2008, van Loon et al. 2008, Verhagen et al. 2004). In addition to cell wall modifications, certain strains can stimulate the defence enzymes phenylalanine ammonia-lyase (PAL), peroxidase (PO) and polyphenol oxidase (PPO) involved in the synthesis of diverse secondary metabolic products such as phytoalexins toxic against fungal spores and hyphae (Chen et al. 2000, van Loon et al. 1998). An example of control through this route is the stimulation of defence responses in gravepine against *Botrytis cinerea* by *P. fluorescens* WCS417r (Verhagen et al. 2010).

Fluorescent pseudomonads constitute a reservoir of strains with a high diversity (Silby et al. 2011) and different colonization capabilities. Besides general studies on the ISR process, current analyses are focused on specific strains isolated from affected soils (Nandakumar et al. 2001) and on the selection of several strains instead of a single one to prepare inoculants that improve ISR response (Saravanakumar et al. 2009).

Concluding Remarks

Plant protection from pathogens by using biocontrol agents is a real necessity. Phytochemicals are hazardous chemicals for humans and the environment, and there is a growing trend in limiting or even banishing their use. Therefore reliable and safe methods are required in order to maintain agronomic production. Fluorescent pseudomonads are one of the best studied groups for this purpose, and are therefore excellent candidates for the formulation of biocontrol agents and products. Although a few biocontrol products based in pseudomonads such as AtEze, Bio-save, BlightBan, Frostban and Spot-Less are already in the USA market, more research is needed in order to know the mechanisms of biocontrol and to improve the formulation and stability of the products. The research presented here shows that no single process is responsible for biocontrol ability of fluorescent pseudomonads and that synergistic effects on the production of antagonistic metabolites and enzymes, rhizosphere colonization and enhancing plant defences are at the base of biocontrol.

Acknowledgments

We thank Dr. Francisca Fernandez-Piñas for reviewing the manuscript and providing helpful comments.

References

Aarons, S., A. Abbas, C. Adams, A. Fenton and F. O'Gara. 2000. A regulatory RNA (PrrB RNA) modulates expression of secondary metabolite genes in *Pseudomonas fluorescens* F113. J. Bacteriol. 182: 3913–3919.

Andersen, J.B., B. Koch, T.H. Nielsen, D. Sorensen, M. Hansen, O. Nybroe, C. Christophersen, J. Sorensen, S. Molin and M. Givskov. 2003. Surface motility in *Pseudomonas* sp. DSS73 is required for efficient biological containment of the root-pathogenic microfungi *Rhizoctonia solani* and *Pythium ultimum*. Microbiology. 149: 37–46.

Anderson, J.P., E. Badruzsaufari, P.M. Schenk, J.M. Manners, O.J. Desmond, C. Ehlert, D.J. Maclean, P.R. Ebert and K. Kazan. 2004. Antagonistic interaction between abscisic acid and jasmonate-ethylene signaling pathways modulates defense gene expression and disease resistance in *Arabidopsis*. The Plant Cell. 16: 3460–3479.

Anjaiah, V., N. Koedam, B. Nowak-Thompson, J.E. Loper, M. Hofte, J.T. Tambong and P. Cornelis. 1998. Involvement of phenazines and anthranilate in the antagonism of *Pseudomonas aeruginosa* PNA1 and Tn5 derivatives toward *Fusarium* spp. and *Pythium* spp. Mol. Plant. Microbe Interact. 11: 847–854.

Bakker, P., D.C.M. Glandorf, M. Viebahn, T.W.M. Ouwens, E. Smit, P. Leeflang, K. Wernars, L.S. Thomashow, J.E. Thomas-Oates and L.C. van Loon. 2002. Effects of *Pseudomonas putida* modified to produce phenazine-1-carboxylic acid and 2, 4-diacetylphloroglucinol on the microflora of field grown wheat. Anton. Leeuw. Int. J.G. 81: 617–624.

Bangera, M.G. and L.S. Thomashow. 1999. Identification and characterization of a gene cluster for synthesis of the polyketide antibiotic 2,4-diacetylphloroglucinol from *Pseudomonas fluorescens* Q2–87. J. Bacteriol. 181: 3155–3163.

Barahona, E., A. Navazo, F. Yousef-Coronado, D. Aguirre de Carcer, F. Martinez-Granero, M. Espinosa-Urgel, M. Martin and R. Rivilla. 2010. Efficient rhizosphere colonization by *Pseudomonas fluorescens* F113 mutants unable to form biofilms on abiotic surfaces. Environ. Microbiol. 12: 3185–3195.

Barahona, E., A. Navazo, F. Martinez-Granero, T. Zea-Bonilla, R.M. Perez-Jimenez, M. Martin and R. Rivilla. 2011. A *Pseudomonas fluorescens* F113 mutant with enhanced competitive colonization ability shows improved biocontrol activity against fungal root pathogens. Appl. Environ. Microbiol. 77: 5412–5419.

Bender, C., V. Rangaswamy and J. Loper. 1999. Polyketide Production by Plant-Associated Pseudomonads. Annu. Rev. Phytopathol. 37: 175–196.

Bolwerk, A., A.L. Lagopodi, A.H.M. Wijfjes, G.E.M. Lamers, T.F.C. Chin-A-Woeng, B.J.J. Lugtenberg and G.V. Bloemberg. 2003. Interactions in the tomato rhizosphere of two *Pseudomonas* biocontrol strains with the phytopathogenic fungus *Fusarium oxysporum* f. sp. *radicis-lycopersici*. Mol. Plant. Microbe Interact. 16: 983–993.

Boutrot, F., C. Segonzac, K.N. Chang, H. Qiao, J.R. Ecker, C. Zipfel and J.P. Rathjen. 2010. Direct transcriptional control of the *Arabidopsis* immune receptor FLS2 by the ethylene-dependent transcription factors EIN3 and EIL1. Proc. Natl. Acad. Sci. U.S.A. 107: 14502–14507.

Bull, C.T., D.M. Weller and L.S. Thomashow. 1991. Relationship between root colonization and suppression of *Gaeumannomyces graminis* var. tritici by *Pseudomonas fluorescens* strain 2–79. Phytopathology. 81: 954–959.

Capdevila, S., F.M. Martínez-Granero, M. Sánchez-Contreras, R. Rivilla and M. Martín. 2004. Analysis of *Pseudomonas fluorescens* F113 genes implicated in flagellar filament synthesis and their role in competitive root colonization. Microbiology. 150: 3889–3897.

Castric, P.A. 1981. The metabolism of hydrogen cyanide by bacteria. *In:* B. Vennesland, E.E. Conn, C.J. Knowles, J. Westley and F. Wissing [eds.]. Cyanide in biology. Academic Press, London. pp. 233–261.

Comolli, J.C. and T.J. Donohue. 2002. *Pseudomonas aeruginosa* RoxR, a response regulator related to *Rhodobacter sphaeroides* PrrA, activates expression of the cyanide-insensitive terminal oxidase. Mol. Microbiol. 45: 755–768.

Couillerot, O., C. Prigent-Combaret, J. Caballero-Mellado and Y. Moënne-Loccoz. 2009. Pseudomonas fluorescens and closely-related fluorescent pseudomonads as biocontrol agents of soil-borne phytopathogens. Lett. Appl. Microbiol. 48: 505–512.

Cuppels, D.A., C.R. Howell, R.D. Stipanovic, A. Stoessl and J.B. Stothers. 1986. Biosynthesis of pyoluteorin—a mixed polyketide-tricarboxilic acid cycle origin demonstrated by [1,2-C-13(2)]acetate incorporation. Z. Naturforsch. C. 41: 532–536.

Cuppels, D.A., R.D. Stipanovic, A. Stoessl and J.B. Stothers. 1987. The constitution and properties of a pyochelin zinc complex. Can. J. Chem. 65: 2126–2130.

Chancey, S.T., D.W. Wood and L.S. Pierson. 1999. Two-component transcriptional regulation of N-acyl-homoserine lactone production in *Pseudomonas aureofaciens*. Appl. Environ. Microbiol. 65: 2294–2299.

Chen, C., R.R. Bélanger, N. Benhamou and T.C. Paulitz. 2000. Defense enzymes induced in cucumber roots by treatment with plant growth-promoting rhizobacteria (PGPR) and *Pythium aphanidermatum*. Physiol. Mol. Plant Pathol. 56: 13–23.

Chin-A-Woeng, T.F.C., G.V. Bloemberg, A.J. van der Bij, K.M.G.F. van der Drift, J. Schripsema, B. Kroon, R.J. Scheffer, C. Keel, P.A.H.M. Bakker, H.V. Tichy, F.J. de Bruijn, J.E. Thomas-Oates and B.J.J. Lugtenberg. 1998. Biocontrol by phenazine-1-carboxamide-producing *Pseudomonas chlororaphis* PCL1391 of tomato root rot caused by *Fusarium oxysporum* f. sp. *radicis-lycopersici*. Mol. Plant. Microbe Interact. 11: 1069–1077.

Chin-A-Woeng, T.F.C., G.V. Bloemberg, I.H.M. Mulders, L.C. Dekkers and B.J.J. Lugtenberg. 2000. Root colonization by phenazine-1-carboxamide-producing bacterium *Pseudomonas chlororaphis* PCL1391 is essential for biocontrol of tomato foot and root rot. Mol. Plant. Microbe Interact. 13: 1340–1345.

Chin-A-Woeng, T.F.C., D. van den Broek, G. de Voer, K. van der Drift, S. Tuinman, J.E. Thomas-Oates, B.J.J. Lugtenberg and G.V. Bloemberg. 2001. Phenazine-1-carboxamide production in the biocontrol strain *Pseudomonas chlororaphis* PCL1391 is regulated by multiple factors secreted into the growth medium. Mol. Plant. Microbe Interact. 14: 969–979.

Chin-A-Woeng, T.F.C., G.V. Bloemberg and B.J.J. Lugtenberg. 2003. Phenazines and their role in biocontrol by *Pseudomonas* bacteria. New Phytol. 157: 503–523.

Chini, A., S. Fonseca, G. Fernandez, B. Adie, J.M. Chico, O. Lorenzo, G. Garcia-Casado, I. Lopez-Vidriero, F.M. Lozano, M.R. Ponce, J.L. Micol and R. Solano. 2007. The JAZ family of repressors is the missing link in jasmonate signalling. Nature. 448: 666–671.

de Bruijn, I., M.J. de Kock, P. de Waard, T.A. van Beek and J.M. Raaijmakers. 2008. Massetolide A biosynthesis in *Pseudomonas fluorescens*. J. Bacteriol. 190: 2777–2789.

de Souza, J.T., C. Arnould, C. Deulvot, P. Lemanceau, V. Gianinazzi-Pearson and J.M. Raaijmakers. 2003. Effect of 2,4-diacetylphloroglucinol on pythium: cellular responses and variation in sensitivity among propagules and species. Phytopathology. 93: 966–975.

De Vleesschauwer, D., M. Djavaheri, P.A. Bakker and M. Hofte. 2008. *Pseudomonas fluorescens* WCS374r-induced systemic resistance in rice against *Magnaporthe oryzae* is based on pseudobactin-mediated priming for a salicylic acid-repressible multifaceted defense response. Plant Physiol. 148: 1996–2012.

de Weert, S., I. Kuiper, E.L. Lagendijk, G.E. Lamers and B.J. Lugtenberg. 2004. Role of chemotaxis toward fusaric acid in colonization of hyphae of *Fusarium oxysporum* f. sp. *radicis-lycopersici* by *Pseudomonas fluorescens* WCS365. Mol. Plant. Microbe Interact. 17: 1185–1191.

de Weger, L.A., C.I.M. van der vlugt, A.H.M. Wijfjes, P.A.H.M. Bakker, B. Schippers and B. Lugtenberg. 1987. Flagella of a plant-growth-stimulating *Pseudomonas fluorescens* strain are required for colonization of potato roots. J. Bacteriol. 169: 2769–2773.

Dekkers, L.C., C.C. Phoelich, L. van der Fits and B.J.J. Lugtenberg. 1998. A site-specific recombinase is required for competitive root colonization by *Pseudomonas fluorescens* WCS365. Proc. Natl. Acad. Sci. U.S.A. 95: 7051–7056.

Dekkers, L.C., I.H.M. Mulders, C.C. Phoelich, T.F.C. Chin-A-Woeng, A.H.M. Wijfjes and B.J.J. Lugtenberg. 2000. The *sss* colonization gene of the tomato-*Fusarium oxysporum F.* sp. *radicis-lycopersici* biocontrol strain *Pseudomonas fluorescens* WCS365 can improve root colonization of other wild-type *Pseudomonas* spp. bacteria. Mol. Plant. Microbe Interact. 13: 1177–1183.

Delany, I.R., U.F. Walsh, I. Ross, A.M. Fenton, D.M. Corkery and F. O'Gara . 2001. Enhancing the biocontrol efficacy of *Pseudomonas fluorescens* F113 by altering the regulation and production of 2,4-diacetylphloroglucinol—Improved *Pseudomonas* biocontrol inoculants. Plant Soil. 232: 195–205.

Dombrecht, B., G.P. Xue, S.J. Sprague, J.A. Kirkegaard, J.J. Ross, J.B. Reid, G.P. Fitt, N. Sewelam, P.M. Schenk, J.M. Manners and K. Kazan. 2007. *MYC2* differentially modulates diverse jasmonate-dependent functions in *Arabidopsis*. The Plant Cell. 19: 2225–2245.

Dong, X. 2004. NPR1, all things considered. Curr. Opin. Plant Biol. 7: 547–552.

Duffy, B.K. and G. Defago. 2000. Controlling instability in *gacS-gacA* regulatory genes during inoculant production of *Pseudomonas fluorescens* biocontrol strains. Appl. Environ. Microbiol. 66: 3142–3150.

Elad, Y. and R. Baker. 1985. Influence of trace amounts of cations and siderophore-producing pseudomonads on chlamydospore germination of *Fusarium oxysporum*. Phytopathology. 75: 1047–1052.

Ellis, R.J., T.M. Timms-Wilson and M.J. Bailey. 2000. Identification of conserved traits in fluorescent pseudomonads with antifungal activity. Environ. Microbiol. 2: 274–284.

Fenton, A.M., P.M. Stephens, J. Crowley, M. O'Callaghan and F. O'Gara. 1992. Exploitation of gene(s) involved in 2,4-diacetylphloroglucinol biosynthesis to confer a new biocontrol capability to a *Pseudomonas* strain. Appl. Environ. Microbiol. 58: 3873–3878.

Flaishman, M.A., Z. Eyal, A. Zilberstein, C. Voisard and D. Haas. 1996. Suppression of *Septoria tritici* blotch and leaf rust of wheat by recombinant cyanide-producing strains of *Pseudomonas putida*. Mol. Plant. Microbe Interact. 9: 642–645.

Frapolli, M., J.F. Pothier, G. Defago and Y. Moenne-Loccoz. 2012. Evolutionary history of synthesis pathway genes for phloroglucinol and cyanide antimicrobials in plant-associated fluorescent pseudomonads. Mol. Phylogenet. Evol. 63: 877–890.

Georgakopoulos, D.G., M. Hendson, N.J. Panopoulos and M.N. Schroth. 1994. Cloning of a phenazine biosynthetic locus of *Pseudomonas aureofaciens* PGS12 and analysis of its expression *in vitro* with the ice nucleation reporter gene. Appl. Environ. Microbiol. 60: 2931–2938.

Glandorf, D.C., P. Verheggen, T. Jansen, J.W. Jorritsma, E. Smit, P. Leeflang, K. Wernars, L.S. Thomashow, E. Laureijs, J.E. Thomas-Oates, P.A. Bakker and L.C. van Loon. 2001. Effect of genetically modified *Pseudomonas putida* WCS358r on the fungal rhizosphere microflora of field-grown wheat. Appl. Environ. Microbiol. 67: 3371–3378.

Gurusiddaiah, S., D.M. Weller, A. Sarkar and R.J. Cook. 1986. Characterization of an antibiotic produced by a strain of *Pseudomonas fluorescens* inhibitory to *Gaeumannomyces graminis* var. *tritici* and *Pythium* spp. Antimicrob. Agents Chemother. 29: 488–495.

Haas, D. and G. Defago. 2005. Biological control of soil-borne pathogens by fluorescent pseudomonads. Nat. Rev. Microbiol. 3: 307–319.

Hammer, P.E., D.S. Hill, S.T. Lam, K.H. VanPee and J.M. Ligon. 1997. Four genes from *Pseudomonas fluorescens* that encode the biosynthesis of pyrrolnitrin. Appl. Environ. Microbiol. 63: 2147–2154.

Hammer, P.E., W. Burd, D.S. Hill, J.M. Ligon and K.H. van Pee. 1999. Conservation of the pyrrolnitrin biosynthetic gene cluster among six pyrrolnitrin-producing strains. FEMS Microbiol. Lett. 180: 39–44.

Hassan, H.M. and I. Fridovich. 1980. Mechanism of the antibiotic action pyocyanine. J. Bacteriol. 141: 156–163.

Hassett, D.J., L. Charniga, K. Bean, D.E. Ohman and M.S. Cohen. 1992. Response of *Pseudomonas aeruginosa* to pyocyanin: Mechanisms of resistance, antioxidant defenses, and demonstration of a manganese-cofactored superoxide-dismutase. Infect. Immun. 60: 328–336.

Heeb, S. and D. Haas. 2001. Regulatory roles of the GacS/GacA two-component system in plant-associated and other Gram-negative bacteria. Mol. Plant. Microbe Interact. 14: 1351–1363.

Henkes, G.J., A. Jousset, M. Bonkowski, M.R. Thorpe, S. Scheu, A. Lanoue, U. Schurr and U.S. Roese. 2011. *Pseudomonas fluorescens* CHA0 maintains carbon delivery to *Fusarium graminearum*-infected roots and prevents reduction in biomass of barley shoots through systemic interactions. J. Exp. Bot. 62: 4337–4344.

Hothersall, J. and C.M. Thomas. 2004. Polyketide antibiotics of *Pseudomonas. In:* J.L. Ramos [ed.]. Pseudomonas, vol. 3. Kluwer Academic/Plenum Publishers, New York. pp. 689–722.

Howell, C.R. and R.D. Stipanovic. 1979. Control of *Rhizoctonia solani* on cotton seedlings with *Pseudomonas fluorescens* and with an antibiotic produced by the bacterium. Phytopathology. 69: 480–482.

Howell, C.R. and R.D. Stipanovic. 1980. Suppression of *Pythium ultimum*-induced damping-off of cotton seedlings by *Pseudomonas fluorescens* and its antibiotic pyoluteorin. Phytopathology. 70: 712–715.

Iavicoli, A., E. Boutet, A. Buchala and J.P. Metraux. 2003. Induced systemic resistance in *Arabidopsis thaliana* in response to root inoculation with *Pseudomonas fluorescens* CHA0. Mol. Plant. Microbe Interact. 16: 851–858.

Jousset, A., L. Rochat, A. Lanoue, M. Bonkowski, C. Keel and S. Scheu. 2011. Plants respond to pathogen infection by enhancing the antifungal gene expression of root-associated bacteria. Mol Plant-Microbe Interact. 24: 352–358.

Kamilova, F., S. Validov, T. Azarova, I. Mulders and B. Lugtenberg. 2005. Enrichment for enhanced competitive plant root tip colonizers selects for a new class of biocontrol bacteria. Environ. Microbiol. 7: 1809–1817.

Kang, B.R., K.Y. Yang, B.H. Cho, T.H. Han, I.S. Kim, M.C. Lee, A.J. Anderson and Y.C. Kim. 2006. Production of indole-3-acetic acid in the plant-beneficial strain *Pseudomonas chlororaphis* O6 is negatively regulated by the global sensor kinase GacS. Curr. Microbiol. 52: 473–476.

Keel, C., U. Schnider, M. Maurhofer, C. Voisard, J. Laville, U. Burger, P. Wirthner, D. Haas and G. Defago. 1992. Suppression of root diseases by *Pseudomonas fluorescens* CHA0: importance of the bacterial secondary metabolite 2,4-diacetylphloroglucinol. Mol. Plant. Microbe Interact. 5: 4–13.

Keel, C., D.M. Weller, A. Natsch, G. Defago, R.J. Cook and L.S. Thomashow. 1996. Conservation of the 2,4-diacetylphloroglucinol biosynthesis locus among fluorescent Pseudomonas strains from diverse geographic locations. Appl. Environ. Microbiol. 62: 552–563.

Kloepper, J.W., J. Leong, M. Teintze and M.N. Schroth. 1980a. Enhanced plant-growth by siderophores produced by plant growth-promoting rhizobacteria. Nature. 286: 885–886.

Kloepper, J.W., J. Leong, M. Teintzeand M.N. Schroth. 1980b. *Pseudomonas* siderophores: a mechanism explaining disease-suppressive soils. Curr. Microbiol. 4: 317–320.

Knoester, M., C.M. Pieterse, J.F. Bol and L.C. Van Loon. 1999. Systemic resistance in Arabidopsis induced by rhizobacteria requires ethylene-dependent signaling at the site of application. Mol. Plant. Microbe Interact. 12: 720–727.

Knowles, C.J. and A.W. Bunch. 1986. Microbial cyanide metabolism. Adv. Microb. Physiol. 27: 73–111.

Koch, B., T.H. Nielsen, D. Sorensen, J.B. Andersen, C. Christophersen, S. Molin, M. Givskov, J. Sorensen and O. Nybroe. 2002. Lipopeptide production in *Pseudomonas* sp. strain DSS73 is regulated by components of sugar beet seed exudate via the Gac two-component regulatory system. Appl. Environ. Microbiol. 68: 4509–4516.

Kraus, J. and J.E. Loper. 1992. Lack of evidence for a role of antifungal metabolite production by *Pseudomonas fluorescens* Pf-5 in biological-control of *Pythium* damping-off of cucumber. Phytopathology. 82: 264–271.

Kraus, J. and J.E. Loper. 1995. Characterization of a genomic region required for production of the antibiotic pyoluteorin by the biological-control agent *Pseudomonas fluorescens* Pf-5. Appl. Environ. Microbiol. 61: 849–854.

Kuiper, I., G.V. Bloemberg and B.J.J. Lugtenberg. 2001. Selection of a plant-bacterium pair as a novel tool for rhizostimulation of polycyclic aromatic hydrocarbon-degrading bacteria. Mol. Plant. Microbe Interact. 14: 1197–1205.

Laville, J., C. Voisard, C. Keel, M. Maurhofer, G. Defago and D. Haas. 1992. Global control in *Pseudomonas fluorescens* mediating antibiotic synthesis and suppression of black root rot of tobacco. Proc. Natl. Acad. Sci. U.S.A. 89: 1562–1566.

Laville, J., C. Blumer, C. Von Schroetter, V. Gaia, G. Defago, C. Keel and D. Haas. 1998. Characterization of the *hcnABC* gene cluster encoding hydrogen cyanide synthase and anaerobic regulation by ANR in the strictly aerobic biocontrol agent *Pseudomonas fluorescens* CHA0. J. Bacteriol. 180: 3187–3196.

Leeman, M., E.M. denOuden, J.A. vanPelt, F.P.M. Dirkx, H. Steijl, P. Bakker and B. Schippers. 1996. Iron availability affects induction of systemic resistance to Fusarium wilt of radish by *Pseudomonas fluorescens*. Phytopathology. 86: 149–155.

Leon-Reyes, A., S.H. Spoel, E.S. De Lange, H. Abe, M. Kobayashi, S. Tsuda, F.F. Millenaar, R.A.M. Welschen, T. Ritsema and C.M.J. Pieterse. 2009. Ethylene Modulates the Role of NONEXPRESSOR OF PATHOGENESIS-RELATED GENES1 in Cross Talk between Salicylate and Jasmonate Signaling. Plant Physiol. 149: 1797–1809.

Levy, E., F.J. Gough, K.D. Berlin, P.W. Guiana and J.T. Smith. 1992. Inhibition of *Septoria tritici* and other phytopathogenic fungi and bacteria by *Pseudomonas fluorescens* and its antibiotics. Plant Pathol. 41: 335–341.

Liao, C.H., D.E. McCallus, J.M. Wells, S.S. Tzean and G.Y. Kang. 1996. The *repD* gene required for production of extracellular enzymes and fluorescent siderophores in *Pseudomonas viridiflava* is an analog of the *gacA* gene of *Pseudomonas syringae*. Can. J. Microbiol. 42: 177–182.

Ligon, J.M., D.S. Hill, P.E. Hammer, N.R. Torkewitz, D. Hofmann, H.J. Kempf and K.H. van Pee. 2000. Natural products with antifungal activity from *Pseudomonas* biocontrol bacteria. Pest Manag. Sci. 56: 688–695.

Lim, H.S. and S.D. Kim. 1995. The role and characterization of beta-1,3-glucanase in biocontrol of *Fusarium solani* by *Pseudomonas stutzeri* YPL-1. J. Microbiol. 33: 295–301.

Lim, H.S., Y.S. Kim and S.D. Kim. 1991. *Pseudomonas-stutzeri* YPL-1 genetic-transformation and antifungal mechanism against *Fusarium solani*, an agent of plant-root rot. Appl. Environ. Microbiol. 57: 510–516.

Loh, J., E.A. Pierson, L.S. Pierson, G. Stacey and A. Chatterjee. 2002. Quorum sensing in plant-associated bacteria. Curr. Opin. Plant Biol. 5: 285–290.

Loper, J.E. and M.D. Henkels. 1997. Availability of iron to *Pseudomonas fluorescens* in rhizosphere and bulk soil evaluated with an ice nucleation reporter gene. Appl. Environ. Microbiol. 63: 99–105.

Loper, J.E. and M.D. Henkels. 1999. Utilization of heterologous siderophores enhances levels of iron available to *Pseudomonas putida* in the rhizosphere. Appl. Environ. Microbiol. 65: 5357–5363.

Lugtenberg, B.J.J., L.A. Deweger and J.W. Bennett. 1991. Microbial stimulation of plant-growth and protection from disease. Curr. Opin. Biotechnol. 2: 457–464.

Lugtenberg, B.J.J., L.V. Kravchenko and M. Simons. 1999. Tomato seed and root exudate sugars: composition, utilization by *Pseudomonas* biocontrol strains and role in rhizosphere colonization. Environ. Microbiol. 1: 439–446.

Mantell, S.H. and B.E.J. Wheeler. 1973. *Rosellinia* and white root rot of *Narcissus* in the Scilly Isles. T. Brit. Mycol. Soc. 60: 23–25.

Markovich, N.A. and G.L. Kononova. 2003. Lytic enzymes of *Trichoderma* and their role in plant defense from fungal diseases: A review. Appl. Biochem. Micro. 39: 341–351.

Martínez-Granero, F., S. Capdevila, M. Sánchez-Contreras, M. Martín and R. Rivilla. 2005. Two site-specific recombinases are implicated in phenotypic variation and competitive rhizosphere colonization in *Pseudomonas fluorescens*. Microbiology. 151: 975–983.

Martínez-Granero, F., R. Rivilla and M. Martin. 2006. Rhizosphere selection of highly motile phenotypic variants of *Pseudomonas fluorescens* with enhanced competitive colonization ability. Appl. Environ. Microbiol. 72: 3429–3434.

Mascher, F., Y. Moenne-Loccoz, U. Schnider-Keel, C. Keel, D. Haas and G. Defago. 2002. Inactivation of the regulatory gene *algU* or *gacA* can affect the ability of biocontrol *Pseudomonas fluorescens* CHA0 to persist as culturable cells in nonsterile soil. Appl. Environ. Microbiol. 68: 2085–2088.

Matsushita, K., M. Yamada, E. Shinagawa, O. Adachi and M. Ameyama. 1983. Membrane-bound respiratory chain of *Pseudomonas aeruginosa* grown aerobically. A KCN-insensitive alternate oxidase chain and its energetics. J. Biochem. (Tokyo). 93: 1137–1144.

Matthijs, S., K.A. Tehrani, G. Laus, R.W. Jackson, R.M. Cooper and P. Cornelis. 2007. Thioquinolobactin, a *Pseudomonas* siderophore with antifungal and anti-*Pythium* activity. Environ. Microbiol. 9: 425–434.

Maurhofer, M., C. Keel, D. Haas and G. Defago. 1994. Pyoluteorin production by *Pseudomonas fluorescens* strain CHA0 is involved in the suppression of *Pythium* damping-off of cress but not of cucumber. Eur. J. Plant Pathol. 100: 221–232.

Maurhofer, M., C. Keel, D. Haas and G. Defago. 1995. Influence of plant species on disease suppression by *Pseudomonas fluorescens* strain CHA0 with enhanced antibiotic production. Plant Pathol. 44: 40–50.

Mavrodi, D.V., V.N. Ksenzenko, R.F. Bonsall, R.J. Cook, A.M. Boronin and L.S. Thomashow. 1998. A seven-gene locus for synthesis is of phenazine-1-carboxylic acid by *Pseudomonas fluorescens* 2–79. J. Bacteriol. 180: 2541–2548.

Mavrodi, O.V., D.V. Mavrodi, D.M. Weller and L.S. Thomashow. 2006. The role of *ptsP*, *orfT*, and *sss* recombinase in root colonization by *Pseudomonas fluorescens* Q8r1-96. Appl. Environ. Microbiol. 7111–7122.

Mazzola, M., R.J. Cook, L.S. Thomashow, D.M. Weller and L.S. Pierson. 1992. Contribution of phenazine antibiotic biosynthesis to the ecological competence of fluorescent pseudomonads in soil habitats. Appl. Environ. Microbiol. 58: 2616–2624.

Michelsen, C.F. and P. Stougaard. 2012. Hydrogen cyanide synthesis and antifungal activity of the biocontrol strain *Pseudomonas fluorescens* In5 from Greenland is highly dependent on growth medium. Can. J. Microbiol. 58: 381–390.

Misaghi, I.J., M.W. Olsen, P.J. Cotty and C.R. Donndelinger. 1988. Fluorescent siderophore-mediated iron deprivation: a contingent biological control mechanism. Soil Biol Biochem. 20: 573–574.

Nandakumar, R., S. Babu, R. Viswanathan, T. Raguchander and R. Samiyappan. 2001. Induction of systemic resistance in rice against sheath blight disease by *Pseudomonas fluorescens*. Soil Biol. Biochem. 33: 603–612.

Naseby, D.C., J.A. Way, N.J. Bainton and J.M. Lynch. 2001. Biocontrol of *Pythium* in the pea rhizosphere by antifungal metabolite producing and non-producing *Pseudomonas* strains. J. Appl. Microbiol. 90: 421–429.

Navazo, A., E. Barahona, M. Redondo-Nieto, F. Martinez-Granero, R. Rivilla and M. Martin. 2009. Three independent signalling pathways repress motility in *Pseudomonas fluorescens* F113. Microb. Biotechnol. 2: 489–498.

Nielsen, M.N., J. Sorensen, J. Fels and H.C. Pedersen. 1998. Secondary metabolite- and endochitinase-dependent antagonism toward plant-pathogenic microfungi of *Pseudomonas fluorescens* isolates from sugar beet rhizosphere. Appl. Environ. Microbiol. 64: 3563–3569.

Nielsen, T.H. and J. Sorensen. 2003. Production of cyclic lipopeptides by *Pseudomonas fluorescens* strains in bulk soil and in the sugar beet rhizosphere. Appl. Environ. Microbiol. 69: 861–868.

Nielsen, T.H., C. Christophersen, U. Anthoni and J. Sorensen. 1999. Viscosinamide, a new cyclic depsipeptide with surfactant and antifungal properties produced by *Pseudomonas fluorescens* DR54. J. Appl. Microbiol. 87: 80–90.

Nielsen, T.H., D. Sorensen, C. Tobiasen, J.B. Andersen, C. Christophersen, M. Givskov and J. Sorensen. 2002. Antibiotic and biosurfactant properties of cyclic lipopeptides produced by fluorescent *Pseudomonas* spp. from the sugar beet rhizosphere. Appl. Environ. Microbiol. 68: 3416–3423.

Nowak-Thompson, B., S.J. Gould, J. Kraus and J.E. Loper. 1994. Production of 2,4-diacetylphloroglucinol by the biocontrol agent *Pseudomonas fluorescens* Pf-5. Can. J. Microbiol. 40: 1064–1066.

Nowak-Thompson, B., S.J. Gould and J.E. Loper. 1997. Identification and sequence analysis of the genes encoding a polyketide synthase required for pyoluteorin biosynthesis in *Pseudomonas fluorescens* Pf-5. Gene. 204: 17–24.

Nybroe, O. and J. Sorensen. 2004. Production of cyclic lipopeptides by fluorescent pseudomonads. *In:* J.L. Ramos [ed.]. Pseudomonas, vol. 3. Kluwer Academic/Plenum Publishers, New York. pp. 147–172.

Pessi, G. and D. Haas. 2000. Transcriptional control of the hydrogen cyanide biosynthetic genes *hcnABC* by the anaerobic regulator ANR and the quorum-sensing regulators LasR and RhlR in *Pseudomonas aeruginosa*. J. Bacteriol. 182: 6940–6949.

Phoebe, C.H., J. Combie, F.G. Albert, K. Van Tran, J. Cabrera, H.J. Correira, Y.H. Guo, J. Lindermuth, N. Rauert, W. Galbraith and C.P. Selitrennikoff. 2001. Extremophilic organisms as an unexplored source of antifungal compounds. J. Antibiot. (Tokyo). 54: 56–65.

Picard, C. and M. Bosco. 2008. Genotypic and phenotypic diversity in populations of plant-probiotic *Pseudomonas* spp. colonizing roots. Naturwissenschaften. 95: 1–16.
Pierson, E.A., D.W. Wood, J.A. Cannon, F.M. Blachere and L.S. Pierson. 1998. Interpopulation signaling via *N*-acyl-homoserine lactones among bacteria in the wheat rhizosphere. Mol. Plant. Microbe Interact. 11: 1078–1084.
Pierson, L.S. and L.S. Thomashow. 1992. Cloning and heterologous expression of the phenazine biosynthetic locus from *Pseudomonas aureofaciens* 30–84. Mol. Plant. Microbe Interact. 5: 330–339.
Pieterse, C.M., S.C. van Wees, J.A. van Pelt, M. Knoester, R. Laan, H. Gerrits, P.J. Weisbeek and L.C. van Loon. 1998. A novel signaling pathway controlling induced systemic resistance in Arabidopsis. Plant Cell. 10: 1571–1580.
Pieterse, C.M.J., S.C.M. van Wees, J. Ton, J.A. van Pelt and L.C. van Loon. 2002. Signalling in Rhizobacteria-Induced Systemic Resistance in *Arabidopsis thaliana*. Plant Biology. 4: 535–544.
Pliego, C., F.M. Cazorla, M.A. Gonzalez-Sanchez, R.M. Perez-Jimenez, A. de Vicente and C. Ramos. 2007. Selection for biocontrol bacteria antagonistic toward *Rosellinia necatrix* by enrichment of competitive avocado root tip colonizers. Res. Microbiol. 158: 463–470.
Pliego, C., S. de Weert, G. Lamers, A. de Vicente, G. Bloemberg, F.M. Cazorla and C. Ramos. 2008. Two similar enhanced root-colonizing *Pseudomonas* strains differ largely in their colonization strategies of avocado roots and *Rosellinia necatrix* hyphae. Environ. Microbiol. 10: 3295–3304.
Pozo, M.J., S. Van Der Ent, L.C. Van Loon and C.M. Pieterse. 2008. Transcription factor MYC2 is involved in priming for enhanced defense during rhizobacteria-induced systemic resistance in *Arabidopsis thaliana*. New Phytol. 180: 511–523.
Prieto, P., C. Navarro-Raya, A. Valverde-Corredor, S.G. Amyotte, K.F. Dobinson and J. Mercado-Blanco. 2009. Colonization process of olive tissues by *Verticillium dahliae* and its *in planta* interaction with the biocontrol root endophyte *Pseudomonas fluorescens* PICF7. Microb. Biotechnol. 2: 499–511.
Raaijmakers, J.M. and D.M. Weller. 2001. Exploiting genotypic diversity of 2,4-diacetylphloroglucinol-producing *Pseudomonas* spp.: characterization of superior root-colonizing *P. fluorescens* strain Q8r1-96. Appl. Environ. Microbiol. 67: 2545–2554.
Raaijmakers, J.M., M. Leeman, M.M.P. Vanoorschot, I. Vandersluis, B. Schippers and P. Bakker. 1995. Dose-response relationships in biological-control of *Fusarium*-wilt of radish by *Pseudomonas* spp. Phytopathology. 85: 1075–1081.
Raaijmakers, J.M., M. Vlami and J.T. de Souza. 2002. Antibiotic production by bacterial biocontrol agents. Anton. Leeuw. Int. J.G. 81: 537–547.
Raaijmakers, J.M., I. de Bruijn and M.J. de Kock. 2006. Cyclic lipopeptide production by plant-associated *Pseudomonas* spp.: diversity, activity, biosynthesis, and regulation. Mol. Plant. Microbe Interact. 19: 699–710.
Ramette, A., Y. Moenne-Loccoz and G. Defago. 2003. Prevalence of fluorescent pseudomonads producing antifungal phloroglucinols and/or hydrogen cyanide in soils naturally suppressive or conducive to tobacco black root rot. FEMS Microbiol. Ecol. 44: 35–43.
Rezzonico, F., M. Zala, C. Keel, B. Duffy, Y. Moenne-Loccoz and G. Defago. 2007. Is the ability of biocontrol fluorescent pseudomonads to produce the antifungal metabolite 2,4-diacetylphloroglucinol really synonymous with higher plant protection? New Phytol. 173: 861–872.
Rosales, A.M., L. Thomashow, R.J. Cook and T.W. Mew. 1995. Isolation and identification of antifungal metabolites produced by rice-associated antagonistic *Pseudomonas* spp. Phytopathology. 85: 1028–1032.
Ross, A.F. 1961. Systemic acquired resistance induced by localized virus infections in plants. Virology. 14: 340–358.
Sacherer, P., G. Defago and D. Haas. 1994. Extracellular protease and phospholipase-C are controlled by the global regulatory gene *gacA* in the biocontrol strain *Pseudomonas fluorescens* CHA0. FEMS Microbiol. Lett. 116: 155–160.

Sánchez-Contreras, M., M. Martín, M. Villacieros, F. O'Gara, I. Bonilla and R. Rivilla. 2002. Phenotypic selection and phase variation occur during alfalfa root colonization by *Pseudomonas fluorescens* F113. J. Bacteriol. 184: 1587–1596.

Saravanakumar, D., N. Lavanya, K. Muthumeena, T. Raguchander and R. Samiyappan. 2009. Fluorescent pseudomonad mixtures mediate disease resistance in rice plants against sheath rot (*Sarocladium oryzae*) disease. Biocontrol. 54: 273–286.

Schnider, U., C. Keel, C. Blumer, J. Troxler, G. Defago and D. Haas. 1995. Amplification of the housekeeping sigma factor in *Pseudomonas fluorescens* CHA0 enhances antibiotic production and improves biocontrol abilities. J. Bacteriol. 177: 5387–5392.

Schroth, M.N. and J.G. Hancock. 1982. Disease-suppressive soil and root-colonizing bacteria. Science. 216: 1376–1381.

Segarra, G., S. Van der Ent, I. Trillas and C.M. Pieterse. 2009. MYB72, a node of convergence in induced systemic resistance triggered by a fungal and a bacterial beneficial microbe. Plant Biology. 11: 90–96.

Selin, C., W.G.D. Fernando and T. de Kievit. 2012. The PhzI/PhzR quorum-sensing system is required for pyrrolnitrin and phenazine production, and exhibits cross-regulation with RpoS in *Pseudomonas chlororaphis* PA23. Microbiology. 158: 896–907.

Sels, J., J. Mathys, B.M. De Coninck, B.P. Cammue and M.F. De Bolle. 2008. Plant pathogenesis-related (PR) proteins: a focus on PR peptides. Plant Physiol. Bioch. 46: 941–950.

Shanahan, P., D.J. O'Sullivan, P. Simpson, J.D. Glennon and F. O'Gara. 1992. Isolation of 2,4-diacetylphloroglucinol from a fluorescent pseudomonad and investigation of physiological parameters influencing its production. Appl. Environ. Microbiol. 58: 353–358.

Shanahan, P., J.D. Glennon, J.J. Crowley, D.F. Donnelly and F. Ogara. 1993. Liquid-chromatographic assay of microbially derived fluoroglucinol antibiotics for establishing the biosynthetic route to production, and the factors affecting their regulation. Anal. Chim. Acta. 272: 271–277.

Sharifi-Tehrani, A., M. Zala, A. Natsch, Y. Moenne-Loccoz and G. Defago. 1998. Biocontrol of soil-borne fungal plant diseases by 2,4-diacetylphloroglucinol-producing fluorescent pseudomonads with different restriction profiles of amplified 16S rDNA. Eur. J. Plant Pathol. 104: 631–643.

Silby, M.W., C. Winstanley, S.A. Godfrey, S.B. Levy and R.W. Jackson. 2011. Pseudomonas genomes: diverse and adaptable. FEMS Microbiol. Rev. 35: 652–680.

Simons, M., H.P. Permentier, L.A. de Weger, C.A. Wijffelman and B.J.J. Lugtenberg. 1997. Amino acid synthesis is necessary for tomato root colonization by *Pseudomonas fluorescens* strain WCS365. Mol. Plant. Microbe Interact. 10: 102–106.

Sorensen, D., T.H. Nielsen, C. Christophersen, J. Sorensen and M. Gajhede. 2001. Cyclic lipoundecapeptide amphisin from *Pseudomonas* sp. strain DSS73. Acta Crystallogr. C. 57: 1123–1124.

Stutz, E.W., G. Defago and H. Kern. 1986. Naturally-occurring fluorescent pseudomonads involved in suppression of black root-rot of tobacco. Phytopathology. 76: 181–185.

Thomashow, L.S. and D.W. Weller. 1988. Role of a phenazine antibiotic from *Pseudomonas fluorescens* in biological control of *Gaeumannomyces graminis* var *tritici*. J. Bacteriol. 170: 3499–3508.

Thomashow, L.S. and D.W. Weller. 1996. Current concepts in the use of introduced bacteria for biological disease control: mechanisms and antifungal metabolites. *In:* G. Stacey and N.T. Keen [eds.]. Plant-Microbe interactions. Chapman and Hall, New York. pp. 236–271.

Thrane, C., T.H. Nielsen, M.N. Nielsen, J. Sorensen and S. Olsson. 2000. Viscosinamide-producing *Pseudomonas fluorescens* DR54 exerts a biocontrol effect on *Pythium ultimum* in sugar beet rhizosphere. FEMS Microbiol. Ecol. 33: 139–146.

Thrane, C., M.N. Nielsen, J. Sorensen and S. Olsson. 2001. *Pseudomonas fluorescens* DR54 reduces sclerotia formation, biomass development, and disease incidence of *Rhizoctonia solani* causing damping-off in sugar beet. Microb. Ecol. 42: 438–445.

Timms-Wilson, T.M., R.J. Ellis, A. Renwick, D.J. Rhodes, D.V. Mavrodi, D.M. Weller, L.S. Thomashow and M.J. Bailey. 2000. Chromosomal insertion of phenazine-1-carboxylic acid biosynthetic pathway enhances efficacy of damping-off disease control by *Pseudomonas fluorescens*. Mol. Plant. Microbe Interact. 13: 1293–1300.

Ton, J., M. De Vos, C. Robben, A. Buchala, J.P. Metraux, L.C. Van Loon and C.M. Pieterse. 2002a. Characterization of *Arabidopsis* enhanced disease susceptibility mutants that are affected in systemically induced resistance. Plant J. 29: 11–21.

Ton, J., J.A. Van Pelt, L.C. Van Loon and C.M.J. Pieterse. 2002b. Differential Effectiveness of Salicylate-Dependent and Jasmonate/Ethylene-Dependent Induced Resistance in *Arabidopsis*. Mol. Plant. Microbe Interact. 15: 27–34.

Tran, H., A. Ficke, T. Asiimwe, M. Hofte and J.M. Raaijmakers. 2007. Role of the cyclic lipopeptide massetolide A in biological control of *Phytophthora infestans* and in colonization of tomato plants by *Pseudomonas fluorescens*. New Phytol. 175: 731–742.

Tripathi, R.K. and D. Gottlieb. 1969. Mechanism of action of antifungal antibiotic pyrrolnitrin. J. Bacteriol. 100: 310–318.

Trutko, S.M., N.P. Golovchenko and V.K. Akimenko. 1979. Study of cyanide-resistant respiration of *Pseudomonas aeruginosa*. Biochemistry-Moscow. 44: 566–573.

Valverde, C., S. Heeb, C. Keel and D. Haas. 2003. RsmY, a small regulatory RNA, is required in concert with RsmZ for GacA-dependent expression of biocontrol traits in *Pseudomonas fluorescens* CHA0. Mol. Microbiol. 50: 1361–1379.

Van der Ent, S., B.W. Verhagen, R. Van Doorn, D. Bakker, M.G. Verlaan, M.J. Pel, R.G. Joosten, M.C. Proveniers, L.C. Van Loon, J. Ton, and C.M. Pieterse. 2008. *MYB72* is required in early signaling steps of rhizobacteria-induced systemic resistance in *Arabidopsis*. Plant Physiol. Bioch. 146: 1293–1304.

Van der Ent, S., S.C.M. Van Wees and C.M.J. Pieterse. 2009. Jasmonate signaling in plant interactions with resistance-inducing beneficial microbes. Phytochemistry. 70: 1581–1588.

van Loon, L. 2007. Plant responses to plant growth-promoting rhizobacteria. Eur. J. Plant Pathol. 119: 243–254.

van Loon, L.C., P.A. Bakker and C.M. Pieterse. 1998. Systemic resistance induced by rhizosphere bacteria. Annu. Rev. Phytopathol. 36: 453–483.

van Loon, L.C., P.A.H.M. Bakker, W.H.W. van der Heijdt, D. Wendehenne and A. Pugin. 2008. Early Responses of Tobacco Suspension Cells to Rhizobacterial Elicitors of Induced Systemic Resistance. Mol. Plant. Microbe Interact. 21: 1609–1621.

van Pee, K.H. and J.M. Ligon. 2000. Biosynthesis of pyrrolnitrin and other phenylpyrrole derivatives by bacteria. Nat. Prod. Rep. 17: 157–164.

VanPeer, R., G.J. Niemann and B. Schipppers. 1991. Induced resistance and phytoalexin accumulation in biological control of fusarium wilt of carnation by *Pseudomonas* sp. strain WCS417r. Phytopathology. 728–734.

Van Wees, S.C., C.M. Pieterse, A. Trijssenaar, Y.A. Van't Westende, F. Hartog and L.C. Van Loon. 1997. Differential induction of systemic resistance in Arabidopsis by biocontrol bacteria. Mol Plant-Microbe Interact. 10: 716–724.

van Wees, S.C., M. Luijendijk, I. Smoorenburg, L.C. van Loon and C.M. Pieterse. 1999. Rhizobacteria-mediated induced systemic resistance (ISR) in Arabidopsis is not associated with a direct effect on expression of known defense-related genes but stimulates the expression of the jasmonate-inducible gene Atvsp upon challenge. Plant Mol. Biol. 41: 537–549.

Verhagen, B.W., J. Glazebrook, T. Zhu, H.S. Chang, L.C. van Loon and C.M. Pieterse. 2004. The transcriptome of rhizobacteria-induced systemic resistance in arabidopsis. Mol. Plant. Microbe Interact. 17: 895–908.

Verhagen, B.W., P. Trotel-Aziz, M. Couderchet, M. Hofte and A. Aziz. 2010. *Pseudomonas* spp.-induced systemic resistance to *Botrytis cinerea* is associated with induction and priming of defence responses in grapevine. J. Exp. Bot. 61: 249–260.

Vincent, M.N., L.A. Harrison, J.M. Brackin, P.A. Kovacevich, P. Mukerji, D.M. Weller and E.A. Pierson. 1991. Genetic analysis of the antifungal activity of a soilborne *Pseudomonas aureofaciens* strain. Appl. Environ. Microbiol. 57: 2928–2934.

Visca, P. 2004. Iron regulation and siderophore signalling in virulence by *Pseudomonas aeruginosa*. *In*: J.L. Ramos [ed.]. Pseudomonas, Vol. 2. Kluwer Academic/Plenum Publishers, New York. pp. 69–123.

Visca, P., G. Colotti, L. Serino, D. Verzili, N. Orsi and E. Chiancone. 1992. Metal regulation of siderophore synthesis in *Pseudomonas aeruginosa* and functional effects of siderophore-metal complexes. Appl. Environ. Microbiol. 58: 2886–2893.

Viswanathan, R. and R. Samiyappan. 2002. Induced systemic resistance by fluorescent pseudomonads against red rot disease of sugarcane caused by *Colletotrichum falcatum*. Crop Prot. 21: 1–10.

Voisard, C., C. Keel, D. Haas and G. Defago. 1989. Cyanide production by *Pseudomonas fluorescens* helps suppress black root-rot of tobacco under gnotobiotic conditions. EMBO J. 8: 351–358.

Walker, T.S., H.P. Bais, E. Grotewold and J.M. Vivanco. 2003. Root exudation and rhizosphere biology. Plant Physiol. 132: 44–51.

Ward, E.R., S.J. Uknes, S.C. Williams, S.S. Dincher, D.L. Wiederhold, D.C. Alexander, P. Ahl-Goy, J.P. Metraux and J.A. Ryals. 1991. Coordinate Gene Activity in Response to Agents that Induce Systemic Acquired Resistance. The Plant Cell. 3: 1085–1094.

Wei, G., J.W. Kloepper and S. Tuzun. 1991. Induction of systemic resistance of cucumber to *Colletotrichum orbiculare* by select strains of plant growth-promopting rhizobacteria. Phytopathology. 1508–1512.

Weller, D.M. 1983. Colonization of wheat roots by a fluorescent pseudomonad suppressive to take-all. Phytopathology. 73: 1548–1553.

Weller, D.M. 2007. *Pseudomonas* biocontrol agents of soilborne pathogens: Looking back over 30 years. Phytopathology. 97: 250–256.

Whistler, C.A., N.A. Corbell, A. Sarniguet, W. Ream and J.E. Loper. 1998. The two-component regulators GacS and GacA influence accumulation of the stationary-phase sigma factor sigma(S) and the stress response in *Pseudomonas fluorescens* Pf-5. J. Bacteriol. 180: 6635–6641.

Wood, D.W., F.C. Gong, M.M. Daykin, P. Williams and L.S. Pierson. 1997. *N*-acyl-homoserine lactone-mediated regulation of phenazine gene expression by *Pseudomonas aureofaciens* 30–84 in the wheat rhizosphere. J. Bacteriol. 179: 7663–7670.

Yan, Z., M.S. Reddy, C.M. Ryu, J.A. McInroy, M. Wilson and J.W. Kloepper. 2002. Induced systemic protection against tomato late blight elicited by plant growth-promoting rhizobacteria. Phytopathology. 92: 1329–1333.

13

Inoculants based in Autochthonous Microorganisms, a Strategy to Optimize Agronomic Performance of Biofertilizers

D. Mulas,[1-3] C.A. Díaz-Alcántara,[2] R. Mulas,[1] I. Marcano,[2] M. Barquero,[1] P. Serrano[1] and F. González-Andrés[1,*]

Introduction

Plant growth has traditionally been considered to be affected by soil as a mineral substrate with no interaction with the crops apart from plant nutrient supply. Nowadays, the scientific community studies the soils as ecosystems in which soil microorganisms play key roles in any transformation (Miransari 2011). Amongst soil microorganisms, those that were initially comprehensively studied were the ones capable of producing

[1]Instituto de Medio Ambiente, Recursos Naturales y Biodiversidad. Universidad de León. Avenida de Portugal, 41. E24071 León. Spain.
[2]Facultad de Ciencias Agronómicas y Veterinarias. Universidad Autónoma de Santo Domingo. Dominican Republic.
[3]Fertiberia S.A. – Departamento de I+D+i. Spain.
*Corresponding author: fernando.gonzalez@unileon.es

plant diseases and pests, thus reducing crop yields. In many cases, these studies focused on developing chemical products to overcome these pests and diseases with a detrimental effect on the soil microbial community as a whole. However, more recently there has been a shift in the research worldwide, considering soil microorganisms from a proactive point of view. This perspective leads to understanding the interactions between soil microorganisms and plants in terms of increasing the plant growth and health, thus directly or indirectly increasing agricultural productivity due to their relatedness with plant fitness (Gutiérrez-Mañero and Ramos-Solano 2010). The following are among their most relevant beneficial effects in agricultural soils: (1) recycling of soil nutrients available in organic form, with special relevance in the nitrogen cycle, (2) enhancing soil nutrient availability and consequently uptake by the plant, (3) alleviating abiotic stresses from the soil on plant growth and yield, (4) controlling pathogenic fungi or bacteria by means of a direct interaction between the beneficial microorganism and the pathogen, or without direct interaction between the plant and the beneficial microorganism, triggering a defending response on the plant against pathogens, (5) improving soil structure by producing different biochemicals (Emmerling et al. 2002, Böhme and Böhme 2006, Arzanesh et al. 2010, Gutiérrez-Mañero and Ramos-Solano 2010, Siddikee et al. 2011).

Nitrogen is after water, the main limiting factor for plant growth (de Felipe 2006). After the discovery of the Haber-Bosch process which made possible the industrial process of N_2 fixation, chemical fertilizers became the fast way of providing crops with this necessary nutrient (Miransari 2011). However, the key importance of N_2 fixation for the biosphere is well known, estimated in 250 Tg (Tg = 10^6 tonnes), of which 87 percent are fixed by prokaryotic organisms through what is called Biological Nitrogen Fixation (BNF) (Arrese-Igor 2010). Approximately 20 percent of the total nitrogen derived from BNF is fixed by free living microorganisms, whereas 80 percent is fixed in symbiosis with plants (de Felipe 2006). In agriculture, BNF presents economic, environmental and agronomic benefits and could be used to a greater extent as an alternative to synthetic fertilizers (Abi-Ghanem et al. 2011). In spite of this, there is an abusive use of nitrogen fertilizers in many agricultural systems worldwide (Good and Beatty 2011), and this causes unfavourable effects including leaching, runoff, greenhouse gases (GHG) emission and eutrophication of aquatic ecosystems (Adesemoye and Kloepper 2009, Adesemoye et al. 2009). A surplus nitrogen balance leads to NH_3, N_2O, NO_3^- and/or NO pollution, whereas a deficit nitrogen balance leads to low soil fertility from depleting soil nutrient pools, resulting in poor crop yield (Bouwman et al. 2005).

Undoubtedly the most important nitrogen fixation process in agriculture is the symbiotic fixation between the genera *Azorhizobium*, *Bradyrhizobium*,

Mesorhizobium, *Rhizobium* and *Ensifer*, collectively known as rhizobia, and the leguminous plants (Velázquez et al. 2011). In 1975 nearly half of the nitrogen needed in agriculture was provided through legume-rhizobial symbiosis (Burns and Hardy 1975), but by 2005, synthetic nitrogen provided three times as much nitrogen to agriculture as BNF on a global basis (Galloway et al. 2008).

Several environmental factors influence the effectiveness of nitrogen fixation in crops and this leads striking variations recorded in total nitrogen fixed by crops (Unkovich and Pate 2000). However as the rhizobia–legume interaction is specific and is reduced to one or few bacterial species for each legume species (Perret et al. 2000, Willems 2006), sometimes is needed to put in contact the bacterium and the plant through inoculation. It is usually thought that the inoculation with rhizobia is needed in legume crops that have never been grown in a particular soil, or for some years (Deaker et al. 2004).

Inoculation in agriculture is an ancient technique. Shortly after Hellriegel's report on nitrogen nutrition in 1886, it was recommended the agronomic practice of transferring soil from one field with high legume yield to another, being the starting point of inoculation, because the soil transfers rhizobial inoculum. During the following years the technique of inoculation with soil was improved, using soil dust in the surroundings of the seed and thereafter a soil paste or muddy water (Smith 1992). It was still in the XIX century, when Nobber and Hiltner in 1896 patented the first inoculant based in a pure culture of rhizobium.

After the description of the plant growth promoting rhizobacteria (PGPR) as soil bacteria that colonize the roots of plants and that enhance plant growth by Kloepper and Schroth (1978), new inoculants based in other bacterial taxa have been developed.

Types of Inoculants in Agriculture and the Concept of Biofertilizer

The beneficial interactions microorganism–plant–environment can be used in agriculture to improve the agronomic performance of crops (Barea et al. 2005a). The PGPR, soil microorganisms that are beneficial for plant growth, are sometimes called "Plant-Probiotic Microorganisms" PPM (Haas and Keel 2003), and are formulated in the inoculants.

The next inoculants classification is based on the main action of the microorganism in the crop, from an agronomic viewpoint.

Inoculants Based in Bacteria that Fix Atmospheric Nitrogen (N_2)

The most extended inoculants in agriculture consist of rhizobia based formulations for legume crops, that fix nitrogen in specialized structures in the legume roots referred to as nodules (Thuita et al. 2012).

Other free living bacteria, fix atmospheric N_2 that can be used to some extent by the crops, improving their growth. The genera *Azotoabacter, Azospirillum, Acetobacter* and *Azolla* are used as inoculants to increase the yield of non legume crops, mainly cereals, especially in conditions of limiting mineral nitrogen added to the soil (Fallik and Okon 1996, Rodríguez-Navarro and Ruiz-Sánchez 2010).

Inoculants Based in Microorganisms that Improve the Plants' Growth by a Complex Mix of Action Modes

According to Glick et al. (2007), the observed improvement in plants' growth as a consequence of the PGPR activity, is the result of the combination and interaction of several modes of action. The main individual actions are listed below:

Improvement of Plants' Nutrition

An important mode of action of PGPR is related with the nutritional state of the plant (Dobbelaere et al. 2003, Van Loon 2007, Singh et al. 2011). First, some bacteria belonging to several taxa frequently found in soils, are able to mobilize inorganic phosphorous to the crops (Dobbelaere et al. 2003). Second, some soil microorganisms produce siderophores, which are low molecular weight compounds excreted by the microorganisms. They join to Fe ions and transform the non-assimilable forms of this cation into assimilable, because plants are able to assimilate the siderophores (Neilands 1981, 1982). In the third place, many studies reported plant increased uptake of nitrogen and some other elements through PGPR inoculation (Sheng and He 2006, Glick et al. 2007). R. Mulas and F. González-Andrés (unpublished) observed that barley plants inoculated with *Rhizobium leguminosarum* showed a higher proportion of nitrogen in the aerial biomass, compared with the uninoculated control.

Arbuscular mycorrhiza (AM) participate in the enhancement of plant growth, mobilizing nutrients to the plant (Adesemoye et al. 2008), especially those nutrients whose molecular species show low mobility in the soil (Barea et al. 2005b, see also Chapter 15 in this volume).

Stimulation of Plant Growth

Some soil bacteria produce phytohormones, the most deeply studied being IAA. The hormones induce a general increase of growth in the plants, not only roots. Frequently it involves an increase of the economic yield of the crop (Dobbelaere et al. 2003, Sheng and He 2006, Glick et al. 2007, Spaepen et al. 2008).

The higher biomass produced by the inoculated plants, together with the increased uptake of nitrogen, has a positive consequence not only in the yield but also in the preservation of the environment from nitrogen loses. In the previously mentioned work of Mulas et al. they observed an enhanced nitrogen uptake by inoculating barley plants with rhizobia, regardless of the origin of the nitrogen, even from mineral fertilizer, improving therefore the apparent efficiency of nitrogen and reducing the risk of contamination.

Overcoming of Plant Stresses

Another important action of PGPR is the synthesis of enzymes that indirectly promote plant growth. The most comprehensively studied case is the production of ACC deaminase, an enzyme which hydrolyses ACC, an immediate precursor of ethylene, so reducing the ethylene production by the plant. The reduction of ethylene levels increases the plants growth under stress situations, therefore improving the plant response to several biotic and abiotic stresses (Dobbelaere 2003, Glick et al. 2007, Nadeem et al. 2007, Contesto et al. 2008, Arshad et al. 2008, Govindasamy et al. 2008). However, determining this enzymatic activity must be optimized in order to obtain liable results, since the technique that uses ACC as the only N source for *in vitro* assays can produce biased results when testing nitrogen-fixing microorganisms (Velázquez, pers. comm.)

The unpublished work of R. Mulas and F. González-Andrés showed a significant increase of yield in barley inoculated with *Rhizobium leguminosarum* under stressing conditions for the plant, consisting on lack of space for the root growth due to the cultivation in pots with limiting space. It seems that the bacteria helped the plant to overcome the stress situation. Unexpectedly, the bacterial strains that increased the plant growth the most as compared with the uninoculated control, were not the same that showed *in vitro* the highest ACC deaminase activity, supporting the theory that the final effect of the PGPR is the result of a mix of complex interrelationships between different modes of action.

Microbial Biocontrol Agents

Some authors have included microorganisms that act by reducing the damage caused by pests and diseases amongst the PGPR. Although by reducing the severe impact of diseases in different ways, PGPR *sensu stricto* must only be considered for those with a direct action on plant nutritional status or plant growth. Microorganisms with a detrimental effect on plant pathogens must be considered as biocontrol agents, since their mechanisms to help plants overcome pests and diseases are different. They are of great interest in the present agricultural scenario of reduction of active substances for plant protection (Dobbelaere et al. 2003, Van Loon 2007, Singh et al. 2011). Microbial biocontrol agents may protect plants against pathogens by direct antagonistic interactions with the pathogen, as well as through induction of host systemic resistance. However, PGPR can indirectly promote the competition against pathogens, such as in the case of production of siderophores to chelate iron, making it unavailable to pathogens, to successfully compete against pathogens for nutrients or specific niches on the root, to synthesize anti-fungal metabolites that suppress the growth of fungal pathogens as for example antibiotics, fungal cell wall lysing enzymes, or hydrogen cyanide (Glick et al. 1995, Bloemberg and Lugtenberg 2001, Persello Cartieaux et al. 2003). For additional information on biocontrol mechanisms, see also Chapter 12 in this volume.

AM protects the plants against abiotic and also biotic stresses (Jeffries et al. 2003, Barea et al. 2005a). It has been demonstrated that plants with a well established symbiotic AM structure have a lower detrimental impact of soil pathogens, due to modifications in the structure of soil microbial community as a consequence of the direct competition in the mychorhizosphera and also to the systemic resistance developed in the plants (Barea et al. 2005a).

The Concept of Biofertilizer

There is not a complete agreement on what must be considered a biofertilizer (Rodríguez-Navarro and Ruíz-Sánchez 2010). The products based on living microorganisms which improve plants nutritional status must be included under the name of biofertilizers. In this sense, the rhizobia-based inoculants for legumes that fix atmospheric nitrogen in root nodules, are undoubtedly biofertilizers. They can be considered "first generation biofertilizers".

Other products based in microorganisms whose main effect is the improvement of plant nutrition and growth, although they do not fix atmospheric N, are also biofertilizers according to the definition. In this case there is also an interaction between the plant and the microorganism but it is not performed in a specific structure as it was in the nodule; instead of this, the interaction occurs in the rizosphere or inside the root tissues and in the last case the microorganisms are called endophytes. They can be collectively named as "second generation biofertilizers". The second generation biofertilizers can be based in a range of soil bacterial taxa. However, the most promising is the beneficial interaction of non legumes with rhizobia, as it has been demonstrated that they promote the plants' growth, and rhizobia are completely safe bacteria for human beings and the environment. The plant growth promoting (PGP) effect of rhizobia in non-legume plants, was demonstrated in cereals like rice, maize or barley (Chabot et al. 1996, Yanni et al. 2001, Gutiérrez-Zamora and Martínez-Romero 2001, Peix et al. 2001, Mishra et al. 2006, Rosenblueth and Martinez-Romero 2006, Mia and Shamsuddin 2010, R. Mulas and F. González-Andrés, unpublished). Currently there is data about the growth promotion effect of rhizobium in vegetables edible as raw leaves such as lettuce (Chabot et al. 1996, Noel et al. 1996), as raw roots such as radishes (Antoun et al. 1998) and as raw fruits (García-Fraile et al. 2012).

Another innovative perspective for soil microorganisms as biofertilizers is their actual content of plant nutrients. Paungfoo-Lonhienne et al. (2010) have found that plants are able to use microbial biomass as nutrient source for plant nutrition when entering into the root as viable cells. However, this approach poses an innovative viewpoint that must be comprehensively studied before describing soil microbes as another nutrient source for crop production.

Other second generation biofertilizers are based in bacteria other than rhizobia, for the inoculation of legumes and non legumes. Among non-rhizobial bacteria species with PGP acitivity the most frequent species belong to the genera *Pseudomonas* and *Bacillus*, although some of the soil diazotrophes like *Azotobacter* or *Azospirillum* also may show other PGP activities apart from providing nitrogen to the crop (Dobbelaere et al. 2003, Hayat et al. 2010, see also Chapter 11 in this volume).

The products based in microorganisms designed for the biocontrol of pathogens or pests, must not be considered as biofertilizers. However, in many cases the mode of action of a microorganism is multiple, for example a bacterium that produces siderophores, thus improving the Fe nutrition of plants when available Fe is limited, simultaneously makes Fe unavailable for plant pathogens (Roy and Chakrabartty 1994), preventing the infection of the pathogen. Moreover, from a general viewpoint the improvement of the phytosanitary situation of the plant has the consequence of a better plant growth.

However, other products are sometimes considered as biofertilizers. For example, compost or other organic fertilizers are rich in living microorganisms, although in this case the microbial community is comprised by a natural population instead of a well known strain or a mix of well known strains. Moreover, some fertilizers which are not based in living microorganisms as for example hydrolyzed proteins have also been considered and designated as a biofertilizer.

The European Case of Legislation Regarding Biofertilizers

The European Union has developed two different legislative tools to regulate the inputs that include the microorganisms in agriculture. The main plant nutrients are nitrogen, phosphorous and potassium. Other chemical elements, such as sulphur, calcium or magnesium are also nutrients, and the microelements include iron, molybdenum, copper, zinc, etc. From a European perspective, a fertilizer is based on inorganic products delivering these elements for plant nutrition. Therefore the European Regulation for fertilizers only considers these elements as nutrients, and there is no mention of microorganisms in any of its articles.

On the other hand, the another important regulation about inputs in agriculture is related to products that reduce, avoid or mitigate the impact on the crops of pests, diseases and weeds. In this case, the European Union has established a list that is renewed periodically in which all products that can be used as plant health products, including microorganisms, are under one of the following categories: approved, under revision or not accepted. However, there is no European law regarding those microorganisms that act as biofertilizers, in fact this term is not described in any law.

Moreover, a recent European regulation is expected to bring a new classification of fertilizer products and will include the term "biostimulant" or similar, under which microorganisms used as inoculants will be able to be regulated. This indicates that microorganisms can be included in this section, but will not imply a different consideration for microorganisms involved in reducing pests and diseases.

Rhizobia Inoculants. Factors Affecting BNF

Many studies have attempted to understand the factors related with the effectiveness of BNF, and from the agronomic viewpoint, the performance of the inoculants. The inoculants based in rhizobia for legumes have been more comprehensively studied, compared with the rest of the inoculants.

Effect of the Plant Cultivar

It has been found that the effect of crop variety and strain-by-variety specific interactions are significant factors in BNF, which in some cases are even more important than the effect of the strain alone (Abi-Ghanem et al. 2011). For example, in lentils (*Lens culinaris* Medik.) Hafeez et al. (2000) found a 42 percent variability in BNF among rhizobial strains, and a 81 percent variability among cultivars of the crop. Abi-Ghanem et al. (2011) found also a significant influence of the cultivar; however, in this case the variability among cultivars was only a 10 percent, compared with a 15 percent of variability among the strains. A strong cultivar × strain interaction on BNF has been also reported in beans (*Phaseolus vulgaris*) (Valverde and Otabbong 1997), peanuts (*Arachis hypogaea* L.) (Wynne et al. 1980) and soybeans (*Glycine max* Linn. Merr.) (Israel 1981). Similarly in peas (*Pisum sativum* L.), Abi-Ghanem et al. (2011) observed a significant effect of the cultivar on the BNF, but surprisingly the effect of the strain was not significant.

Effect of Abiotic Stresses

The main abiotic stresses that can affect legume growth or establishment and function of the legume–rhizobia symbiosis are temperature, water, salinity, sodicity, acidity and nutrient disorders (Chalk et al. 2010, see also Chapters 4 and 5 in this volume). However, the response of the symbiosis to a particular stress depends on several factors, as the legume cultivar, inoculant, climatic conditions, and the duration, timing and severity of the stress (Chalk et al. 2010).

Soil temperature has a marked effect in the inoculant performance, because it influences several aspects related with the survival of rhizobia and the nodulation process (Chalk et al. 2010). The soil temperature affects the bacterial survival in the soil after inoculation, as was demonstrated in the classical experiment of Brockwell et al. (1987). High temperatures usually affect the survival of rhizobia, but they are more resistant to dry heat than to moist heat (Wilkins 1967 in Deaker et al. 2004). The resistance to high temperatures is an inherent characteristic of the strain, as it has been demonstrated by a strain × temperature interaction (Montanez et al. 1995). One temptation is therefore the selection of strains *in vitro* for tolerance in high temperature, but this is not usually linked with the nodulation and BNF capacity under similar conditions (Hungría and Vargas 2000, Giller 2001). Temperature also affects the nodulation process, probably the root infection process is the most affected component by high temperatures. Once the nodules are formed, high temperature may also affect nodule function (Hungría and Vargas 2000). Besides this, temperature can also affect the response to inoculation through an effect on competitive performance, these are the interactions occurring in the soil microbial community. Finally, the temperature greatly affects the macrosymbiont, which is the plant, and therefore affects the proportional dependence of biomass N on N_2 fixation (Montanez et al. 1995).

Shortage of water affects plant and rhizobial growth, and is a major cause of nodulation failure and low BNF (Hungria and Vargas 2000). However, the review article of Chalk et al. (2010) found conflicting reports in the literature regarding the effect of drought in the proportional dependence of the biomass N on N_2 fixation. They compiled several studies that demonstrated a negative impact of drought stress on the symbiotic performance and the response to inoculation: for example, Bremer et al. (1988, 1989) in lentil and peas and Kirda et al. (1989) and Hungria and Vargas (2000) for soybean. Conversely, working with the common bean (*Phaseolus vulgaris* L.) Pena-Cabrales and Castellanos (1993) observed that the proportional dependence of the biomass N on N_2 fixation did not differ between treatment regarding different water shortages. However, for the same crop Calvache and Reichardt (1996) observed that when the water shortage was in the early vegetative growth there was no significant effect in the symbiotic performance, but the same stress in flowering or grain filling reduced the dependence of the biomass N on N_2 fixation.

Different rhizobia strains usually show different response *in vitro* to salinity stress (Mulas 2010). However, some studies (Rao et al. 2002) showed that the plant × bacterium interactions under salt stress do not depend on the salt tolerance of the bacterium. Therefore they suggested a selection based in legume cultivars. For a detailed review on the mechanisms of adaptation of the rhizobium-legume symbiosis to salinity, see Chapter 5.

According to the review by Hungría and Vargas (2000), low soil pH affects the growth of rhizobia (Coventry and Evans 1989, Campo 1995), of the host legume (Andrews et al. 1973, Kim et al. 1985) and the symbiosis (Brady et al. 1990, Campo 1995) due to the increased Al and Mn toxicity and reduced Ca, Mg and P supply (Jayasundara et al. 1998). Agronomic research has been based on the effect of different quantities of lime applications to neutralize the pH, and the results obtained in terms of nitrogen fixation or proportional dependence of the biomass N on N_2 fixation were very variable, and in most of the occasions related with environmental factors like rainfall, or the mineral nitrogen content in the soil (which inhibits the BNF) (compilation of Chalk et al. 2010). Not only low pH, but also alkalinity above pH 8.5 limits inoculant strain survival (Howieson and Ballard 2004).

Nutritional disorders in the plant also serve as important constraints in the function of the legume–rhizobium symbiosis (Chalk et al. 2010). Phosphorous is the limiting nutrient for legume BNF in most of the plants, and usually the P requirement of legumes is generally higher when their growth depends on the nitrogen derived from atmospheric fixation rather than on ammonium or nitrate (Drevon et al. 2011). However, the response of nodulation parameters to P strongly depends on the plant variety (Attar et al. 2012).

Effect of the So-called "Helper" Bacteria

It has been known since the 1970's that some microorganisms enhance a variety of legume-rhizobia symbiosis (Vessey 2003). The effects of PGPR coinoculation on legume symbiosis include increases in nodule number and/or nodule weight, and in some cases an enhancement of nitrogen fixation or plant nitrogen content. For example, a recent work of Fox et al. (2011) showed that coinoculation with *Ensifer* (*Sinorhizobium*) *medicae* WSM419 and *Pseudomonas fluorescens* WSM3457 enhanced the early nodule initiation and development in the model plant *Medicago truncatula*, which can be a significant advantage for the success of nodulation in field grown legumes under environmental stressors detrimental for the establishment of a successful symbiosis.

Competition for Nodule Ocupancy

The most important challenge to maximize BNF in inoculated legumes is to prevent rhizobia resident in the soil from dominating nodule occupation (Yates et al. 2011). Usually, the native strains are numerically dominant and they are well adapted to their niche (Sadowsky and Graham 1998), but often they are not optimally matched for nitrogen fixation with the introduced legume (Yates et al. 2011). Such adapted microbes are said to form "barriers" that compromise the ability of introduced inoculants strains to occupy nodules (Sadowsky and Graham 1998). This is the main problem for the success of inoculants, but also an opportunity for the development of inoculants based on effective autochthonous strains.

Effect of the Formulation of the Rhizobia

The formulation determines the potential success of inoculants, frequently being the most common barrier to the commercialization of the product (Fages 1992). The strain must survive in the inoculant formulation during storage; therefore, a suitable rhizobial carrier should have a high water holding capacity to protect against desiccation and a lack of toxic compounds to the bacteria. At the same time, these materials should be non-toxic in nature, easily sterilized, manufactured and handled in the field, chemically and physically uniform, provide slow release of bacteria for long periods in the soil, and inexpensive (Smith 1992, Rebah 2002, Bashan 2005). In addition to the intrinsic properties of the carrier, the factors that significantly influence the viability of rhizobia during storage of the product are: (1) the method of preparation, mainly its sterility by gamma radiation (Khavazi et al. 2007); (2) the final moisture content (Roughley and Vincent 1967); (3) the use of packaging material which allows gas exchange (Roughley 1968) and (4) the storage temperature and time of inoculum (Roughley 1968).

Inoculants are commercially available as solid products, powder, produced from peat, or as granular form, or liquid inoculants using broth medium (Stephens and Rask 2000, Rebah et al. 2002). However, survival in liquid inoculants is worse because bacteria lack carrier protection (Tittabutr et al. 2007, Albareda et al. 2008). Powder inoculants, applied to the seed prior to sowing, are improved by the use of adhesives and polymers to ensure contact between the rhizobia and the legume seed coat, to reduce rhizobial desiccation (Denardin and Freire 2000, Deaker et al. 2004). A large amount of granular formulations are also evaluated using clay as a carrier, mineral sands, charcoal, vermiculite, perlite, ground rock phosphate, calcium sulphate, talc, polyacrylamide gel, alginate beads, polythene beads and others (Brockwell et al. 1980, Daza et al. 2000, Deaker et al. 2004) and have shown certain advantages over the powdered form, particularly when it is necessary to apply fungicides or insecticides in the seed (Denton et al. 2009) or when natural toxic seed coat exudates affect the survival of the bacteria (Deaker et al. 2004).

The inoculant quality standards vary according to the legislation in each country. Most established minimum values of viable cell counts of rhizobia ranging from 5×10^7 to 1×10^9 rhizobium cells/g of inoculants product (Lupwayi et al. 2000). But there is another way to evaluate the quality of a commercial inoculant, based on the minimum number of rhizobia per seed, thus, it considers the number of viable cells per g or ml of inoculums and the application rate recommended by the manufacturer. The minimum number of rhizobia per seed stipulated is 10^3 rhizobia per small seed, 10^4 per medium seed, and 10^5 per large seed (Ben Rebah 2007).

Evidence of the Advantages of Native Strains for Inoculants' Formulation

The competition of the infective but ineffective native rhizobia against the inoculant can be of considerable economic consequence (Yates et al. 2011). Therefore, the use of a selected and effective strain as an inoculant in agriculture can become a complete failure in a given agrosystem, if the strain in the inoculant is not able to colonize enough the nodules. For example, in soybean, Thies et al. (1991) reported that significant increases in yield were observed only when no less than a doubling of nodule mass and 66 percent nodule occupancy were reached. It is generally accepted that background populations of rhizobia strongly influence the response to inoculation and that nodulation effects are attenuated as background nodulation from resident rhizobia increases (Thies et al. 1991, Rice and Olsen 1992, Denton et al. 2009).

For this reason, several attempts have been made for managing nodule occupancy by rhizobial strains. For this, Barram and Bromfield (1997) suggested a deep understanding of the following points: (1) the dynamics of the soil population of rhizobia, (2) the abiotic and biotic factors involved in the survival and development of rhizobia populations, (3) the role of the legume host, and (4) the genetic basis of successful nodulating strains. In spite of the research already done, the last two points still remain unclear, because of the high variability of the experimental outcomes, to some extent due to the lack of standardization for discussing them (Yates et al. 2011). A topic intensively studied is whether or not the legume and rhizobium selectively influence the strain that successfully nodulates, but at the moment this is a difficult question to answer (Yates et al. 2011). There is some evidence of selective nodulation, for example in the symbiosis between *Trifolium* sp. and *Rhizobium leguminosarum* bv. *trifolii*, but further research on this phenomenon is still needed (Yates et al. 2011). In absence of a stronger fundamental support, several theories have been drawn for competition in nodulation by variably numerous and effective rhizobial strains in the soil (Sessitsch et al. 2002). A strongly supported theory is that rhizobial genotypes are differentially adapted to soil conditions, and it is this adaptation (rather than the relationship with the legume) that primarily determines the outcome of competition for nodule occupancy (Sprent 2001, Howieson et al. 2000, Sessitch et al. 2002). From the agronomic viewpoint, this theory opens a chief strategy for the selection of strains to design an inoculant, which is the isolation and assessment of efficient strains from local sites of production for each specific legume (Mostasso et al. 2002).

The importance of this strategy is sufficiently supported by agronomic results. Rodríguez-Blanco et al. (2010) working with *Trifolium* species in the prairies of Uruguay demonstrated that it may be possible to isolate efficient, competitive, and genetically different native rhizobial strains to be used in pastures.

In soybean, Melchiorre et al. (2011) compared the symbiotic performance of native *Bradyrhizobium* strains isolated from soybean in several locations, with the *Bradyrhizobium japonicum* strain officially recommended in Argentina for inoculant formulation, and they found that a consortium of the strains isolated from at least one location showed better symbiotic parameters than the official one. The isolation was made from commercial soybean fields that had been neither inoculated nor supplied with nitrogen fertilizer for at least 5 yr, and the isolated strains were those that showed the highest nodule occupancy.

Probably the most outstanding result is with the common bean, a promiscuous species that may establish symbiosis with a high number of bacterial taxa. The common bean has a reputation of lacking effective nodulation and poor nitrogen fixation (Rodríguez-Navarro et al. 2000). A lack of response to inoculation when non-native strains are used in the inoculant has also been observed (Rodríguez-Navarro et al. 2000, Daza et al. 2000). Mostasso et al. (2002) in their work with *R. tropici*, observed that the type strain CIAT899^{T} only reached 80 percent of the yield obtained with the best native strain. Conversely, the suitability of native strains as inoculants for the local common bean has been demonstrated for *Rhizobium tropici* in America (Hungría et al. 2000, 2003, Mostasso et al. 2002), for *R. gallicum* in Africa (Mrabet et al. 2005), and for *R. leguminosarum* in Castilla y León-Spain (Mulas et al. 2011). In all the mentioned works it was demonstrated that the grain yield obtained by inoculating with autochthonous elite strains was similar, or even higher, when compared to a crop fully supplied with nitrogen as mineral fertilizer.

Specific agronomic results have proven the good performance of autochthonous rhizobia regarding competition against native microorganisms in the soil. The work by Mostasso et al. (2002), showed a similar yield when inoculating the common bean with autochthonous rhizobia and when supplying nitrogen as a mineral fertilizer even in soils with high populations of soil borne rhizobia (10^4 rhizobia per g of soil). The same conclusion was reached by Mrabet et al. (2005) after inoculation of the common bean with native *R. gallicum* strains in Tunisian soils. They additionally observed high nodule occupancy by the introduced strain, even in a soil which showed a MPN value of 10^3 resident rhizobia per g of soil. Mulas et al. (2011) observed the same yield in the inoculated treatments and in the nitrogen supplied control in two different soils in León province, the first with less than 10 rhizobia nodulating the common bean per gram of soil, and the second with 3.1×10^4 nodulating rhizobia per gram of soil.

Several attempts have been undertaken to improve the nitrogen fixation efficiency of rhizobial strains by genetic engineering (Sessitsch et al. 2002). However, the release of genetically improved strains is often restricted by national regulations, so in our opinion, the use of native bacteria has advantages with respect to the use of transformed bacteria. In a society which values natural products, the use of selected but wild organisms substituting partially agrochemicals, can give an added value to the food produced in such way.

Strategy for the Development of a Successful Inoculant Based in Autochthonous Microorganisms

The key point for the development of an inoculant based in autochthonous microorganisms is to optimize the triple interaction (plant genotype × bacterium genotype × environment) proposed by Sessitsch et al. (2002). According to these authors, in a non limiting environment, we can consider the nitrogen fixation as a product of the interaction between the genotype of the plant Gp, and the bacterium Gb, viz G^2. However, the environment (E) is usually limiting, as a consequence of both biotic and abiotic components, thus the net nitrogen fixation can be represented as $G^2 \times E$. E can impact on either Gp or Gb. G^2 can be assessed in a controlled environment minimizing the interaction with E, for example in hydroponic media with an adequate supply of nutrients except nitrogen (when selecting bacteria for their nitrogen fixation capacity). Accepting the hypothesis that plants have the capacity to select a bacterium from a pool of strains, this may be the end point of the first phase in the research. The second phase is the introduction of the inoculated crop in real agricultural soils. To achieve a good agronomic performance of the inoculant, it has to survive in the soil, and establish effective symbiosis with the plant. Supporting the view that Gb × E more strongly impacts nitrogen fixation, then the next steps of strains selection research inevitably must graduate from the controlled environment to the field environment. In this phase, substandard inoculant strains poorly adapted to their target environment are discarded. The last point, which is not in the scope of this chapter, is breeding of the plant for enhanced nitrogen fixation in legumes (Gp × E).

The sequential process of selection of autochthonous bacteria for inoculation can be seen in Fig. 13.1. The first step, which is the isolation of strains from the soil, is also the most important, as at this point a wide range of superior strains as possible are needed, to have enough variability to select the final candidates for the biofertilizer.

Melchiorre et al. (2011) developed rhizobia inoculants for soybean in Argentina, and they successfully proved that to catch rhizobial strains for the inoculant, the best strategy was to isolate native strains from nodules

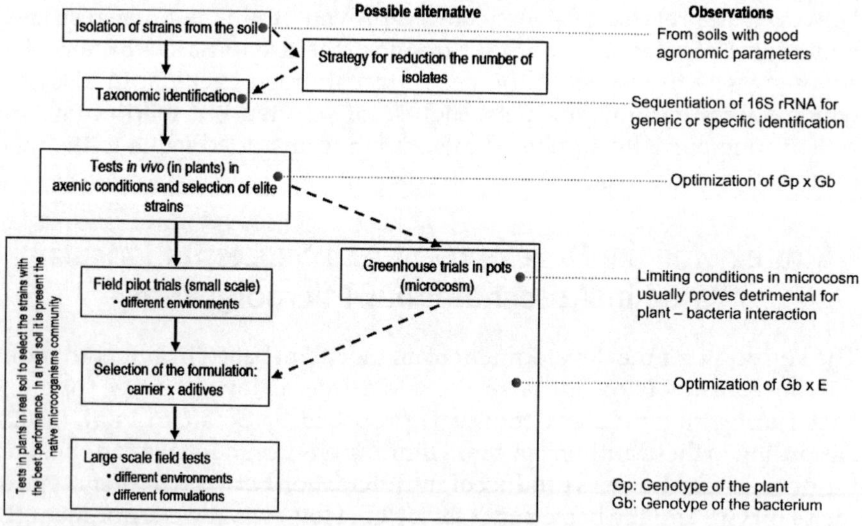

Figure 13.1 Sequential process of selection of autochthonous bacteria for the development of a biofertilizer. Note: Gp: means plant genotype, Gb: bacterium genotype, E: environment.

in non-inoculated and non-nitrogen fertilized fields, that maintain high competitiveness and efficiency in providing adequate nitrogen levels to the plants, and therefore high yields. They isolated bacteria from nodules in fields selected by the criterion that the crop did not show decreases in yields even though they had been neither supplied with nitrogen fertilizer nor inoculated for at least 5 yr. A similar strategy was used by Mulas et al. (2011) in the common bean in Spain. In this case, the criterion for this selection was a long tradition of cultivation and good agronomic practices carried out by the farmer.

An important point before the release of an inoculant or even before the manipulation of microorganisms is the taxonomic identification. This is because inside the plant tissues of a healthy plant or even inside a functional nodule, it is possible to find a great variety of microorganisms, most of them beneficial, but some of them as possible pathogens for animals, humans or crops (García-Fraile et al. 2012, see also Chapter 10 in this volume).

The next main step is to optimize the interaction between the plant and the bacterium (Gp × Gb), but sometimes an intermediate step to reduce the number of isolates to be tested is necessary. For example, Melchiorre et al. (2011) developed a *Bradyrhizobium*-based inoculant and proposed a previous grouping of the strains from each location, based on the pattern obtained for each strain with rep-PCR fingerprinting, using cluster analysis based on the Dice coefficient to calculate similarity, and the UPGMA for clustering. Then they selected the cluster formed by the highest number of nodules, and that showed 70 percent or higher of a similarity coefficient. From such cluster, a "strain pool" formed by four strains that had shown highest within-cluster distance was used to evaluate their symbiotic behaviour in a greenhouse assay.

In the case of working with other PGPR, there are two options: to isolate endophytic bacteria, or to isolate rhizospheric ones. In both cases, the number of cultivable bacteria that can be isolated can become very high, and in such a case it is important to develop an efficient protocol to reduce the number of isolates before the assays in plant. In the case of endophytic bacteria, after the isolation of pure cultures from the interior of a surface-sterilized root, isolates are commonly tested for different PGP activities *in vitro*, like IAA production, ACC deaminase activity, inorganic phosphorous solubilization, or siderophores production. Those strains with significant rates of one or several of these activities would be selected for tests in the plant, depending of the main object of the future biofertilizer. However, some researchers have demonstrated that a relevant *in vitro* activity in properties as for example IAA production or ACC deaminase activity, it is not necessarily related with a clear PGP activity *in vivo* (Marcano et al. 2011, Smyth et al. 2011, R. Mulas and F. González-Andrés, unpublished). For this reason, it has been suggested to taxonomically identify the isolates instead of testing their PGP capacities *in vitro*, and then to test in plants the isolates belonging to taxa that have been showed good PGP activity (Dr. Borriss, Humboldt University, pers. comm.), as for example *Bacillus amyloliquefaciens* (Kloepper et al. 2004, Araujo et al. 2005, Idriss et al. 2007).

In the case of rhizospheric bacteria, as the number of isolates can become extremely high, it would be desirable to make the isolation in a specific media, in order to only isolate bacteria with specific PGP activities *in vitro*. The most frequent is to isolate bacteria for inorganic phosphorus solubilization, or ACC deaminase activity.

After the pre-selection of the isolated strains, the following step is the optimization between the genotype of the plant and the bacterium (Gp × Gb). The assay must be carried out under axenic and hydroponic conditions, and all the materials used must be sterilized, including the recipients, substratum, and the watering solution, it is very important to surface-sterilize the seeds to avoid seed-borne microorganisms. Exfoliated vermiculite is usually the best substratum, and the irrigation is made with a sterile N-Free solution, as for example Rigaud and Puppo (1975) solution in the case of rhizobia inoculants. It is important that watering of the substrate is done from the bottom to avoid washing the inoculated bacteria. The recipient for the plants' growth depends on the size of the seed, and ranges from test tubes to 1-l capacity plastic glasses. The typical treatment in rhizobia inoculants is the following: inoculation with the strains to be tested, a non inoculated control, and a non inoculated control with nitrogen added as a mineral salt to fulfil the plant demand. The experimental unit must have enough number of plants, and an adequate number could be around 15. The inoculation must be carried out with a pure culture. Plants must be grown for a period of time which varies with the species, but the time must be enough for the formation of the nodules and they must be functional to appreciate the differences in the plants' growth. Around 30 to 45 d could be adequate, as a longer time increases the risks of uncontrolled infections. For the plants' growth the best option is a growth chamber with controlled photoperiod, thermoperiod and humidity. At the end of the experiment the most important parameters to measure are: aerial fresh and dry biomass, roots dry biomass and number and dry biomass of nodules. To calculate nitrogen fixation effectiveness, the concentration of N Kjeldahl (percent) in the aerial biomass and its total content are calculated. Usually, the following indexes are calculated (Beck et al. 1993):

$$\text{Specific strain effectiveness (\%)} = 100 \times \frac{Ns - No}{Nn - No}$$

$$\text{Nitrogen derived from the atmosphere (\%)} = 100 \times \frac{Ns - No}{No}$$

Ns: nitrogen in the aerial biomass in plants inoculated with the specific strain

No: nitrogen in the aerial biomass in the non-inoculated control

Nn: nitrogen in the aerial biomass in the nitrogen-supplemented, non-inoculated control

The final step is the optimization of Gb × E, and the general recommendation for the strains selection is a graduate testing from the controlled environment to the field environment, which usually includes a microcosm test before the field test (Sessitsch et al. 2002) (Fig. 13.1). However, in our experience (D. Mulas, C.A. Diaz-Alcántara, F. González-Andrés) with rhizobia based inoculants, the microcosm experiment usually gives biased results, because of detrimental environmental conditions generated in the pots that usually badly affects the inoculated treatment. The abnormally high temperatures reached in the pots could be the main reason why the inoculated treatment is clearly worse than the control with nitrogen, due to the pernicious effect of high temperatures in nodules formation and nitrogen fixation process as indicated above in this chapter. Table 13.1 presents the results of a microcosm experiment with autochthonous rhizobia strains isolated from the common bean in the Dominican Republic. The strains tested in microcosm, were firstly selected under axenic conditions (Díaz-Alcántara et al. 2010), then in microcosm, and finally in the field. In microcosm, significant differences between treatments were observed,

Table 13.1 Average values of the dependent variables aerial biomass and seeds yield in a microcosm trial of the common bean (*Phaseolus vulgaris* L.) cv. José Veta inoculated with autochthonous rhizobium strains isolated from mountainous agrosystems in the Dominican Republic. The experimental design was a randomized complete block, with 15 plants per treatment and three repetitions. The independent variables were the soil (three different soils) and the inoculation treatment. The data presented are the average for the three soils, in each inoculation treatment. The mean values followed by same letter did not differ significantly ($p<0.05$; Tukey test).

Inoculation treatment (strain or control)	Origin of the strain (province)	Aerial biomass ($g\ plant^{-1}$)		Seed yield ($g\ plant^{-1}$)	
ESC1029	Elías Piña	6.4	ab	3.6	a
ESC1120	Elías Piña	6.5	ab	3.7	a
VLP1091	La Vega	6.6	ab	3.9	a
VRI2090	La Vega	7.1	b	4.1	a
OAC1020	San José de Ocoa	7.0	b	4.2	ab
Control without N		5.8	a	3.5	a
Control with N		8.7	c	5.0	b
Mean squares		34.3		10.35	
p value		0.000		0.008	

and the strains VRI2090 and OAC1020 produced significantly higher aerial biomass than the non-fertilized control (Table 13.1), and the seed yield produced by OAC1020 did not differ from the control fertilized with N. However, the result obtained in microcosm with the inoculated treatment was clearly lower than that obtained in the nitrogen-fertilized uninoculated control, and this is in conflict with the results obtained in the field experiment with the same strains (C.A. Diaz-Alcántara and F. González-Andrés, unpublished), where the inoculated treatments in some environments exceeded the yield of the nitrogen fertilized control, and in other environments the yield was equivalent in the inoculated and in the fertilized treatments. Similarly, Mulas et al. (2011) demonstrated in field experiments that it is possible to replace nitrogen fertilization by inoculation with autochthonous strains in the common bean in León (Spain), even when the results obtained in the microcosm trials with the same strains were erratic (Mulas 2010).

On the basis of the erratic results of the microcosm trials, our recommendation is to test the strains with better performance under axenic conditions, directly in the field in small plots (Fig. 13.2C). For biofertilizers other than rhizobium for legumes, in our experience the detrimental conditions of the microcosm also affect the interaction between the plant and the bacterium, so the microcosm trials are not recommended. Either in the microcosm trials, if in spite of the evidences they are used, or in field trials, several environments must be tested, comprising the agroecological conditions of the farms where the biofertilizer will be used. In the case of the microcosm, the different environments will correspond to different soils, and in the case of the field, to different locations.

The last step will be large-scale field trials in several locations or environments. In this step, several formulations (carrier × additives) will be tested. This is the last step before the release of a successful inoculant to the market.

The Inoculants Based in Autochthonous Microorganisms as a Source for Local Development

The success of autochthonous microorganisms as inoculants for crops is an opportunity to create local companies that, providing they reach the minimum level of business, can be profitable and contribute to the local development.

Important efforts in research have been made, not only for the development of first-generation biofertilizers based in autochthonous bacteria, but also for the development of second-generation biofertilizers in several crops. For example, in the Dominican Republic, Marcano et al. (2011) isolated bacteria from surface-sterilized plant roots of organic

Figure 13.2 A. Nodules formed in the common bean's (*Phaseolus vulgaris* L.) roots, after inoculation under axenic conditions of seedlings with *Rhizobium leguminosarum* strains isolated from commercial common bean fields in León–Spain. B. *Phaseolus vulgaris* L. plants inoculated under axenic conditions and in the absence of mineral nitrogen, with different strains isolated from commercial common beans fields in León–Spain. The effectiveness of each strain is evidenced by the colour of the plant and its growth. C. Field pilot trials (small scale) to select, out of several strains pre-selected in axenic conditions, those with better adaptation to the environment. D. *Rhizobium leguminosarum* colonizing the surface of the rizosphere of barley (*Hordeum vulgare* L.). Bacteria have been transformed with a fluorescent GFP. E. Biocontrol assay of *Fusarium oxysporum* f. sp. phaseoli *in vitro* (in the centre of the Petri dish) with two different bacterial strains isolated from surface sterilized roots of the common bean and pepper (*Capsicun annuum* L.). F. *In vivo* assay of biocontrol of *Fusarium oxysporum* f. sp. phaseoli in the common bean. The four plants on the left side are the non-inoculated control, the four on the right side were inoculated with the pathogen, and the four in the middle were inoculated with the pathogen and a strain of *Pseudomonas brassicacearum* subsp. *neoaurantiaca* isolated from common bean.

Color image of this figure appears in the color plate section at the end of the book.

banana plantations of cv. Cavendish (*Musa acuminata* group AAA), in four areas of the country with high production of organic banana (Mao, Azua, Montecristi and Dajabón provinces). The isolated bacteria mostly belonged to the genera *Pseudomonas*, in the species *P. taiwanensis* and *P. mosselii*, but they also found some species of other genera like *Bacillus*, *Serratia*, *Citrobacter* and *Rhizobium*, although the last belongs to the plant pathogen *Rhizobium radiobacter*. Several isolates of *P. taiwanensis* and *P. mosselii* were assayed in plants. They used to inoculate banana plantlets produced *in vitro*, in order to improve the production of plants in the nursery, during the acclimatization and subsequent phases before the plant was transplanted in the field. Some strains produced a more vigorous and faster growth. Moreover, root inoculation with some of the strains at the time of the transplant to the field produced a better growth and a significant degree of systemic resistance against the attack of the fungi *Mycosphaerella fijiensis* Morelet (I. Marcano and F. González-Andrés, unpublished).

In another work, M. Barquero and F. González-Andrés (unpublished) characterized and tested in plants the PGP effect of bacteria isolated from surface-sterilized roots of the common bean and pepper (*Capsicun annuum* L.) in Castilla y León (Spain). From 198 isolates that were tested *in vitro* for PGP activities and biocontrol of *Fusarium oxysporum* f. sp. phaseoli and *Phytophthora capsici*, 40 isolates with relevant activities *in vitro* were identified by 16S rRNA gene sequencing, and belonged to 14 different species of the genera *Bacillus, Enterobacter, Rhizobium, Pseudomonas* and *Achromobacter*. Amongst them, the species *Pseudomonas corrugata* was discarded because it is a plant pathogen, and also *Bacillus cereus* because of its phylogenetic proximity to the pathogen *Bacillus anthracis*. Some strains which produced an increase of the roots' growth under axenic conditions, and a control of phytopatogenic fungi, are currently being tested in agricultural soils. For example, one strain of *Achromobacter marplatenses* and one strain of *Rhizobium massiliae* significantly promoted the growth of pepper roots, and one strain of *Pseudomonas brassicacearum* subsp. *neoaurantiaca* promoted the growth of common bean roots. Moreover, *Bacillus amyloliquefaciens* subsp. *plantarum* and *Pseudomonas brassicacearum* subsp. *neoaurantiaca* completely controlled the attack of *Fusarium oxysporum* f. sp. phaseoli in the common bean.

The inoculants based in bacteria or mycorrhiza with an effect of the plants' phytofortification can be commercialized as biofertilizers. More difficult is the registration of biocontrol agents, because of the regulations and testing procedures, that include costly and long-lasting field and laboratory tests.

The use of biofertilizers is especially important in developing countries, where the costs of fertilizers and phytosanitary products are very expensive. As a consequence, agricultural soils are usually under oligotrophic conditions, and in such situations the effect of biofertilizers is very significant. For example, in the case of legumes it is cheaper and usually more effective to ensure BNF, compared to the application of nitrogen fertilizers (Chianu et al. 2011). For this reason, biofertilizers are very common in countries with a less developed chemical industry.

The technological needs for the isolation and selection of autochthonous microorganisms to create biofertilizers, concerning economic aspects and technical complexity, are affordable in most of the cases for the scientific community in developing countries. In such conditions, it is possible for them to develop specific products optimized for particular environmental conditions, giving the opportunity to produce their own agricultural inputs.

Provided that there is continuous supply of electricity, the laboratory appliances are very simple. The equipment needed for sterilization of materials and culture media, incubators with temperature control, and an air-flow bench. General equipment for the analysis of soils and plant biomass is also necessary. The most difficult is to create axenic conditions for growing plants under hydroponic conditions, in order to select the strains with better interaction with a specific crop and even variety. The best is of course a growth chamber with automatic regulation of photoperiod and thermoperiod, but a clean and well controlled greenhouse can also be used. The next phase of the selection in real soil harbours less technological requirements.

With minimal equipment, a research group in the Dominican Republic has developed an effective inoculant for *Cajanus cajan* (L.) millsp., a pulse legume which is basic for the human consumption on the island. The inoculant was based in autochthonous bacteria isolated from inside root nodules grown in real farms. They belonged to an unknown lineage of *Bradyrhizobium* (C.A. Diaz-Alcántara, D. Mulas and F. González-Andrés, unpublished).

Once effective strains have been selected, the production process of inoculants is based in an aerobic fermentation, following the formulation process. However, maintaining axenic conditions during the fermentation and formulation processes is the most difficult in order to produce a quality inoculant and this is the main problem faced by inoculants produced in developing countries (D. Mulas, unpublished).

Acknowledgements

The original data presented in this chapter, corresponds to the results of research projects financially supported by *Junta de Castilla y León*, Spain (research projects: LE025A07 and LE029A10-2), *Agencia Española de Cooperación Internacional para el Desarrollo* (AECID) (research projects A/7473/07, A/4824/06, A/023132/09, A/030020/10 and A1/035364/11), and by the company Fertiberia S.A.

References

Abi-Ghanem, R., L. Carpenter-Boggs and J.L. Smith. 2011. Cultivar effects on nitrogen fixation in peas and lentils. Biology and Fertility of Soils. 47: 115–120.

Adesemoye, A.O. and J.W. Kloepper. 2009. Plant-microbes interactions in enhanced fertilizer-use efficiency. Applied Microbiology and Biotechnology. 85(1): 1–12.

Adesemoye, A.O., H.A. Torbert and J.W. Kloepper. 2008. Enhanced plant nutrient use efficiency with PGPR and AMF in an integrated nutrient management system. Canadian Journal of Microbiology. 54(10): 876–886.

Adesemoye, A.O., H.A. Torbert and J.W. Kloepper. 2009. Plant Growth Promoting Rhizobacteria Allow Reduced Application Rates of Chemical Fertilizers. Microbial Ecology. 58(4): 921–929.

Albareda, M., D.N. Rodríguez-Navarro, M. Camacho and F.J. Temprano. 2008. Alternatives to peat as a carrier for rhizobia inoculants: Solid and liquid formulations. Soil Biology and Biochemistry. 40: 2771–2779.

Andrews, C.S., A.D. Johnson and R.L. Sandland. 1973. Effect of aluminium on the growth and chemical composition of some tropical and temperate pasture legumes. Aust. J. Agric. Res. 24: 325–339.

Antoun, H., C.J. Beauchamp, N. Goussard, R. Chabot and R. Lalande. 1998. Potential of *Rhizobium* and *Bradyrhizobium* species as growth promoting rhizobacteria on non-legumes: effect on radishes (*Raphanus sativus* L.). Plant Soil. 204: 57–67.

Araujo, F.F., A.A. Henning and M. Hungria. 2005. Phytohormones and antibiotics produced by *Bacillus subtilis* and their effects on seed pathogenic fungi and on soybean root development. World J. Microbiol. Biotech. 21: 1639–1645.

Arrese-Igor, C. 2010. Biological Nitrogen Fixation. *In*: A. Gonzalez-Fontes, A. Gárate and I. Bonilla [eds.]. Agricultural Sciences Topics in Modern Agriculture. Studium Press LLC Houton. Texas. pp. 233–255.

Arshad, M., B. Shaharoona and T. Mahmood. 2008. Inoculation with *Pseudomonas* spp. containing ACC-deaminase partially eliminates the effects of drought stress on growth, yield, and ripening of pea (*Pisum sativum* L.). Pedosphere. 18(5): 611–620.

Arzanesh, M.H., H.A. Alikhani, K. Khavazi, H.A. Rahimian and M. Miransari. 2010. Wheat (*Triticum aestivum* L.) growth enhancement by *Azospirillum* spp. under drought stress. World J. Microbiol. Biotechnol. 27: 197–205.

Attar, H.A., D. Blavet, E.M. Selim, M.T. Abdelhamid and J.J. Drevon. 2012. Relationship between phosphorus status and nitrogen fixation by common beans (*Phaseolus vulgaris* L.) under drip irrigation. International Journal of Environmental Science and Technology. Is phosphorous efficiency low for symbiotic nitrogen fixation. 17th International Congress on Nitrogen Fixation. Frementale. Western Australia. 9(1): 1–13.

Barea, J.M., M.J. Pozo, R. Azcón and C. Azcón-Aguilar. 2005a. Microbial co-operation in the rhizosphere. Journal of Experimental Botany. 56(417): 1761–1778.

Barea, J.M., R. Azcón and C. Azcón-Aguilar. 2005b. Interactions between mycorrhizal fungi and bacteria to improve plant nutrient cycling and soil structure. *In*: F. Buscot, S. Varma [eds.]. Microorganisms in soils: roles in genesis and functions. Springer-Verlag. Heidelberg, Germany. pp. 195–212.

Barran, L.R. and E.S.P. Bromfield. 1997. Competition among rhizobia for nodulation of legumes. *In*: B.D. McKersie and D.C.W. Brown [eds.]. Biotechnology and the Improvement of Forage Legumes. University Press, Cambridge. pp. 343–374.

Bashan, Y. 2005. Plant Growth-Promoting. Encyclopedia of soils in the environment. *In:* D. Hillel [ed.]. Encyclopedia of soils in the environment. Elsevier, Oxford, UK. pp. 103–115.

Beck, D.P., L.A. Materon and F. Afandi. 1993. Practical Rhizobium-Legume technology manual. Technical Manual No. 19. International Center for Agricultural Research in the Dry Areas. Aleppo, Syria.

Ben Rebah, F., D. Prévost, A. Yezza and R.D. Tyagi. 2007. Agro-industrial waste materials and wastewater sludge for rhizobial inoculant production: a review. Bioresource technology. 98: 3535–3546.

Bloemberg, G.V. and B.J.J. Lugtenberg. 2001. Molecular basis of plant growth promotion and biocontrol by rhizobacteria. Curr. Opin. Plant. Biol. 4: 343–350.

Böhme, L. and F. Böhme. 2006. Soil microbiological and biochemical properties affected by plant growth and different long-term fertilization. Europ. J. Soil. Biol. 42: 1–12.

Bouwman, A.F., G. Drecht and K.W. Hoek. 2005. Surface N balance and reactive N loss to the environment from global intensive agricultural production systems for the period. 1970–2030. Science in China 48: 1–13.

Brady, D.J., C.H. Hecht-Buchholz, C.D. Asher and D.G. Edwards. 1990. Effects of low activities of aluminium on soybean (*Glycine max*). Early growth and nodulation. *In*: M.L. Beusichem [ed.]. Plant Nutrition—Physiology and Applications. Kluwer Academic Publishers. Dordrecht. pp. 329–334.

Bremer, E., R.J. Rennie and D.A. Rennie. 1988. Dinitrogen fixation of lentil, field pea and fababean under dryland conditions. Can. J. Soil. Sci. 68: 553–562.

Bremer, E., C. Kessel and R. Karamanos. 1989. Inoculant, phosphorus and nitrogen responses of lentil. Can. J. Plant. Sci. 69: 691–701.

Brockwell, J., R.R. Gault, D.L. Chase, F.W. Hely, M. Zorin and E.J. Corbin. 1980. An appraisal of practical alternatives to legume seed inoculation: field experiments on seed bed inoculation with solid and liquid inoculants. Australian Journal of Agricultural Research. 31: 47–60.

Brockwell, J., R.J. Roughley and D.F. Herridge. 1987. Population dynamics of *Rhizobium japonicum* strains used to inoculate three successive crops of soybean. Aust. J. Agric. Res. 38: 61–74.

Burns, R.C. and R.W. Hardy. 1975. Nitrogen fixation in bacteria and higher plants. Springer. New York.

Calvache, A.M. and K. Reichardt. 1996. Water deficit at different growth stages for common bean (*Phaseolus vulgaris* L.) cv. Imbabello and nitrogen use efficiency. Scientia Agricola. 53: 343–353.

Campo, R.J. 1995. Residual effects of aluminium on *Bradyrhizobium japonicum* in defined medium and soil solution. Ph.D. Thesis. The University of Reading. Reading.

Chabot, R., H. Antoun and M.P. Cescas. 1996. Growth promotion of maize and lettuce by phosphate-solubilizing *Rhizobium leguminosarum* biovar *phaseoli*. Plant Soil. 184: 311–321.

Chalk, P.M., B.J.R. Alves, R.N. Boddey and S. Urquiaga. 2010. Integrated effects of abiotic stresses on inoculant performance, legume growth and symbiotic dependence estimated by 15N dilution. Plant Soil. 328: 1–16.

Chianu, J.N., E.M. Nkonya, F.S. Mairura, J.N. Chianu and F.K. Akinnifesi. 2011. Biological nitrogen fixation and socioeconomic factors for legume production in sub-Saharan Africa. Agronomy For Sustainable Development. 31 (1): 139–154.

Contesto, C., G. Desbrosses, C. Lefoulon, G. Bena, F. Borel, M. Galland, L. Gamet, F. Varoquaux and B. Touraine. 2008. Effects of rhizobacterial ACC deaminase activity on Arabidopsis indicate that ethylene mediates local root responses to plant growth-promoting rhizobacteria. Plant Science. 175(1–2): 178–189.

Coventry, D.R. and J. Evans. 1989. Symbiotic nitrogen fixation and soil acidity. *In*: A.D. Robson [ed.]. Soil Acidity and Plant Growth. Academic Press, Sydney. pp. 103–137.

Daza, A., C. Santamaría, D.N. Rodríguez-Navarro, M. Camacho, R. Orive and F. Temprano. 2000. Perlite as carrier for bacterial inoculants. Soil Biology and Biochemistry. 32: 567–572.

De Felipe, M.R. 2006. Fijación de dinitrógeno atmosférico en vida libre. *In:* E.J. Bedmar, J. González, C. Lluch and B. Rodelas [eds.]. Fijación de Nitrógeno: Fundamentos y Aplicaciones. Sociedad Española de Fijación de Nitrógeno (SEFIN). Spain. pp. 9–16.

Deaker, R., R.J. Roughley and I.R. Kennedy. 2004. Legume seed inoculation technology a review. Soil Biology and Biochemistry. 36: 1275–1288.

Denardin, N.D. and J.R.J. Freire. 2000. Assessment of polymers for the formulation of legume inoculants. World Journal of Microbiology And Biotechnology. 16(1): 215–217.

Denton, M.D., D.J. Pearce, R.A. Ballard, M.C. Hannah, L.A. Mutch, S. Norng and J.F. Slattery. 2009. A multi-site field evaluation of granular inoculants for legume nodulation. Soil Biology and Biochemistry. 41 (12): 2508–2516.

Díaz-Alcántara, C.A. 2010. Aislamiento, caracterización y selección de rhizobia autóctonos que nodulan habichuela roja (*Phaseolus vulgaris* L.) en la Repúlbica Dominicana. PhD Thesis. Universidad de León. Spain.

Dobbelaere, S., J. Vanderleyden and Y. Okon. 2003. Plant Growth Promoting Effects of Diazotrophs in the Rhizosphere. Critical Reviews in Plant Sciences. 22 (2): 107–149.

Drevon, J.J., H. Attar, A. Bargaz, M. Faghire, C. Ghoulam, M. Lazali, M. Ounan and P. Rodiño. 2011. Is phosphorus efficiency low for symbiotic nitrogen fixation? 17th International Congress on Nitrogen Fixation. Fremantle, Western Australia.

Emmerling, C., M. Schloter, A. Hartmann and E. Kandeler. 2002. Functional diversity of soils organisms a review of recent research activities in Germany. J. Plant. Nutr. Soil Sci. 165: 408–420.

Fages, F. 1992. An industrial view of Azospirillum inoculants: formulation and application technology. Symbiosis. 13: 15–26.

Fallik, E. and Y. Okon. 1996. Inoculants of *Azospirillum brasilense*: biomass production, survival and growth promotion of *Setaria italica* and *Zea mays*. Soil Biology Biochemistry. 28: 123–126.

Fox, S.L., G.W. O'Hara and L. Bräu. 2011. Enhanced nodulation and symbiotic effectiveness of *Medicago truncatula* when co-inoculated with *Pseudomonas fluorescens*. WSM3457 and *Ensifer (Sinorhizobium) medicae* WSM419. Plant Soil. 348: 245–254.

Galloway, J.N., A.R. Townsend, J.W. Erisman, M. Bekunda, Z. Cai, J.R. Freney, L.A. Martinelli, S.P. Seitzinger and M.A. Sutton. 2008. Transformation of the nitrogen cycle: recent trends, questions, and potential solutions. Science. 320: 889–892.

García-Fraile, P., L. Carro, M. Robledo, M.H. Ramírez-Bahena, J.D. Flores-Félix, M.T. Fernández, P.F. Mateos, R. Rivas, J.M. Igual, E. Martínez-Molina, Á. Peix and E. Velázquez. 2012. *Rhizobium* promotes non-legumes growth and quality in several production steps: towards a biofertilization of edible raw vegetables healthy for humans. PLOS one. In press.

Giller, K.E. 2001. Nitrogen fixation in tropical cropping systems, 2nd edn. Cabi, Wallingford.

Glick, B.R., D.M. Karaturovíc and P.C. Newell. 1995. A novel procedure for rapid isolation of plant growth-promoting pseudomonads. Can. J. Microbiol. 41: 533–536.

Glick, B.R., B. Todorovic, J. Czarny, Z. Cheng, J. Duan and B. McConkey. 2007. Promotion of plant growth by bacterial ACC deaminase. Crit. Rev. Plant Sci. 26: 227–242.

Good, A.G. and P.H. Beatty. 2011. Fertilizing Nature: A Tragedy of Excess in the Commons. Plos Biology. 9(8): 9.

Govindasamy, V., M. Senthilkumar, K. Gaikwad and K. Annapurna. 2008. Isolation and characterization of ACC deaminase gene from two plant growth-promoting rhizobacteria. Current Microbiology. 57(4): 312–317.

Gutiérrez-Mañero, F.J. and B. Ramos-Solano. 2010. Bacteria and Agriculture Agricultural Sciences: Topics in Modern Agriculture. Studium Press LLC.

Gutiérrez-Zamora, M.L. and E. Martínez-Romero. 2001. Natural endophytic association between *Rhizobium etli* and maize (*Zea mays* L.). J. Biotechnol. 91: 117–126.

Haas, D. and C. Keel. 2003. Regulation of antibiotic production in root-colonizing *Peudomonas* spp. and relevance for biological control of plant disease, Annual Review of Phytopathology. 41 pp. 117–153.

Hafeez, F.Y., N.H. Shah and K.A. Malik. 2000. Field evaluation of lentil cultivars inoculated with *Rhizobium leguminosarum* bv. *viciae* strains for nitrogen fixation using nitrogen-15 isotope dilution. Biology and Fertility of Soils. 31: 65–69.

Hayat, R., S. Ali, U. Amara, R. Khalid and I. Ahmed. 2010. Soil beneficial bacteria and their role in plant growth promotion: a review. Annals of Microbiology. 60(4): 579–598.

Howieson, J.G. and R. Ballard. 2004. Optimising the legume symbiosis in stressful and competitive environments within southern Australia. Soil Biol. Biochem. 36: 1261–1273.

Howieson, J.G., G.W. O'Hara, and S.J. Carr. 2000. Changing roles for legumes in Mediterranean agriculture: Developments from an Australian perspective. Field Crops Res. 65: 107–122.

Hungria, M. and M.A.T. Vargas. 2000. Environmental factors affecting N_2 fixation in grain legumes in the tropics with an emphasis on Brazil. Field Crops Res. 65: 151–164.

Hungria, M., D.S. Andrade, L.M.O. Chueire, A. Probanza, F.J. Gutiérrez-Mañero and M. Megías. 2000. Isolation and characterization of new efficient and competitive bean (*Phaseolus vulgaris* L.) rhizobia from Brazil. Soil Biology and Biochemistry. 32: 1515–1528.

Hungria, M., R.J. Campo and I.C. Mendes. 2003. Benefits of inoculation of the common bean (*Phaseolus vulgaris*) crop with efficient and competitive *Rhizobium tropici* strains. Biology and Fertility of Soils. 39: 88–93.

Idriss, E.E., D.J. Iglesias, M. Talon and R. Borriss. 2007. Tryptophan-dependent production of indole-3-acetic acid (IAA) affects level of plant growth promotion by *Bacillus amyloliquefaciens* FZB42. MPMI. 20: 619–626.

Israel, D.W. 1981. Cultivar and Rhizobium strain effects on nitrogen fixation and remobilization by soybeans. Agron. J. 73: 509–516.

Jayasundara, H.P.S., B.D. Thomson and C. Tang. 1998. Responses of cool season grain legumes to soil abiotic stresses. Adv. Agron. 63: 77–151.

Jeffries, P., S. Gianinazzi, S. Perotto, K. Turnau and J.M. Barea. 2003. The contribution of arbuscular mycorrhizal fungi in sustainable maintenance of plant health and soil fertility. Biology and Fertility of Soil. 37: 1–16.

Khavazi, K., F. Rejali, P. Seguin and M. Miransari. 2007. Effects of carrier, sterilisation method, and incubation on survival of *Bradyrhizobium japonicum* in soybean (*Glycine max* L.) inoculants. Enzyme and Microbial Technology. 41: 780–784.

Kim, M.M., C.J. Asher, D.G. Edwards and R.A. Date. 1985. Aluminium toxicity: effects on growth and nodulation of subterranean clover. Proceedings of the 15th International Grasslands Congress. Kentucky. pp. 501–503.

Kirda, C., S.K.A. Danso and F. Zapata. 1989. Temporal water stress effects on nodulation, nitrogen accumulation and growth of soybean. Plant Soil. 120: 49–55.

Kloepper, J.W. and M.N. Schroth. 1978. Plant growth-promoting rhizobacteria on radishes. Proc. of the 4th Internat. Conf. on Plant Pathogenic Bacter. Station de Pathologie Vegetale et Phytobacteriologie. INRA, Angers, France. 2: 879–882.

Kloepper, J.W., C.M. Ryu and S. Zhang. 2004. Induced systemic resistance and promotion of plant growth by *Bacillus* spp. Phytopathol. 94: 1259–1266.

Lupwayi, N.Z., P.E. Olsen, E.S. Sande, H.H. Keyser, M.M., Collins, P.W. Singleton and W.A. Rice. 2000. Inoculant quality and its evaluation. Field Crops Research. 65: 259–270.

Marcano, I., C.A. Diaz Alcantara, B. Urbano, E. Velazquez and F. Gonzalez-Andres. 2011. *In vitro* plant growth promotion properties of endophitic bacteria isolated from banana (*Musa* sp.) roots in Dominican Republic. 17th International Congress on Nitrogen Fixation. Fremantle, Western Australia.

Melchiorre, M., M.J. de Luca, G. Gonzalez-Anta, P. Suarez, C. Lopez, R. Lascano and R.W. Racca. 2011. Evaluation of bradyrhizobia strains isolated from field-grown soybean plants in Argentina as improved inoculants. Biology and Fertility of Soils. 47: 81–89.

Mia, M.A.B. and Z.H. Shamsuddin. 2010. Rhizobium as a crop enhancer and biofertilizer for increased cereal production. African J. of Biotechnology. 9(37): 6001–6009.

Miransari, M. 2011. Soil microbes and plant fertilization. Appl. Microbiol. Biotechnol. 92: 875–885.

Mishra, R.P., R.K. Singh, H.K. Jaiswal, V. Kumar and S. Maurya. 2006. *Rhizobium*-mediated induction of phenolics and plant growth promotion in rice (*Oryza sativa* L.). Curr. Microbiol. 52: 383–389.

Montanez, A., S.K.A. Danso and G. Hardarson. 1995. The effect of temperature on nodulation and nitrogen fixation by five *Bradyrhizobium japonicum* strains. Appl. Soil. Ecol. 2: 165–174.

Mostasso, L., F.L. Mostasso, B.G. Dias, M.A.T. Vargas and M. Hungría. 2002. Selection of bean (*Phaseolus vulgaris* L.) rhizobial strains for the Brazilian Cerrados. Field Crops Research. 73: 121–132.

Mrabet, M., R. Mhamdi, F. Tajini, R. Tiwari, M.Trabelsi and M.E. Aouani. 2005. Competitiveness and symbiotic effectiveness of *R. gallicum* strain isolated from root nodule of *Phaseolus vulgaris*. European Journal of Agronomy. 22: 209–216.

Mulas, D. 2010. Caracterización y selección de cepas autóctonas de Rhizobium para la biofertilización de la alubia (*Phaseolus vulgaris* L.) en la Indicación Geográfica Protegida (IGP) "Alubia de La Bañeza-León". PhD Thesis. Universidad de León. Spain.

Mulas, D., P. García-Fraile, L. Carro, M.H. Ramírez-Bahena, P. Casquero, E. Velázquez and F. González-Andrés. 2011. Distribution and efficiency of *Rhizobium leguminosarum* strains nodulating *Phaseolus vulgaris* in Northern Spanish soils: Selection of native strains that replace conventional N fertilization, Soil Biology and Biochemistry. 43(11): 2283–2293.

Nadeem, S.M., Z.A. Zahir, M. Naveed and M. Arshad. 2007. Preliminary investigations on inducing salt tolerance in maize through inoculation with rhizobacteria containing ACC deaminase activity. Canadian Journal of Microbiology. 53(10): 1141–1149.

Neilands, J.B. 1981. Microbial iron compounds. Ann. Rev. Biochem. 50: 715–731

Neilands, J.B. 1982. Microbial envelope proteins related to iron. Ann. Rev. Biochem. 36: 285–309.

Noel, T.C., C. Sheng, C.K. Yost, R.P. Pharis and M.F. Hynes. 1996. *Rhizobium leguminosarum* as a plant growth-promoting rhizobacterium: direct growth promotion of canola and lettuce. Can. J. Microbiol. 42: 279–283.

Paungfoo-Lonhienne, C, D. Rentsch, S. Robatzek, R.I. Webb, E. Sagulenko, T. Näsholm, S. Schmidt and T.G. Lonhienne. 2010. Turning the Table: Plants Consume Microbes as a Source of Nutrients. PLoS ONE 5(7): e11915.

Peix, A., A.A. Rivas-Boyero, P.F. Mateos, C. Rodríguez-Barrueco, E. Martínez-Molina and E. Velázquez. 2001. Growth promotion of chickpea and barley by a phosphate solubilizing strain of *Mesorhizobium mediterraneum* under growth chamber conditions. Soil Biol. Biochem. 33: 103–110.

Pena-Cabriales, J.J. and J.Z. Castellanos. 1993. Effects of water stress on N_2 fixation and grain yield of *Phaseolus vulgaris* L. Plant Soil. 152: 151–155.

Perret, X., C. Staehelin and H.P. Spaink. 2000. Molecular basis of symbiotic promiscuity. Microbiol Mol. Biol. 64: 180–201.

Persello Cartieaux, F., L. Nussaume and C. Robaglia. 2003. Tales from the underground: molecular plant–rhizobacteria interactions. Plant. Cell. Environ. 26: 189–199.

Rao, D.L.N., K.E. Giller, A.R. Yeo and T.J. Flowers. 2002. The effects of salinity and sodicity upon nodulation and nitrogen fixation in chickpea (*Cicer arietinum*). Ann. Bot. 89: 563–570.

Rebah, F., R.D. Tyagi and D. Prévost. 2002. Wastewater sludge as a substrate for growth and carrier for rhizobia: the effect of storage conditions on survival of *Sinorhizobium meliloti*. Bioresource technology. 83: 145–51.

Rice, W.A. and P.E. Olsen. 1992. Effects of inoculation method and size of *Rhizobium meliloti* population in the soil on nodulation of alfalfa. Canadian Journal of Soil Science. 72: 57–67.

Rigaud, J. and A. Puppo. 1975. Indole-3-Acetic Acid cataboslism by soybean bacterioids. Journal of General Microbiology. 88: 223–228.

Rodríguez-Blanco. A., M. Sicardi and L. Frioni. 2010. Competition for nodule occupancy between introduced and native strains of *Rhizobium leguminosarum* biovar *trifolii*. Biology and Fertility of Soils. 46(4): 419–425.

Rodríguez-Navarro, D.N., A.M. Buendía, M. Camacho, M. Lucas and C. Santamaría. 2000. Characterization of *Rhizobium* spp. bean isolates from southwest of Spain. Soil Biology and Biochemistry. 32: 1601–1613.

Rodríguez-Navarro, D.N. and J.E.C. Ruíz-Sánchez. 2010. Biofertilisers. *In:* A. Gonzalez-Fontes, A. Gárate and I. Bonilla [eds.]. Agricultual Sciences Topics in Modern Agriculture. Studium Press LLC Houton. Texas. pp. 79–102.

Rose, M.T., R. Deaker, S. Potard, T. Cuc Kim Thi, V. Nga Thuy and I.R. Kennedy. 2011. The survival of plant growth promoting microorganisms in peat inoculant as measured by selective plate counting and enzyme-linked immunoassay. World Journal of Microbiology and Biotechnology. 27(7): 1649–1659.

Rosenblueth, M. and E. Martínez-Romero. 2006. Bacterial endophytes and their interactions with hosts. Molecular Plant-Microbe Interactions. 19(8): 827–837.

Roughley, R. and J. Vincent. 1967. Growth and survival of *Rhizobium* spp. in peat culture. Journal of Applied Bacteriology. 30: 362–376.

Roughley, R. 1968. Some factors influencing the growth and survival of root nodule bacteria in peat culture. Journal of Applied Bacteriology. 31: 259–265.

Roy, N., P. Bhattacharyya and P.K. Chakrabartty. 1994. Iron acquisition during growth in an iron-deficient medium by *Rhizobium* spp. isolated from *Cicer arietinum*. Microbiology UK. 140: 2811–2820.

Sadowsky, M.J. and P.H. Graham. 1998. Soil biology of the Rhizobiaceae. *In*: H.P. Spaink, A. Kondorosi and P.J.J. Hooykaas [eds.]. Molecular biology of model plant-associated bacteria. Kluwer Academic Publishers. Dordrecht. pp. 155–172.

Sessitsch, A., J.G. Howieson, X. Pret, H. Antoun and E. Martínez-Romero. 2002. Advances in Rhizobium research. Critical Reviews in Plant Sciences. 21(4): 323–378.

Sheng, X.F. and L.Y. He. 2006. Solubilization of potassium-bearing minerals by a wild-type strain of *Bacillus edaphicus* and its mutants and increased potassium uptake by wheat. Can. J. Microbiol. 52: 66–72.

Siddikee, M.A., B.R. Glick, P.S. Chauhan, W.J. Yim and T. Sa. 2011. Enhancement of growth and salt tolerance of red pepper seedlings (*Capsicum annuum* L.) by regulating stress ethylene synthesis with halotolerant bacteria containing 1-aminocyclopropane-1-carboxylic acid deaminase activity. Plant Physiology And Biochemistry. 49(4): 427–434.

Singh, J.S., V.C. Pandey and D.P. Singh. 2011. Efficient soil microorganisms: A new dimension for sustainable agriculture and environmental development. Agriculture, Ecosystems and Environment. 140(3-4): 339–353.

Smith, R.S. 1992. Legume inoculant formulation and application. Canadian Journal of Microbiology. 38(6): 485–492.

Smyth, E.M., J. McCarthy, R. Nevin, M.R. Khan, J.M. Dow, F. O'Gara and F.M. Doohan. 2011. *In vitro* analyses are not reliable predictors of the plant growth promotion capability of bacteria; a *Pseudomonas fluorescens* strain that promotes the growth and yield of wheat. J. Appl. Microbiol. 111(3): 683–92.

Spaepen, S., S. Dobbelaere, A. Croonenborghs and J. Vanderleyden. 2008. Effects of *Azospirillum brasilense* indole-3-acetic acid production on inoculated wheat plants. Plant and Soil. 312(1-2): 15–23.

Sprent, J.I. 2001. Nodulation in Legumes. Royal Botanical Gardens. Kew. UK.

Stephens, J.H.G. and H.M. Rask. 2000. Inoculant production and formulation. Field Crops Research. 65: 249–258.

Thies, J.E., P.W. Singleton and B.B. Bohlool. 1991. Modeling symbiotic performance of introduced rhizobia in the field by use of indices of indigenous population size and nitrogen status of the soil. Applied and Environmental Microbiology. 57: 29–37.

Thuita, M., P. Pypers, L. Herrmann, R.J. Okalebo, C. Othieno, E. Muema and D. Lesueur. 2012. Commercial rhizobial inoculants significantly enhance growth and nitrogen fixation of a promiscuous soybean variety in Kenyan soils. Biology and Fertility of soils. 48: 87–96.

Tittabutr, P., P. Payakapong, W. Teaumroong, P.W. Singleton and N. Boonkerd. 2007. Growth, survival and field performance of bradyrhizobial liquid inoculant formulations with polymeric additives. ScienceAsia. 33: 69–77.

Unkovich, M.J. and J.S. Pate. 2000. An appraisal of recent field measurements of symbiotic N_2 fixation by annual legumes. Field Crops Research. 65(2-3): 211–228.

Valverde, G. and E. Otabbong. 1997. Evaluation of N_2-fixation measured by the 15N-dilution and N-difference methods in Nicaraguan and Ecuadorian *Phaseolus vulgaris* L. plants inoculated with *Rhizobium leguminoserum*. Acta Agric. Scand. B. 47: 71–80.

Van Loon, L.C. 2007. Plant responses to plant growth-promoting rhizobacteria. European Journal of Plant Pathology. 119(3): 243–254.

Velázquez, E., P. García-Fraile, M.H. Ramírez-Bahena, A. Peix and R. Rivas. 2011. Proteobacteria forming nitrogen fixing symbiosis with higher plants. *In*: M.L. Sezenna [ed.] Proteobacteria: Phylogeny, Metabolic Diversity and Ecological Effects 1st edn. Nova Science Publishers Inc. New York, U.S.A. pp. 37–56.

Vessey, J.K. 2003. Plant growth promoting rhizobacteria as biofertilizers. Plant Soil. 255: 571–586.

Wilkins, J. 1967. The effects of high temperature on certain root-nodule bacteria. Australian Journal of Agricultural Research. 18: 299–304.

Willems, A. 2006. The taxonomy of rhizobia: an overview. Plant Soil. 287: 3–14.

Wynne, J.C., G.H. Elkan, C.M. Meisner, T.J. Scheewis and J.M. Ligon. 1980. Greenhouse evaluation of strains of Rhizobium for peanuts. Agron. J. 72: 645–649.

Yanni, Y., R. Rizk, F. Abd-El Fattah, A. Squartini, V. Corich, F. de Bruijn, J. Rademaker, J. Maya-Flores, P. Ostrom, M. Vega-Hernández, R. Hollingsworth, E. Martínez-Molina, P. Mateos, E. Velázquez, J. Wopereis, E. Triplett, U. Umali-García, B. Rolfe, J.K. Ladha, J. Hilland and F.B. Dazzo. 2001. The beneficial plant growth-promoting association of *Rhizobium leguminosarum* bv. *trifolii* with rice roots. Australian Journal of Plant Physiology. 28: 845–870.

Yates, R.J., J.G. Howieson, W.G. Reeve and G.W. O'Hara. 2011. A re-appraisal of the biology and terminology describing rhizobial strain success in nodule occupancy of legumes in agriculture. Plant Soil. 348: 255–267.

14

Engineering the Rhizosphere of Legumes for Improved Bioremediation

E. Pajuelo,[1] *I.D. Rodríguez-Llorente,*[1] *A. Lafuente,*[1] *J. Delgadillo,*[1,3] *P. Pérez-Palacios,*[1] *B. Doukkali,*[1] *M.A. Chamber*[2] and *M.A. Caviedes*[1,*]

Introduction

Over the last few decades, environmental pollution and resources depletion threaten sustainable social and economic development. Especially in developing countries, more attention has been paid to fast economic growth and people's well-being, than to the environmental consequences of a rapid, and many times not well planned, economic development (Prasad 2011). As a result, large amounts of toxic wastes have been dispersed all over the world, the consequence of the increase in world population,

[1]Departamento de Microbiología y Parasitología, Facultad de Farmacia, Universidad de Sevilla, C/Profesor García González, 2, 41012-Sevilla.
[2]Centro IFAPA Las Torres-Tomejil. Junta de Andalucía. Carretera Sevilla-Cazalla, Km. 12,2. Apartado Oficial, 41200-Alcalá del Río, Sevilla.
[3]Present address: Área de Microbiología. Colegio de Post-graduados. Campus de Montecillo. Carretera Federal México-Texcoco, Montecillo, México C.P. 56230, Km. 36.5
*Corresponding author: caviedes@us.es

industrialization and urbanization, unsafe residues' disposal, agricultural practices, oil refineries, mining and smelting, etc. Global pollution has harmful consequences on natural resources, biodiversity and human health, contributing to the incidence of cancer, degenerative diseases and endocrine disorders (USEPA, Environmental Protection Agency, http://www.usepa.gov; World Health Organization, http://www.who.int).

Restrictive environmental regulations are being approved in order to prevent and control pollution, and caring for human health and environmental protection (USEPA, http://www.epa.gov). Current physico-chemical technologies for remediation of contaminated sites include *in situ* technologies (including solidification, stabilization, soil vapour extraction, chemical treatment, etc.), and *ex situ* technologies (such as soil washing, solvent extraction, incineration, landfill, etc.), depending on the nature and extension of the pollution.

The development of innovative and cost-effective solutions to decontaminate polluted environments has become one of the main focuses of environmental research. Bioremediation has emerged in the past two decades as an effective innovative technology for the treatment of a wide variety of contaminants (Kumar et al. 2011, Prasad 2011). The principle of bioremediation lies in utilizing the capacities of biological diversity for the complete elimination of pollutants, and when not possible, their immobilization or the reduction of the toxicity. Living organisms have developed a number of strategies to cope with environmental pollutants, both organic and inorganic compounds (Díaz 2004, Silver and Phung 2005, Shröeder and Collins 2011). In particular, bacteria have developed strategies for obtaining energy from virtually every organic compound, both in aerobic or anaerobic conditions. Horizontal transfer of genes for xenobiotics degradation among soil microbial populations is facilitated by the fact that these genes are usually found in mobile genetic elements, such as transposons and plasmids, and allows the rapid adaptation of microorganisms to new pollutants (Díaz 2004).

Plants have also developed mechanisms to cope with environmental contaminants, both organics and inorganics. The use of plants in bioremediation is known as phytoremediation.

Phytoremediation

Phytoremediation is a solar-driven process that offers an environmentally friendly, aesthetic and low-cost option for environmental restoration, compared to most conventional cleanup techniques (Pilon-Smits 2005, Vangronsveld et al. 2009). Phytoremediation includes several different techniques, any one of them is most appropriate to a determinate kind of pollutants.

Metal Phytoextraction

Phytoextraction is a process useful for the removal of metals or metalloids, that takes advantage of the unusual ability of some plants to (hyper-) accumulate these elements in their shoots. At the time of disposal, metals are concentrated in the much smaller volume of the plant matter than in the initially contaminated soil or sediment, and the residue can be disposed off adequately. A relatively small group of hyperaccumulator plants is capable of tolerating very high concentrations of one (or several) metals and sequestering them in their shoot tissues at high concentrations (Baker 1981). In recent years, major scientific progress has been made in understanding the physiological mechanisms of metal uptake and transport in these plants (Verbruggen et al. 2009, Krämer 2010). The major processes involved in hyperaccumulation of trace metals include: (a) enhanced uptake of metals towards the root symplast by plasma membrane proteins, (b) chelation of metals in the cytoplasm with various ligands, (c) limited transport of metal-chelates into the vacuoles of the roots, (d) enhanced upload of metal-chelates to the xylem by elevated expression of xylem-loading transporters in the roots, and (e) sequestration of metals into epidermal vacuoles of the leaves by tonoplast-located transporters (Krämer 2010, Sarma 2011). Evidence is clear against a higher amount of phytochelatins in hyperaccumulators. By contrast, metals are mostly chelated to weaker O-ligands, mainly like organic acids, instead of stronger S-ligands in hyperaccumulators (Küpper et al. 2004, Callahan et al. 2006). Histidine and nicotinamide are important chelators in Ni hyperaccumulators (Krämer et al. 1996, Mari et al. 2006). The control of oxidative stress is also a crucial mechanism for metal hypertolerance and hyperaccumulation (Srivastava et al. 2005, Krämer 2010). Up to date, over 500 hyperaccumulator plants (about 0.2 percent of all angiosperms) are described (Baker 1981, Krämer 2010, Sarma 2011). Only a few legumes, belonging to the genus *Astragalus*, are hyperaccumulators of Se, including *A. sinicus, A. bisulcatus* and *A. racemosus*, which are able to accumulate between 5,600 ppm and 10,000 ppm Se in the shoots (Freeman et al. 2006, Pilon-Smits and Quinn 2010). Furthermore, *Lupinus albus* has been reported to accumulate over 3,000 ppm Zn in polluted acidic soils (Pastor et al. 2003).

Metal Phytostabilization

Metal phytostabilization is a different strategy that pursues immobilization of metals in the soil, preventing dispersion and entrance in the food chain (Mendez and Maier 2008). Phytostabilization appears to be a convenient and less expensive method that can immediately reduce the negative impacts of metal pollution. It is also used in large polluted areas or inaccessible sites where other remediation procedures are impossible. This strategy uses metal excluders, i.e., plants with high metal tolerance, and in which the resistance mechanism consists in decreasing metal availability in the soil and restricting metal movement into the shoot (Baker 1981). Legumes accumulate metals mainly in roots (Pajuelo et al. 2007), so they are useful in phytostabilization. Usually, metal resistant plant ecotypes can be collected at polluted sites, for instance mine sites, whereas ecotypes from non-polluted sites may be non-tolerant (Mahieu et al. 2011). An ideal plant for phytostabilization must tolerate metals and accumulate them mainly in roots, displaying very low translocation factors (TFs). Since the technique does not pursue pollutants removal, but stabilization, periodic monitoring is required. Many plants have been proposed for metal immobilization, including trees, shrubs or herbs, like metal-tolerant Poaceae, Compositae, Boraginaceae, Plantaginaceae and Cyperaceae (Mendez and Maier 2008). Metal-tolerant legume plants can be also introduced, for instance *Lupinus* sp. (Vázquez et al. 2006, Dary et al. 2010), *Cytisus* sp. (Moreno-Jiménez et al. 2009), *Retama spherocarpa* (Moreno-Jiménez et al. 2012), *Sesbania rostrata* (Yang et al. 1997), *Anthyllis vulneraria* (Mahieu et al. 2011) or even trees such as *Acacia* sp. (Maldonado-Magaña et al. 2011). These plants have an important role in increasing the nitrogen content in soils, due to their symbiotic interaction with rhizobia (Graham and Vance 2003, Olroyd and Downie 2008). The nodulation process is extremely sensitive to even low concentrations of pollutants (Castro et al. 1997, Broos et al. 2004, Kopittke et al. 2007, Wani et al. 2007a, Pajuelo et al. 2008, Lafuente et al. 2010). Nevertheless, nodulation in polluted soils has been observed when tolerant *Rhizobium* strains are present (Wani et al. 2007b, 2008, Dary et al. 2010, Mahieu et al. 2011). Primary colonizers can favour the growth of other new plants, thus increasing the biodiversity and the regeneration of polluted soils, together with the immobilization of metals due to the vegetal cover (Lei and Duan 2008).

Metal Phytovolatilization

This technique is useful when metals are previously converted into volatile forms. Legumes of the genus *Astragalus*, like *A. sinicus*, *A. bisulcatus*, *A. racemosus*, can accumulate and volatilize Se (Freeman et al. 2006, Pilon-Smits and Quinn 2010). Organics, in particular, volatile organic compounds (VOCs) like TCE (tetrachloroethylene) can be also volatilized (Doty et al. 2003).

Phytodegradation of Organic Compounds

Plants have the inherent ability to detoxify xenobiotics. Since plants are photosynthetic organisms, compared to microorganisms, they generally lack the complete catabolic pathway for the entire degradation of these compounds. In higher plants, xenobiotics can be partially transformed through several biochemical processes (Sanderman 1994, Gerhardt et al. 2009), which can be grouped into three main phases. i) phase I or conversion, which is an activation step that creates a reactive group in the compound, with the participation of enzymes such as cytochrome P450 monooxygenases. P450 monooxygenases is a large family of enzymes that catalyze most of the oxidation steps in plant secondary metabolism, and at the same time, act on xenobiotics through N-, O-, and S-dealkylation, aromatic and aliphatic hydroxylation, epoxidation, peroxidation, oxidative desulphuration, sulphoxidation or reduction (Siminszky 2006), ii) phase II or conjugation, is a step in which the activated xenobiotic is conjugated to pre-existing plant ligands such as glutathione, glucose, malonate, aminoacids, etc. The enzymes involved in these reactions are glutathione-S-transferases (Brazier-Hicks et al. 2008), glucosyl C- and O-transferases, (Gandía-Herrero et al. 2008), manolyl-transferases, etc., and iii) phase III or compartmentalization/sequestration, in this step, the complex xenobiotic-ligand is transported to vacuoles (soluble complex) or to cell walls (insoluble complex) with the participation of specific membrane transporters.

Plants can metabolize a wide range of xenobiotics, including halogenated compounds, chloroaromatics, PCBs, PHAs, pesticides, etc. For example, *Glycine max* is able to detoxify the explosive TNT through a reduction of the nitro group catalyzed by a nitroreductase (Adamia et al. 2006). Also alfalfa, in synergy with rhizosphere microorganisms, can detoxify the toxic gasoline additive MTEB (Zhang et al. 2000). In some cases, the incomplete degradation of xenobiotics creates toxic intermediates, which could be even more toxic than the original compounds. These intermediates can accumulate in plant tissues, so there are also concerns of the potential for introducing contaminants into the food chain. The question of how to dispose plants that accumulate xenobiotics is also a serious concern, so that the feasibility of phytoremediation as an approach to remediate environmental contamination still needs to be answered (Gerhardt et al. 2009). For instance, leaching of contaminants has been reported due to loss of leaves in poplars trees after remediation of explosives (Yoon et al. 2006). In other cases, some xenobiotics are lethal to legumes, therefore is impossible to use these plants in phytoremediation. For instance, legumes were shown to be more sensitive to petroleum hydrocarbons, PCBs and TNT than grasses in several pots studies (Dzantor et al. 2000).

Phytoremediation Using Trees

When using phytoremediation, it is important to select not only a plant that is capable of degrading the pollutant, but also one that will grow well in a contaminated environment. Ideal plants for phytoremediation should develop a deep and big root system, allowing greater contact with the pollutant, and hence, an improved phytoremediation. In this sense, trees can develop root systems to a great depth, so the idea is to move as much water through them as possible so that they take up as much of the contaminants as possible. With water, trees take in pollutants from the soils. Once the heavy metals are absorbed, they are sequestered in the plants' leaves and/or roots. Any organic compounds that are absorbed are metabolized. Several trees have been used for phytoremediation, including willows, poplars, *Eucalyptus* sp., etc. (Capuana 2011). Among them, some legume trees have been used, particularly in arid or semi-arid regions, including *Leucaena leucocephala* (Doty et al. 2003), *Mimosa pudica* (Chen et al. 2008), *Acacia farnesiana* (Maldonado-Magaña et al. 2011), etc. These trees can phytoremediate metals like Pb (Maldonado-Magaña et al. 2011) and volatile organics like TCE, EDB or MTEB (Doty et al. 2003, Chen et al. 2008). An added value to the phytoremediation is the ultimate utilization of its by-products (Ghosh and Singh 2005). Trees from phytoremediation processes can give additional benefits such as (i) generation of plant biomass for paper or wood production, (ii) as an energy source for industry, and (iii) use of soils for carbon farming by increasing carbon sequestration.

A summary of legumes proposed in phytoremediation is given in Table 14.1, where the different remediation mechanisms are indicated.

Rhizoremediation

Rhizoremediation refers to the removal of pollutants from contaminated sites by mutual interaction of plant roots and rhizosphere microorganisms, and it is considered as the most evolved process of bioremediation (Glick 2003, Kuiper et al. 2004, Gerhardt et al. 2009, Wenzel 2009). Rhizoremediation is an attractive process, since the plant root offers a large surface for microbial colonization and allows the microorganisms to move deeper into soil layers, and hence, increases the contact of detoxifying microorganisms and soil pollutants (Chaudhry et al. 2005, Zhuang et al. 2007, Wenzel 2009).

In the rhizoremediation process, every partner has a role. Plants release a complex array of organic compounds in the rhizosphere (aminoacids, sugars, organic acids, phenolics and secondary metabolites, and high molecular mass biomolecules, like cellulose, lignin, mucilage and proteins) (Bais et al. 2006, Shukla et al. 2011), that serve as carbon (C) sources for rhizosphere microbial communities, including fungi and bacteria. Up to 20 percent of C fixed by a plant may be released from its roots (Hinsinger et al. 2005). As a consequence, the rhizosphere shows a higher microbial density (10^2 to 10^4 fold) compared to the neighbouring bulk soil. Root exudates also affect the availability of pollutants in the rhizosphere by production of biosurfactants

Table 14.1 Legumes proposed in phytoremediation of metals and xenobiotics.

Toxic substance	Legume plant	Phytoremediation process	Reference
Metals			
Multi-metal polluted soils (As, Cd, Cu, Pb, Zn)	*Lupinus luteus* (inoculated)	Phytostabilization of metals in mine soils	Dary et al. 2010, El Aafi et al. 2012
As, Cd	*Lupinus albus*	Phytostabilization of metals	Vázquez et al. 2006
Ni, Zn	*Vigna unguiculata* (nodulated)	Stimulation of growth, nodulation, phytostabilization	Wani et al. 2007b
Cr	*Cicer arietinum* (nodulated)	Cr Phytostabilization	Wani et al. 2008
Pb, Zn and Cd	*Anthyllis vulneraria ecotype metallicolous*	Nodulation by metal-resistant *Mesohizobium metallidurans*, establishment of a plant cover in mine sites, phytostabilization	Mahieu et al. 2011
Mine soils	*Cytisus striatus* *C. oromediterraneus*	Metal phytostabilization of mine soils	Moreno-Jiménez et al. 2009
Mine soil supplemented with pyrite sludge	*Retama sphaerocarpa*	Metal phytostabilization	Moreno-Jiménez et al. 2012
Pb	*Acacia farnesiana* (nodulated)	Metal phytostabilization	Maldonado-Magaña et al. 2011
Se	*Astragalus sinicus*, *A. bisulcatus*, *A. racemosus*	Se Hyperaccumulation, phytovolatilization of Se in the forms of MSe and DMSe	Freeman et al. 2006
Organics			
TCE (trichloro-ethylene)	*Leucaena leucocephala*	Uptake and degradation "in planta", detection of dichloroethanol	Doty et al. 2003
EDB (ethylene-dibromide)	*Leucaena leucocephala*	Uptake and degradation "in planta", detection of bromide ion	Doty et al. 2003
MTEB (methyl-terc-butyl-eter)	*Medicago sativa* (alfalfa)	Phytotransformation and phytovolatilization. Detection of a intermediate in plants (MTA)	Zhang et al. 2000
2,4-D (herbicide)	*Trifolium pratense*	Rhizodegradation	Shaw and Burns 2004
2,4-D	*Pissum sativum*	Degradation by Endophytes (*Pseudomonas putida*)	Germaine et al. 2006
TNT (explosive)	*Glycine max*	Rapid uptake, phytotransformation: nitro-group reduction by nitroreductase	Adamia et al. 2006
Hexachlorocyclohexane	*Cytisus striatus*	Degradation by rhizosphere microorganisms	Kidd et al. 2008

(Read et al. 2003, Gao et al. 2010). Rhizosphere microorganisms play a crucial role on the adaptation of plants to environmental stresses, like the presence of pollutants (Burd et al. 2000, Khan 2005, Khan et al. 2009). The effect of rhizosphere microorganisms can be envisioned from a dual perspective: on the one hand, rhizosphere microorganisms display a range of plant growth promoting (PGP) activities, for example, secretion of siderophores, phosphate solubilization, nitrogen-fixation, plant defence against pathogens (biocontrol) (Han et al. 2005), and stimulation of plant growth by secreting phytohormones, like auxins, by inhibiting ethylene accumulation via the expression of aminocyclopropane deaminase activity, etc. (Bernard 2005, Avis et al. 2008). The plant growth promoting (PGP) activity of rhizosphere microorganisms alleviates the toxicity caused by pollutants, increases root length and biomass, enhances plant yield and biomass, improves soil quality, increases the content of organic matter and the content of nitrogen in soils (Khan 2005, Wani et al. 2008, Khan et al. 2009, Dary et al. 2010). On the other hand, in polluted environments, rhizosphere populations affect the mobility and availability of pollutants in the surroundings of plant roots. For instance, metal mobility and availability can be increased by the presence of siderophores, organic acids, aminoacids, etc., secreted by rhizosphere microorganisms, being the metal more available for plant uptake (Whiting et al. 2001, Chen et al. 2005, Zaidi et al. 2006, Sheng and Xia 2006). In other cases, the activity of rhizosphere microorganisms can immobilize metals in the rhizosphere, contributing to metal phytostabilization (Dary et al. 2010, El Aafi et al. 2012). Concerning organic contaminants, microbial communities foster the degradation of xenobiotics in the rhizosphere of legumes, for instance PHA (Liste and Prutz 2006), herbicides (Shaw and Burns 2004) and chloronitroaromatics (Liu et al. 2007). Enhanced biodegradation of MTBE was observed in the rhizosphere of alfalfa plants compared to unplanted controls (Zhang et al. 2000). In other experiment, degradation of hexachlorocyclohexane was observed by rhizosphere microorganisms of the legume of *Cytisus striatus* (Kidd et al. 2008). Another effect is the increase of pollutants' availability by production of biosurfactants, like rhamnolipids (Bento et al. 2005). Legumes are plants able to establish interactions with a high number of bacteria and mycorrhiza. Besides rhizobia, which are the most studied symbionts of legumes, many other alpha-proteobacteria (including *Ochrobactrum cytisi* or *Methylobacterium*) and beta-proteobacteria (like *Burkholderia, Ralstonia* and *Cupriavidus*) (revised in Rivas et al. 2009) can nodulate legumes. Some examples of rhizoremediation of metals and organics using legumes are revised in Table 14.2. For example, with metals, inoculation of legumes with metal-resistant rhizobia alleviates phytotoxicity at the time that decreases metal content in plant shoots, helping phytostabilization (Reichman 2007, Wani et al. 2008, Dary et al. 2010, El Aafi et al. 2012). There are also several

Table 14.2 Examples of Legume-microorganisms symbioses proposed in rhizoremediation.

Plant	Inoculant	Toxic compound	Rhizoremediation effect	Reference
	Rhizobia			
Glycine max	*Bradyrhizobium* sp.	As	38% increased plant biomass, no effect on As accumulation	Reichman 2007
Lupinus luteus	*Bradyrhizobium* sp. + *Ochrobactrum cytisi* + *Pseudomonas fluorescens/*	As, Cd, Cu, Zn, Pb	Alleviation of phytotoxicity, Plant growth promotion, nodulation and phytostabilization	Dary et al. 2010
Cicer arietinum	*Mesorhizobium* sp.	Cr	Diminution of toxicity, enhanced plant growth, nodulation and phytostabilization	Wani et al. 2008
Mimosa pudica	Metal-resistant *Cupriavidus tawanensis*	Pb, Cu and Cd	Increased metal uptake by *M. pudica*, Accumulation in roots and nodules, low translocation to shoots	Chen et al. 2008
Galega	*Rhizobium galegae* + *Pseudomonas*	Oil-contaminated soil	Degradation of oil by the *Pseudomonas* strain, nodulation by *Rhizobium*	Lindström et al. 2003
Medicago sativa	*Rhizobium meliloti* + *Flavobacterium* sp.	PCBs	Enhanced degradation of PCB in the rhizosphere, *Flavobacterium* sp. seems to perform the degradation	Xu et al. 2010
	Other bacteria			
Pissum sativum	PHA-degrading *Bacillus, Staphyloccocus, Micrococcus, Stenotrophomonas, Pseudomonas, Cellulomonas)*	Oil hydrocarbons (PHAs and PCHs)	Plant roots increase oil extractability, degradation by rhizosphere bacteria	Liste and Prutz 2006
Trifolium pratense	α-Proteobacteria belonging to the *Bradyrhizobium-Agromonas-Nitrobacter-Afipia* cluster	2,4-D (herbicide)	Complete degradation (mineralization) of 2,4-D in the rhizosphere	Shaw and Burns 2004
Medicago sativa	*Comamonas* sp. CNB-1	Chloronitro-aromatics	Strong ability for degradation of 4-CNB in the rhizosphere	Liu et al. 2007
	Mycorrhiza			
Pisum sativum	*Glomus intraradices*	Cd	Buffering stress caused by Cd	Rivera-Becerril et al. 2002
	Endophytes			
Pisum sativum	*Pseudomonas putida*	2,4-D	Colonization of tissues and rhizosphere. Enhanced degradation of 2,4-D, no detected in plant aerial tissues	Germaine et al. 2006

reports on the degradation of xenobiotics in the rhizosphere of legumes, for instance, polycyclic aromatic hydrocarbons and petroleum hydrocarbons (Liste and Prutz 2006).

The symbiotic interaction of legumes with mycorrhiza is also of critical importance to the nutrient status of these plants. Arbuscular mycorrhiza fungi (AMF) are known to promote plant growth and health by improving mineral nutrition, particularly phosphorous, or increasing resistance or tolerance to biotic and abiotic stresses (Göhre and Paszkowski 2006, Audet and Charest 2007, see also Chapter 15 in this volume). The interaction between plants and AMF applied to phytoremediation has been termed Mycorrhizoremediation (Khan 2006). Some works report the diminution of the stress caused by heavy metals in mycorrhizal plants (see Chapter 15 for further review).

Also, endophytes play a crucial role in the degradation of toxic organics within legumes. For example, degradation of the herbicide 2,4-D by endophyte *Pseudomonas putida* in alfalfa has been reported (Germaine et al. 2006).

Engineering the Rhizosphere of Legumes for Improved Rhizoremediation

The rhizoremediation process using legumes and suitable rhizosphere microorganisms can be done with different levels of human intervention. *Natural attenuation* means almost no human action. Although this process can occur naturally, it is slow and it can be speeded up by different kinds of techniques, including biostimulation, bioaugmentation, or introduction of genetically modified organisms (GMOs) (Darrah et al. 2006). Some recent proposals using legumes will be reviewed.

Biostimulation

The term "biostimulation" is often used to describe the addition of electron acceptors, electron donors or nutrients to stimulate naturally occurring microbial populations, in order to enhance the activity of indigenous microbial degraders or to promote co-metabolism (Miller 2010). The primary advantage of biostimulation is that bioremediation is undertaken by already present native microorganisms that are well-suited to the subsurface environment, and are well distributed spatially within the subsurface, compared to the introduction of non-indigenous bacteria (inoculation), which is inconvenient because of the short survival of the inoculum (Tyagi et al. 2011).

Different substances have been used for stimulating microbial populations for improved bioremediation of soils. When metals are the pollutants, usually soils display a low content of organic matter and nitrogen. A carbon source (for example, in the forms of melasses, sludge, organic compost, manure, etc.), could be added as an electron donor and carbon source. Ideal amendments should fulfil determinate C:N ratios (from 12:1 to 20:1). Different products have been used as organic amendments, including wood chips, straw, biosolids, composted municipal waste or manure (Munshower 1994). The establishment of a vegetal cover on polluted sites usually is impeded in the absence of organic amendments (Mendez and Maier 2008).

On the contrary, if metal phytoextraction is pursued, the addition of chelating agents increases metal availability and hence, plant uptake. Several substances have been used as chelators in field experiments, including EDTA, DTPA, NTA and a number of biodegradable, low-molecular weight, organic acids (citric acid, ascorbic acid, oxalic acid, salicylic acid), sulphur amendments, chitosan, etc. (Kayser et al. 2000, Chen et al. 2004). Leaching of metals and dispersion in the environment is the inconvenient use of chelants, with the possibility of groundwater contamination (Sun et al. 2001, Nowack et al. 2006). Ethylen-diamine succinate acid (EDDS) was proposed as an environmentally-safe chelator for lead phytoextraction (Grcman et al. 2003).

For soils polluted with organic compounds, availability can reduce the rate of phytodegradation and rhizodegradation. The addition of amendments (sewage sludge, N + P, organic acids, EDTA, agricultural compost) can increase availability of pyrene, herbicides, etc. (Perrin-Ganier et al. 2001, White et al. 2003, Kobayashi et al. 2008).

Bioaugmentation

The term "bioaugmentation" refers to the enrichment of the native rhizosphere population by inoculation with indigenous microorganisms previously isolated from the same rhizosphere. This practice guarantees the competitiveness of the inoculum, since native microorganisms are suited to particular rhizosphere conditions (Tyagi et al. 2011).

Several studies have been produced on the use of bioaugmentation at the legume rhizosphere level. Many works demonstrated the protection exerted by indigenous rhizosphere bacteria on legumes roots in metal polluted soils (Lebeau et al. 2008). Dary et al. (2010) reported the protection of *Lupinus* plants in an *in situ* experiment at a contaminated site after a mine spill. Inoculation with *Rhizobium* and metal-resistant PGPRs (in particular, strains of *Pseudomonas* sp. and *Ochrobactrum cytisi*) isolated from the same contaminated site, enhanced plant biomass and nitrogen content, and at the same time decreased metal concentrations in plant tissues, improving metal phytostabilization (between 50-80 percent reduction in metal accumulation). Native bacterial strains showed high resistance towards an array of metals (multiresistance), which is a desired characteristic in sites polluted by a mix of metals, as was the case of this study (Dary et al. 2010). Similar results were obtained with plants of *Vigna unguiculata* grown in soils polluted by Ni and Zn. Inoculation with a metal-resistant *Bradyrhizobium sp.* increased plant biomass, nodulation and nitrogen content (40–50 percent), and seed yield, and decreased phytotoxicity and metal (Ni and Zn) accumulation (Wani et al. 2007b). Analogously, a chromium-reducing and metal-resistant *Mesorhizobium* strain alleviates Cr toxicity in its host chickpea plants (Wani et al. 2008). Different mechanisms could account for this protection at the rhizosphere level, as biosorption at the plant roots and to the bacterial exopolysaccharides, bioprecipitation, complexation, redox changes, etc. (Silver and Phung 2005, Nies and Silver 2007). Bioaugmentation may be used as well for enhanced degradation of organics in the legume rhizosphere. For example, increased degradation of PCB was reported by alfalfa plants after bioaugmentation using a PCB-degrading *Rhizobium* strain. However, in this case, a *Flavobacterium* sp. strain, rather than *Rhizobium*, may have substantially contributed to the degradation of PCBs (Xu et al. 2010). In order to guarantee the efficiency of the nodulation in polluted soils, resistant and very competitive rhizobia must be used (Wani et al. 2007b, Pajuelo et al. 2008). Otherwise, other resistant genera (for example *Ochrobactrum*) can nodulate legumes in polluted soils more competitively (Pajuelo et al. 2010). However, since they do not produce nitrogen-fixing nodules, this does not contribute to the enrichment of nitrogen. As concluded from 16S rRNA gene analysis, shifting in the microbial populations are produced during the bioremediation process (Singer et al. 2003, Rosario et al. 2007).

Genetically Modified Legumes

The generation of transgenic plants with improved phytoremediation capacities is a focus of active research. A large number of transgenic approaches have been developed for phytoremediation of heavy metals (Eapen and Souza 2005, Kortba et al. 2009) and organics (Eapen et al. 2007, Maceck et al. 2008, Abhilash et al. 2009), not only in model plants such as *Arabidopsis thaliana*, but also in other plants more suitable for phytoremediation (tobacco, *Nicotiana glauca, Brassica, Alysum,* etc.). However, only a few reports have been produced on the generation of transgenic legumes for enhanced bioremediation, in spite of the existence of very efficient transformation protocols for model and other legumes like *Medicago truncatula* and *Lotus japonicus* (Thyjkaer et al. 1998, Kamaté et al. 2000). For example, transgenic alfalfa plants have been generated by expression of a modified bacterial atrazine chlorohydrolase (*atzA*) gene. These plants showed a 50-fold increase resistance to the herbicide atrazine, being able to tolerate up to 10 mg L^{-1}, when wild type plants only tolerated 0.2 mg L^{-1}. The enzyme was fully active in plants, being able to dechlorinate atrazine to hydroxi-atrazine, a non-herbicidal metabolite (Wang et al. 2005).

More recently, a different approach has been generated in our laboratory. Composite plants of *Medicago truncatula* were obtained by infecting roots with *Agrobacterium rhizogenes* carrying a metallothionein *mt4a* gene. When these plants were grown in the presence of Cu, transgenic hairy roots showed increased length and branching, and developed a higher number of nodules, as compared to wild type plants. The insertion of the *mt4a* gene in transgenic roots alleviated plant phytotoxicity and increased nitrogen content in plants (Pérez-Palacios, unpublished data).

GMROs (Genetically Modified Rhizosphere Microorganisms)

Much more abundant are the examples of genetically modified microorganisms. Genetically modified *Pseudomonas* or *Rhizobium* were generated with enhanced capacity to degrade PCBs (Toure et al. 2003), TCE (Yee et al. 1998) or dioxin-like compounds (Saiki et al. 2003) in the rhizosphere of plants. The *bph* operon from *Burkholderia xenovorans* LB400, responsible for PCB degradation, was chromosomally inserted into a spontaneous rifampicin-resistant mutant of *Pseudomonas* F113 (Brazil et al. 1995). More recently, the *bph* operon was cloned under the control of a strongly inducible (*nod* box) promoter from *Sinorhizobium meliloti*, and the cassette was subsequently chromosomally integrated into *P. fluorescens* strain F113 (Villacieros et al. 2005). This *Pseudomonas fluorescens* strain was initially isolated from the rhizosphere of sugar beet, and it has an additional advantage, since it was found to be an excellent root colonizator of other plants including tomato and willow, and also legumes like pea and alfalfa (Power et al. 2011). The GMO can perform the degradation of PCB in a very efficient manner in the rhizosphere of an array of plants. Insertion and expression of the arsenic-resistant operon *arsRDABC* into *Pseudomonas* F113rifPCB strain has allowed this strain growing, colonizing the root and degrading biphenyl in the presence of sodium arsenate (Ryan et al. 2007). This is an important task in soils polluted by organics and metals simultaneously, which seems to be the case in more than 75 percent of the contaminated sites (Glass 1998).

In an interesting approach, a genetically modified *Mesorhizobium huakuii* strain was developed by expressing a metallothionein under the control of a nodule-specific *pnifH* promoter in the bacterial surface. Inoculation of *Astragalus* with this GMO allowed the accumulation of cadmium in nodules and the protection of the plant towards this metal (Sriprang et al. 2002). Better results were reported by expression of two genes in rhizobia; a metallothionein and a phytochelatin synthase (Ike et al. 2007).

Recently, a genetically modified *Sinorhizobium medicae* strain has been generated by expressing the copper resistance genes *copAB* from *Pseudomonas fluorescens*. These two genes are part of the copper resistance operon (*copABCD*), and constitute the minimum genetic unit able to confer copper resistance (Cánovas et al. 2003). The genes were cloned under the control of a nodulation specific promoter, *pnifH*. The resulting *Sinorhizobium* strain did not show and enhanced copper resistance "*in vitro*". However, when inoculated on alfalfa plants, it was able to alleviate the phytotoxicity caused by Cu on plants. Furthermore, copper was selectively accumulated in the central tissue of efficient nodules, reaching similar levels to macronutrients Ca and K (Delgadillo 2011) (Fig. 14.1).

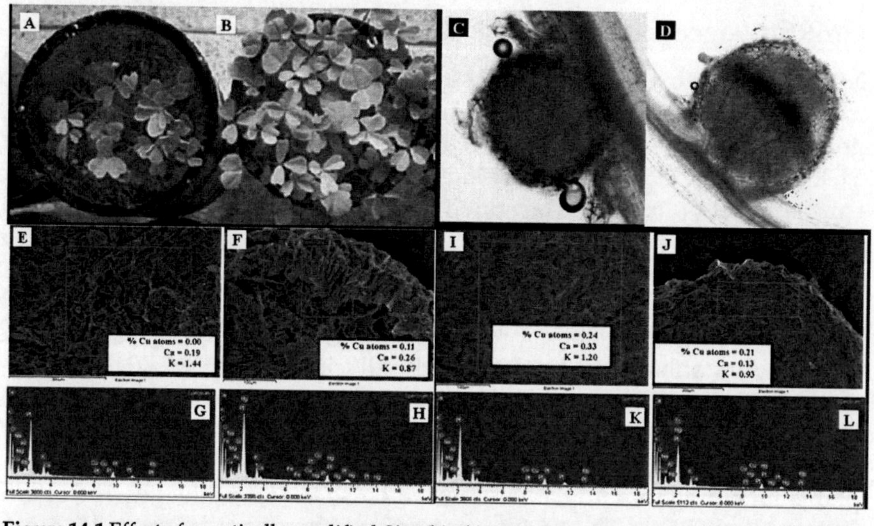

Figure 14.1 Effect of genetically modified *Sinorhizobium medicae* resistant to copper on *Medicago truncatula* growth and nodulation under Cu stress. Plants were grown in copper 200 μM and inoculated with wild type *Sinorhizobium medicae* MA11 or genetically modified *S. medicae* (*copAB*) containing the copper resistance genes *copAB* from *Pseudomonas fluorescens* under the control of the *pnifH* promoter. (A, B) Aspect of plants grown in 200 μCu, (A) non-inoculated, (B) inoculated with *S. medicae* (*copAB*). (C, D) Aspects of nodules stained with X-gal, (C) nodules formed by wild type *S. medicae* MA11, (D) nodules formed by *S. medicae* (*copAB*), (note the dark cortex in C). (E, F) Accumulation of Cu determined by SEM-EDX in nodules formed by the wild type strain, (E) central tissue, (F) cortex. (G, H) SEM-EDX spectra of the tissues shown in E, F, respectively. (I, J) Accumulation of Cu determined by SEM-EDX in nodules formed by the modified strain, (I) central tissue, (J) cortex. (K, L) SEM-EDX spectra of the tissues shown in I, J, respectively. (Note the increase in Cu accumulation from not detected to 0.24 units in nodules formed by the genetically modified strain).

Color image of this figure appears in the color plate section at the end of the book.

The risk of gene transfer in the rhizosphere must be evaluated. Different strategies have been designed for containing or minimizing gene transfer (Ronchel and Ramos 2001, Davison 2005), especially by inserting the transgenes into the bacterial chromosome or encapsulating the microorganisms (Power et al. 2011). Some recent studies done under field conditions reveal that repeated introduction of modified rhizosphere bacteria affects microbial population in the rhizosphere, with low or no effect on bulk soil beyond it (Viebahn et al. 2003, de Cárcer et al. 2007).

GMEs (Genetically Modified Endophytes)

Endophytes have been also modified to improve phytoremediation (Doty 2008, Weyens et al. 2009). A study with high diffusion was the construction of an endophyte *Burkholderia cepacia* strain, termed Endegrader, carrying the plasmid pTOL from *Pseudomonas,* which contains the operon for the complete degradation (mineralization) of toluene. This strain was inoculated in *Lupinus* plants, allowing complete mineralization of toluene *in planta* (Barac et al. 2004). Horizontal transfer of these genes to indigenous endophytes of poplar was shown to occur, which improves the degradation of volatile organic compounds at a higher rate by trees (Tagavi et al. 2005). In another work, inoculation of pea plants with an endophyte able to degrade the herbicide 2,4-D, diminished the signs of phytotoxicity at the time that increased the ability to remove the chemical from the soil (Xu et al. 2010).

Endophytes were also engineered with metal resistance determinants. For instance, a strain of *B. cepacia* was engineered with the *ncc-nre* (Ni resistance operon from *Ralstonia eutropha*) and inoculated on plants of *Lupinus luteus*. In this case, an increase of Ni accumulation in plant roots was reported (Lodewyckx et al. 2001). More recently, endophytes were generated with increased metal (Ni) resistance and with the capacity to degrade small organic compounds (toluene and tetrachloroethylene). The strain harbouring both traits, used in combination with the host plant (*Lupinus*), resulted in decreased Ni and toluene phytotoxicity, degradation of toluene and accumulation of Ni in roots (phytostabilization) (Weyens et al. 2011).

Conclusions and Perspectives

Phytoremediation has emerged in the last decades as an attractive low-cost alternative for soil decontamination. In particular, rhizoremediation is a promising strategy. Among the different plants with potential in rhizoremediation, legumes present several advantages; plant-microbe interactions with rhizobia, many other bacteria and mycorrhiza; pioneer plants; known agronomy, abundance of characterized ecotypes; moderate tolerance to metals and some organic compounds; possibility of co-cropping with more remediative species. There are many studies at the bench/ green house scale, and a few works are also available on field studies. It is necessary to increase our current knowledge on the use of legume-*Rhizobium* symbioses in "real" polluted soils (Fig. 14.2). It could be also interesting to create a platform that agglutinates all the work done on bioremediation using legume plants.

For the establishment of "real phytoremediation projects" using legumes, tailored remediation projects can be designed for specific sites, with the participation of all the stakeholders involved (scientists, owners, end-users, consultants and remediators, small-to medium sized enterprises, environmental regulators and auditors) (Hage et al. 2010). As these projects are affordable, opportunities for value creation must be considered (Ghosh and Singh, 2005). Some of them are presented in Fig. 14.2 (regenerated soil for new uses, non-food plants like ornamentals, fibres, paper, wood, education and training in environmental agriculture and biotechnology, etc.).

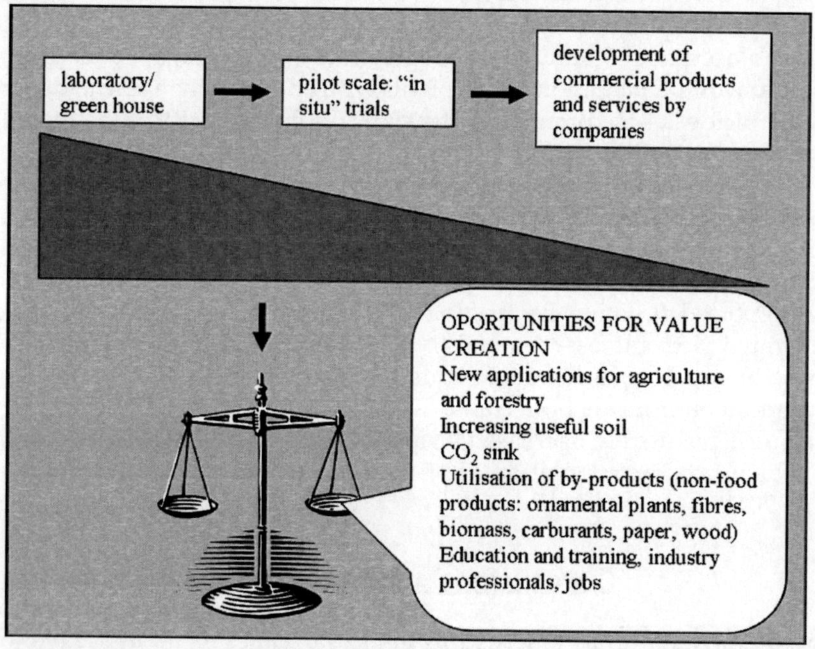

Figure 14.2 Current scenario in our knowledge of the results of phytoremediation projects towards commercial application. The figure depicts our current knowledge on the use of legume-microorganisms interactions in rhizoremediation of pollutants: many laboratory/ greenhouse studies are available and very few *in situ* "real phytoremediation projects". In order to change this scenario, the opportunities for value creation must be considered.

Acknowledgements

This work has been financed by a project of the Ministerio de Economía y Competitividad and FEDER (project number BIO 2009 7766) and project of Junta de Andalucía (reference number RNM-07274)

References

Abhilash, P.C., S. Jamil and N. Singh. 2009. Transgenic plants for enhanced biodegradation and phytoremediation of organic xenobiotics. Biotechnol. Adv. 27: 474–488.

Adamia, G., M. Ghoghoberidze, D. Graves, G. Khatisashvili, G. Kvesitadze, E. Lomidze et al. 2006. Absortion, distribution and transformation of TNT in higher plants. Ecotoxicol. Environ. Saf. 64: 136–145.

Audet, P. and C. Charest. 2007. Dynamics of arbuscular mycorrhyzal symbiosis in heavy metal phytoremediation: Meta-analytical and conceptual perspectives. Environ. Pollut. 147(3): 609–614.

Avis, T.J., V. Gravel, H. Antoun and R.J. Tweddell. 2008. Multifaceted beneficial effects of rhizosphere microorganisms on plant health and productivity. Soil Biol. Biochem. 40: 1733–1740.

Bais, H.P., T.L. Weir, L.G. Perry, S. Gilroy and J.M. Vivanco. 2006. The Role of Root Exudates in Rhizosphere Interactions with Plants and Other Organisms. Ann. Rev. Plant Biol. 57: 233–266.

Baker, A.J.M. 1981. Accumulators and excluders strategies in the response of plants to heavy metals. J. Plant. Nutr. 3(-4): 643–654.

Barac, T., S. Taghavi, B. Borremans, A. Provoost, L. Oeyen, J.V. Colpaert et al. 2004. Engineered endophytic bacteria improve phytoremediation of water soluble, volatile, organic pollutants. Nature Biotechnol. 22: 583–588.

Bento, F.M., F.A. de Oliveira Camargo, B.C. Okeke and W.T. Frankenberger. 2005. Diversity of biosurfactant producing microorganisms isolated from soils contaminated with diesel soil. Microbiol. Res. 160: 249–255.

Bernard, R.G. 2005. Modulation of plant ethylene levels by the bacterial enzyme ACC deaminase. FEMS Microbiol. Lett. 251: 1–7.

Brazier-Hicks, M., K.M. Evans, O.D. Cunningham, D.R. Hodgson, P.G. Steel and R. Edwards. 2008. Catabolism of glutathione conjugates in *Arabidopsis thaliana*. Role in metabolic reactivation of the herbicide safener fenclorim. J. Biol. Chem. 283: 21102–21112.

Brazil, G.M., L. Kenefick, M. Callanan, A. Haro, V. de Lorenzo, D.N. Dowling et al. 1995. Construction of a rhizosphere pseudomonad with potential to degrade polychlorinated biphenyls and detection of *bph* gene expression in the rhizosphere. Appl. Environ. Microbiol. 61: 1946–1952.

Broos, K., M. Uyttebroek, J. Mertens and E. Smolders. 2004. A survey of symbiotic nitrogen fixation by white clover grown on metal contaminated soils. Soil Biol. Biochem. 36: 633–640.

Burd, G.I., D.G. Dixon and B.R. Glick. 2000. Plant growth promoting bacteria that decrease heavy metal toxicity in plants. *Can. J. Microbiol.* 46(3): 237–245.

Callahan, D.L., A.J.M. Baker, S.D. Kolev and A.G. Wedd. 2006. Metal ion ligands in hyperaccumulating plants. J. Biol. Inorg. Chem. 11: 2–12.

Cánovas, D., I. Cases and V. de Lorenzo. 2003. Heavy metal tolerante and metal homeostasis in pseudomonas putida as revealed by complete genome analysis. Environ. Microbiol. 5: 1242–1256.

Capuana, M. 2011. Heavy metals and woody plants - biotechnologies for phytoremediation. iForest 4: 7–15. URL: http://www.sisef.it/ iforest/show.php?id=555.

Castro, I.V., E.M. Ferreira and S.P. McGrath. 1997. Effectiveness and genetic diversity of *Rhizobium leguminosarum* bv. *trifolii* isolates in Portuguese soils polluted by industrial effluents. Soil Biol. Biochem. 29: 1209–1213.

Chaudhry, Q., M. Blom-Zandstra, S. Gupta and E.J. Joner. 2005. Utilising the synergy between plants and rhizosphere microorganisms to enhance breakdown of organic pollutants in the environment. Environ. Sci. Pollut. Res. 12: 34–48.

Chen, W.M., C.H. Wu, E.K. James and J.S. Chang. 2008. Metal biosorption capability of *Cupriavidus taiwanensis* and its effects on heavy metal removal by nodulated *Mimosa pudica*. J. Haz. Mat. 151: 364–371.

Chen, Y., X. Li and Z. Shen. 2004. Leaching and uptake of heavy metals by ten different species of plants during an EDTA-assisted phytoextraction process. Chemosphere. 57: 187–196.

Chen, Y.X., Y.P. Wang, Q. Lin and Y.M. Luo. 2005. Effect of copper-tolerant rhizosphere bacteria on mobility of copper ions and copper accumulation by *Elsholtzia splendens*. Environ. Int. 31: 861–866.

Darrah, P.R., D.L. Jones, G.J.D. Kirk and T. Roose. 2006. Modeling the rhizosphere: a review of methods for up scaling to the whole plant scale. Eur. J. Soil Sci. 57: 13–25.

Dary, M., M.A. Chamber-Pérez, A.J. Palomares and E. Pajuelo. 2010. "*in situ*" phytostabilisation of heavy metal polluted soils using *Lupinus luteus* inoculated with metal resistant plant-growth promoting rhizobacteria. J. Haz. Mat. 177: 323–330.

Davison, J. 2005. Risk mitigation of genetically modified bacteria and plants designed for bioremediation. J. Ind. Microbiol. Biotechnol. 32: 639–650.

de Cárcer, A.D., M. Martín, M. Mackova, T. Macek, U. Karlson and R. Rivilla. 2007. The introduction of genetically modified microorganisms designed for rhizoremediation induces changes on native bacteria in the rhizosphere but not in the surrounding soil. ISME J. 1: 215–223.

Delgadillo, J. 2011. Ingeniería genética de la rizosfera de *Medicago*. Aplicaciones en fitorremediación de suelos contaminados por cobre. PhD Thesis, Univ. of Seville. Spain.

Díaz, E. 2004. Bacterial degradation of aromatic pollutants: a paradigm of metabolic versatility. Int. Microbiol. 7: 173–180.

Doty, S.L., Q.T. Shang, A.M. Wilson, A.L. Moore, L.A. Newman, S.E. Strand et al. 2003. Metabolism of the soil and ground water contaminants ethylene-dibromide and trichloroethylene by the tropical leguminous tree *Leucaena leucocephala*. Water Res. 37: 441–449.

Doty, S.L. 2008. Enhancing phytoremediation through the use of transgenic plants and entophytes. New Phytol. 179: 318–333.

Dzantor, E.K., T. Chekol and L.R. Vough. 2000. Feasibility of using forage grasses and legumes for phytoremediation of organic pollutants. J. Environ. Sci. Health, Part A. 35: 1645–1661.

Eapen, S. and S.F. De Souza. 2005. Prospect of genetic engineering of plants for phytoremediation of toxic metals. Biotechnol. Adv. 23: 97–116.

Eapen, S., S. Singh and S.F. D'Souza. 2007. Advances in development of transgenic plants for remediation of xenobiotic pollutants. Biotechnol. Adv. 25: 442–451.

El Aafi, N., F. Brhada, M. Dary, A. Filali Maltouf and E. Pajuelo. 2012. Rhizostabilization of metals in soils using *Lupinus Luteus* inoculated with the metal resistant rhizobacterium *Serratia* sp. MSMC541. Int. J. Phytorem. 14: 261–274.

Freeman, J.L., L.H. Zhang, M.A. Marcus, S. Fakra, S.P. McGrath and E.A.H. Pilon-Smits. 2006. Spatial imaging, speciation, and quantification of selenium in the hyperaccumulator plants *Astragalus bisulcatus* and *Stanleya pinnata*. Plant Physiol. 142: 124–134.

Gandía-Herrero, F., A. Lorenz, T. Larson, I.A. Graham, J. Bowles, E.L. Rylott et al. 2008. Detoxification of the explosive 2,4,6- trinitrotoluene in *Arabidopsis*: discovery of bifunctional O and C-glucosyltransferases. Plant J. 56: 963–974.

Gao, Y., L. Ren, W. Ling, S. Gong, B. Sun and Y. Zhang. 2010. Desorption of phenanthrene and pyrene in soils by root exudates. Biores. Technol. 101: 1159–1165.

Gerhardt, K.E., X.D. Huang, B.R. Glick and B.M. Greenberg. 2009. Phytoremediation and rhizoremediation of organic soil contaminants: potential and challenges. Plant Sci. 176: 20–30.

Germaine, K.J., X. Liu, G.G. Cabellos, J.P. Hogan, D. Ryan, and D.N. Dowling. 2006. Bacterial endophyte-enhanced phytoremediation of the organochlorine herbicide 2,4-dichlorophenoxyacetic acid. FEMS Microbiol. Ecol. 57: 302–310.

Ghosh, M. and S.P. Singh. 2005. A review on phytoremediation of heavy metals and utilization of its byproducts. Appl. Ecol. Environ. Res. 3: 1–18.

Glass, D.J. 1998. The United States market for phytoremediation. A.D. Glass Associates, Inc. Needham, Mass.

Glick, B.R. 2003. Phytoremediation: synergistic use of plants and bacteria to clean up the environment. Biotechnol. Adv. 21: 239–244.

Göhre, V. and U. Paszkowski. 2006. Contribution of the arbuscular mycorrhizal symbiosis to heavy metal phytoremediation. Planta. 223: 1115–1122.

Graham, P.H. and C.P. Vance. 2003. Legumes: importance and constraints to greater use. Plant Physiol. 131: 872–877.

Grcman, H., D. Vodnik, S. Velikonja-Bolta and D. Lestan. 2003. Ethylenediamine-dissuccinate as a new chelate for environmentally safe enhanced lead phytoextraction. J. Environ. Qual. 32: 500–506.

Hage, M., P. Leroy P and A.C. Petersen. 2010. Stakeholder participation in environmental knowledge production. Futures. 42: 254–264.

Han, J.G., L. Sun, X.Z. Dong, Z.Q. Cai, X.L. Sun, H.L. Yang et al. 2005. Characterization of a novel plant growth-promoting bacteria strain *Delftia tsuruhatensis* HR4 both as a diazotroph and a potential biocontrol agent against various plant pathogens. Sys. Appl. Microbiol. 28: 66–76.

Hinsinger, P., G.R. Gobran, P.J. Gregory and W.W. Wenzel. 2005. Rhizosphere geometry and heterogeneity arising from root-mediated physical and chemical processes. New Phytol. 168: 293–303.

Ike, A., R. Sriprang, H. Ono and YM. Mitsuo. 2007. Bioremediation of cadmium contaminated soil using symbiosis between leguminous plant and recombinant rhizobia with the *MTL4* and the *PCS* genes. Chemosphere. 66: 1670–1676.

Kamaté, K., I.D. Rodríguez-Llorente, M. Scholte, P. Durand, P. Ratet, E. Kondorosi, A. Kondorosi and T.H. Trinh. 2000. Transformation of floral organs with GFP in *Medicago truncatula*. Plant Cell Rep. 19: 647–653.

Kayser, A., K. Wenger, A. Keller, W. Attinger, H.R. Felix, S.K. Gupta et al. 2000. Enhancement of phytoextraction of Zn, Cd and Cu from calcareous soil: the use of NTA and sulphur amendments. Environ. Sci. Technol. 34: 1778–1783.

Khan, A.G. 2005. Role of soil microbes in the rhizospheres of plants growing on trace metal contaminated soils in phytoremediation. J. Trace Elem. Med. Biol. 18: 355–364.

Khan, A.G. 2006. Mycorrhizoremediation, an enhanced form of phytoremediation. J. Zhejiang Univ. Science B. 7: 503–514.

Khan, M.S., A. Zaidi, P.A. Wani and M. Oves. 2009. Role of plant growth promoting rhizobacteria in the remediation of metal contaminated soils. Environ. Chem. Lett. 7: 1–19.

Kidd, P.S., A. Prieto-Fernández and C. Monterroso. 2008. Rhizosphere microbial community and hexachlorocyclohexane degradative potential in contrasting plant species. Plant Soil, 302: 233–247.

Kobayashi, T., R.R. Navarro, K. Tatsumi and Y. Iimura. 2008. Influence of compost amendment on pyrene availability to two subspecies of *Cucurbita pepo*. Sci. Total Environ. 404: 1–9.

Kopittke, P.M., P.J. Dart and N.W. Menzies. 2007. Toxic effects of low concentrations of Cu on nodulation of cowpea (*Vigna unguiculata*). Environ Pollut. 145: 309–315.

Kortba, P., J. Naymanova, T. Macek, T. Ruml and M. Mackova. 2009. Genetically modified plants in phytoremediation of heavy metals and metalloids in soil and sediment pollution. Biotechnol. Adv. 27: 799–810.

Krämer, U. 2010. Metal hyperaccumulation in plants. Ann. Rev. Plant Biol. 61: 517–534.

Krämer, U., J.D. Cotter-Howells, J.M. Charnock, A.J.M. Baker and J.A.C. Smith. 1996. Free histidine as a metal chelator in plants that accumulate nickel. Nature. 379: 635–638.

Kumar, A., B.S. Bisht, V.D. Joshi and T. Dhewa. 2011. Review on Bioremediation of Polluted Environment: A Management Tool. International Journal of Environmental Sciences. 1: 1079–1093.

Kuiper, I., E.L. Lagendijk, G.V. Bloemberg and B.J.J. Lugtenberg. 2004. Rhizoremediation: a beneficial plant–microbe interaction. Mol. Plant Microb. Interact. 17: 6–15.

Küpper, H., A. Mijovilovich, W. Meyer-Klauck and P.M.H. Kroneck. 2004. Tissue- and age-dependent differences in the complexation of cadmium and zinc in the Cd/Zn hyperaccumulator *Thlaspi caerulescens* (Ganges ecotype) revealed by x-ray absorption spectroscopy. Plant Physiol. 134: 748–757.

Lafuente, A., E. Pajuelo, M.A. Caviedes and I.D. Rodríguez-Llorente. 2010. Reduced nodulation in alfalfa induced by arsenic correlates with altered expression of early nodulins. J. Plant Physiol. 167: 286–291.

Lebeau, T., A. Braud and K. Jézéquel. 2008. Performance of bioaugmentation-assisted phytoextraction applied to metal contaminated soils: A review. Environ. Pollut. 153: 497–522.

Lei, D. and C. Duan. 2008. Restoration potential of pioneer plants growing on lead-zinc mine tailings in Lanping, southwest China. J. Environ. Sci. 20: 1202–1209.

Lindström, K., M.M. Jussila, H. Hintsa, A. Kaksonen, L. Mokelke, K. Mäkeläinen et al. 2003. Potential of the *Galega–Rhizobium galegae* System for Bioremediation of Oil-Contaminated Soil. Food Technol. Biotechnol. 41: 11–16.

Liste, H.H. and I. Prutz. 2006. Plant performance, dioxygenase-expressing rhizosphere bacteria, and biodegradation of weathered hydrocarbons in contaminated soil. Chemosphere. 62: 1411–1420.

Liu, L., C.Y. Jiang, X.Y. Liu, J.F. Wu, J.G. Han and S.J. Liu. 2007. Plant-microbe association for rhizoremediation of chloronitroaromatic pollutants with *Comamonas* sp. strain CNB-1. Environ. Microbiol. 9: 465–473.

Lodewyckx, C., S. Taghavi, M. Mergeay, J. Vangronsveld, H. Clijsters and D. van der Lelie. 2001. The effect of recombinant heavy metal resistant endophytic bacteria on heavy metal uptake by their host plant. Int. J. Phytorem. 3: 173–187.

Macek, T., P. Kotrba, A. Svatos, M. Novakova, K. Demnerova and M. Mackova. 2008. Novel roles for genetically modified plants in environmental protection. Trends Biotechnol. 26: 146–152.

Mahieu, S., H. Frérot, C. Vidal, A. Galiana, K. Heulin, L. Maure, B. Brunel, C. Lefèbvre, J. Escarré and J.C. Cleyet-Marel. 2011. *Anthyllis vulneraria/Mesorhizobium metallidurans*, an efficient symbiotic nitrogen fixing association able to grow in mine tailings highly contaminated by Zn, Pb and Cd. Plant Soil. 342: 405–417.

Maldonado-Magaña, A., E. Favela-Torres, F. Rivera-Cabrera and T.L. Volke-Sepulveda. 2011. Lead bioaccumulation in *Acacia farnesiana* and its effect on lipid peroxidation and glutathione production. Plant Soil. 339: 377–389.

Mari, S., D. Gendre, K. Pianelli, L. Ouerdane, R. Lobinski, J.F. Briat, M. Lebrun and P. Czernic. 2006. Root-to-shoot long-distance circulation of nicotianamine and nicotianamine–nickel chelates in the metal hyperaccumulator *Thlaspi caerulescens*. J. Exp. Bot. 57: 4111–4122.

Miller, H. 2010. Biostimulation as a Form of Bioremediation of Soil Pollutants. MMG445 Basic Biotechnology eJournal. 6: 1.

Mendez, M.O. and R.M. Maier. 2008. Phytostabilization of Mine Tailings in Arid and Semiarid Environments—An Emerging Remediation Technology. Environ. Health Perspect. 116: 278–283.

Moreno-Jiménez, E., J.M. Peñalosa, R. Manzano, R.O. Carpena-Ruiz, R. Gamarra and E. Esteban. 2009. Heavy metals distribution in soils surrounding an abandoned mine in NW Madrid (Spain) and their transference to wild flora. J. Haz. Mat. 162: 854–859.

Moreno-Jiménez, E., E. Esteban, R.O. Carpena-Ruiz, M.C. Lobo and J.M. Peñalosa. 2012. Phytostabilisation with Mediterranean shrubs and liming improved soil quality in a pot experiment with a pyrite mine soil. J. Haz. Mat. 201–202: 52–59.

Munshower, F.F. 1994. Practical Handbook of Disturbed Land Revegetation. Lewis Publishing. Boca Raton, Florida.

Nies, D.H. and S. Silver [eds.]. 2007. Molecular microbiology of heavy metals. Springer Verlag Heidelberg.

Nowack, B., R. Schulin and B.H. Robinson. 2006. Critical assessment of chelant-enhanced metal phytoextraction. Environm. Sci. Technol. 17: 5225–5232.

Oldroyd, G.E.D. and J.A. Downie. 2008. Coordinating nodule morphogenesis with rhizobial infection in legumes. Annu. Rev. Plant Biol. 59: 519–546.

Pajuelo, E., J.A. Carrasco, L.C. Romero, M.A. Chamber and C. Gotor. 2007. Evaluation of the metal phytoextraction potential of crop legumes. Regulation of the expression of O-acetylserine- (thyol)-lyase under metal stress. Plant Biol. 9: 672–681.

Pajuelo, E., I.D. Rodríguez-Llorente, M. Dary and A.J. Palomares. 2008. Toxics effects of arsenic on *Sinorhizobium-Medicago sativa* symbiotic interaction. Environ Pollut. 154: 203–211.

Pajuelo, E., B. Doukkali, F. Pérez-Paniagua, I.D. Rodríguez-Llorente and M.A. Caviedes. 2010. Ultraestructura de nódulos no fijadores de nitrógeno inducidos por *Ochrobactrum cytisi* en plantas leguminosas. X Reunión Nacional del Metabolismo del Nitrógeno. Benalauría, Málaga (Spain).

Pastor, J., A.J. Hernández, N. Prieto and M. Fernández-Pascual. 2003. Accumulating behaviour of *Lupinus albus* L. growing in a normal and a decalcified calcic luvisol polluted witn Zn. J. Plant Physiol. 160: 1457–1465.

Perrin-Ganier, C., F. Schiavon, J.L. Morel and M. Schiavon. 2001. Effect of sludge-amendment or nutrition addition on the biodegradation of the herbicide isoproturon in soil. Chemosphere. 44: 887–892.

Pilon-Smits, E.A.H. 2005. Phytoremediation. Annu. Rev. Plant Biol. 56: 15–39.

Pilon-Smits, E.A.H. and C.F. Quinn. 2010. Selenium metabolism in plants. *In:* R. Hell and R.R. Mendel [eds.]. Cell Biology of Metals and Nutrients Springer Verlag Berlin Heidelberg. pp. 225–241.

Power, B., X. Liu, K.J. Germaine, D. Ryan, D. Braziland and D.N. Dowling. 2011. Alginate beads as a storage, delivery and containment system for genetically modified PCB degrader and PCB biosensor derivatives of *Pseudomonas fluorescens* F113. J. Appl. Microbiol. 110: 1351–1358.

Prasad, M.N.V. 2011. A State-of-the-Art report on Bioremediation: its Applications to Contaminated Sites in India. Published by the Ministry of Environment and Forest, Government of India. India.

Read, D.B., A.G. Bengough, P.J. Gregory, J.W. Crawford, D. Robinson, C.M. Scrimgeouret al. 2003. Plant roots release phospholipids surfactants that modify the physical and chemical properties of soil. New Phytol. 157: 315–326.

Reichman, S.M.A. 2007. The potential use of the legume–*Rhizobium* symbiosis for the remediation of arsenic contaminated sites. Soil Biol. 39: 2587–2593.

Rivera-Becerril, F., C. Calantzis, K. Turnau, J.P. Caussanel, A.A. Belimov, S. Gianinazzi et al. 2002. Cadmium accumulation and buffering of cadmium-induced stress by arbuscular mycorrhiza in three *Pisum sativum* L. genotypes. J. Exp. Bot. 53: 1177–1185.

Rivera-Becerril, F, A. Metwally, F. Martin-Laurent, D. Van Tuinen, K.J. Dietz, S. Gianinazzi and V. Gianinazzi-Pearson. 2005. Molecular responses to cadmium in roots of *Pissum sativum* L. Water Air Soil Pollut. 168: 171–186.

Rivas, R., P. García-Fraile and E. Velázquez. 2009. Taxonomy of bacteria nodulating legumes. Microbiol. Insights. 2: 51–69.

Ronchel, M.C. and J.L. Ramos. 2001. Dual system to reinforce biological containment of recombinant bacteria designed for rhizoremediation. Appl. Environ. Microbiol. 67: 2649–2656.

Rosario, K., S.L. Iverson, D.A. Henderson, S. Chartrand, C. Mckeon, E.P. Glenn et al. 2007. Bacterial community changes during plant establishment at the San Pedro River mine tailings site. J. Environ. Qual. 36: 1249–1259.

Ryan, R.P., D. Ryan and D.N. Dowling. 2007. Plant protection by the recombinant, root-colonizing *Pseudomonas fluorescens* F113rifPCB strain expressing arsenic resistance: improving rhizoremediation. Lett. Appl. Microbiol. 45: 668–674.

Saiki, Y., H. Habe, T. Yuuki, M. Ikeda, T. Yoshida, H. Nojiri et al. 2003. Rhizoremediation of dioxine-like compounds by a recombinant *Rhizobium tropici* strain expressing carbazole 1,9a-dioxigenase constitutively. Biosci. Biotechnol. Biochem. 67: 1144–1148.

Sanderman, H. 1994. Higher plant metabolism of xenobiotics: the 'green liver' concept. Pharmacogenetics. 4: 225–41.

Sarma, H. 2011. Metal Hyperaccumulation in Plants: A Review Focusing on Phytoremediation Technology. J. Environ. Sci. Technol. 4: 118–138.

Schröder, P. and C.D. Collins [eds.]. 2011. Organic Xenobiotics and Plants: From Mode of Action to Ecophysiology. Plant Ecophysiology 8. Springer Verlag New York and Business Media B.V.

Shaw, L.J. and R.G. Burns. 2004. Enhanced mineralization of [U14-C] 2,4-dichlorophenoxy-acetic acid in the soil from the rhizosphere of *Trifolium pratense*. Appl. Environ. Microbiol. 70: 4766–4774.

Sheng, X.F. and J.J. Xia. 2006. Improvement of rape (*Brassica napus*) plant growth and cadmium uptake by cadmium resistant bacteria. Chemosphere. 64: 1036–1042.

Silver, S. and L.T. Phung. 2005. A bacterial view of the periodic table: genes and proteins for toxic inorganic ions. J. Ind. Microbiol. Biotechnol. 32: 587–605.

Siminszky, B. 2006. Plant cytochrome P450-mediated herbicide metabolism. Phytochem. Rev. 5: 445–458.

Singer, C.A., D. Smith, W.A. Jury, K. Hathuc and D.E. Crowley. 2003. Impact of the plant rhizosphere and augmentation on remediation of polychlorinated biphenyl contaminated soil. Environ. Toxicol. Chem. 22: 1998–2004.

Sriprang, R., M. Hayashi, M. Yamashita, H. Ono, K. Saeki and Y. Murooka. 2002. A novel bioremediation system for heavy metals using the symbiosis between leguminous plant and genetically engineered rhizobia. J. Biotechnol. 99: 279–293.

Srivastava, M., L.Q. Ma and N. Singh. 2005. Antioxidant responses of hyperaccumulator and sensitive fern species to arsenic. J. Exp. Bot. 56: 1335–1342.

Shukla, K.P., S. Sharma, N.K. Singh, V. Singh, K. Tiwari and S. Singh. 2011. Nature and role of root exudates: Efficacy in bioremediation. African J. Biotechnol. 10: 9717–9724.

Sun, B., F.J. Zhao, E. Lombi and S.P. McGrath. 2001. Leaching of heavy metals from contaminated soils using EDTA. Environ. Pollut. 113: 111–120.

Taghavi. S., T. Barac, B. Greenberg, B. Borremans, J. Vangronsveld and D. van der Lelie. 2005. Horizontal gene transfer to endogenous endophytic bacteria from Poplar improves phytoremediation of toluene. Appl. Environ. Microbiol. 71: 8500–8505.

Thyjkaer, T., L. Schauser, D. Danielsen, J. Finneman and J. Stougaard. 1998. Transgenic plants *Agrobacterium*-mediated transformation of the diploid legume *Lotus japonicus*. *In:* Cell Biol: A Laboratory handbook. 2nd edn., vol. 3, Acadmic Press, San Francisco. pp. 514–520.

Toure O., Y.Q. Chen and S.K. Dutta. 2003. *Sinorhizobium meliloti* electrotransporant containing *ortho*dechlorination gene shows enhanced PCB dechlorination. Fresenius Environ. Bull. 12: 320–322.

Tyagi, M., M. Manuela, R. Da Fonseca and C.C.C.R. De Carvalho. 2011. Bioaugmentation and biostimulation strategies to improve the effectiveness of bioremediation processes. Biodegradation. 22: 231–241.

USEPA, US Environmental Protection Agency, http//www.epa.gov.

Van Aken, B. 2009. Transgenic plants for enhanced phytoremediation of toxic explosives. Curr. Opin. Biotechnol. 20: 231–236.

Vangronsveld, J., R. Herzig, N. Weyens, J. Boulet, K. Adriaensen, A. Ruttens et al. 2009. Phytoremediation of contaminated soils and groundwater: lessons from the field. Environ. Sci. Pollut. Res. 16: 765–794.

Vázquez, S., R. Agha, A. Granado, M.J. Sarro, E. Esteban, J.M. Peñalosa et al. 2006. Use of white lupin plant for phytostabilization of Cd and As polluted acid soil. Water Air Soil Pollut. 177: 349–365.

Verbruggen, N., C. Hermans and H. Schat. 2009. Molecular mechanisms of metal hyperaccumulation in plants. New Phytol. 181: 759–776.

Viebahn, M., D.C. Glandorf, T.W. Ouwens, E. Smit, P. Leeflang, K. Wernars et al. 2003. Repeated introduction of genetically modified *Pseudomonas putida* WCS358r without intensified effects on the indigenous microflora of field-grown wheat. Appl. Environ. Microbiol. 69: 3110–3118.

Villacieros, M., C. Whelan, M. Mackova, J. Molgaard, M. Sanchez-Contreras, J. Lloret et al. 2005. Polychlorinated biphenyl rhizoremediation by *Pseudomonas fluorescens* F113 derivatives, using a *Sinorhizobium meliloti nod* system to drive *bph* gene expression. Appl. Environ. Microbiol. 71: 2687–2694.

Wang, L., D.A. Samac, N. Shapir, L.P. Wackett, C.P. Vance, N.E. Olszewski and M.J. Sadowsky. 2005. Biodegradation of atrazine in transgenic plants expressing a modified bacterial atrazine chlorohydrolase (*atzA*) gene. Plant Biotechnol. J. 3: 475–486.

Wani, P.A., M.S. Khan and A. Zaidi. 2007a. Cadmium, chromium and copper in greengram plants. Agron. Sustain. Develop. 27: 145–153.

Wani, P.A., M.S. Khan and A. Zaidi. 2007b. Effect of metal tolerant plant growth promoting *Bradyrhizobium* sp. (*Vigna*) on growth, symbiosis, seed yield and metal uptake by greengram plants. Chemosphere. 70: 36–45.

Wani, P.A., M.S. Khan and A. Zaidi. 2008. Chromium reducing and plant growth promoting *Mesorhizobium* improves chickpea growth in chromium amended soil. Biotechnol. Lett. 30: 159–163.

Wenzel, W.W. 2009. Rhizosphere processes and management in plant-assisted bioremediation (phytoremediation) of soils. Plant Soil. 321: 385–408.

Weyens, N., D. van der Lelie, S. Taghavi and J. Vangronsveld. 2009. Phytoremediation: plant–endophyte partnerships take the challenge. Curr. Opin. Biotechnol. 20: 248–254.

Weyens, N., S. Truyens, E. Saenen, J. Boulet, J. Dupae, S. Taghavi et al. 2011. Endophytes and Their Potential to Deal with Co-Contamination of Organic Contaminants (Toluene) and Toxic Metals (Nickel) During Phytoremediation. Int. J. Phytorem. 13: 244–255.

White, J.C., M.J.I. Mattina, W.Y. Lee, B.D. Eitzer and W. Iannucci-Berger. 2003. Role of organic acids in enhancing the desorption and uptake of weathered p,p′-DDE by *Cucurbita pepo*. Environ. Pollut. 124: 71–80.

Whiting, S.N., De Souza, M. and N. Terry. 2001. Rhizosphere bacteria mobilize Zn for hyperaccumulator *Thlaspi caerulescens*. Environ. Sci. Technol. 35: 3144–3150.

WHO, World Health Organization, http://www.who.int.

Xu, L., Y. Teng, Z.G. Li, J.M. Norton and Y.M. Luo. 2010. Enhanced removal of polychlorinated biphenyls from alfalfa rhizosphere soil in a field study: the impact of a rhizobial inoculum. Sci. Total Environ. 408: 1007–1013.

Yang, Z.Y., J.G. Yuan, G.R. Xin, H.T. Chang and M.H. Wong. 1997. Germination, growth and nodulation of *Sesbania rostrata* grown in Pb/Zn mine tailings. Environ. Manage. 21: 617–622.

Yee, D.C., J.A. Maynard and T.K. Wood. 1998. Rhizoremediation of trichloroethylene by a recombinant, root-colonizing *Pseudomonas fluorescens* strain expressing toluene ortho-monooxygenese constitutively. Appl. Environ. Microbiol. 64: 112–118.

Yoon, J.M., B. Van Aken and J.L. Schnoor. 2006. Leaching of contaminated leaves following uptake and phytoremediation of RDX, HMX, and TNT by poplar. Int. J. Phytorem. 8: 81–94.

Zaidi, S., S. Usmani, B.R. Singh and J. Musarrat. 2006. Significance of *Bacillus subtilis* strain SJ 101 as a bioinoculant for concurrent plant growth promotion and nickel accumulation in *Brassica juncea*. Chemosphere. 64: 991–997.

Zhang, Q., L.C. Davis and L.E. Erickson. 2000. An experimental study of phytoremediation of methyl-tert-butyl ether (MTBE) in ground water. J. Haz. Subs. Res. 2: 4.1–4.19.

Zhuang, X., J. Chen, H. Shin and Z. Bai. 2009. New advances in plant growth promoting rhizobacteria for bioremediation. Environ. Int. 33: 406–413.

15

Arbuscular Mycorrhizas and their Significance in Promoting Soil-Plant System Sustainability against Environmental Stresses

J.M. Barea, M.J. Pozo, J.A. López-Ráez, R. Aroca, J.M. Ruíz-Lozano, N. Ferrol, R. Azcón* and *C. Azcón-Aguilar*

Introduction

The stability and productivity of agro-ecosystems is largely dependent on maintaining the quality of soil resources through a sustainable soil management, which aims to conserve its productive capacity, minimizing energy and resource use by optimizing the rate of turnover and recycling of nutrients (Altieri 2004). These are fundamental issues not only for a sustainable production of healthy crops, or to ensure ecosystem self-sustainability, but also to prevent erosion and to minimize negative environmental impacts (Buscot 2005).

Department of Soil Microbiology and Symbiotic Systems, Estación Experimental del Zaidín, CSIC, Granada.
*Corresponding author: josemiguel.barea@eez.csic.es

Sustainability of the plant-soil systems can be achieved when utilization of mineral resources by plants is balanced by an efficient biogeochemical cycling, thereby avoiding nutrients to be rapidly depleted. This means that agricultural productivity can be maintained with a reduction of fertilizer inputs and that natural plant communities can exist in a stable form for prolonged periods of time (Jeffries and Barea 2012). To achieve sustainability both environmental pollution and depletion of agricultural and forestry resources have to be prevented, the structure and diversity of natural plant communities should be maintained, and the susceptibility of soils to erosion reduced (Klironomos 2002).

Diverse chemical, physical and biological factors are involved in the framework of plant-soil interactions responsible for a sustainable ecosystem functioning (Barea et al. 2005a). The biological components are based on diverse genetic and functional groups of soil microbial populations able to carry out critical ecosystem functions such as the biogeochemical cycling of mineral nutrients, organic matter decomposition and the formation and maintenance of soil structure, key issues in a sustainable production scenario (Barea et al. 2005b, Chaudhary et al. 2009, Richardson et al. 2009). Among the beneficial microbes, mycorrhizal fungi are recognized as one of the most influential group of soil biota in the context of ecosystem sustainability once they establish mutualistic plant-fungus symbioses, so-called mycorrhizas (Jeffries and Barea 2012). Mycorrhizal associations are formed by most vascular plant species on Earth and can be found in almost all terrestrial ecosystems worldwide (Smith and Read 2008, Brundrett 2009), being universally accepted that they are fundamental to improve plant performance and soil quality (Jeffries et al. 2003).

Mycorrhizal functioning is based on the exchange of nutrients and energy between both the plant and fungal partners (Brundrett 2002). A variety of mycorrhizal types are formed, depending on the plant and fungal taxa involved. However, arbuscular mycorrhizal (AM) symbiosis is the most common and over 70 percent of plant species are capable of forming these associations (Smith and Read 2008, Brundrett 2009). This chapter will focus on AM symbiosis but the importance and ecological meaning of other mycorrhizal types are reviewed in this book (Chapter 17).

AM fungi, included in the phylum Glomeromycota (Schüßler et al. 2001), are ubiquitous soil-borne fungi, whose origin and divergence have been dated back over 450 million years (Redecker et al. 2000, Honrubia 2009, Schübler and Walker 2011). Pioneering observations, followed by morphological and phylogenetic (molecular) studies, suggested that AM symbioses played a key role in land colonization by plants (Pirozynski and Malloch 1975), and that primitive roots developed in association with AM fungi and co-evolved with them to build up the mycorrhizal root systems of extant vascular plants (Brundrett 2002). As a consequence of this co-evolution the AM relationship became an integral component of plant ecology in both natural and agricultural ecosystems (Brundrett 2002).

Nowadays it is well accepted that AM associations, which helped plants to thrive in hostile environments such as those prior to their origin and during their evolution, continue helping plants to develop in stressed environments (Barea et al. 2011). This is fundamental because adverse conditions of different origin, particularly exacerbated in the current scenario of global change, generate a great array of stress situations which interferes with the stability of both natural and agricultural systems. Plants must be able to cope with these stresses. Consequently, adaptive strategies able to increase their resilience to overcome negative impacts should be developed. AM establishment can be considered as one of these adaptive strategies, crucial in maintaining ecosystem sustainability, through activities able to confer the plant an increased tolerance to environmental constraints (Barea et al. 2011, Jeffries and Barea 2012).

In this chapter we describe how the AM symbiosis results fundamentally in maintaining ecosystem sustainability. Accordingly, we first summarize key information on the establishment, functioning and significance of AM symbiosis in the soil-plant system. Then the physiological and molecular mechanisms responsible for the increased tolerance of mycorrhizal plants to a variety of environmental stresses, either biotic or abiotic, and their implications for plant performance are analyzed. We later discuss how to explore and manage the diversity of AM fungal populations from different habitats. Finally, examples where the appropriate management of the AM symbiosis can help restore disturbed environments as an adaptive strategy to improve their resilience against adversity are given. The trends of this thematic area will be outlined to suggest what research is needed in the future.

Arbuscular Mycorrhiza: Establishment and Functioning

The information generated during the last years on the cellular and molecular events that took place during AM establishment and the eco-physiological and molecular components of AM functioning have recently been reviewed (Bonfante and Genre 2008, Parniske 2008, Smith and Read 2008; Gianinazzi-Pearson et al. 2009). Accordingly, only the main well-established conclusions from these review articles are considered and critically summarized here.

AM Establishment

AM fungi can colonize plant roots from three main types of soil-based propagules: spores, fragments of mycorrhizal roots and extraradical hyphae, all of them producing more or less a well-developed mycelial network expanding in the soil. Establishment and functioning of the AM symbiosis requires a high degree of coordination between the two partners based on a finely regulated molecular dialogue (Hause et al. 2007, López-Ráez et al. 2010). The AM fungi-plant communication starts in the rhizosphere with the production and exudation of certain signaling molecules—strigolactones—by the host plant that are recognized by AM fungi (Bouwmeester et al. 2007, López-Ráez et al. 2011b). As signaling cues in the rhizosphere, strigolactones are perceived by the AM fungus engaging its metabolism and giving rise to the so-called pre-symbiotic stage. This stage is characterized by a profuse hyphal branching, thus increasing the probability to contact the root and establish symbiosis (Akiyama et al. 2005). It has been suggested that they might also act as the chemo attractant that directs the hyphal growth to the roots (Sbrana and Giovannetti 2005). Conversely, when the AM fungus starts to branch in the vicinity of the root, plants perceive diffusible fungal signals, called "Myc factors", that induce symbiosis-specific responses in the host root, even in the absence of any physical contact (Parniske 2008, Genre and Bonfante 2010). The chemical structure of these Myc factors was identified as lipochito-oligosaccharides with structural similarities with rhizobial Nod factors (Maillet et al. 2011).

When finally a hypha contacts the plant root, it adheres to epidermal cells forming a characteristic fungal structure called appressorium (also called hyphopodium). This event marks the initiation of the symbiotic phase. From the appressorium a penetrating hypha is formed, which reaches the root cortex by following, in most cases, an intracellular route across epidermal cells. The plant cell actively prepares a pre-penetration apparatus to guide the fungus into the cell (Genre et al. 2005, 2008). Colonization then proceeds to produce the characteristic tree-like structures, called "arbuscules" that the fungus develops within the root cortical cells. They consist of highly ramified hyphae with very fine terminal tips. The arbuscules, which give the name to the symbiosis, are structures where most nutrient exchange between the fungus and the plant is thought to occur (Smith and Read 2008). Vesicles, potentially important fungal storage structures, are also developed by some AM fungal species.

Following root colonization AM fungi form extensive mycelial networks outside the root—the extraradical mycelium, ERM—where spores are developed completing their life cycle. The ERM result in a tri-dimensional structure specialized in the acquisition of mineral nutrients from soil, particularly those whose ionic forms have poor mobility or are present in low concentration in the soil solution, as it is the case with phosphate and ammonia (Barea et al. 2005a). The ERM produced by different fungi have quite varied characteristics, in terms of size and architecture. For example, there exist important differences in hyphal diameter (usually in the range of 2–20 μm) and extend from the roots and able to absorb and translocate nutrients (at a distance up to 25 cm) among individual AM fungi (Smith and Smith 2011). Hyphal length densities in field soils are also variable and range from 3 to 14 m/g depending at least in part on the identity of the AM fungus. They are, in any case, much higher than the root length densities of associated plants (Li et al. 2008), emphasizing how effectively AM fungi can explore soil.

AM Functioning

Through the activities of the interlinked and extensive soil ERM, AM fungi affect the distribution and movement of nutrients within the soil ecosystem (Richardson et al. 2009). The major flux is the transfer of carbon from plant to fungus (and thereby to the soil) and the reciprocal movement of phosphate and ammonium from fungus to plant. In addition to the uptake of nutrients, the AM symbiosis improves plant performance through increased protection against environmental stresses, whether they be biotic (e.g., pathogen attack) or abiotic (e.g., drought, salinity, heavy metals toxicity or presence of organic pollutants), and also enhances soil structure through the formation of hydro-stable aggregates necessary for good soil tilth (Barea et al. 2005a,b, Turnau et al. 2006, Ferrol and Pérez-Tienda 2009, González-Guerrero et al. 2009, Ruíz-Lozano and Aroca 2010, Pozo et al. 2010, Azcón and Barea 2010, López-Ráez et al. 2012). The role of AM fungi in water uptake, and at improving plant tolerance to environmental stresses is discussed later in this chapter. Here we focus on the effects of extraradical AM hyphae on plant nutrient acquisition and in the formation of stable soil aggregates.

Phosphorus (P) availability is the most limiting factor for crop yield in 30–40 percent of the world's arable soils (Barea et al. 2008). The role of AM fungi in phosphate uptake has been well documented for a long time (see Smith and Read 2008). The use of ^{32}P-based isotope dilution approaches has allowed to ascertain and to quantify the AM fungal contribution to plant P acquisition (Barea 2010) and it has been shown that the majority of P taken up by plants comes via the fungal partner (Smith and Smith 2012). The AM fungi take up P from the same pool of soluble ions as do roots, and thus act as an extension to the root system. There is evidence that phosphatase activity is higher in the rhizosphere soil around AM-plants than around non-mycorrhizal roots but there is no clear evidence if this is a fungal-mediated phenomenon that allows alternative P sources to be accessed (Jeffries and Barea 2012).

The acquisition of N via the fungal partner can also be considerable (Leigh et al. 2009), despite the high N demand by the fungi themselves and that the global pool of N in AM mycelia is at least as big as that in fine roots (Hodge and Fitter 2010). While the mycorrhizal activity at improving

N_2-fixation represents a considerable contribution to N inputs in legume species (Azcón and Barea 2010), the possibilities that the mycorrhizal hyphae uses N sources less available to non-mycorrhizal plants was suggested sometime ago (Hodge et al. 2001). To investigate such an AM contribution to N acquisition by plants, the isotope ^{15}N was used to measure the apparent plant available N pool size, i.e., the A_N value of the soil for AM and non-mycorrhizal plants. Several studies, as reviewed by Barea et al. (2005a), demonstrated that the A_N value for mycorrhizal plants is higher than that for non-AM controls. This suggests that the AM mycelium is accessing N forms which are sparingly available for non-mycorrhizal plants. In spite of this, the role played by AM fungi in N acquisition from organic N sources, the dominant form of N in most soils remains controversial (Hodge et al. 2010).

In addition to P and N, AM fungi are able to improve assimilation of other nutrients. When plants are growing in soils deficient in low mobility micronutrients such as Zn, Cu or Fe, their uptake by plants is often increased by the formation of AM symbioses (Liu et al. 2000). As in the case of P and N, this has been attributed to the capacity of the external mycelium to exploit larger volumes of soil. By contrast, it has also been observed that AM fungi are able to confer increased metal tolerance to host plants when they are exposed to supra-optimal concentrations of metals. Therefore, it has been postulated that AM function as a "buffer" that protects the plant against damaging alterations in metal levels in the soil (González-Guerrero et al. 2009).

The analysis of the biochemical and molecular mechanisms involved in nutrient transport processes in AM associations, and in the bidirectional nutrient exchange between symbionts is a matter of recent and current interest but will not be considered here. The interested reader is referred to some of the reviews on the topic (see Barea et al. 2008, Ferrol and Pérez-Tienda 2009, Franken 2010, Harrison et al. 2010, Pérez-Tienda et al. 2011 for details and references).

The AM hyphal network is thus vital for plant nutrition in natural ecosystems and for sustainable crop production in low-input agricultural situations. Many studies have also shown that inter-plant bridges formed by AM fungi can provide channels for direct nutrient transfer between AM fungi associated with different plants (Azcón and Barea 2010). It is clear that the AM hyphal network is essential for the continued cycling of nutrients within the plant community, and once it is lost, nutrient sequestration or leaching will occur at a faster rate than in its presence (Jeffries and Barea 2012).

In addition to the content in available nutrients, physical properties are fundamental for soil quality, with soil structure being one of the most influential factors (Buscot 2005). Soil structure depends on the aggregation status of soil particles and a well-aggregated soil ensures appropriate soil plant water relations, good aeration, root penetrability and organic matter accumulation (Miller and Jastrow 2000). AM hyphae are involved in the formation of stable soil aggregates and, in consequence, in soil conservation and in promoting good soil tilth (Rillig and Mummey 2006, Wright et al. 2007, Kohler et al. 2010, Barea et al. 2011). In addition to the ERM effect entangling soil particles, glomalin-related glyco-proteins produced by AM fungi seem to be involved in the initiation and stabilization of water-stable soil aggregates due to its glue-like hydrophobic nature (Miller and Jastrow 2000, Rillig and Mummey 2006, Wright et al. 2007, Bedini et al. 2009, Curaqueo et al. 2011).

Physiological and Molecular Basis of AM Functioning, and their Implications in Promoting Soil-Plant System Sustainability against Environmental Stresses

The ecological, physiological and molecular components of AM functioning and their implications at enabling the plant to cope more effectively with cultural or environmental stress, either biotic or abiotic, have been the subject of diverse experimental and review studies during the last decade (Barea et al. 2013). The main conclusions from recent articles specialized in the different thematic areas involved are considered and critically summarized here. The selected topics of interest include:

Priming Plant Defenses against Pathogens

Root colonization by AM fungi can improve plant resistance/tolerance to biotic stresses. Although this bioprotection has been widely described in different plant systems, some of the underlying mechanisms remain largely unknown but experimental evidence supports the activation of plant defenses in the AM-mediated plant protection (Azcón-Aguilar and Barea 1996, Goicoechea et al. 2010). During AM establishment modulation of plant defense responses occurs upon recognition of AM fungi to achieve a functional symbiosis and a mild but effective activation of the plant immune responses may occur, both in local and in systemic tissues. This leads to a primed state of the plant that allows a more efficient activation of defense mechanisms in response to attacks by potential enemies (Pozo et al. 2009, 2010, Jung et al. 2012).

Modulation of Plant Defense Responses in Mycorrhizal Plants

The establishment of mycorrhizas implies an alteration in the roots in the level of plant hormones related to defense. For example, changes in jasmonates (JA), salicylic acid (SA) and ethylene (ET) have been described in tomato roots colonized by different *Glomus* species (López-Ráez et al. 2010). As a consequence of this altered balance, the regulation of plant defenses may have an impact on potential attackers. Accumulation of reactive oxygen species and transcripts of certain defense related genes, activation of phenylpropanoid metabolism and accumulation of specific isoforms of hydrolytic enzymes such as chitinases and glucanases has been reported in mycorrhizal roots. Although these reactions are mostly localized, suggesting a role in AM establishment or control of the symbiosis (Pozo et al. 2009), they may contribute to the induction of resistance against certain root pathogens.

Concerning above-ground effects, transcriptional regulation of defense-related genes and accumulation of insect anti-feedant compounds have been reported in shoots of mycorrhizal plants (Gange 2006, Liu et al. 2007). Defense-related genes were among those with altered expression levels, and the authors correlated this finding with increased resistance to shoot pathogens. Furthermore, the volatile blends released by AM plants can be more attractive to aphid parasitoids than those from non-mycorrhizal, as shown in tomato plants (Guerrieri et al. 2004). These results indicate that not only direct, but also indirect plant defense mechanisms may be modulated in mycorrhizal plants. All together, experimental evidences confirm systemic modulation of plant defenses in AM. This modulation may explain the pattern of enhanced resistance/susceptibility of AM plants to diverse enemies on the basis of the different signaling pathways involved in the plant response to particular attackers. In addition, it would explain the fact that AM can modulate the effectiveness of chemically induced plant resistance (Sonnemann et al. 2005).

Priming of Defense Mechanisms in Mycorrhizal Plants

With the exception of limited activation of plant defenses discussed above, a direct activation of defenses has not been observed in AM plants. Despite the vital character of defense responses, constitutive expression of defenses is too costly for the plant in the absence of challenging attackers. Thus, beneficial microorganisms have developed the ability of enhancing resistance not through a direct activation of defenses but through priming of the defense mechanisms (Pozo et al. 2009). A rapid and strong activation of defense mechanisms is crucial for success in controlling attackers. Accordingly, preconditioning of plant tissues for a quick and more effective activation of defenses upon attack has important ecological fitness benefits and seems to be a common feature of the plant´s immune system (Conrath et al. 2006). This boost of basal defenses, known as priming, seems to be successfully triggered by certain beneficial microorganisms including AM fungi (Pozo and Azcón-Aguilar 2007).

Several studies point to priming as a main mechanism operating in Micorrhiza Induced Resistance (MIR), as indicated by the stronger defense reactions triggered in the mycorrhizal plant upon attack (Pozo et al. 2009, 2010). To investigate whether mycorrhizal colonization leads to priming of defenses in aerial tissues, the response of non-mycorrhizal plants or plants colonized by either *Glomus mosseae* or *Glomus intraradices* to the application in the shoots of different defense-related stimuli were compared. Gene expression and enzymatic activities were monitored in a time-course experiment after shoot treatment with JA, ET and SA. Transcript profiling of leaves of mycorrhizal and non-mycorrhizal plants 24 hr upon treatment with JA indicated a stronger induction in mycorrhizal plants of JA regulated genes, including typical defense-related JA responsive genes supporting a prominent role of priming for JA-dependent responses in AM-induced resistance (Pozo et al. 2009, 2010).

Signaling Pathways Involved

Plant defense mechanisms are tightly regulated through an interconnected network of signaling pathways in which JA, ET and SA play major roles. Priming is often manifested as a sensitization of the tissues to one or some of the signaling molecules (Conrath et al. 2006). SA coordinates defense mechanisms that are generally effective against biotrophic pathogens, whereas JA regulates wounding responses and resistance against necrotrophs (Glazebrook 2005). Nevertheless, there is some overlap in their spectrum of efficiency, especially on pathogens with intermediate lifestyles (Pozo et al. 2004). Insect feeding guilds also determine the response they trigger in the plant. Generalist chewing insects, but not phloem-feeding ones, cause wounding and trigger JA-regulated responses (De Vos et al. 2005). These signaling pathways are not independent, intensive interactions ranging from synergism to antagonism shape a complex regulation network, in which trade-offs between SA and JA pathways are well documented (Koornneef and Pieterse 2008). Mycorrhizal roots are related with increased levels of endogenous JA (Hause et al. 2007). The increase probably occurs when connected to a fully AM establishment. Elevated levels of basal JA production could be related to the increased resistance of mycorrhizal plants to certain pathogens and insects. In shoots, increased sensitivity to the hormone, rather than an increase in its level, seems to underlay the primed defense capacity observed in AM plants. Thereby, a partial suppression of some SA-dependent defense responses in the plant is compensated by an enhancement of these JA-regulated. This would result in priming of JA-dependent defense mechanisms (Pozo and Azcón-Aguilar 2007), a defense regulation model which is coherent with the spectrum of effectiveness described for MIR: increased susceptibility to biotrophs, and increased resistance to necrotrophs and generalist chewing insects (Pozo et al. 2009, 2010, Jung et al. 2012).

Agricultural Weed Control

In addition to the already mentioned role of strigolactones as signaling molecules for AM symbiosis establishment, they also act in the rhizosphere as detection signals for root parasitic plants of the *Orobanchaceae*, including the genera *Striga*, *Orobanche* and *Phelipanche* (Bouwmeester et al. 2007, López-Ráez et al. 2011a). These parasitic weeds are some of the most damaging agricultural pests worldwide affecting important crops such as rice, maize, legumes and tomato, causing up to 70 percent of yield losses (Parker 2009). Their seeds surrounding the roots perceive the strigolactones produced by the host plant and germinate, thus starting their lifecycle (López-Ráez et al. 2009). The fact that most of the lifecycle of these weeds occurs belowground, causes great difficulty in diagnosing the infection and usually only after irreversible damage has already been caused to the crop.

Currently, several approaches such as hand weeding, crop rotation, sanitation, fumigation, solarization and improvement of soil fertility are being used to control root parasites without desirable success. Moreover, the most efficient control method—fumigation—is environmentally hazardous. Therefore, new and more environmentally-friendly methods for a more effective control against these agricultural pests should focus on the initial steps in the host-parasite interaction, particularly based on the germination stage induced by strigolactones (López-Ráez et al. 2009).

The fact that strigolactones play a dual role in the rhizosphere as signaling molecules for both AM fungi and root parasitic plants opens the possibility of potential interactions in both biosystems and the use of AM symbiosis as a strategy for pest management. In this sense, it was shown that AM colonization in maize and sorghum induced a lower infection by the parasitic weed *Striga hermonthica* in the field (Lendzemo et al. 2005). These authors proposed that this reduced infection was caused, at least partially, by a reduction in the production and/or exudation of strigolactones in the mycorrhizal plants. Similarly, AM-colonized pea plants showed less stimulatory capacity of germination for seeds of *Orobanche* and *Phelipanche* species than non-colonized plants (Fernández-Aparicio et al. 2010), again suggesting a reduction in strigolactones. More recently, this AM-induced reduction in strigolactone production has been confirmed in tomato and showed that a well established colonization is required for such a reduction. AM symbiosis led to a reduction in the germination of seeds of the root parasite *Phelipanche ramosa*. When the root exudates were analyzed by liquid chromatography tandem mass spectrometry, a correlation between the germination activity and the levels of strigolactones was analytically demonstrated (López-Ráez et al. 2011a). These results observed in maize, sorghum, pea and tomato indicate that reduction in strigolactone production induced by AM symbiosis is a conserved phenomenon across the plant kingdom. Since this beneficial association is established by most plant species worldwide, it can potentially be used as an environmentally-friendly biocontrol strategy for economically important crops that suffer from these root parasites worldwide. Interestingly, these crops could take advantage of all the other well-known benefits of the AM symbiosis, such as positive effect on plant fitness and tolerance/resistance against biotic and abiotic stresses. However, since strigolactones are also AM hyphal branching factors and are involved in plant architecture, the consequences for the AM fungal community in the soil and possible unwanted side-effects on plant architecture should be carefully evaluated before following this approach.

In addition to the root parasitic plants, it has been suggested that certain AM fungi may suppress growth of other aggressive agricultural weeds such as *Chenopodium album* and *Echinocloa crus-galli*, which cause between 10 to 30 percent of crop yield losses every year (Rinaudo et al. 2010). The authors reported that the presence of AM fungi reduced total weed biomass, while sunflower benefited from AM symbiosis via enhanced phosphorus nutrition. Overall, these observations indicate that the use or stimulation of AM fungi in agro-ecosystems may suppress some aggressive weeds and a possible applicability of the AM symbiosis in weed control is suggested, an agricultural practice in the context of sustainability issues.

Plant Response to Osmotic Stresses

In the last decades the land areas of the world becoming arid or semi-arid have increased progressively, rising drought and salinity events that plants need to cope with (Porcel et al. 2012, see also Chapters 4 and 5 in this volume). Both stresses (drought and salinity) have an osmotic component which causes dehydration of plant tissues. Indeed, both stresses cause a diminution of the soil water potential by increasing solutes concentration (salinity) or by diminishing water availability (drought). Both stresses also cause stomatal closure in order to limit water loss (Aroca et al. 2012). So, drought and salinity stresses share a common osmotic component and some of the plant responses are common, besides salinity also causes ionic imbalance and toxic effects (Porcel et al. 2012), which will not be discussed here. Plant mechanisms to cope with osmotic stresses include fine regulation of water uptake capacity and transpiration rate and activation of the antioxidant machinery to overcome the overproduction of reactive oxygen species (ROS) caused by the stress (Aroca et al. 2012, Porcel et al. 2012). These two mechanisms (keeping water and ROS balance) may be ameliorated by the establishment of the AM symbiosis (Ruiz-Lozano 2003, Porcel et al. 2012).

AM Ameliorates Plant Water Balance

It has been extensively found that AM plants showed better leaf water status than non-AM plants under osmotic stress conditions (Porcel and Ruiz-Lozano 2004, Aroca et al. 2007, Porcel et al. 2012). However, when AM plants are bigger than non-AM ones and the growing substrate is limited, the former may have worse leaf water status because the limited water availability in a pot is depleted faster (Ruiz-Sánchez et al. 2011). The better water status of AM plants under osmotic stress conditions is mainly caused by the direct transfer of water from the fungal hyphae to the plant roots. Recently, Ruth et al. (2011) estimated that about 20 percent of root water uptake taken by roots of AM plants is caused by the presence of mycorrhizal mycelium. The higher root water uptake capacity of AM plants is caused by more soil volume explored by the AM roots plus fungal mycelium, and because AM mycelium can reach water reservoirs in the soil not accessible to plant roots because of their bigger size (Ruth et al. 2011). Consequently, AM plants have more chances to take water from soil under osmotic stress conditions. However, to take water efficiently from soils affected by osmotic stresses it is also necessary that the roots (and the mycelium) decrease their osmotic potential below the osmotic potential of the soil. The AM symbiosis has been shown to improve the plant osmotic adjustment by accumulation of different compounds such as proline, sugars, free amino acids, etc., although this effect may differ according to the plant tissue considered (Bheemareddy and Lakshman 2011, Sheng et al. 2011). In this context, Porcel and Ruiz-Lozano (2004) and Ruiz-Lozano et al. (2011) found that non-AM plants accumulated more proline in shoots than AM plants. In contrast, in roots, AM plants subjected to drought accumulated more proline than non-AM plants. This suggests that in root tissues AM plants accumulate more proline in order to cope with the low water potential of drying soil and to keep a water potential gradient favoring water flow into the roots.

Root water uptake depends on root hydraulic conductivity (L) values, which depend ultimately on aquaporin functioning (Aroca et al. 2012). Aquaporins are membrane intrinsic proteins that allow water and other small neutral molecules to pass through them following an osmotic gradient (see Maurel et al. 2008). Thus, *Arabidopsis* plants lacking the expression of one aquaporin gene showed less L than wild type plants (Postaire et al. 2010). Both L and aquaporins are regulated by AM symbiosis under osmotic stress conditions. Indeed, L decreased less or even remained unchanged in AM plants under osmotic stress conditions, together with a different regulation of the expression of aquaporin genes (Aroca et al. 2007). The modulation of particular aquaporins by the AM symbiosis results in a better regulation of plant water status and contributes to global plant resistance to the stressful conditions, as evidenced by better growth and water status of AM plants under conditions of water deficit. As a result, the better water uptake capacity of AM plants allows them to also have higher transpiration rates and hence higher photosynthetic rates under osmotic stress conditions (Porcel et al. 2012).

AM Ameliorates Plant ROS Balance

Since osmotic stresses induce stomatal closure, they also cause an oxidative stress because of the enhancement of ROS (Porcel et al. 2012). This ROS over-production is caused when the stomata are closed there is not enough CO_2 available to be fixed by the enzyme Rubisco, and reduction power is accumulated in excess. Such an excess of reduction power causes an inhibition of the electron transport from photosystem II to photosystem I, and the excess of electrons are trapped by oxygen molecules generating ROS (Miller et al. 2010). Plants have several enzymatic and non-enzymatic mechanisms to eliminate ROS (for details see Miller et al. 2010, Porcel et al. 2012). Several studies found that AM plants show less oxidative damage than non-AM plants under osmotic stress conditions (Porcel and Ruiz-Lozano 2004, Ruiz-Sánchez et al. 2010, Porcel et al. 2012), in some cases because of a higher activity of antioxidant enzymes (Porcel et al. 2012), or a higher concentration of antioxidant compounds (Ruiz-Sánchez et al. 2010). However, since AM plants are able to overcome water deficit induced by osmotic stresses and show higher stomatal conductance than non-AM plants, they also generate less ROS under osmotic stress conditions (Porcel and Ruiz-Lozano 2004). Hence, AM fungi enhance osmotic tolerance of the host plants by several mechanisms or by avoiding water deficit induced by osmotic stresses and/or by increasing antioxidant capacity. It is evident that more research is needed in order to fully understand the involvement of AM symbiosis in tolerating osmotic stresses.

Mechanisms Underlying Tolerance to Heavy Metals

Soil heavy metal (HM) pollution is one of the major problems that negatively affect environmental health. Excessive accumulation of metals in the soil threatens ecosystem sustainability by altering above- and below-ground community structure, often through severe loss of floristic diversity due to removal of the more HM sensitive components of the community. Although considerable variability in plant responses to AM inoculation has been observed in contaminated soils, the potential of AM fungi to buffer HM-stress has been demonstrated in a number of studies (reviewed by Göhre and Paszkowski 2006 and Hildebrandt et al. 2007). This alleviative influence can be attributed not only to AM-mediated nutritional effects, but also to the impact of AM fungi on metal distribution at the soil–fungus–plant interface. Immobilization of metals on both extra- and intraradical fungal structures has been shown (Kaldorf et al. 1999, Joner et al. 2000), thus providing a plausible explanation for repeated enhancement of the barrier for metal translocation from the roots to the shoots of inoculated plants as reported.

HM have been reported to reduce or delay AM colonization. However, even in highly contaminated soils, AM fungal propagules never disappear completely, and whenever colonization occurs, even to a small extent, it induces beneficial effects on the host plant (Gamalero et al. 2009). Heavy metal-tolerant AM fungi have been isolated from polluted soils and these indigenous AM fungal populations have been reported to cope better with HM-toxicity than those not exposed to such long-term selection pressure (del Val et al. 1999). In order to persist in environments with high metal content, AM fungi should have evolved a series of strategies to restrict entry of toxic metal species and to keep intracellular metal homeostasis.

In general, mechanisms of HM tolerance in fungi include reduction of metal uptake and/or increased efflux, metal immobilization (e.g., cell-wall adsorption), extracellular metal sequestration by, e.g., exo-polysaccharides or other extracellular metabolites, intracellular chelation by, e.g., metallothioneins or phytochelatins, and metal localization/sequestration within vacuoles (Gadd 2007). These mechanisms seem to be used in AM fungi as well (reviewed by Ferrol et al. 2009 and González-Guerrero et al. 2009). Accumulation of metal ions on the cell wall has been shown to be an important mechanism leading to metal immobilization by AM fungi (Joner et al. 2000, González-Guerrero et al. 2008). Chitin, the main component of the fungal cell wall, displays free amino and hydroxyl groups that can be very efficient in binding toxic metals. Moreover, AM fungi produce glomalin (glomalin-related soil protein, GRSP), first believed to be a hydrophobin, later identified as likely a 60-kDa heat shock protein homolog (Gadkar and Rillig 2006) that strongly and irreversibly sequesters metal such as Cu, Cd and Zn (González-Chávez et al. 2004, Cornejo et al. 2008) therefore contributing to metal stabilization in the soil.

The fungal vacuoles also seem to play an important role in the regulation of cytosolic metal ion concentrations and the detoxification of potentially toxic metals. Through the activity of specific metal transporters, such as the vacuolar Zn transporter GintZnT1 or the ABC transporter GintABC1 (González-Guerrero et al. 2005, 2010), excess metal is translocated to the vacuoles, where it would cause less damage. Simultaneously, excess metal would be sequestered in the cytosol by the metallothioneins GmarMT1 or GintMT1, which are cysteine-rich polypeptides that confer increased tolerance to Cd and Cu in *Gigaspora margarita* and *Glomus intraradices*, respectively (Lanfranco et al. 2002, González-Guerrero et al. 2007). Additionally, the fungus struggles to reduce the free radicals produced by heavy metals by activating its enzymatic and non-enzymatic antioxidant systems (reviewed by Ferrol et al. 2009).

In addition to metal immobilization by the fungus, increased HM tolerance of mycorrhizal plants may be related to extensive changes in gene expression and protein synthesis induced by the symbiosis itself. Some studies suggest that plant mechanisms responsible for HM tolerance are down-regulated in mycorrhizal plants. For example, while Cd induced up-regulation of a metallothionein gene and induction of gluthation reductase activity in roots of non-mycorrhizal pea plants, no effect was observed in mycorrhizal roots (Rivera-Becerril et al. 2005). Down-regulation of some heavy metal transporters was observed in mycorrhizal roots of several plant species when grown under toxic metal conditions (Burleigh et al. 2003, Ouziad et al. 2005). Additionally, it has been found that AM fungi modulate the leaf transcriptome of a poplar clone grown on a HM-contaminated soil (Cicatelli et al. 2012) and that changes induced by the AM symbiosis on the shoot proteome of *Medicago* are modulated during Cd stress (Aloui et al. 2011). The proteomic suggests that metal toxicity escape in shoots of mycorrhizotrophic plants is supported by a mobilization of the defense mechanisms at the expense of the photosynthesis-dependent symbiotic sucrose sink.

Exploring and Managing the Diversity of AM Fungal Populations in Agro-ecosystems

In AM ecological research it is fundamental to have information on the individual fungi that are actually functional within an ecosystem since differential benefits can be conferred to each plant host by different AM fungi. Even more, different isolates of the same species differ in their effectiveness in conferring benefits to similar host plants (Jeffries and Barea 2012). Because of the necessity to identify and discriminate particular individuals within any AM fungal population, analyzing this sub-set of soil biodiversity is a key issue in rhizosphere ecology. Exploring AM fungal diversity is important per se and is used as a marker for assessing the impact of perturbations over a target agro-ecosystem. The reasons for maintaining AM diversity in agriculture is that mixed crops will benefit more from a mixed population of AM symbionts. In non-agricultural situations, as plant diversity has been related to AM diversity, maintenance of a sustainable mixed plant population depends on the maintenance of a diverse AM fungi population and vice versa (Maherali and Klironomos 2007). Thus, it is important to recognize the impact of mycorrhizal fungi on plant community composition and functioning is a key issue in ecosystem dynamic and that the ad hoc research over the last few decades has led to the conclusion that the activity and diversity of mycorrhizal fungi is a key mechanism in linking biodiversity and ecosystem functioning (Read 1998).

On the other hand, and because of an increasing demand for low-input agriculture has resulted in a greater interest in the manipulation and use of beneficial soil microorganisms, management of the native population of AM fungi is recognized as a sustainable strategy in agriculture or for the restoration of natural ecosystems, because it can reduce the use of chemicals and energy in agriculture leading to a more economical and sustainable production, while minimizing environmental degradation (Jeffries and Barea 2012). These biological interventions are becoming more attractive as the use of chemicals for fumigation and disease control is progressively discouraged and fertilizers have become more expensive (Atkinson 2009).

Accordingly, recent knowledge on the analysis of AM fungal diversity and on AM technological developments is summarized here.

Exploring AM Fungus Diversity and Inoculum and Inoculation Technologies

Diversity Analysis of AM Fungal Populations

Earlier studies concerning the analysis of the diversity of AM fungal communities were based on the morphology, wall characters and ontogeny of their large multinucleated spores. However, molecular tools are now available for a challenging dissection of AM fungal population dynamics (Robinson-Boyer et al. 2009). For molecular identification, the PCR-amplified rRNA gene fragments of the spores and/or the mycelia from AM fungi are usually subjected to cloning, fingerprinting and sequencing (Hempel et al. 2007, Öpik et al. 2008, Toljander et al. 2008, Rosendahl et al. 2009, Alguacil et al. 2009, Sonjak et al. 2009, Sánchez-Castro et al. 2012a,b). Alternative molecular tools now exist to quantitatively analyze the effect of environment, management or inoculation of soils on more diverse AM fungal communities. For example, q-PCR can be used for simultaneous specific and quantitative investigations of particular taxa of AM fungi in roots and soils colonized by several taxa (Gamper et al. 2008, König et al. 2010). In addition, new techniques of high-throughput sequencing (e.g., pyrosequencing) are now being used for AM fungi (Lumini et al. 2010). Despite the advancement in molecular techniques, the identification approaches employed for AM fungi based on morphological characteristics are still valid and used, being considered complementary to the molecular methods (Morton 2008). A lack of relationship between genetic and functional diversity has been shown (Munkvold et al. 2004, Ehinger et al. 2009).

Molecular techniques able to discriminate individual AM fungi within roots have allowed us to corroborate that AM fungal taxa exhibit little host specificity, whilst a single plant can be colonized by many different AM fungal taxa within the same root. These molecular approaches also help to corroborate that a certain degree of host preference (functional compatibility) occurs and this has been shown to play an important role in regulating diversity, stability and productivity of natural ecosystems (Smith and Read 2008).

AM Inoculants' Production and Inoculation Techniques

Once the importance of the AM symbiosis in sustainable strategies of agriculture or restoration of natural ecosystems is recognized, a decision must be made as to whether the native population of AM fungi suffices as the starting material from which to develop a sustainable system. If not, it will be necessary to augment the native species with inoculum of exotic or indigenous isolates, as it is the case with other beneficial soil microorganisms.

Accordingly, novel agro-biotechnological approaches include the use of microbial inoculants but while technology for the production of inexpensive rhizobial and free-living beneficial bacteria is commercially available, constraints on the production of AM inocula and the development of inoculation techniques have limited the exploitation of the AM potential. This is a key issue for sustainable management of either agricultural or natural ecosystems (Azcón and Barea 2010).

The difficulty in culturing obligate symbionts such as AM fungi in the absence of their host plant is a major obstacle for massive inoculum productions (Baar 2008). Despite these problems, the beneficial effects of AM fungi on plant growth have led to their development as bioinoculants for forestry, agriculture and horticulture (Ijdo et al. 2011) and several companies worldwide are producing AM inoculum products which are now commercially available (Gianinazzi and Vosátka 2004, Vosátka et al. 2008, Ijdo et al. 2011). Selection of the appropriate AM fungi is however a key step (Estaún et al. 2002), while specific procedures are required to multiply AM-fungi and to produce high quality inocula (von Alten et al. 2002). Recent developments in AM-inoculum production systems range from nursery plots (Koltai et al. 2008, Cuenca et al. 2008) to *in vitro* monoxenic root organ cultures (Bago and Cano 2005). The resulting materials (spores, hyphae, root fragments, etc.) are added to different carriers, resulting in a wide range of formulations, including encapsulation, to be applied at an agronomical scale using different application methods (Jaizme-Vega et al. 2003, Cuenca et al. 2008, Vosátka et al. 2008), including hydro-seeding (Estaún et al. 2007). It is indeed a matter of discussion whether "generic products" , containing several AM fungi, potentially suitable for a range of applications, are more appropriate for the market than those with precise formulations and AM fungi specifically tuned to particular end-uses (Smith and Read 2008).

Among other companies producing AM inocula in Spain, two *spin-off*, related to the Spanish Thematic Network on Beneficial Plant-Microbial Interactions, have emerged to produce AM inocula. One of them, MYCOVITRO (www.mycovitro.com), from EEZ-CSIC, produces (*in vitro* as monoxenics root cultures) and commercializes inoculants globally known as "glomigel", specifics for each cropping system/ecosystem. The other, Thader-Biotechnology (www.thaderbiotechnology.com), from University of Murcia, which produce and commercialize several microbial (mainly fungi)-based formulated, known as AGROMIC BASE, as an AM fungal inoculants for horticulture and fruit-culture.

Inoculation at broad scale in highly developed farming systems, however has many constraints. To overcome this, management of indigenous populations is a currently viable option (Brito et al. 2008). However, at relatively small-scale (nursery production), AM inoculation is feasible and advantageous. Inoculation of seedlings is potentially a good method for establishing selected fungi in roots before potting on or planting-out into the field. Inoculation is appropriate where transplanting is part of the normal production system, as is the case with horticulture, including plantation crops. If possible it is best to use indigenous isolates that are already adapted to the prevailing conditions at the field site (Requena et al. 2001, Pelligrino et al. 2011), thus, the concept of using native inoculants rather than exotic ones is now widely accepted. Introduction of exotic inoculum may only be necessary if a native population is either absent or has a low inoculum potential, or if the native fungi are ineffective for the crop species that are being planted (Jeffries and Barea 2012).

Managing AM Fungi for the Restoration of Degraded Ecosystems

Ecosystem degradation is one of the most serious problems in the world, affecting plant community structure and functionality and in some geographic areas can have a high risk of desertization, a complex and dynamic process which is claiming several hundred million hectares annually (Jeffries and Barea 2012). Human activities can cause or accelerate desertification and the loss of most plant species and their corresponding symbionts (Barea et al. 2007).

As a result of degradation/desertification processes, disturbance of natural plant communities is often accompanied, or preceded by, loss of physical, chemical and biological soil properties, such as soil structure, plant nutrient availability, organic matter content, microbial activity, etc. (Jeffries and Barea 2012). These traits are fundamental for soil quality, particularly soil structure stabilization (Miller and Jastrow 2000, Buscot 2005), and thus limit re-establishment of the natural plant cover. In particular, desertification causes a sharp decrease in the numbers and diversity of AM fungal spores while soil disturbance is also responsible for the destruction of the AM hyphal network, thereby affecting inoculum potential (Cuenca et al. 1998, Requena et al. 2001, Allen 2007, Barea et al. 2011). Thus, consideration of the AM symbiosis is usually recommended in the establishment of plants in degraded soils, particularly in arid and semi-arid ecosystems where they would enable greater plant tolerance of environmental stresses characteristic of these ecosystems including nutrient deficiency, drought and soil disturbance (Goicoechea et al. 2004, Allen 2007, Barea et al. 2011).

The semiarid Southeast Spain could be a representative scenario for prospecting and applying AM fungi to improve functioning of plant communities, the most arid region of continental Europe. The area is characterized by a warm dry Mediterranean climate where scarce and irregular rainfall, a long dry and hot summer and man-mediated degradation activities may synergistically act as driving-forces able to promote desertification processes (Requena et al. 2001). The ecological and functional impact of mycorrhizas in semiarid ecosystems of Southeast Spain have been investigated for more than 20 yr through basic, strategic and applied studies addressed at ascertaining how the activity and diversity of mycorrhizal fungi affect plant community composition, structure and dynamics in the region. These studies have recently been reviewed (Barea et al. 2011) and the main conclusions can be summarized as follows.

Disturbance of the target semi-arid ecosystems caused by global climate change or other stresses decreases the density and diversity of AM fungal populations, yet some mycorrhizal propagules remain, suggesting adaptation to the situation within the stressed area. Many field experiments have demonstrated that using native plant species, inoculated with a managed community of indigenous AM fungi, is a successful biotechnological tool to aid the integral recovery of degraded semi-arid Southeast Iberian ecosystems. Inoculation with indigenous AM fungi has resulted in improvement of both plant performance and the physicochemical properties in the soil around native plant species, whereas exotic fungi were effective only during the first year after transplanting. This is the way for restoration and maintenance of a self-sustaining ecosystem.

Field experiments have shown that tailored inoculation with AM fungi in combination with composted organic residues, using indigenous shrub/tree species as test plants, improved plant establishment, enhanced enzymatic activities involved in C, N and P cycling, and increased soil aggregate stability.

New information on the specificity of mycorrhizal fungus-plant interactions in patchy environments has been reported, indicating a strong influence of mycorrhizal fungi on plant populations and community dynamics in arid ecosystems. Isotopic techniques based on approaches using $\delta^{18}O$, $\delta^{13}C$ and $\delta^{15}N$ measurements have proved useful in determining how mycorrhizal inoculation affects some eco-physiological responses related to water-use efficiency and nutrient acquisition by plants.

Recovering Endangered Flora

Ecosystems worldwide can include large numbers of endangered plant species, some of them threatened with extinction (Bothe et al. 2010). Although extinction is a natural biological process, it is reckoned that the current rate of plant species extinction is between a hundred and a thousand times faster than the average historical extinction rate for the Earth (Pimm et al. 1995). As it is widely accepted, this accelerated extinction rate is linked almost exclusively to the activities of human beings including overexploitation of species, alterations in their natural habitats or environmental changes, threats which are exacerbated in the current scenario of global warming and the subsequent climate change (Stork 2010). According to the International Union for Conservation of Nature (IUCN) Red List, these constraints are particularly relevant for tropical forests and high mountain ecosystems, where the numbers of plant species included in the catalogues of endangered flora are currently increasing.

In the context of managing mycorrhizal symbioses to exploit their role in promoting soil-plant systems sustainability against environmental stresses, special attention has been devoted to ascertain their potential significance in protecting endangered plants and habitats in high mountain ecosystems. This is an important issue because high mountains, which often bear plant diversities richer than those in their surrounding lowlands, usually include large numbers of threatened species (Körner 2003). Plants growing in high mountain habitats have developed numerous adaptations imposed by the harsh conditions derived from the increase in altitude (low temperatures, intense winds, short growing season, nutrient-poor soils, etc.). These adaptations are manifested in different life strategies and physiological processes (Grabherr et al. 2003). As most plants depend on mycorrhizas to thrive, particularly in fragile and stressed environments, the establishment of mycorrhizal associations was suggested to be one of these adaptive strategies for endangered plants to survive and reproduce in high mountain ecosystems (Fuchs and Haselwandter 2004). Accordingly, diverse studies have been carried out trying to corroborate whether mycorrhiza formation help endangered plants to perform in degraded high mountain ecosystems and contribute to the regeneration of their natural habitats (see Azcón-Aguilar et al. 2012 for references).

Since many endangered species in high mountains form AM associations (Zubek et al. 2008), the potential role of AM fungi in protecting endangered plants in these environments has been especially investigated (Zubek et al. 2009, Bothe et al. 2010, Navarro-Fernández et al. 2011, Azcón-Aguilar et al. 2012). In one of these studies (Bothe et al. 2010) it was shown that AM colonization affects vegetative growth and sexual reproduction of endangered plant species by increasing the number of inflorescences, fruit and seed production, and offspring vigor. In another study, the response of endangered plant species to inoculation with AM fungi was investigated with the aim of improving the *ex situ* propagation of the threatened taxa and the success of their reintroduction in their natural habitats (Zubek et al. 2009). They concluded that AM fungal application can improve performance of the target species and suggested a beneficial influence of AM fungi on the acclimatization of the reintroduced endangered species in nature. In a series of studies carried out in Sierra de Baza Natural Park in Granada, Spain, Navarro-Fernández et al. (2011) found that the AM symbiosis was critical for improving the growth of *Thymus granatensis*, an endemic and threatened plant species from dolomitic soils. These soils have supra-optimal Mg and Ca concentrations and a low water holding capacity, characteristics which restrict the number of plants able to survive in these environments and their reproduction. The AM fungal community adapted to these dolomitic soils was able to contribute to the tolerance mechanisms of *T. granatensis* plants to high Mg concentrations under the prevalent drought conditions.

The significance of AM fungi on the conservation of endangered flora is also being investigated in the high Mediterranean mountains of the Sierra Nevada National Park, Granada, Spain (Azcón-Aguilar et al. 2012). Sierra Nevada constitutes an exceptional refuge for flora and one of the enclaves with a higher biodiversity level of the European continent. Seedlings from a number of selected endangered species were nursery produced and inoculated with specific mixes of suitable autochthonous AM fungi previously isolated from the target environments.

In conclusion, research carried out in the field of AM conservational management supports that native AM fungi could be used as a tool for the propagation of endangered species for restoration purposes, and for the *in situ* and *ex situ* conservation of these plant species and their natural habitats. Additionally, this research also aims at the conservation of AM fungal population diversity in threatened areas worldwide. In this context,

Turrini and Giovannetti (2012) suggest the use of protected areas (national parks and other nature reserves) as a basis for a strategic perspective for the *in situ* conservation of AM fungi, alongwith the conservation of their endangered plant species.

Mycorrhizosphere Interactions to Improve Plant Development under Stress Situations

Managing AM fungi and their interactions with ecosystem beneficial microorganisms (mycorrhizosphere tailoring) is nowadays recognized as a feasible biotechnological tool to improve plant growth and health, and soil quality as a sustainable practice in agriculture or restoration of natural ecosystems (Barea et al. 2005b). Many co-inoculation experiments using selected AM fungi and rhizosphere microorganisms and their ecological impact have recently been reported (Barea et al. 2013). The main conclusions from some of these experiments are summarized below with emphasis on the ecological impacts of interactions related to: (1) biofertilization in nutrient deficient soils; (2) abiotic stress alleviation; and (3) restoration of degraded ecosystems.

Mycorrhizosphere Interactions to Improve Plant Growth in Nutrient Deficient Soils

Soil microorganisms that increase the amount of nutrients available to plants (biofertilizers) include rhizobial bacteria and phosphate-solubilizing bacteria and fungi. The interactions of these beneficial rhizosphere microorganisms with AM fungi in the so-called mycorrhizosphere are relevant in stressed soils with a low availablity of nutrients.

The widespread presence of the AM symbiosis in legumes and its role in improving nodulation and N_2 fixation by legume-rhizobia associations are both universally recognized processes, as based on the supply of P by the AM fungi to satisfy the high P-demand of symbiotic N_2 fixation. This information has recently been reviewed (Azcón and Barea 2010, Barea et al. 2013), and only a summarized analysis is given here. Methodologies based on the use of ^{15}N-enriched inorganic fertilizer allow to ascertain and quantify the amount of N which is actually fixed by legume-rhizobia consortia in a particular situation and to measure the contribution of the AM symbiosis to the process. A lower $^{15}N/^{14}N$ ratio in the shoots of rhizobia-inoculated AM plants with respect to those achieved by the same rhizobial strain in non-mycorrhizal plants was found. This indicated an enhancement of the N_2 fixation rates (an increase in ^{14}N from the atmosphere), as induced by the AM activity (Barea et al. 2005a). The ecological impact and benefit from using two biological systems to supply the plant with the two major nutrients, N and P, in nutrient-deficient, stressed soils, is self-evident. In addition, the ^{15}N-based isotopic techniques have been also used to measure N-transfer in mixed cropping where legumes are usually involved. Since the AM mycelia can link different plant species growing nearby, and help overlap the pool of available nutrients for the intercropped plant species, the N released into the overlapping mycorrhizospheres by legume root exudation, or by nodule decay, can result in nitrogen available for non-fixing plants (Azcón and Barea 2010).

The interactions between AM fungi and phosphate-solubilizing-microorganisms (PSM) are relevant to P cycling and plant nutrition, particularly in P-deficient soils. Because the Pi made available by PSM, acting on sparingly-soluble P sources, has limited diffusion in soil solution, the already available Pi may not reach the root surface. However, AM fungi could tap the phosphate ions solubilized by the PSM and translocate them into plant roots (Barea 2010). The microbial interaction of AM fungi and PSM has been tested in experiments using ^{32}P-tracer methodologies (Barea et al. 2007). Upon adding a small amount of ^{32}P to label the exchangeable soil P pool, the isotopic composition, or "specific activity" ($SA = {}^{32}P/{}^{31}P$ quotient), was determined in plant tissues. It was found that dual inoculation reduced the SA of the host plant, indicating that these plants acquired P from sources which were not directly available to non-inoculated or singly-inoculated plants. Microbial inoculation improved biomass production and P accumulation in plants, demonstrating the interactive effects of PSM and AM-fungi on P capture, cycling and supply in a tailored mycorrhizosphere (Barea et al. 2008).

Multi-microbial interactions, including those between locally isolated AM fungi, PSB, and *Azospirillum*, have also been reported, which indicate that microorganisms can act synergistically when co-inoculated (Azcón and Barea 2010).

Mycorrhizosphere Interactions in Phytoremediation of Heavy Metal Contaminated Soils

AM fungi improve phytoremediation of soils contaminated with heavy metals, radionuclides or polycyclic aromatic hydrocarbons (Leyval et al. 2002). Most phytoremediation assays involving mycorrhizosphere interactions concern HM and different strategies of phytoremediation, have been investigated (Turnau et al. 2006, Ruíz-Lozano and Azcón 2011). These studies mostly concentrated on Zn, Cu, Cd, Pb or Ni.

Interactions between rhizobacteria and AM-fungi have been investigated in diverse experiments to ascertain whether they are able to benefit phytoremediation (Azcón et al. 2009a,b, 2010). The main achievements resulting from these experiments using HM-multiple contaminated soils and *Trifolium* as test plants were: (i) a number of bacteria and the AM-fungi were isolated from a HM-contaminated soil, and identified by 16S rRNA or 18S rRNA gene sequencing, respectively; (ii) the target bacteria were able to accumulate large amounts of metals; (iii) co-inoculation with a HM-adapted autochthonous bacteria and AM fungi increased biomass, N and P content as compared to non-inoculated plants, and also enhanced the establishment of symbiotic structures (nodule number and AM colonization), which were negatively affected as the level of HM in soil increased; (iv) dual inoculation lowered HM concentrations in *Trifolium* plants, inferring a phytostabilization-based activity, however, as the total HM content in plant shoots was higher in dually-inoculated plants, due to the effect on biomass accumulation, a possible phytoextraction activity was suggested; and (v) inoculated HM-adapted bacteria increased dehydrogenase, phosphatase and β-gluconase activities, and auxin production, in the mycorrhizosphere, indicating an enhancement of microbial activities related to plant development. The physiological/biochemical mechanisms by which the tested bacterial isolates enhanced phytoremediation activity in AM plants include: (i) improved rooting, and AM formation and functioning; (ii) enhanced microbial activity in the mycorrhizosphere; and (iii) accumulation of metals in the root-soil environment, thus avoiding their transfer to the trophic chain, or to aquifers (Ruíz-Lozano and Azcón 2011).

Inoculation of autochthonous AM fungi and bacteria, together with the application of treated agrowaste residue, changed the bacterial community structure and enhanced phytoextraction to remediate heavy metals contaminated soils (Azcón et al. 2009a). An enhancement of antioxidant activities in plants inoculated with AM fungi and bacteria, and agrowaste residue, was observed (Azcón et al. 2009b). Such a mycorrhizosphere effect seems to help plants to limit oxidative damage to bio-molecules in response to metal stress.

Mycorrhizosphere Interaction for Restoring Degraded Ecosystems

As desertification causes disturbance of plant-microbe symbioses, the recovery of populations of AM fungi and rhizobial bacteria is essential to the integral restoration of a degraded area (Barea et al. 2011, Jeffries and Barea 2012). Accordingly, management of AM fungi, together with rhizosphere bacteria, was proposed for the integral restoration of degraded ecosystems. A model experiment in this context was carried out in a desertified semi-arid ecosystem with *Anthyllis cytisoides*, a drought-tolerant legume, as the test plant (Requena et al. 2001). *Anthyllis* seedlings inoculated with indigenous rhizobia and AM fungi were transplanted to field plots for a five-year-trial. The tailored mycorrhizosphere enhanced seedling survival and growth, P-acquisition, N-fixation, and N-transfer from N-fixing to associated non-fixing species in the natural succession. The improvement in the physical-chemical properties in the soil around the *Anthyllis* plants was shown by the increased levels of N, organic matter and number of hydro-stable soil aggregates. Glomalin-related glycol-proteins, produced by the external hyphae of AM fungi, seen to be involved in the initiation and stabilization of water-stable soil aggregates, due to its glue-like hydrophobic nature (Miller and Jastrow 2000, Rillig and Mummey 2006, Bedini et al. 2009).

Future Trends

As the current scenario of climate change is threatening the stability of both natural and agricultural systems, research efforts must be addressed to potentiate the AM role as an adaptive strategy to confer plants an increased tolerance to environmental stresses. Therefore, both basic and applied research needs to be implemented. On the one hand, we need to increase our knowledge on the underlying mechanisms involved in an increased tolerance of mycorrhizal plants to either biotic or abiotic stresses. Future functional genomics (including transcriptomics, proteomics and metabolomics) developments will be useful to identify the genes expressed in AM plants that play key roles in processes such as nutrient mobilization, abiotic stress alleviation or suppression of plant diseases. On the other hand, the specific management of AM fungi should be one of the main objectives of applied studies in the future and the use of AM inoculants must take into account the importance of retaining AM diversity in the rhizosphere to achieve realistic and effective biotechnological applications in agricultural and natural systems.

Accordingly, mycorrhizal technology should be a component in sustainable agricultural strategies in the future, since application of AM fungi can reduce fertilizer and energy inputs yet promoting healthy plant growth, particularly in a world of depleting non-renewable resources. The application of AM inoculants is likely to become even more important due to the agro-ecological threats of agrochemicals, which are urgently required to be reduced, and even avoided, to increase food quality, sustainable food production and environmental protection. Further studies must address the consequences of the co-operation between microbes in the mycorhizosphere under field conditions to assess their ecological impacts as a viable alternative for agriculture, horticulture, and revegetation of degraded ecosystems.

With regard to natural ecosystem conservation, in spite of many achievements that have been reached investigating the ecological and functional role of mycorrhizas in the restoration of degraded areas, we still lack a comprehensive view of the mycorrhizal potentiality in propelling ecosystem dynamics and activities by improving the composition, diversity, structure and functionality of stress-adapted plant communities.

Acknowledgements

We thank support from the Spanish National Research Program (R & D+i)-European Union (FEDER) through different grants which financed research on the different topics covered in this chapter: CGL2009-08825/BOS, AGL2009-07691, AGL2011-25403, AGL2009-12530-C02-02, AGL2009-08868; and from the Andalucian (Spain) Government, PAIDI (R & D+i) Program: P07-CVI-02952, P11-CVI-07107, CVI-7640.

References

Akiyama, K., K. Matsuzaki and H. Hayashi. 2005. Plant sesquiterpenes induce hyphal branching in arbuscular mycorrhizal fungi. Nature. 435: 824–827.

Alguacil, M.M., A. Roldán and M.P. Torres. 2009. Assessing the diversity of AM fungi in arid gypsophilous plant communities. Environmental Microbiology. 11: 2649–2659.

Aloui, A., G. Recorbet, F. Robert, B. Schoefs, M. Bertrand, C. Henry, V. Gianinazzi-Pearson, E. Dumas-Gaudot and S. Aschi-Smiti. 2011. Arbuscular mycorrhizal symbiosis elicits shoot proteome changes that are modified during cadmium stress alleviation in *Medicago truncatula*. Bmc Plant Biology. 11.

Altieri, M.A. 2004. Linking ecologists and traditional farmers in the search for sustainable agriculture. Frontiers in Ecology and the Environment. 2: 35–42.

Allen, M.F. 2007. Mycorrhizal fungi: Highways for water and nutrients in arid soils. Vadose Zone Journal. 6: 291–297.

Aroca, R., R. Porcel and J.M. Ruíz-Lozano. 2007. How does arbuscular mycorrhizal symbiosis regulate root hydraulic properties and plasma membrane aquaporins in *Phaseolus vulgaris* under drought, cold or salinity stresses? New Phytologist. 173: 808–816.

Aroca, R., R. Porcel and J. Manuel Ruiz-Lozano. 2012. Regulation of root water uptake under abiotic stress conditions. Journal of Experimental Botany. 63: 43–57.

Atkinson, D. 2009. Soil microbial resources and agricultural policies. *In:* C. Azcón-Aguilar, J.M. Barea, S. Gianinazzi and V. Gianinazzi-Pearson [eds.]. Mycorrhizas Functional Processes and Ecological Impact. Springer-Verlag, Berlin, Heidelberg. pp. 33–45.

Azcón-Aguilar, C. and J.M. Barea. 1996. Arbuscular mycorrhizas and biological control of soil-borne plant pathogens—an overview of the mechanisms involved. Mycorrhiza. 6: 457–464.

Azcón-Aguilar, C., J. Palenzuela, N. Ferrol, F. Oehl and J.M. Barea. 2012. Mycorrhizal status and arbuscular mycorrizal fungal diversity of endangered plant species in the Sierra Nevada National Park. *In:* R. Duponnois and M. Hafidi [eds.]. The Mycorrhizal Symbiosis in Mediterranean Environment: Importance in Ecosystem Stability and in Soil Rehabilitation Strategies. Editions Nova Science Publishers, New York. pp. 49–70.

Azcón, R. and J.M. Barea. 2010. Mycorrhizosphere interactions for legume improvement. *In:* M.S. Khan, A. Zaidi and J. Musarrat [eds.]. Microbes for Legume Improvement Springer-Verlag, Vienna, New York. pp. 237–271.

Azcón, R., A. Medina, A. Roldán, B. Biró and A. Vivas. 2009a. Significance of treated agrowaste residue and autochthonous inoculates (Arbuscular mycorrhizal fungi and *Bacillus cereus*) on bacterial community structure and phytoextraction to remediate heavy metals contaminated soils. Chemosphere. 75: 327–334.

Azcón, R. M.D. Perálvarez, B. Biró, A. Roldán and J.M. Ruíz-Lozano. 2009b. Antioxidant activities and metal acquisition in mycorrhizal plants growing in a heavy-metal multicontaminated soil amended with treated lignocellulosic agrowaste. Applied Soil Ecology. 41: 168–177.

Azcón, R., M.D. Perálvarez, A. Roldán and J.M. Barea. 2010. Arbuscular mycorrhizal fungi, *Bacillus cereus*, and *Candida parapsilosis* from a multicontaminated soil alleviate metal toxicity in plants. Microbial Ecology. 59: 668–677.

Baar, J. 2008. From production to application of arbuscular mycorrhizal fungi in agricultural systems: requirements and needs. *In:* A. Varma [ed.]. Mycorrhiza: State of the Art, Genetics and Molecular Biology, Eco-Function, Biotechnology, Eco-Physiology, Structure and Systematics. 3rd edn. Springer-Verlag, Berlin, Heidelberg, Germany. pp. 361–373.

Bago, B. and C. Cano. 2005. Breaking myths on arbuscular mycorrhizas *in vitro* biology. *In:* S. Declerck, F.G. Strullu and J.A. Fortin [eds.]. *In Vitro* Culture of Mycorrhizas, vol. 4. Soil Biology. Springer-Verlag, Berlin, Heidelberg. pp. 111–138.

Barea, J.M. 2010. Mycorrhizas and agricultural fertility. *In:* A. González-Fontes, A. Gárate and I. Bonilla [eds.]. Agricultural Sciences: Topics in Modern Agriculture. Studium Press. Houston, Texas, USA. pp. 257–274.

Barea, J.M., R. Azcón and C. Azcón-Aguilar. 2005a. Interactions between mycorrhizal fungi and bacteria to improve plant nutrient cycling and soil structure. *In:* F. Buscot and A. Varma [eds.]. Microorganisms in Soils: Roles in Genesis and Functions. Springer-Verlag, Berlin, Heidelbert. pp. 195–212.

Barea, J.M., M.J. Pozo, R. Azcón and C. Azcón-Aguilar. 2005b. Microbial co-operation in the rhizosphere. Journal of Experimental Botany. 56: 1761–1778.

Barea, J.M., J. Palenzuela, P. Cornejo, I. Sánchez, C. Navarro, P.B. Quiñones, R. Azcón, N. Ferrol and C. Azcón-Aguilar. 2007. Significado, diversidad e impacto de los hongos de las micorrizas arbusculares en ambientes mediterráneos. *In:* J.M. Barea-Azcón, M. Moleón, R. Travesí, E. Ballesteros, J.M. Luzón and J.M. Tierno [eds.]. Biodiversidad y Conservación de Fauna y Flora en Ambientes Mediterráneos. Sociedad Granatense de Historia Natural, Granada, España. pp. 155–185.

Barea, J.M., N. Ferrol, C. Azcón-Aguilar and R. Azcón. 2008. Mycorrhizal symbioses. *In:* P.J. White and J.P. Hammond [eds.]. The Ecophysiology of Plant-Phosphorus Interactions. Series: Plant Ecophysiology, vol. 7 Springer, Dordrecht. pp. 143–163.

Barea, J.M., J. Palenzuela, P. Cornejo, I. Sánchez-Castro, C. Navarro-Fernández, A. Lopéz-García, B. Estrada, R. Azcón, N. Ferrol and C. Azcón-Aguilar. 2011. Ecological and functional roles of mycorrhizas in semi-arid ecosystems of Southeast Spain. Journal of Arid Environments. 75: 1292–1301.

Barea, J.M., M.J. Pozo, R. Azcón and C. Azcón-Aguilar. 2013. Microbial Interactions in the Rhizosphere. *In:* F. de Bruijn [ed.]. Molecular Microbial Ecology of the Rhizosphere. Wiley-Blackwell, USA. pp. 29–44.

Bedini, S., E. Pellegrino, L. Avio, S. Pellegrini, P. Bazzoffi, E. Argese and M. Giovannetti. 2009. Changes in soil aggregation and glomalin-related soil protein content as affected by the arbuscular mycorrhizal fungal species *Glomus mosseae* and *Glomus intraradices*. Soil Biology & Biochemistry. 41: 1491–1496.

Bheemareddy, V.S. and H.C. Lakshman. 2011. Effect of AM fungus *Glomus fasciculatum* on metabolite accumulation in four varieties of *Triticum aestivum* L. under short-term water stress. Vegetos. 24: 41–49.

Blanca, G., M.R. López Onieva, J. Lorite, M.J. Matínez Lirola, J. Molero Mesa, S. Quintas, M. Ruíz Girela, M.A. Varo and S. Vidal. 2002. Flora Amenazada y Endémica de Sierra Nevada Editorial Universidad de Granada, Consejería de Medio Ambiente, Junta de Andalucía, Granada.

Bonfante, P. and A. Genre. 2008. Plants and arbuscular mycorrhizal fungi: an evolutionary-developmental perspective. Trends in Plant Science. 13: 492–498.

Bothe, H., K. Turnau and M. Regvar. 2010. The potential role of arbuscular mycorrhizal fungi in protecting endangered plants and habitats. Mycorrhiza. 20: 445–457.

Bouwmeester, H.J., C. Roux, J.A. López-Ráez and G. Becard. 2007. Rhizosphere communication of plants, parasitic plants and AM fungi. Trends in Plant Science. 12: 224–230.

Brito, I., M.J. Goss, M. de Carvalho, D. van Tuinen and P.M. Antunes. 2008. Agronomic management of indigenous mycorrhizas. *In:* A. Varma [ed.]. Mycorrhiza: State of the Art, Genetics and Molecular Biology, Eco-Function, Biotechnology, Eco-Physiology, Structure and Systematics. 3rd edn. Springer-Verlag, Berlin, Heidelberg, Germany. pp. 375–402.

Brundrett, M.C. 2002. Coevolution of roots and mycorrhizas of land plants. New Phytologist. 154: 275–304.

Brundrett, M.C. 2009. Mycorrhizal associations and other means of nutrition of vascular plants: understanding the global diversity of host plants by resolving conflicting information and developing reliable means of diagnosis. Plant and Soil. 320: 37–77.

Burleigh, S.H., B.K. Kristensen and I.E. Bechmann. 2003. A plasma membrane zinc transporter from *Medicago truncatula* is up-regulated in roots by Zn fertilization, yet down-regulated by arbuscular mycorrhizal colonization. Plant Molecular Biology. 52: 1077–1088.

Buscot, F. 2005. What are soils? *In:* F. Buscot and S. Varma [eds.]. Microorganisms in soils: roles in genesis and functions. Springer-Verlag. Heidelberg, Germany. pp. 3–18.

Cicatelli, A., G. Lingua, V. Todeschini, S. Biondi, P. Torrigiani and S. Castiglione. 2012. Arbuscular mycorrhizal fungi modulate the leaf transcriptome of a *Populus alba* L. clone grown on a zinc and copper-contaminated soil. Environmental and Experimental Botany. 75: 25–35.

Conrath, U., G.J.M. Beckers, V. Flors, P. García-Agustín, G. Jakab, F. Mauch, M.A. Newman, C.M.J. Pieterse, B. Poinssot, M.J. Pozo, A. Pugin, U. Schaffrath, J. Ton, D. Wendehenne, L. Zimmerli and B. Mauch-Mani. 2006. Priming: Getting ready for battle. Molecular Plant-Microbe Interactions. 19: 1062–1071.

Cornejo, P., S. Meiera, G. Borie, M. Rillig and F. Borie. 2008. Glomalin-related soil protein in a Mediterranean ecosystem affected by a copper smelter and its contribution to Cu and Zn sequestration. Science of the Total Environment. 406: 154–160.

Cuenca, G., Z. De Andrade and G. Escalante. 1998. Diversity of Glomalean spores from natural, disturbed and revegetated communities growing on nutrient-poor tropical soils. Soil Biology & Biochemistry. 30: 711–719.

Cuenca, G., A. Cáceres and M.G. González. 2008. AM inoculation in tropical agriculture: field results. *In:* A. Varma [ed.]. Mycorrhiza: State of the Art, Genetics and Molecular Biology, Eco-Function, Biotechnology, Eco-Physiology, Structure and Systematics. 3rd edn. Springer-Verlag, Berlin, Heidelberg, Germany. pp. 403–417.

Curaqueo, G., J.M. Barea, E. Acevedo, R. Rubio, P. Cornejo and F. Borie. 2011. Effects of different tillage system on arbuscular mycorrhizal fungal propagules and physical properties in a Mediterranean agroecosystem in central Chile. Soil & Tillage Research 113: 11–18.

Chaudhary, V.B., M.A. Bowker, T.E. O'Dell, J.B. Grace, A.E. Redman, M.C. Rillig and N.C. Johnson. 2009. Untangling the biological contributions to soil stability in semiarid shrublands. Ecological Applications. 19: 110–122.

De Vos, M., V.R. Van Oosten, R.M.P. Van Poecke, J.A. Van Pelt, M.J. Pozo, M.J. Mueller, A.J. Buchala, J.P. Metraux, L.C. Van Loon, M. Dicke and C.M.J. Pieterse. 2005. Signal signature and transcriptome changes of Arabidopsis during pathogen and insect attack. Molecular Plant-Microbe Interactions. 18: 923–937.

del Val, C., J.M. Barea and C. Azcón-Aguilar. 1999. Assessing the tolerance to heavy metals of arbuscular mycorrhizal fungi isolated from sewage sludge-contaminated soils. Applied Soil Ecology. 11: 261–269.

Ehinger, M., A.M. Koch and I.R. Sanders. 2009. Changes in arbuscular mycorrhizal fungal phenotypes and genotypes in response to plant species identity and phosphorus concentration. New Phytologist. 184: 412–423.

Estaún, V., A. Camprubí and E.J. Joner. 2002. Selecting arbuscular mycorrhizal fungi for field application. *In:* S. Gianinazzi, H. Schüepp, J.M. Barea and K. Haselwandter [eds.]. Mycorrhiza Technology in Agriculture, from Genes to Bioproducts. Birkhauser Verlag. Basel, Switzerland. pp. 249–259.

Estaún, V., S. Vicente, C. Calvet, A. Camprubí and M. Busquets. 2007. Integration of arbuscular mycorrhiza inoculation in hydroseeding technology. Effects on plant growth and inter-species competition. Land Degradation & Development. 18: 621–630.

Fernández-Aparicio, M., J.M. García-Garrido, J.A. Ocampo and D. Rubiales. 2010. Colonisation of field pea roots by arbuscular mycorrhizal fungi reduces *Orobanche* and *Phelipanche* species seed germination. Weed Research. 50: 262–268.

Ferrol, N., M. González-Guerrero, A. Valderas, K. Benabdellah and C. Azcón-Aguilar. 2009. Survival strategies of arbuscular mycorrhizal fungi in Cu-polluted environments. Phytochemistry Reviews. 8: 551–559.

Ferrol, N. and J. Pérez-Tienda. 2009. Coordinated nutrient exchange in arbuscular mycorrhiza. *In:* C. Azcón-Aguilar, J.M. Barea, S. Gianinazzi and V. Gianinazzi-Pearson [eds.]. Mycorrhizas Functional Processes and Ecological Impact. Springer-Verlag, Berlin, Heidelberg. pp. 73–87.

Franken, P. 2010. Molecular-physiological aspects of the AM symbiosis post penetration. *In:* H. Koltai and Y. Kapulnik [eds.]. Arbuscular Mycorrhizas: Physiology and Function. Springer, The Netherlands. pp. 93–116.

Fuchs, B. and K. Haselwandter. 2004. Red list plants: colonization by arbuscular mycorrhizal fungi and dark septate endophytes. Mycorrhiza. 14: 277–281.

Gadd, G.M. 2007. Geomycology: biogeochemical transformations of rocks, minerals, metals and radionuclides by fungi, bioweathering and bioremediation. Mycological Research. 111: 3–49.

Gadkar, V. and M.C. Rillig. 2006. The arbuscular mycorrhizal fungal protein glomalin is a putative homolog of heat shock protein 60. Fems Microbiology Letters. 263: 93–101.

Gamalero, E., G. Lingua, G. Berta and B.R. Glick. 2009. Beneficial role of plant growth promoting bacteria and arbuscular mycorrhizal fungi on plant responses to heavy metal stress. Canadian Journal of Microbiology. 55: 501–514.

Gamper, H.A., J.P.W. Young, D.L. Jones and A. Hodge. 2008. Real-time PCR and microscopy: Are the two methods measuring the same unit of arbuscular mycorrhizal fungal abundance? Fungal Genetics and Biology. 45: 581–596.

Gange, A. 2006. Insect-mycorrhizal interactions: patterns, processes, and consequences. *In:* T. Ohgushi, T. Craig and P. Price [eds.]. Indirect interaction webs: Nontrophic linkages through induced plant traits. University Press, Cambridge. pp. 124–144.

Genre, A. and P. Bonfante. 2010. The making of symbiotic cells in arbuscular mycorrhizal roots. *In:* H. Koltai and Y. Kapulnik [eds.]. Arbuscular Mycorrhizas: Physiology and Function. Springer, The Netherlands. pp. 57–71.

Genre, A., M. Chabaud, T. Timmers, P. Bonfante and D.G. Barker. 2005. Arbuscular mycorrhizal fungi elicit a novel intracellular apparatus in *Medicago truncatula* root epidermal cells before infection. Plant Cell. 17: 3489–3499.

Genre, A., M. Chabaud, A. Faccio, D.G. Barker and P. Bonfante. 2008. Prepenetration apparatus assembly precedes and predicts the colonization patterns of arbuscular mycorrhizal fungi within the root cortex of both *Medicago truncatula* and *Daucus carota*. Plant Cell. 20: 1407–1420.

Gianinazzi, S. and M. Vosátka. 2004. Inoculum of arbuscular mycorrhizal fungi for production systems: science meets business. Canadian Journal of Botany-Revue Canadienne de Botanique. 82: 1264–1271.

Gianinazzi-Pearson, V., M. Tollot and P.M.A. Seddas. 2009. Dissection of genetic cell programmes driving early arbuscular mycorrhiza interactions. *In:* C. Azcón-Aguilar, J.M. Barea, S. Gianinazzi and V. Gianinazzi-Pearson [eds.]. Mycorrhizas Functional Processes and Ecological Impact. Springer-Verlag, Berlin, Heidelberg. pp. 33–45.

Glazebrook, J. 2005. Contrasting mechanisms of defense against biotrophic and necrotrophic pathogens. Annual Review of Phytopathology. 43: 205–227.

Göhre, V. and U. Paszkowski. 2006. Contribution of the arbuscular mycorrhizal symbiosis to heavy metal phytoremediation. Planta. 223: 1115–1122.

Goicoechea, N., S. Merino and M. Sánchez-Díaz. 2004. Contribution of arbuscular mycorrhizal fungi (AMF) to the adaptations exhibited by the deciduous shrub *Anthyllis cytisoides* L. under water deficit. Physiologia Plantarum. 122: 453–464.

Goicoechea, N., I. Garmendia, M. Sánchez-Díaz and J. Aguirreolea. 2010. Arbuscular mycorrhizal fungi (AMF) as bioprotector agents against wilt induced by *Verticillium* spp. in pepper. Spanish Journal of Agricultural Research. 8: S25–S42.

González-Chávez, M.C., R. Carrillo-González, S.F. Wright and K.A. Nichols. 2004. The role of glomalin, a protein produced by arbuscular mycorrhizal fungi, in sequestering potentially toxic elements. Environmental Pollution. 130: 317–323.

González-Guerrero, M., C. Azcón-Aguilar, M. Mooney, A. Valderas, C.W. MacDiarmid, D.J. Eide and N. Ferrol. 2005. Characterization of a *Glomus intraradices* gene encoding a putative Zn transporter of the cation diffusion facilitator family. Fungal Genetics and Biology. 42: 130–140.

González-Guerrero, M., C. Cano, C. Azcón-Aguilar and N. Ferrol. 2007. GintMT1 encodes a functional metallothionein in *Glomus intraradices* that responds to oxidative stress. Mycorrhiza. 17: 327–335.

González-Guerrero, M., L.H. Melville, N. Ferrol, J.N.A. Lott, C. Azcón-Aguilar and R.L. Peterson. 2008. Ultrastructural localization of heavy metals in the extraradical mycelium and spores of the arbuscular mycorrhizal fungus *Glomus intraradices*. Canadian Journal of Microbiology. 54: 103–110.

González-Guerrero, M., K. Benabdellah, N. Ferrol and C. Azcón-Aguilar. 2009. Mechanisms underlying heavy metal tolerance in arbuscular mycorrhizas. *In:* C. Azcón-Aguilar, J.M. Barea, S. Gianinazzi and V. Gianinazzi-Pearson [eds.]. Mycorrhizas Functional Processes and Ecological Impact. Springer-Verlag, Berlin, Heidelberg. pp. 107–122.

González-Guerrero, M., K. Benabdellah, A. Valderas, C. Azcón-Aguilar and N. Ferrol. 2010. GintABC1 encodes a putative ABC transporter of the MRP subfamily induced by Cu, Cd, and oxidative stress in *Glomus intraradices*. Mycorrhiza. 20: 137–146.

Grabherr, G., L. Nagy and D. Thompson. 2003. An outline of Europe's alpine areas. *In:* L. Nagy, G. Grabherr, Ch. Körner and D.B.A. Thompson [eds.]. Alpine Biodiversity in Europe. Ecological Series vol. 167. Springer, Berlin, Heidelberg. pp. 3–12.

Guerrieri, E., G. Lingua, M.C. Digilio, N. Massa and G. Berta. 2004. Do interactions between plant roots and the rhizosphere affect parasitoid behaviour? Ecological Entomology. 29: 753–756.

Harrison, M.J., N. Pumplin, F.J. Breuillin, R.D. Noar and H.J. Park. 2010. Phosphate transporters in arbuscular mycorrhizal symbiosis. *In:* H. Koltai and Y. Kapulnik [eds.]. Arbuscular Mycorrhizas: Physiology and Function. Springer, The Netherlands. pp. 117–135.

Hause, B., C. Mrosk, S. Isayenkov and D. Strack. 2007. Jasmonates in arbuscular mycorrhizal interactions. Phytochemistry. 68: 101–110.

Hempel, S., C. Renker and F. Buscot. 2007. Differences in the species composition of arbuscular mycorrhizal fungi in spore, root and soil communities in a grassland ecosystem. Environmental Microbiology. 9: 1930–1938.

Hildebrandt, U., M. Regvar and H. Bothe. 2007. Arbuscular mycorrhiza and heavy metal tolerance. Phytochemistry. 68: 139–146.

Hodge, A. and A.H. Fitter. 2010. Substantial nitrogen acquisition by arbuscular mycorrhizal fungi from organic material has implications for N cycling. Proceedings of the National Academy of Sciences of the United States of America. 107: 13754–13759.

Hodge, A., C.D. Campbell and A.H. Fitter. 2001. An arbuscular mycorrhizal fungus accelerates decomposition and acquires nitrogen directly from organic material. Nature. 413: 297–299.

Hodge, A., T. Helgason and A.H. Fitter. 2010. Nutritional ecology of arbuscular mycorrhizal fungi. Fungal Ecology. 3: 267–273.

Honrubia, M. 2009. The Mycorrhizae: a plant-fungus relation that has existed for more than 400 million years. Anales del Jardin Botanico de Madrid. 66: 133–144.

Ijdo, M., S. Cranenbrouck and S. Declerck. 2011. Methods for large-scale production of AM fungi: past, present, and future. Mycorrhiza. 21: 1–16.

Jaizme-Vega, M.C., A.S. Rodríguez-Romero, C.M. Hermoso and S. Declerck. 2003. Growth of micropropagated bananas colonized by root-organ culture produced arbuscular mycorrhizal fungi entrapped in Ca-alginate beads. Plant and Soil. 254: 329–335.

Jeffries, P. and J.M. Barea. 2012. Arbuscular Mycorrhiza—a key component of sustainable plant-soil ecosystems. *In:* B. Hock [ed.]. The Mycota. Springer-Verlag, Berlin, Heidelberg. pp. 51–75.

Jeffries, P., S. Gianinazzi, S. Perotto, K. Turnau and J.M. Barea. 2003. The contribution of arbuscular mycorrhizal fungi in sustainable maintenance of plant health and soil fertility. Biology and Fertility of Soils. 37: 1–16.

Joner, E.J., R. Briones and C. Leyval. 2000. Metal-binding capacity of arbuscular mycorrhizal mycelium. Plant and Soil. 226: 227–234.

Jung, S., A. Martínez-Medina, J.A. López-Ráez and M.J. Pozo. 2012. Mycorrhiza-induced resistance and priming of plant defences. Journal of Chemical Ecology. 38: 651–664.

Kaldorf, M., A.J. Kuhn, W.H. Schroder, U. Hildebrandt and H. Bothe. 1999. Selective element deposits in maize colonized by a heavy metal tolerance conferring arbuscular mycorrhizal fungus. Journal of Plant Physiology. 154: 718–728.

Klironomos, J.N. 2002. Feedback with soil biota contributes to plant rarity and invasiveness in communities. Nature. 417: 67–70.

Kohler, J., F. Caravaca and A. Roldán. 2010. An AM fungus and a PGPR intensify the adverse effects of salinity on the stability of rhizosphere soil aggregates of *Lactuca sativa*. Soil Biology & Biochemistry. 42: 429–434.

Koltai, H., D. Meir, E. Shlomo, N. Resnick, O. Ziv, S. Wininger, B. Ben-Dor and Y. Kapulnik. 2008. Exploiting arbuscular mycorrhizal technology in different cropping systems under greenhouse conditions in semi-arid regions. *In:* C. Kubota and M. Kacira [eds.]. Book series Acta Horticulturae. Proceedings of the International Workshop on Greenhouse Environmental Control and Crop Production in Semi-Arid Regions, vol 797. ISHS publishes, Tucson, AZ, USA. pp. 223–228.

König, S., T. Wubet, C.F. Dormann, S. Hempel, C. Renker and F. Buscot. 2010. Taqman real-time PCR assays to assess arbuscular mycorrhizal responses to field manipulation of grassland biodiversity: effects of soil characteristics, plant species richness, and functional traits. Applied and Environmental Microbiology. 76: 3765–3775.

Koornneef, A. and C.M.J. Pieterse. 2008. Cross talk in defense signaling. Plant Physiology. 146: 839–844.

Körner, C. 2003. Alpine Plant Life—Functional Plant Ecology of High Mountain Ecosystems. 2nd edn. Springer, Heidelberg.

Lanfranco, L., A. Bolchi, E.C. Ros, S. Ottonello and P. Bonfante. 2002. Differential expression of a metallothionein gene during the presymbiotic versus the symbiotic phase of an arbuscular mycorrhizal fungus. Plant Physiology. 130: 58–67.

Leigh, J., A. Hodge and A.H. Fitter. 2009. Arbuscular mycorrhizal fungi can transfer substantial amounts of nitrogen to their host plant from organic material. New Phytologist. 181: 199–207.

Lendzemo, V.W., T.W. Kuyper, M.J. Kropff and A. van Ast. 2005. Field inoculation with arbuscular mycorrhizal fungi reduces Striga hermonthica performance on cereal crops and has the potential to contribute to integrated Striga management. Field Crops Research. 91: 51–61.

Leyval, C., E.J. Joner, C. del Val and K. Haselwandter. 2002. Potential of arbuscular mycorrhizal fungi for bioremediation. *In:* S. Gianinazzi, H. Schüepp, J.M. Barea and K. Haselwandter [eds.]. Mycorrhizal Technology in Agriculture. Birkhäuser Verlag, Basel, Switzerland. pp. 175–186.

Li, H.Y., S.E. Smith, K. Ophel-Keller, R.E. Holloway and F.A. Smith. 2008. Naturally occurring arbuscular mycorrhizal fungi can replace direct P uptake by wheat when roots cannot access added P fertiliser. Functional Plant Biology. 35: 124–130.

Liu, A., C. Hamel, R.I. Hamilton, B.L. Ma and D.L. Smith. 2000. Acquisition of Cu, Zn, Mn and Fe by mycorrhizal maize (*Zea mays* L.) grown in soil at different P and micronutrient levels. Mycorrhiza. 9: 331–336.

Liu, J., I. Maldonado-Mendoza, M. López-Meyer, F. Cheung, C.D. Town and M.J. Harrison. 2007. Arbuscular mycorrhizal symbiosis is accompanied by local and systemic alterations in gene expression and an increase in disease resistance in the shoots. Plant Journal. 50: 529–544.

López-Ráez, J.A., R. Matusova, C. Cardoso, M. Jamil, T. Charnikhova, W. Kohlen, C. Ruyter-Spira, F. Verstappen and H. Bouwmeester. 2009. Strigolactones: ecological significance and use as a target for parasitic plant control. Pest Management Science. 65: 471–477.

López-Ráez, J.A., A. Verhage, I. Fernández, J.M. García, C. Azcón-Aguilar, V. Flors and M.J. Pozo. 2010. Hormonal and transcriptional profiles highlight common and differential host responses to arbuscular mycorrhizal fungi and the regulation of the oxylipin pathway. Journal of Experimental Botany. 61: 2589–2601.

López-Ráez, J.A., T. Charnikhova, I. Fernández, H. Bouwmeester and M.J. Pozo. 2011a. Arbuscular mycorrhizal symbiosis decreases strigolactone production in tomato. Journal of Plant Physiology. 168: 294–297.

López-Ráez, J.A., M.J. Pozo and J.M. García-Garrido. 2011b. Strigolactones: a cry for help in the rhizosphere. Botany-Botanique. 89: 513–522.

López-Ráez, J.A., H. Bouwmeester and M.J. Pozo. 2012. Communication in the rhizosphere, a target for pest management. *In:* E. Lichtfouse [ed.]. Sustainable Agriculture Reviews, vol 8. Agroecology and Strategies for Climate Change, vol. 8. Sustainable Agriculture Reviews. Springer. The Netherlands. pp. 109–133.

Lumini, E., A. Orgiazzi, R. Borriello, P. Bonfante and V. Bianciotto. 2010. Disclosing arbuscular mycorrhizal fungal biodiversity in soil through a land-use gradient using a pyrosequencing approach. Environmental Microbiology. 12: 2165–2179.

Maherali, H. and J.N. Klironomos. 2007. Influence of Phylogeny on fungal community assembly and ecosystem functioning. Science. 316: 1746–1748.

Maillet, F., V. Poinsot, O. Andre, V. Puech-Pages, A. Haouy, M. Gueunier, L. Cromer, D. Giraudet, D. Formey, A. Niebel, E.A. Martínez, H. Driguez, G. Bécard and J. Denarie. 2011. Fungal lipochitooligosaccharide symbiotic signals in arbuscular mycorrhiza. Nature. 469: 58–64.

Maurel, C., L. Verdoucq, D.-T. Luu and V. Santoni. 2008. Plant aquaporins: Membrane channels with multiple integrated functions. Annual Review of Plant Biology. 59: 595–624.

Miller, G., N. Suzuki, S. Ciftci-Yilmaz and R. Mittler. 2010. Reactive oxygen species homeostasis and signalling during drought and salinity stresses. Plant Cell and Environment. 33: 453–467.

Miller, R.M. and J.D. Jastrow. 2000. Mycorrhizal fungi influence soil structure. *In:* Y. Kapulnik and D.D. Douds [eds.]. Arbuscular Mycorrhizas: Physiology and Function. Kluwer Academic Publishers, Dordrecht, The Netherlands. pp. 3–18.

Morton, J.B. 2009. Reconciliation of conflicting phenotypic and rRNA gene phylogenies of fungi in glomeromycota based on underlying patterns and processes. *In:* C. Azcón-Aguilar, J.M. Barea, S. Gianinazzi and V. Gianinazzi-Pearson [eds.]. Mycorrhizas Functional Processes and Ecological Impact. Springer-Verlag, Berlin, Heidelberg. pp. 137–154.

Munkvold, L., R. Kjøller, M. Vestberg, S. Rosendahl and I. Jakobsen. 2004. High functional diversity within species of arbuscular mycorrhizal fungi. New Phytologist. 164: 357–364.

Navarro-Fernández, C.M., R. Aroca and J.M. Barea. 2011. Influence of arbuscular mycorrhizal fungi and water regime on the development of endemic *Thymus* species in dolomitic soils. Applied Soil Ecology. 48: 31–37.

Öpik, M., Ü. Saks, J. Kennedy and T. Daniell. 2008. Global diversity patterns of arbuscular mycorrhizal fungi-community composition and links with functionality. *In:* A. Varma [ed.]. Mycorrhiza: State of the Art, Genetics and Molecular Biology, Eco-Function,

Biotechnology, Eco-Physiology, Structure and Systematics. 3rd edn. Springer-Verlag, Berlin, Heidelberg, Germany. pp. 89–111.

Ouziad, F., U. Hildebrandt, E. Schmelzer and H. Bothe. 2005. Differential gene expressions in arbuscular mycorrhizal-colonized tomato grown under heavy metal stress. Journal of Plant Physiology. 162: 634–649.

Parker, C. 2009. Observations on the current status of *Orobanche* and *Striga* problems worldwide. Pest Management Science. 65: 453–459.

Parniske, M. 2008. Arbuscular mycorrhiza: the mother of plant root endosymbioses. Nature Reviews Microbiology. 6: 763–775.

Pellegrino, E., S. Bedini, L. Avio, E. Bonari and M. Giovannetti. 2011. Field inoculation effectiveness of native and exotic arbuscular mycorrhizal fungi in a Mediterranean agricultural soil. Soil Biology & Biochemistry. 43: 367–376.

Pérez-Tienda, J., P.S. Testillano, R. Balestrini, V. Fiorilli, C. Azcón-Aguilar and N. Ferrol. 2011. GintAMT2, a new member of the ammonium transporter family in the arbuscular mycorrhizal fungus *Glomus intraradices*. Fungal Genetics and Biology. 48: 1044–1055.

Pimm, S.L., G.J. Russell, J.L. Gittleman and T.M. Brooks. 1995. The future of biodiversity. Science. 269: 347–350.

Pirozynski, K.A. and D.W. Malloch. 1975. Origin of land plants: a matter of mycotropism. BioSystems. 6: 153–164.

Porcel, R. and J.M. Ruíz-Lozano. 2004. Arbuscular mycorrhizal influence on leaf water potential, solute accumulation, and oxidative stress in soybean plants subjected to drought stress. Journal of Experimental Botany. 55: 1743–1750.

Porcel, R., R. Aroca and J. Manuel Ruíz-Lozano. 2012. Salinity stress alleviation using arbuscular mycorrhizal fungi. A review. Agronomy for Sustainable Development. 32: 181–200.

Postaire, O., C. Tournaire-Roux, A. Grondin, Y. Boursiac, R. Morillon, A.R. Schaeffner and C. Maurel. 2010. A PIP1 aquaporin contributes to hydrostatic pressure-induced water transport in both the root and rosette of *Arabidopsis*. Plant Physiology. 152: 1418–1430.

Pozo, M.J. and C. Azcón-Aguilar. 2007. Unraveling mycorrhiza-induced resistance. Current Opinion in Plant Biology. 10: 393–398.

Pozo, M., L. Van Loon and C. Pieterse. 2004. Jasmonates—Signals in plant-microbe interactions. Journal of Plant Growth Regulation. 23: 211–222.

Pozo, M.J., A. Verhage, J. García-Andrade, J.M. García and C. Azcón-Aguilar. 2009. Priming plant defence against pathogens by arbuscular mycorrhizal fungi. *In:* C. Azcón-Aguilar, J.M. Barea, S. Gianinazzi and V. Gianinazzi-Pearson [eds.]. Mycorrhizas Functional Processes and Ecological Impact. Springer-Verlag, Berlin, Heidelberg. pp. 123–135.

Pozo, M.J., S.C. Jung, J.A. López-Ráez and C. Azcón-Aguilar. 2010. Impact of arbuscular mycorrhizal symbiosis on plant response to biotec stress: The role of plant defence mechanisms. *In:* H. Koltai and Y. Kapulnik [eds.]. Arbuscular Mycorrhizas: Physiology and Function. Springer, The Netherlands. pp. 193–207.

Read, D. 1998. Biodiversity—Plants on the web. Nature. 396: 22–23.

Redecker, D., R. Kodner and L.E. Graham. 2000. Glomalean fungi from the Ordovician. Science. 289: 1920–1921.

Requena, N., E. Pérez-Solis, C. Azcón-Aguilar, P. Jeffries and J.M. Barea. 2001. Management of indigenous plant-microbe symbioses aids restoration of desertified ecosystems. Applied and Environmental Microbiology. 67: 495–498.

Richardson, A.E., J.M. Barea, A.M. McNeill and C. Prigent-Combaret. 2009. Acquisition of phosphorus and nitrogen in the rhizosphere and plant growth promotion by microorganisms. Plant and Soil. 321: 305–339.

Rillig, M.C. and D.L. Mummey. 2006. Mycorrhizas and soil structure. New Phytologist. 171: 41–53.

Rinaudo, V., P. Barberi, M. Giovannetti and M.G.A. van der Heijden. 2010. Mycorrhizal fungi suppress aggressive agricultural weeds. Plant and Soil. 333: 7–20.

Rivera-Becerril, F., D. van Tuinen, F. Martin-Laurent, A. Metwally, K.J. Dietz, S. Gianinazzi and V. Gianinazzi-Pearson. 2005. Molecular changes in *Pisum sativum* L. roots during arbuscular mycorrhiza buffering of cadmium stress. Mycorrhiza. 16: 51–60.

Robinson-Boyer, L., I. Grzyb and P. Jeffries. 2009. Shifting the balance from qualitative to quantitative analysis of arbuscular mycorrhizal communities in field soils. Fungal Ecology. 2: 1–9.

Rosendahl, S., P. McGee and J.B. Morton. 2009. Lack of global population genetic differentiation in the arbuscular mycorrhizal fungus *Glomus mosseae* suggests a recent range expansion which may have coincided with the spread of agriculture. Molecular Ecology. 18: 4316–4329.

Ruíz-Lozano, J.M. 2003. Arbuscular mycorrhizal symbiosis and alleviation of osmotic stress. New perspectives for molecular studies. Mycorrhiza. 13: 309–317.

Ruíz-Lozano, J.M. and R. Aroca. 2010. Host response to osmotic stresses: stomatal behaviour and water use efficiency of arbuscular mycorrhizal plants. *In:* H. Koltai and Y. Kapulnik [eds.]. Arbuscular Mycorrhizas: Physiology and Function, 2nd edn. Springer Science+Business Media B.V. , Dordrecht, The Netherlands. pp. 239–256.

Ruíz-Lozano, J.M. and R. Azcón. 2011. *Brevibacillus*, arbuscular mycorrhizae and remediation of metal toxicity in agricultural soils. *In:* N.A. Logan and P. de Vos [eds.]. Endospore-forming Soil Bacteria, Soil Biology. Springer-Verlag. Berlin, Heidelberg, Germany. pp. 235–258.

Ruíz-Lozano, J.M., M.C. Peralvarez, R. Aroca and R. Azcon. 2011. The application of a treated sugar beet waste residue to soil modifies the responses of mycorrhizal and non mycorrhizal lettuce plants to drought stress. Plant and Soil. 346: 153–166.

Ruíz-Sánchez, M., E. Armada, Y. Munoz, I.E. García de Salamone, R. Aroca, J.M. Ruíz-Lozano and R. Azcón. 2011. *Azospirillum* and arbuscular mycorrhizal colonization enhance rice growth and physiological traits under well-watered and drought conditions. Journal of Plant Physiology. 168: 1031–1037.

Ruíz-Sánchez, M., R. Aroca, Y. Munoz, R. Polon and J.M. Ruíz-Lozano. 2010. The arbuscular mycorrhizal symbiosis enhances the photosynthetic efficiency and the antioxidative response of rice plants subjected to drought stress. Journal of Plant Physiology. 167: 862–869.

Ruth, B., M. Khalvati and U. Schmidhalter. 2011. Quantification of mycorrhizal water uptake via high-resolution on-line water content sensors. Plant and Soil. 342: 459–468.

Sánchez-Castro, I., N. Ferrol and J.M. Barea. 2012a. Analyzing the community composition of arbuscular mycorrhizal fungi colonizing the roots of representative shrubland species in a Mediterranean ecosystem. Journal of Arid Environments. 80: 1–9.

Sánchez-Castro, I., N. Ferrol, P. Cornejo and J.M. Barea. 2012b. Temporal dynamics of arbuscular mycorrhizal fungi colonizing roots of representative shrub species in a semi-arid Mediterranean ecosystem. Mycorrhiza. (in press).

Sbrana, C. and M. Giovannetti. 2005. Chemotropism in the arbuscular mycorrhizal fungus *Glomus mosseae*. Mycorrhiza. 15: 539–545.

Schüßler, A. and C. Walker. 2011. Evolution of the 'plant-symbiotic' fungal phylum, Glomeromycota. *In:* S. Pöggeler and J. Wöstemeyer [eds.]. Evolution of fungi and fungal-like organisms Springer-Verlag, Berlin Heidelberg. pp. 163–185.

Schüßler, A., H. Gehrig, D. Schwarzott and C. Walker. 2001. Analysis of partial *Glomales* SSU rRNA gene sequences: implications for primer design and phylogeny. Mycological Research. 105: 5–15.

Sheng, M., M. Tang, F. Zhang and Y. Huang. 2011. Influence of arbuscular mycorrhiza on organic solutes in maize leaves under salt stress. Mycorrhiza. 21: 423–430.

Smith, S.E. and D.J. Read. 2008. Mycorrhizal Symbiosis. 3rd edn. Elsevier, Academic Press, New York.

Smith, S.E. and F.A. Smith. 2011. Roles of arbuscular mycorrhizas in plant nutrition and growth: new paradigms from cellular to ecosystem scales. Annual Review of Plant Biology. 62: 227–250.

Smith, S.E. and F.A. Smith. 2012. Fresh perspectives on the roles of arbuscular mycorrhizal fungi in plant nutrition and growth. Mycologia. 104: 1–13.

Sonjak, S., T. Beguiristain, C. Leyval and M. Regvar. 2009. Temporal temperature gradient gel electrophoresis (TTGE) analysis of arbuscular mycorrhizal fungi associated with selected plants from saline and metal polluted environments. Plant and Soil. 314: 25–34.

Sonnemann, I., N.M. Streicher and V. Wolters. 2005. Root associated organisms modify the effectiveness of chemically induced resistance in barley. Soil Biology & Biochemistry. 37: 1837–1842.

Stork, N.E. 2010. Re-assessing current extinction rates. Biodiversity and Conservation. 19: 357–371.

Toljander, J.F., J.C. Santos-González, A. Tehler and R.D. Finlay. 2008. Community analysis of arbuscular mycorrhizal fungi and bacteria in the maize mycorrhizosphere in a long-term fertilization trial. FEMS Microbiology Ecology. 65: 323–338.

Turnau, K., A. Jurkiewicz, G. Língua, J.M. Barea and V. Gianinazzi-Pearson. 2006. Role of arbuscular mycorrhiza and associated microorganisms in phytoremediation of heavy metal-polluted sites. *In:* M.N.V. Prasad, K.S. Sajwan and R. Naidu [eds.]. Trace Elements in the Environment. Biogeochemistry, Biotechnology and Bioremediation. CRC/Taylor & Francis, Boca Raton, Florida. pp. 235–252.

Turrini, A. and M. Giovannetti. 2012. Arbuscular mycorrhizal fungi in national parks, nature reserves and protected areas worldwide: a strategic perspective for their *in situ* conservation. Mycorrhiza. 22: 81–97.

von Alten, H., B. Blal, J.C. Dodd, F. Feldman and M. Vosátka. 2002. Quality control of arbuscular mycorrhizal fungi inoculum in Europe. *In:* S. Gianinazzi, H. Schüepp, J.M. Barea and K. Haselwandter [eds.]. Mycorrhiza Technology in Agriculture, from Genes to Bioproducts. Birkhauser Verlag. Basel, Switzerland. pp. 281–296.

Vosátka, M., J. Albrechtová and R. Patten. 2008. The international marked development for mycorrhizal technology. *In:* A. Varma [ed.]. Mycorrhiza: State of the Art, Genetics and Molecular Biology, Eco-Function, Biotechnology, Eco-Physiology, Structure and Systematics. 3rd edn. Springer-Verlag, Berlin, Heidelberg, Germany. pp. 419–438.

Wright, S.F., V.S. Green and M.A. Cavigelli. 2007. Glomalin in aggregate size classes from three different farming systems. Soil & Tillage Research. 94: 546–549.

Zubek, S., K. Turnau and J. Blaszkowski. 2008. Arbuscular mycorrhiza of endemic and endangered plants from the Tatra Mts. Acta Societatis Botanicorum Poloniae. 77: 149–156.

Zubek, S., K. Turnau, M. Tsimilli-Michael and R.J. Strasser. 2009. Response of endangered plant species to inoculation with arbuscular mycorrhizal fungi and soil bacteria. Mycorrhiza. 19: 113–123.

16

Can Arbuscular Mycorrhizal Fungi (AMF) be Effective Tools for Improving the Nutritional Quality of Crops? Findings from a Worldwide Consumed Vegetable: Lettuce

M. Baslam, I. Pascual, M. Sánchez-Díaz and *N. Goicoechea**

Introduction

The Role of Fruits and Vegetables on Human Health

Several studies have demonstrated a positive correlation between the consumption of fresh fruits and vegetables and human health. In fact, epidemiological research works have shown that food exerts a direct effect on health so that there is a close correlation between dietary habits and disease risk. These beneficial effects of fruits and vegetables are believed to be due, among others, to their contents of fiber, ascorbic acid, carotenoids

Departamento de Biología Vegetal, Sección Biología Vegetal (Unidad Asociada al CSIC, EEAD, Zaragoza e ICVV, Logroño), Facultades de Ciencias y Farmacia, Universidad de Navarra, C/ Irunlarrea 1, 31008—Pamplona, Spain.
*Corresponding author: niegoi@unav.es

and polyphenols that may protect key biological constituents such as lipoproteins, membranes, and DNA (Szeto et al. 2004).

As described by Kendall et al. (2010), the dietary pattern of humans has shifted towards a diet in which refined grains, meat, added fats and sugar are common. In contrast, vegetable protein and fiber intake is low. This change in diet together with a sedentary lifestyle is largely responsible for the increased prevalence of obesity and chronic diseases including diabetes, heart disease and cancer.

Ascorbic acid is one of the most important water soluble vitamins and it is essential for collagen, carnitine and neurotransmitters biosynthesis. In contrast with most plants and animals, which synthesize ascorbic acid for their own requirement, humans cannot synthesize ascorbic acid due to the lack of the enzyme L-gulonolactone oxidase. This is the reason why ascorbic acid must be supplemented through food in humans (see Naidu 2003 for a review). Ascorbic acid is widely distributed in fresh fruits and vegetables.

Carotenoids are a widespread group of naturally occurring fat-soluble pigments (yellow, orange or red) that are synthesized by plants and microorganisms but not by animals (Rao and Rao 2007). Carotenoids are especially abundant in yellow-orange fruits and vegetables and in dark green, leafy vegetables (Mangels et al. 1993). Outer leaves or skin of fruits and vegetables usually contain higher levels of carotenoids than inner parts of the plants (Hart and Scott 1995). In plant cells, carotenoids are mainly present in lipid membranes or stored in plastids (West and Poortvliet 1993, Mangels et al. 1993) and play indispensable roles in photosynthesis as accessory pigments for light harvesting and in the prevention of photo-oxidative damage (Demmig-Adams and Adams 1996). In human beings, carotenoids can play important roles due to their provitamin A activity and antioxidant potential, which may enhance the function of the immune system and inhibit the development of some types of cancers (Maiani et al. 2009).

Natural phenolic compounds are secondary metabolites and a major class of antioxidants in plants (You et al. 2011). Anthocyanins are a type of phenolic compounds of increasing interest for human diet. Together with chlorophylls and carotenoids, anthocyanins are some of the most common pigments in the Angiospermae. They are water-soluble plant pigments particularly evident in fruit and flower tissues, where they are responsible for a diverse range of red, blue and purple colors (Jaganath and Crozier 2010). The most common connotation with pigmented flower petals and fruits is the attraction of animals both for pollination and seed dispersal. However, anthocyanins play a crucial role in increasing the antioxidant response of plants to both biotic and abiotic stress factors (Stintzing and Carle 2004). Apart from their physiological roles in plants, anthocyanins are regarded as important components in human nutrition due to their antioxidant potential (Ross and Kasum 2002). In fact, anthocyanins are regarded as "supporting human defense mechanisms" when excessive radical production may surpass the antioxidant capacity provided by enzymes and other antioxidant compounds, such as glutathione, tocopherol or ascorbic acid (Stintzing and Carle 2004).

Relevance of Lettuce (*Lactuca sativa* L.) for Human Diet and Health

Lettuce (*Lactuca sativa* L.) is an annual herbaceous plant belonging to the Asteraceae family. Humans have had a long history of cultivation of lettuce for domestic needs. This vegetable is native to the Mediterranean region and its cultivation may have started in Egypt as early as 4,500 BC, perhaps initially for the edible oil extracted from its seeds. Salad lettuce was popular with the Ancient Greeks and Romans (Harlan 1986). Lettuce is a minimally processed food-product and its use has continued to increase in salads, sandwiches and fast foods because of its long storage life, good quality and its perception as a healthy food (Ferreres et al. 1997, Mou 2009). Lettuce is also the most used food crop for the so called 'Fourth Range' of vegetables. The term originally meant fresh, cleaned, possibly chopped and mixed vegetables ready to be seasoned and eaten (Borghi 2003). These vegetables are widely accepted by consumers because they are easy to prepare for eating.

Lettuce, a cool season crop, is widely grown in all the continents, especially in temperate and subtropical regions, and is a major food crop within the European Union. According to FAOSTAT (FAO Statistics Division 2011), the production quantity of lettuce and chicory in Spain, France, Germany and Greece was, respectively, 1,000,000, 430,000, 320,000 and 80,000 tons in 2009. In 2010 the total production quantity of lettuce and chicory in the European Union achieved 3,023,174 tons, being in Italy (843,344 tons) and in Spain (809,200 tons) the main producing countries.

The nutrient content of this vegetable includes useful amounts of several important components, such as phenolics, vitamins A and C, calcium and iron as well as to its fiber content (Romani et al. 2002, Nicolle et al. 2004a, b). In folk medicine, lettuce seeds are used in powder form for the treatment of rhinitis, asthma and cough, and its decoction is used for the treatment of insomnia and as a sedative; moreover, the oil extracted from the seeds has shown some analgesic effect after external application on the head (Said et al. 1996).

The nutrient content of lettuce, however, is affected or conditioned by genetic differences between the types of lettuce, environmental factors and the interactions between both factors (Mou 2009). According to Llorach et al. (2008), caffeic acid derivatives are the predominant phenolics in green varieties of lettuce (iceberg, romaine and continental), while flavonols were detected in greater quantities in red-leaf lettuces (Red Oak Leaf and Lollo Rosso). These authors also measured higher antioxidant activity in red than in green varieties of lettuce. In a recent study, Mulabagal et al. (2010) suggested that the lack of anthocyanins in green lettuce (North Star variety) could explain its lower biological activity compared with that of the red lettuce var. Cherokee. In addition to genetic differences, nutrient content of lettuce may also be affected by environmental factors (Mou 2009). Light promotes the synthesis of carotenoids and generally the optimal temperature for carotenogenesis in plants is relatively low (Gross 1991). Recently, Boo et al. (2011) found that low temperatures lead to an increase of anthocyanins in lettuce plants and that day/night temperatures of 20/13°C enhanced the activities of the enzymes polyphenol oxidase (PPO) and phenylalanine ammonia-lyase (PAL).

How to Improve the Nutritional Quality of Lettuce?

As described by Mou (2009), despite the efforts from public health organizations, increasing the consumption of fruits and vegetables is often hard to achieve. In this context, enhancing the nutritional levels of vegetables would improve the nutrient intake without requiring an increase in consumption. This same author discusses two strategies used for improving the nutritional quality in lettuce: (i) a conventional plant breeding and (ii) the application of Biotechnology. According to Mou (2009), a breeding program must start with an assessment of the existing genetic variability to find sources of high nutritional value. Some of the main challenges for the breeding programs would be (i) the lack of analytical capabilities in most crop improvement programs and (ii) the need of carefully designed and executed experiments in order to avoid great environmental influences on the tested nutritional parameters. When selecting lettuces, it has been found that the varieties or cultivars with high amounts of chlorophylls also have important quantities of β-carotene and lutein (Mou 2005). There has also been considerable interest in the improvement of the nutritional quality of lettuce through biotechnology. Transformation of lettuce has been routinely accomplished by using *Agrobacterium tumefaciens* or electroporation. However, the commercialization of transgenic lettuce may depend on progress in transgene expression, public acceptance, economic and marketing challenges, intellectual property issues and risk assessment (Mou 2009).

All these difficulties impel us to look for other alternatives to improve the nutritional quality of lettuce. In fact, some studies have focused on enhancing growth and nutritional value of different varieties of lettuce by applying supplemental light of selected wavelengths in greenhouses. For example, Li and Kubota (2009) found that supplemental blue, ultraviolet-A (UV-A) or red lights can induce, respectively, the accumulation of anthocyanins, carotenoids and phenolics in baby leaf lettuce grown under white light. Likewise, Caldwell and Britz (2006) studied the effects of supplemental UV radiation on the levels of carotenoids and chlorophylls in several green and red leaf lettuces cultivated in greenhouses. Other authors (Lizarazo et al. 2010) observed that the quantity of α-tocopherol (vitamin E) in leaves of lettuce notably increased during ontogeny and, consequently, they suggested that the content of vitamin E could be easily enhanced in this crop by selecting the appropriate harvesting time, which would avoid the use of genetically-modified plants.

In this context, the application of arbuscular mycorrhizal fungi (AMF) could be another alternative to both conventional plant breeding and transgenic lettuce for enhancing the amount of some compounds of nutritional interest without a risk for human health and respectful toward the environment. Mycorrhizal fungi colonize the roots of over 90 percent of plant species mostly to the mutual benefit of both plant host and fungus (Smith and Read 1997, see also Chapter 15 in this volume), being arbuscular mycorrhizas the type of mycorrhizal symbiosis formed by the majority of crop and horticultural plants, including lettuce. The association between arbuscular mycorrhizal fungi (AMF) and plant roots develops in two functional phases (Smith and Read 1997): the extraradical phase extending from the root into the soil and the intraradical phase with intercellular hyphae and specialized intracellular structures called "arbuscules". The external mycelium transfers water and minerals from soil to roots. Arbuscules are the structures where exchanges of carbon to the fungus and nutrient and water to the host plant take place. The intraradical phase sometimes includes the "vesicles", structures considered to be mainly for storage. The establishment of this mutualistic association involves a continuous cellular and molecular dialogue between the mycorrhizal fungus and the host plant (Bonfante-Fasolo 1984, Balestrini and Lanfranco 2006) that includes the activation of the antioxidant (Garmendia et al. 2004a), phenylpropanoid (Azcón-Aguilar et al. 2002) or carotenoid metabolic pathways (Strack and Fester 2006). It is becoming evident that AM symbiosis can stimulate the synthesis of secondary metabolites, which can both increase plant tolerance to abiotic and biotic stresses and enhance the accumulation of antioxidant compounds potentially beneficial to human health in plant tissues (Seeram 2008). However, only a few analyses have targeted final crop products (leaves, roots or fruits) of mycorrhizal plants used in food or in medicinal remedies (Gianinazzi et al. 2010).

The Application of Arbuscular Mycorrhizal Fungi (AMF) for Improving the Nutritional Quality of Lettuce: Insights from a Case Study in the Greenhouse

The results obtained in research work carried out with greenhouse-grown lettuce inoculated or not with AMF are focused here. The findings with lettuce are compared with those obtained with other crops used in human diet or for medicinal purposes.

Most studies with lettuce plants included two cultivars, Batavia Rubia Munguía and Maravilla de Verano, belonging to the variety Capitata. One experimental design also included the variety Longifolia, commonly known as Cogollos de Tudela. These three types of lettuce are mainly consumed as salads and are highly commercialized in Europe. Batavia lettuce, extensively cultivated in greenhouses, is classified as a loose-leaf lettuce or semi-heading lettuce. Batavia lettuce have an excellent shelf life, maintaining their crispness from the time they are harvested until the time they are consumed. Batavia Rubia Munguía (BRM) has yellow-green leaves, with very ruffled borders and a consistent, crispy texture. It develops a round, dense head. Maravilla de Verano (MV) is adapted to be grown all year around, especially in summer due to its resistance to high temperatures. It has rounded and very broad leaves with a crunchy texture, green color and red pigmentation, especially in the borders of the most ruffled leaves. It develops a good size and a firm head. Cogollos de Tudela (CT) is a delicious variety of fan-leafed lettuce that received its name because it is grown in the city of Tudela and the surrounding municipalities in Navarra, in the North of Spain. It is around 10 centimeters long and is characterized by its long, thick, strong leaves, ranging in color from the bright green on the outer leaves to the yellow in the inner ones. Its head is dense and erect.

Different species of AMF were applied to lettuce plants: (1) *Glomus fasciculatum* (Taxter *sensu* Gerd.) Gerdemann and Trappe that came from a pot culture with alfalfa (*Medicago sativa* L.) as host plant or (2) the commercial inoculum AEGIS Endo Gránulo commercialized by Atens (Tarragona, Spain). The commercial formulation of AMF was a mixture of *Glomus intraradices* (Schenck and Smith) (actually *Rhizophagus intraradices* (Schenck and Smith) Walker & Schüβler comb. nov.) and *Glomus mosseae* (Nicol. and Gerd.) Gerd. and Trappe (actually *Funneliformis mosseae* (Nicol. and Gerd.) Walker & Schüβler comb. nov.) that contained around 100 spores and other infective propagules per gram of product.

The Effectiveness of AMF for Improving the Nutritional Quality of Greenhouse-grown Lettuce depends on both the Type of Lettuce and Fungal Species

Growth of lettuce clearly benefited from the association of plants with AMF, being a mixture of *G. mosseae* and *G. intraradices* (the commercial inoculum) more effective than *G. fasciculatum* in enhancing biomass production of shoots and roots, regardless of the variety or cultivar of lettuce (Baslam et al. 2011a). The greater size achieved by mycorrhizal lettuce plants compared with non-inoculated controls had a subsequent dilution effect on the concentrations of some mineral nutrients and antioxidant compounds of interest for the human diet. Nevertheless, the levels of Zn, Cu, Fe, anthocyanins, carotenoids and, to a lesser extent, phenolics, appeared in higher concentrations (expressed on a wet basis) in mycorrhizal than in non-mycorrhizal lettuce plants. This would mean that cultivation of mycorrhizal lettuce would allow improving the intake of those compounds without requiring an increase in consumption. However, the beneficial effect of AMF on nutrient quality of lettuce varied among types of mycorrhizal inocula applied, cultivars of lettuce and between outer and inner leaves.

When analyzed the levels of mineral nutrients in both outer and inner leaves, Baslam et al. (2011a) found that the dilution effect produced by increased size contributed to the similar concentrations of Zn in leaves of mycorrhizal lettuce plants BRM and MV compared with their respective non-inoculated controls. However, the application of AMF stimulated the accumulation of Zn in outer leaves of CT and sharply enhanced the concentrations of Cu in both outer and inner leaves of BRM. In CT and MV only *G. fasciculatum* was clearly effective in enhancing Cu levels in the inner leaves (Baslam et al. 2011a). In a previous study performed with tomato, Cavagnaro et al. (2006) observed a significant enhancement of Zn in fruits of mycorrhizal plants. Castellanos-Morales et al. (2010) found that strawberry fruits of plants colonized by *G. intraradices* accumulated higher quantities of K and Cu. As described by Loladze (2002), concentrations of several essential elements in modern crops are insufficient for optimal human nutrition, thus contributing to the enormous "hidden hunger" problem. Among these elements insufficiently consumed in developed countries are Fe, Cu and Zn. Therefore, increased accumulation of these elements in mycorrhizal plants constitutes a strong reason for considering the application of AMF in programs focused on the improvement of the quality of crops.

When expressed as total contents per plant, all types of lettuce plants associated with AMF showed significantly higher quantities of total carotenoids (Fig. 16.1A), total soluble phenolics, anthocyanins (Fig. 16.1B) and total ascorbate (ASC) (Fig. 16.1C) than their respective non-inoculated

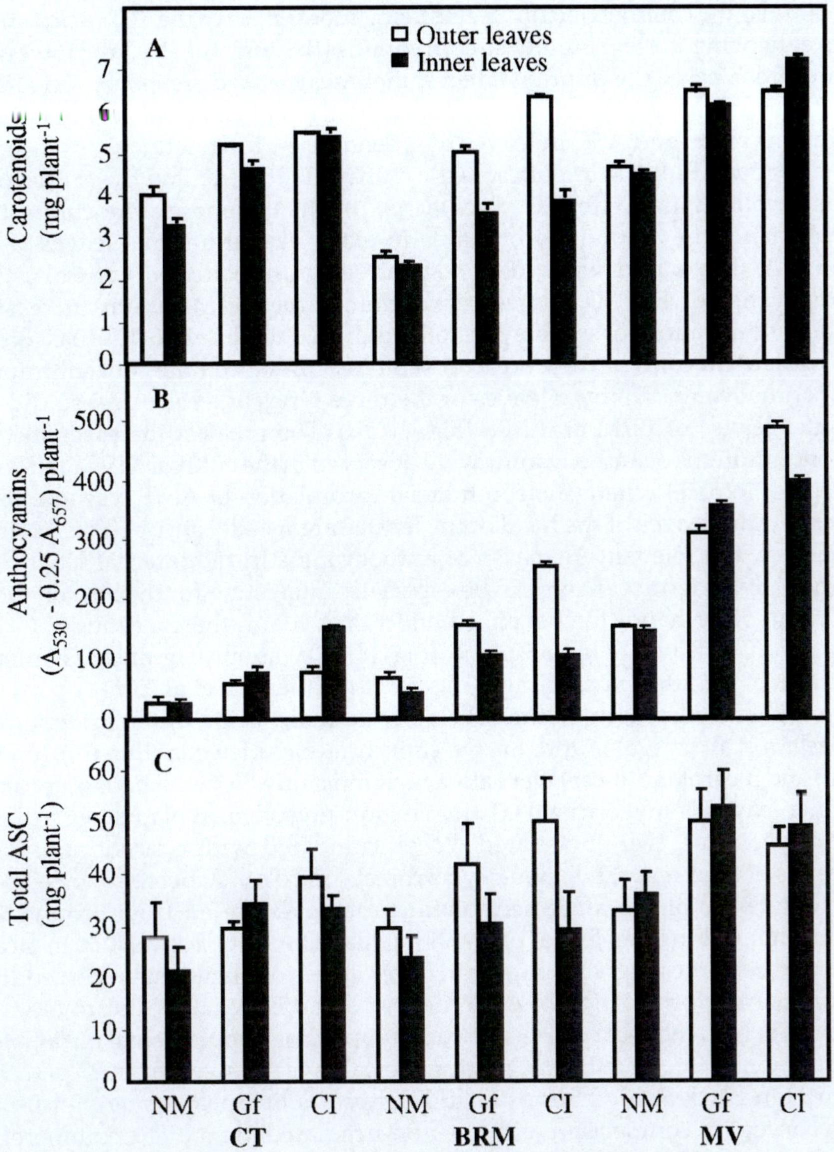

Figure 16.1 Total content of carotenoids (A), anthocyanins (B) and total ascorbate (ASC) (C) in outer (white histograms) and inner leaves (black histograms) of lettuce Cogollos de Tudela (CT), Batavia Rubia Munguía (BRM) and Maravilla de Verano (MV) non-inoculated with mycorrhizal fungi (NM), inoculated with *Glomus fasciculatum* (Gf) or inoculated with the commercial inoculum (CI). Values are means ± SE. Total content of carotenoids and ASC are expressed as mg plant^{-1}. Data on anthocyanins are expressed as optical density (OD) (A_{530}–$0.25A_{657}$) plant^{-1}.

controls, the commercial inoculums being more effective than *G. fasciculatum* in enhancing the levels of these compounds (Baslam et al. 2011a). However, on a fresh basis, the dilution effect sometimes masked the beneficial effect of mycorrhizal symbiosis on the concentrations of carotenoids, phenolics, anthocyanins and ASC in leaves of greenhouse-grown lettuce.

Batavia Rubia Munguía was the cultivar that benefited the most from the application of either *G. fasciculatum* or the commercial inoculum for enhancing the amount carotenoids in external and internal leaves (Fig. 16.2A). Compared with their respective non-inoculated controls, the application of AMF (*G. fasciculatum* or the commercial inoculum) increased the concentrations of soluble phenolics only in outer leaves of CT (Baslam et al. 2011a). In contrast, mycorrhizal symbiosis enhanced the concentrations of anthocyanins in inner leaves of the three types of lettuce and so did in outer leaves of BRM and MV (Fig. 16.2B). The greatest increases in the concentrations of anthocyanins were observed in the cultivar MV (a red-leaf type of lettuce) when the commercial formulation of AMF was applied. Since outer leaves of the head of the lettuce are usually stripped off during harvest, the relevant increases of anthocyanins in the internal leaves of mycorrhizal lettuce plants can be especially interesting for the human diet. On the other hand, the beneficial effect of AMF on the concentrations of total ASC in leaves of lettuce plants (Fig. 16.2C) was only significant in inner leaves of CT after inoculating *G. fasciculatum* (Baslam et al. 2011a).

When determining the levels of the main individual carotenoids, Baslam et al. (2012a) found that the concentrations of neoxanthin (N), lutein (L) and β-carotene (β-car) were always significantly higher in both outer and inner leaves of mycorrhizal CT than in non-mycorrhizal plants, regardless the species of AMF inoculated. When compared with concentrations in leaves of their respective non-mycorrhizal controls (100 percent), increases in the levels of the aforementioned carotenoids varied from around 24 percent for N in inner leaves of plants inoculated with *G. fasciculatum* to more than three times higher for β-car in outer leaves of plants that received the commercial formulation of AMF. Caldwell and Britz (2006) improved the levels of N, L and β-car of green lettuce by applying supplemental ultraviolet radiation in the greenhouse, getting increases that ranged from 27 percent for N in Black-Seeded Simpson and Simpson Elite to 46 percent for β-car in concept in comparison with the non-irradiated plants. The commercial inoculum was more effective than *G. fasciculatum* for enhancing the levels of violaxanthin (V) in outer leaves of CT. In contrast, *G. fasciculatum* was more effective than the commercial inoculum for inducing the accumulation of zeaxanthin (Z) in both types of leaves of this variety of lettuce. In the cultivar BRM the inoculation with *G. fasciculatum* resulted in enhanced

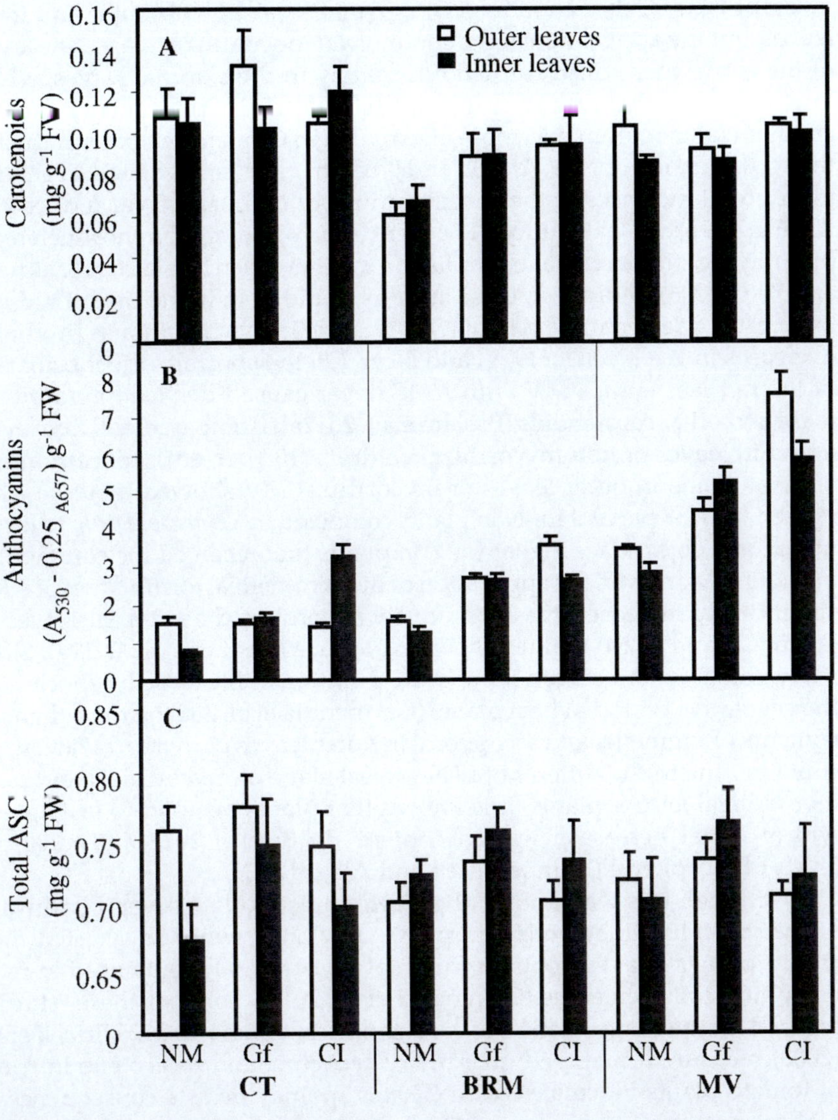

Figure 16.2 Concentrations of carotenoids (A), anthocyanins (B) and total ascorbate (ASC) (C) in outer (white histograms) and inner leaves (black histograms) of lettuce Cogollos de Tudela (CT), Batavia Rubia Munguía (BRM) and Maravilla de Verano (MV) non-inoculated with mycorrhizal fungi (NM), inoculated with *Glomus fasciculatum* (Gf) or inoculated with the commercial inoculum (CI). Values are means ± SE. Concentrations of carotenoids and ASC are expressed as mg g^{-1} FW. Data on anthocyanins are expressed as optical density (OD) (A_{530} $-0.25A_{657}$) g^{-1} FW. FW = fresh weight.

levels of N, V, antheraxanthin (A), Z, L and β-car in both outer and inner leaves, but the application of the commercial inoculum increased the levels of the above mentioned carotenoids mainly in the external leaves, which are stripped off during harvest (Baslam et al. 2012a). On the other hand, the commercial inoculum was more effective than *G. fasciculatum* for inducing the accumulation of N, V, A, Z, L, and β-car in outer leaves of MV. In general terms, total increases in the concentrations of N, L, and β-car induced by AMF in the green leaf lettuce BRM were clearly higher than those achieved by applying supplemental UV radiation to other green leaf lettuce (Caldwell and Britz 2006). Moreover, these authors found that, in the case of red leaf lettuce, the use of supplemental UV radiation in the greenhouse produced decreases in the levels of N, L, and β-car. On the contrary, the inoculation of the red leaf lettuce MV with AMF never caused declined amounts of these and other carotenoids (Baslam et al. 2012a). Compared with the levels found in leaves of non-mycorrhizal controls (100 percent), concentrations of carotenoids in outer leaves of mycorrhizal MV showed increases that ranged from 37 percent for N in plants colonized by *G. fasciculatum* to levels more than eight times higher for Z in plants that received the commercial inoculum. Moreover, the application of the commercial formulation of AMF significantly increased the levels of tocopherol in the external leaves of green (CT and BRM) and red (MV) leaf lettuce (Baslam et al. 2012a). Since mycorrhizal symbiosis can modify the duration of life cycle by shortening the vegetative period of host plants (Garmendia et al. 2004b), the enhanced contents of pigments and tocopherol in outer leaves of mycorrhizal lettuce could be due to the different phenological stage of mycorrhizal and non-mycorrhizal lettuce plants. It is known that plants respond to leaf ageing with an initial increase in pigment content (Pantin et al. 2012) and enhanced levels of tocopherol (Munné-Bosch and Alegre 2004).

Although it is well-known that AMF can stimulate several pathways of the secondary metabolism in plants, few analyses have targeted final crop products (leaves, roots or fruits) of mycorrhizal plants used in food or in medicinal remedies (Gianinazzi et al. 2010). Some of these studies focused on the accumulation of carotenoids in tomato fruits. Ulrichs et al. (2008) measured increased quantity of lycopene and β-carotene in fruits of tomato plants inoculated with *Glomus* sp. may be as a consequence of higher photosynthesis in mycorrhizal plants. More recently, Giovannetti et al. (2012) have found that *G. intraradices* can improve nutraceutical value of tomato fruits without the production of mutagenic compounds.

On the other hand, the effect of AMF on the content of secondary compounds is contradictory in the literature because it depends on plant genotype, fungal species and fertilization regime. For example, when the influence of AM symbiosis on the levels of phenolics was analyzed, Toussaint et al. (2007) showed the potential of three AMF to enhance the production of phenolic compounds with antioxidant activity in sweet basil leaves. In contrast, Geneva et al. (2010) observed decreased concentrations of total phenols in leaves of *Salvia officinalis* associated with *G. intraradices*. Castellanos-Morales et al. (2010) found that strawberry fruits of plants inoculated with *G. intraradices* had greater amount of phenolic compounds than non-mycorrhizal plants; however, differences between fruits of mycorrhizal and non-mycorrhizal plants varied according to the nitrogen level applied to plants. Grape plants inoculated with AMF (*Glomus* sp.) accumulated higher quantities of quercitin in leaves than non-mycorrhizal grape plants, but the response depended on the grape genotype (Eftekhari et al. 2012).

The Source of Phosphorus (P) Fertilization Applied to Plants can Modulate the Effect of AMF on Nutritional Quality of Greenhouse-grown Lettuce

As elucidated before, association of green and red-leaf cultivars of lettuce with AMF induced the accumulation of carotenoids, anthocyanins or some mineral nutrients (e.g., Cu and Fe) in leaves (Baslam et al. 2011a; Baslam et al. 2012a). In these studies lettuce plants received a water-soluble source of phosphorus (P) as a modified Hewitt's nutrient solution. However, the water-soluble sources of P as fertilizers for plants can exhibit rather low efficiency when applied in acidic and calcareous soils as a consequence of the precipitation of water-insoluble iron, aluminum or calcium phosphates (Tunesi et al. 1999, Delgado and Torrent 2000). Consequently, new fertilizers that include P sources of limited water solubility but high solubility in organic acids released to the rhizosphere by plant roots have recently been produced and commercialized (Erro et al. 2011). Therefore, a new study was performed to clarify if the application of P sources differed in water-solubility (modified Hewitt's nutrient solution (H), single superphosphate (SSP) and a water-insoluble (WI) source of P) can influence the capacity of AMF for enhancing nutritional quality of lettuce (Baslam et al. 2011b). This aspect may have relevant importance for the potential use of AMF for the cultivation of lettuce not only in greenhouses but also in the field. WI was the water-insoluble fraction of a "rhizosphere controlled fertilizer" (RFC) (Erro et al. 2011).

Growth of lettuce plants (cv. BRM and MV) was significantly improved by AMF, regardless of the cultivar of lettuce and type of P fertilizer applied. Despite the dilution effect (especially significant in the cultivar MV) due to the greater size reached by mycorrhizal lettuce, symbiosis improved, in most cases, the nutritional quality of this crop, although the effect of AMF was dependent on both cultivar of lettuce and source of P nutrition (Baslam et al. 2011b). When SSP was applied, BRM was the cultivar which most benefited from mycorrhizal symbiosis, with the levels of Cu, Fe (Table 16.1), non-structural sugars (soluble sugars and starch), proteins, phenolics (Figs. 16.3 and 16.4) and total ASC being most enhanced when compared with those of non-mycorrhizal plants (Baslam et al. 2011b). When the modified Hewitt's nutrient solution was used, mycorrhizal symbiosis improved the contents of non-structural sugars (Fig. 16.3A) and total ASC in BRM; increased the levels of chlorophylls and anthocyanins in MV (Fig. 16.4C)

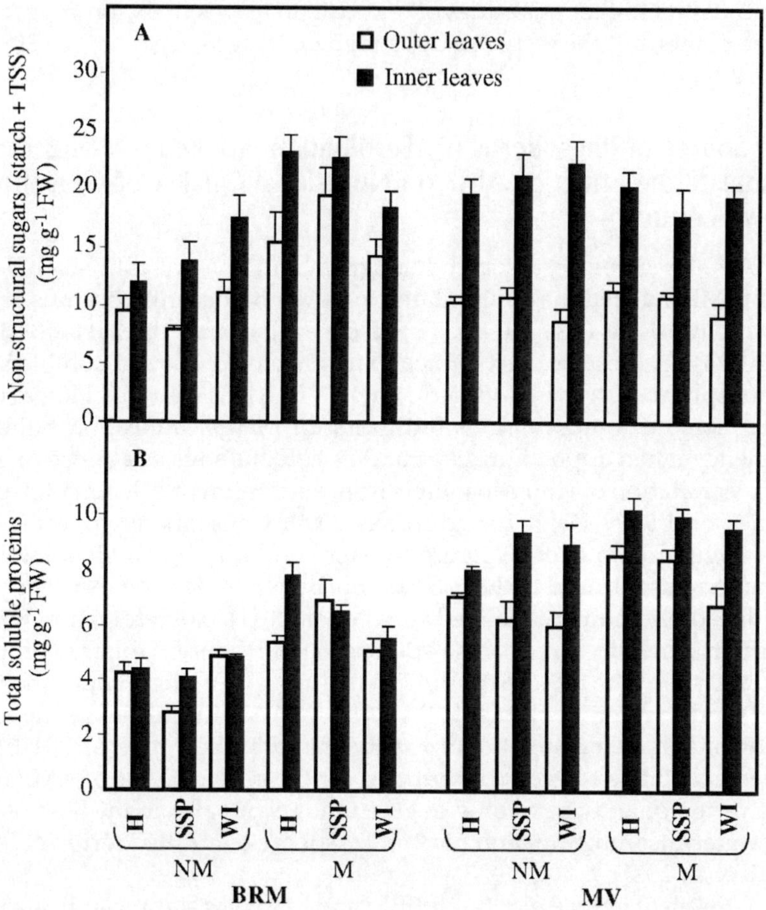

Figure 16.3 Concentrations (mg g^{-1} FW) of non-structural sugars (starch + total soluble sugars, TSS) (A) and total soluble proteins (B) in outer (white histograms) and inner leaves (black histograms) of lettuce Batavia Rubia Munguía (BRM) and Maravilla de Verano (MV), non-mycorrhizal (NM) or mycorrhizal (M), and fertilized with either modified Hewitt's nutrient solution (H), single superphosphate (SSP) or the water-insoluble (WI) fraction of a "rhizosphere controlled fertilizer" (RCF). Values are means ± SE. FW = fresh weight.

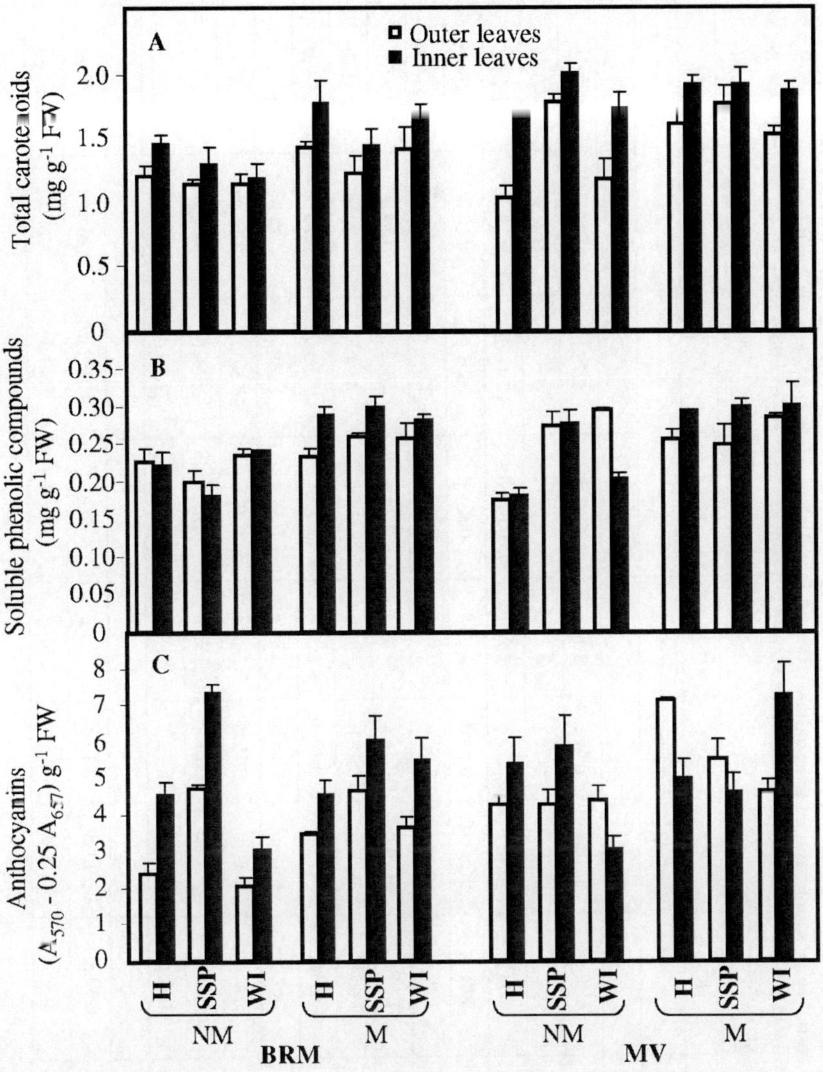

Figure 16.4 Concentrations of total carotenoids (A), soluble phenolic compounds (B) and anthocyanins (C) in outer (white histograms) and inner leaves (black histograms) of lettuce Batavia Rubia Munguía (BRM) and Maravilla de Verano (MV), non-mycorrhizal (NM) or mycorrhizal (M), and fertilized with either modified Hewitt's nutrient solution (H), single superphosphate (SSP) or the water-insoluble (WI) fraction of a "rhizosphere controlled fertilizer" (RCF). Values are means ± SE. Concentrations of carotenoids and phenolic compounds are expressed as mg g^{-1} FW. Concentrations of anthocyanins are expressed as optical density (OD) (A_{530}–$0.25A_{657}$) g^{-1} FW. FW = fresh weight.

Table 16.1 Concentrations of mineral nutrients in outer and inner leaves of Batavia Rubia Munguía (BRM) and Maravilla de Verano (MV). Data are means of three replications. Within each column and cultivar values followed by the same letter did not differ significantly ($P \leq 0.05$). NM = non-mycorrhizal; M= plants inoculated with the commercial inoculum; H = modified Hewitt's nutrient solution; SSP = single superphosphate; WI = water-insoluble fraction of a "rhizosphere controlled fertilizer".

Treatments			K	P	S	N	Cu	Fe	Mo	Zn
			(mg g^{-1} FW)				(ppm)			
BRM										
NM	H	Outer leaves	6.42 c	0.13 h	0.17 c	3.16 b	0.55 d	5.75 h	0.21 b	7.57 dc
		Inner leaves	4.34 ef	0.19 d	0.16 c	2.31 d	0.57 d	5.00 h	0.21 b	3.95 f
	SSP	Outer leaves	7.68 b	0.16 f	0.19 b	2.78 c	0.62 cd	17.41 de	0.19 b	8.24 d
		Inner leaves	4.01 f	0.20 cd	0.14 d	2.08 e	0.63 cd	9.01 g	0.10 d	4.09 ef
	WI	Outer leaves	7.69 b	0.16 f	0.18 b	2.82 c	0.76 bc	22.94 c	0.23 b	18.30 a
		Inner leaves	4.64 e	0.24 b	0.16 c	1.87 f	2.25 a	75.13 a	0.19 b	9.85 c
M	H	Outer leaves	7.65 b	0.15 g	0.18 b	3.36 a	0.53 d	4.94 h	0.22 b	7.31 dc
		Inner leaves	4.69 de	0.21 c	0.17 c	2.05 e	0.75 b	18.66 d	0.14 c	4.34 ef
	SSP	Outer leaves	7.62 b	0.17 ef	0.18 b	2.64 c	0.56 d	12.49 f	0.24 b	5.90 c
		Inner leaves	5.07 d	0.26 a	0.19 b	2.17 e	0.78 b	17.53 de	0.10 d	5.05 e
	WI	Outer leaves	8.54 a	0.17 e	0.21 a	3.33 a	0.75 bc	25.50 b	0.32 a	14.35 b
		Inner leaves	3.49 g	0.17 e	0.12 e	1.47 f	0.61 d	16.19 e	0.07 d	3.81 f
MV										
NM	H	Outer leaves	6.52 c	0.13 d	0.15 c	2.21 c	0.56 abc	8.98 bc	0.15 abc	5.38 cd
		Inner leaves	3.77 f	0.19 b	0.18 b	1.95 d	0.66 ab	5.48 d	0.10 d	5.33 cd

	SSP	Outer leaves	5.39 d	0.12 d	0.14 c	1.58 c	0.42 d	14.09 a	0.13 bcd	3.59 g
		Inner leaves	3.96 ef	0.22 a	0.19 a	2.12 d	0.61 abc	9.47 bc	0.10 cd	3.85 fg
	WI	Outer leaves	7.89 a	0.15 c	0.18 b	2.56 a	0.68 a	16.15 a	0.19 a	8.44 a
		Inner leaves	4.00 ef	0.21 a	0.19 a	1.96 d	0.49 cd	8.11 cd	0.09 d	4.72 d
M	H	Outer leaves	6.85 b	0.14 c	0.18 b	2.31 b	0.64 ab	10.50 b	0.17 ab	5.01 de
		Inner leaves	4.06 ef	0.21 a	0.20 a	2.20 c	0.54 abc	7.64 cd	0.14 bcd	4.85 def
	SSP	Outer leaves	5.18 e	0.11 d	0.13 d	1.32 f	0.38 d	6.13 d	0.11 cd	4.03 efg
		Inner leaves	4.11 ef	0.21 a	0.18 b	1.88 e	0.48 cd	5.78 d	0.09 d	4.18 efg
	WI	Outer leaves	6.99 b	0.15 c	0.18 b	2.00 d	0.52 b	10.51 b	0.18 ab	6.83 b
		Inner leaves	4.23 e	0.21 a	0.20 a	2.03 d	0.56 abc	8.23 bc	0.12 cd	6.14 bc

and enhanced the accumulation of proteins, carotenoids and phenolics in both BRM and MV (Figs. 16.3 and 16.4) (Baslam et al. 2011b). Finally, when the WI was the source of P, mycorrhizal BRM showed increased quantities of K, S, N, Fe and Mo (Table 16.1) and total ASC; mycorrhizal MV accumulated enhanced amounts of Zn (Table 16.1), and both cultivars had increased contents of chlorophylls, carotenoids, phenolics and anthocyanins (Fig. 16.4) when associated with AMF (Baslam et al. 2011b). In mycorrhizal BRM, the enhanced levels of non-structural carbohydrates (starch and soluble sugars) in shoots, compared with those of non-mycorrhizal plants, could partially explain the increased concentrations of some secondary metabolites (carotenoids, phenolics or anthocyanins) in leaves. In MV, the association of plants with AMF favored the accumulation of soluble proteins, some of which could presumably be key enzymes of secondary metabolite pathways.

The Application of AMF can be more Beneficial for the Nutritional Quality of Greenhouse-grown Lettuce when Plants are Cultivated under Restricted Water supply

Cultivation of lettuce requires frequent irrigation for better growth and development because this crop plant lacks a deep root system. Although growing lettuce in dry soil conditions promotes bolting, soil moisture contents ranging from 50 to 75 percent of field capacity (FC) allow lettuce plants to produce similar biomass than plants under optimal water supply (Gallardo et al. 1996, Tsabedze and Wahome 2010). These studies, however, do not show any data concerning the effect of suboptimal irrigation on the nutritional quality of lettuce, and other studies dealing with the loss of lettuce quality under conditions that favor the wilting phenomenon (Agüero et al. 2011) are focused on the postharvest period of this crop vegetable. When lettuce is combined with AMF, fungal hyphae can contribute to water uptake (Ruiz-Lozano and Azcón 1995), *G. intraradices* and *G. mosseae* are some of the most effective AMF in taking soil water (Marulanda et al. 2003). In addition, the establishment of this mutualistic association involves a continuous cellular and molecular dialogue between the mycorrhizal fungus and the host plant (Balestrini and Lanfranco 2006) that includes the activation and/or the synthesis of secondary metabolites, which can both increase plant tolerance to abiotic and biotic stresses and enhance the accumulation of antioxidant compounds potentially beneficial to human health in plant tissues (Seeram 2008). However, biochemical and physiological responses of mycorrhizal plants undergoing drought not only depend on the type of AMF and/or host plant but also on both the degree and the duration of the water deficit (Goicoechea et al. 2004). For this reason, Baslam and Goicoechea (2012) applied three different types of suboptimal irrigation to lettuce plants associated or not with AMF: a soil water content equivalent to 2/3 of field capacity (2/3 FC) that was maintained from planting until harvesting, a soil water content equivalent to 1/2 of field capacity (1/2 FC) from planting until harvesting and a cyclic water deficit (CD) that included five cycles of drought.

Leaves of both non-mycorrhizal and mycorrhizal lettuce plants (cv. BRM and MV) maintained control values of relative water content (RWC) when subjected to moderate water limitation (2/3 FC) and also in the cultivar MV, a water regime equivalent to 1/2 FC was applied (Baslam and Goicoechea 2012). This finding is relevant because one of the main features of lettuce quality is the high amount of water in its tissues (Agüero et al. 2008).

The association of lettuce plants with AMF significantly enhanced the concentrations of total chlorophylls in inner leaves of both cultivars (BRM and MV) regardless of the water irrigation regime (Baslam and Goicoechea 2012). Similarly, Zuccarini (2007) measured greater contents of chlorophylls in lettuce inoculated with a mixture of *G. mosseae*, *G. intraradices* and *G. coronatum* than in non-mycorrhizal controls. The increased levels of chlorophylls in mycorrhizal lettuce presumably contributed to higher photosynthetic rates in plants associated with AMF, thus benefiting both growth of host plants and the development and functionality of the symbiosis. In fact, shoot biomass in mycorrhizal lettuce plants (BRM and MV) subjected to moderate and maintained water deficit (2/3 FC) were similar to those of non-mycorrhizal plants grown under well-watered conditions (Baslam and Goicoechea 2012).

Mycorrhizal symbiosis also favored the accumulation of total carotenoids in both outer and inner leaves of BRM and MV under either optimal irrigation or restricted water supplies compared with their respective non-mycorrhizal controls (Baslam and Goicoechea 2012). When mycorrhizal plants subjected to different types and degrees of water deficits were compared, Baslam and Goicoechea (2012) found that prolonged water restrictions (2/3 FC and 1/2 FC) were the most effective in enhancing the concentrations of carotenoids in external leaves of BRM. Any type of drought increased the quantities of carotenoids in inner leaves of MV.

The imposition of water deficits to non-mycorrhizal plants only induced a slight accumulation of soluble phenolic compounds in the outer leaves of the cultivar BRM (Baslam and Goicoechea 2012). In contrast, Oh et al. (2009) found that the activity of the enzyme phenylalanine ammonia-lyase (PAL), involved in the biosynthesis of many phenolic compounds, increased in lettuce cv. Baronet exposed to water stress. Activation of the phenylpropanoid pathway may be dependent on both the type and degree of water deficit applied and the genotype of lettuce. When associated with AMF, water restrictions increased the levels of these secondary metabolites in the inner leaves of BRM and, in a lesser extent, in the internal leaves of MV (Baslam and Goicoechea 2012).

In non-mycorrhizal lettuce plants, the concentrations of anthocyanins in leaves of BRM increased when subjected to any type and degree of drought. In non-mycorrhizal MV, only the application of cyclic drought produced an enhancement in the levels of anthocyanins in the inner leaves. On the other hand, the association of plants with AMF clearly enhanced the levels of anthocyanins in both varieties of lettuce, the increases being especially significant in plants undergoing moderate and maintained water deficit (2/3 FC) (Baslam and Goicoechea 2012).

Oh et al. (2009) concluded that the adaptation of lettuce plants to water stress was associated to the activation of the enzyme L-galactose dehydrogenase (L-GalDH) implied in the biosynthesis of ascorbic acid. In the study performed by Baslam and Goicoechea (2012), however, neither of the drought regimes that were applied had significant effects on the concentrations of the total ASC pool in the two cultivars of lettuce, whether they were associated or not with AMF.

The findings of Baslam and Goicoechea (2012) suggested that mycorrhizal symbiosis can not only improve the quality of lettuce but may also allow reducing irrigation without subsequent detrimental effects on lettuce production. The possible presence of some microorganisms accompanying the mycorrhizal population could also contribute to enhance the beneficial effect of AMF on the nutritional quality of greenhouse-grown lettuce. In this context, Kohler et al. (2008) proposed the potential use of a plant-growth-promoting rhizobacterium (PGPR) together with AMF as an inoculant to alleviate the oxidative damage of lettuce plants produced under water stress.

Future Perspectives: What will be the Effectiveness of AMF as Enhancers of Growth and Quality of Crops in a Scenario of Changing Climate?

Levels of atmospheric CO_2 have been constantly increasing since the industrial revolution due to anthropogenic activities, including burning of fossil fuels, deforestation and intensive animal husbandry. The atmospheric CO_2 rising is expected to double within the next 100 yr (Watson et al. 1990). The enhanced CO_2 concentration increases the potential net photosynthesis in C3 plants, such as lettuce and other crops (Drake et al. 1997) and therefore can improve yield (Oliveira et al. 2010). However, when the synthesis of carbohydrates in plants exposed to elevated CO_2 exceeds the capacity to produce new sinks, the plants decrease their photosynthetic rate so as to balance source activity and sink capacity (Thomas and Strain 1991). Several studies performed under air CO_2 enrichment have highlighted how sink development is restricted under N limiting conditions (Aranjuelo et al. 2009). Apart from N, P supply can also be crucial in determining the response of plants to elevated CO_2 (Campbell and Sage 2006).

In the context of rising atmospheric CO_2, AMF are predicted to be important in defining plant responses to elevated CO_2 concentrations. As elevated CO_2 affects photosynthetic rates, it would also influence C supply to AMF colonizing roots (Jifon et al. 2002). In turn, AMF may diminish the plant nutrient limitation associated with enhanced photosynthetic rates (Treseder and Allen 2000). In fact, lower concentrations of P in tissues of plants when grown under elevated CO_2 can be alleviated by the formation of AMF and any improvement in plant N nutrition resulting from the formation of AMF may be also important in determining plant responses to atmospheric CO_2 enrichment (Cavagnaro et al. 2011). For example, Gavito et al. (2000) found that elevated CO_2 and AMF independently enhanced shoot growth in *Pisum sativum*, being additive effects of both factors.

When linked high concentrations of atmospheric CO_2 with plant stoichiometry and quality of human nutrition, Loladze (2002) concluded that elevated CO_2 could both reduce the ratio between nutritional and caloric values of crops and intensify the already acute problem of micronutrient malnutrition (e.g., Fe, Zn, I) in the developing world. As AMF can favor the uptake of nutrients with low mobility, such as Cu, Fe or Zn (Clark and Zeto 2000), it was reasonable to hypothesize that mycorrhizal symbiosis could also mitigate the presumed imbalance of essential elements in crops exposed to elevated CO_2 and consumed in the human diet. However, Baslam et al. (2012b) found that the accumulation of minerals and nutraceuticals in leaves of greenhouse-grown lettuce induced by AMF may diminish or even disappear when plants are exposed to elevate CO_2 as a consequence of different partitioning of resources at ambient and under elevated CO_2. Under elevated CO_2, a relevant quantity of photoassimilates seemed to be used for improving shoot growth and spreading mycorrhizal colonization in roots in detriment to the secondary metabolism.

Enhanced atmospheric CO_2 will not be the only factor in a changing climate scenario. Climate change could also increase plant water stress and the agricultural areas under drought, affecting crop production (IPCC 2007a). In addition, CO_2 is the most important anthropogenic greenhouse gas, and if its emissions continue at high levels, temperature is predicted to increase between 1.8 and 4.0°C (IPCC 2007b). Therefore, studies focused on the effects of rising atmospheric CO_2 and its interactions with water deficit and/or increased temperature on growth and nutritional quality of crops, as well as on the capacity of AMF for enhancing both growth and nutritional quality of vegetables and fruits consumed in the human diet, would be of great interest.

Concluding Remarks

It is not known if it is possible to enhance the nutritional quality of lettuce and other crops just by increasing P fertilization to non-mycorrhizal plants. However the use of AMF as biofertilizers to counteract the application of extra fertilization could promote agriculture in the future, based on the implementation of practices that maintain resilience of the ecosystem services (Gianinazzi et al. 2010). Moreover, as concluded by Bonfante and Genre (2010), "in the context of modern agriculture, unraveling the contribution of mycorrhizal fungi to the nutritional quality of edible plant organs becomes a priority".

Acknowledgments

The authors wish to thank Dr. Idoia Garmendia (University of Alicante, Spain) and Dr. Javier Erro (International—R&D Roullier Group Timac—Agro) for analyses of mineral nutrients, Dr. José María García-Mina (International—R&D Roullier Group Timac-Agro) for his contribution in the study focused on the influence of phosphorus sources on the effectiveness of mycorrhizal fungi for improving the nutritional quality of lettuce, Dr. Raquel Esteban and Dr. José Ignacio García-Plazaola for analysis of carotenoids and tocopherol, Adriana Hernández (Atens, Tarragona, Spain) for providing the commercial inoculum, as well as Mónica Oyarzun and Amadeo Urdiáin for technical support. Asociación de Amigos de la Universidad de Navarra (ADA) and International-R&D Roullier Group Timac-Agro provided financial support. The authors are also very grateful to ACS Publications for permission to reprint the following material: Figs. 16.1 and 16.2 are reprinted (adapted) with permission from Journal of Agricultural and Food Chemistry 59: 5504–5515, Copyright 2011, American Chemical Society; Table 16.1, Figs. 16.3 and 16.4 are reprinted (adapted) with permission from Journal of Agricultural and Food Chemistry 59: 11129–11140, Copyright 2011, American Chemical Society.

References

Agüero, M.V., M.V. Barg, A. Yommi, A. Camelo and S.I. Roura. 2008. Postharvest changes in water status and chlorophyll content of lettuce (*Lactuca sativa* L.) and their relationship with overall visual quality. J. Food Sci. 73: 47–55.

Agüero, M.V., A.G. Ponce, M.R. Moreira and S.I. Roura. 2011. Lettuce quality loss under conditions that favor the wilting phenomenon. Postharv. Biol. Technol. 59: 124–131.

Aranjuelo, I., J.J. Irigoyen, S. Nogués and M. Sánchez-Díaz. 2009. Elevated CO_2 and water-availability effect on gas exchange and nodule development in N_2-fixing alfalfa plants. Environ. Exp. Bot. 65: 18–26.

Azcón-Aguilar, C., M.C. Jaizme-Vega and C. Calvet. 2002. The contribution of arbuscular mycorrhizal fungi for bioremediation. *In:* S. Gianinazzi, H. Schüepp, J.M. Barea and K. Haselwandter [eds.]. Mycorrhizal Technology in Agriculture. From Genes to Bioproducts. Birkhäuser Verlag. Berlin, Germany. pp. 187–197.

Balestrini, R. and L. Lanfranco. 2006. Fungal and plant gene expression in arbuscular mycorrhizal symbiosis. Mycorrhiza. 16: 509–524.

Baslam, M. and N. Goicoechea. 2012. Water deficit improved the capacity of arbuscular mycorrhizal fungi (AMF) for inducing the accumulation of antioxidant compounds in lettuce leaves. Mycorrhiza. 22: 347–359.

Baslam, M., I. Garmendia and N. Goicoechea. 2011a. Arbuscular mycorrhizal fungi (AMF) improved growth and nutritional quality of greenhouse grown lettuce. J. Agric. Food Chem. 59: 5504–5515.

Baslam, M., I. Pascual, M. Sánchez-Díaz, J. Erro, J.M. García-Mina and N. Goicoechea. 2011b. Improvement of nutritional quality of greenhouse-grown lettuce by arbuscular mycorrhizal fungi is conditioned by the source of phosphorus nutrition. J. Agric. Food Chem. 59: 11129–11140.

Baslam, M., R. Esteban, J.I. García-Plazaola and N. Goicoechea. 2012a. Effectiveness of arbuscular mycorrhizal fungi (AMF) for inducing the accumulation of major carotenoids, chlorophylls and tocopherol in green and red leaf lettuces. Appl. Microbiol. Biotechnol. doi: 10.1007/s00253-012-4526-x.

Baslam, M., I. Garmendia and N. Goicoechea. 2012b. Elevated CO_2 may impair the beneficial effect of arbuscular mycorrhizal fungi on the mineral and phytochemical quality of lettuce. Ann. Appl. Biol. 161: 180–191.

Bonfante-Fasolo, P. 1984. Anatomy and morphology of VA mycorrhizae. *In:* C.L. Powell and D.J. Bagyaraj [eds.]. VA Mycorrhiza. CRC Press. Boca Raton, Florida, U.S.A. pp. 35–46.

Bonfante, P. and A. Genre. 2010. Mechanisms underlying beneficial plant-fungus interactions in mycorrhizal symbiosis. Nat. Commun. 1: 48.

Boo, H.-O., B.-G. Heo, S. Gorinstein and S.-U. Chon. 2011. Positive effects of temperature and growth conditions on enzymatic and antioxidant status in lettuce plants. Plant Sci. 181: 479–484.

Borghi, S. 2003. Special: IV range (vegetables). Colture Protette. 32: 21–43.

Caldwell, C.R. and S.J. Britz. 2006. Effect of supplemental ultraviolet radiation on the carotenoid and chlorophyll composition of green house-grown leaf lettuce (*Lactuca sativa* L.) cultivars. J. Food Compos. Anal. 19: 637–644.

Campbell, C.D. and R.F. Sage. 2006. Interactions between the effects of atmospheric CO_2 content and P nutrition on phosynthesis in white lupin (*Lupinus alba* L.). Plant Cell Environ. 29: 844–853.

Castellanos-Morales, V., J. Villegas, S. Wendelin, H. Vierheiling, E. Reinhard and R. Cárdenas-Navarro. 2010. Root colonisation by the arbuscular mycorrhizal fungus *Glomus intraradices* alters the quality of strawberry fruits (*Fragaria x ananassa* Duch.) at different nitrogen levels. J. Sci. Food Agric. 90: 1774–1782.

Cavagnaro, T.R., L.E. Jackson, J. Six, H. Ferris, S. Goyal, D. Asami and K.M. Scow. 2006. Arbuscular mycorrhizas, microbial communities, nutrient availability, and soil aggregates in organic tomato production. Plant Soil. 282: 209–225.

Cavagnaro, T.R., R.M. Gleadow and R.E. Miller. 2011. Plant nutrient acquisition and utilization in a high carbon dioxide world. Funct. Plant Biol. 38: 87–96.

Clark, R.B. and S.K. Zeto. 2000. Mineral acquisition by arbuscular mycorrhizal plants. J. Plant Nutr. 23: 867–902.

Delgado, A. and J. Torrent. 2000. Phosphorus forms and desorption patterns in heavily fertilized calcareous and limed acid soils. Soil Sci. Soc. Am. J. 64: 2031–2037.

Demmig-Adams, B. and W.W. Adams III. 1996. Chlorophyll and carotenoid composition in leaves of *Euonymus kiautschovicus* acclimated to different degrees of light stress in the field. Aust. J. Plant Physiol. 23: 649–659.

Drake, B.G., M.A. González-Meler and S.P. Long. 1997. More efficient plants: a consequence of rising atmospheric CO_2. Annu. Rev. Plant Physiol. Plant Mol. Biol. 48: 609–639.

Eftekhari, M., M. Alizadeh and P. Ebrahimi. 2012. Evaluation of the total phenolics and quercetin content of foliage in mycorrhizal grape (*Vitis vinifera* L.) varieties and effect of postharvest drying on quercetin yield. Ind. Crop. Prod. 38: 160–165.

Erro, J., R. Baigorri, J.-C. Yvin and J.M. García-Mina. 2011. ^{31}P NMR characterization and efficiency of new types of water-insoluble phosphate fertilizers to supply plant-available phosphorus in diverse soil types. J. Agric. Food Chem. 59: 1900–1908.

Ferreres, F., M.I. Gil, M. Castaner and F.A. Tomás-Barberán. 1997. Phenolic metabolites in red pigmented lettuce. Changes with minimal processing and cold storage. J. Agric. Food Chem. 45: 4249–4254.

Gallardo, M., L.E. Jackson, K. Schullback, R. Snyder, R. Thompson and L. Wyland. 1996. Production and water use in lettuces under variable water supply. Irrig. Sci. 16: 125–137.

Garmendia, I., N. Goicoechea and J. Aguirreolea. 2004a. Antioxidant metabolism in asymptomatic leaves of *Verticillium*-infected pepper associated with an arbuscular mycorrhizal fungus. J. Phytopathol. 152: 593–599.

Garmendia, I., N. Goicoechea and J. Aguirreolea. 2004b. Effectiveness of three *Glomus* species in protecting pepper (*Capsicum annuum* L.) against verticillium wilt. Biol. Control. 31: 296–305.

Gavito, M.E., P.S. Curtis, T.N. Mikkelsen and I. Jakobsen. 2000. Atmospheric CO_2 and mycorrhiza effects on biomass allocation and nutrient uptake of nodulated pea (*Pisum sativum* L.) plants. J. Exp. Bot. 51: 1931–1938.

Geneva, M.P., I.V. Stancheva, M.M. Boychinova, N.H. Mincheva and P.A. Yonova. 2010. Effects of foliar fertilization and arbuscular mycorrhizal colonization on *Salvia officinalis* L. growth, antioxidant capacity, and essential oil composition. J. Sci. Food Agric. 90: 696–702.

Gianinazzi, S., A. Gollotte, M.-N. Binet, D. van Tuinen, D. Redecker and D. Wipf. 2010. Agroecology: the key role of arbuscular mycorrhizas in ecosystem services. Mycorrhiza. 20: 519–530.

Giovannetti, M., L. Avio, R. Barale, N. Ceccarelli, R. Cristofani, A. Lezzi, F. Mignolli, P. Picciarelli, B. Pinto, D. Reali, C. Sbrana and R. Scarpato. 2012. Nutraceutical value and safety of tomato fruits produced by mycorrhizal plants. Brit. J. Nutr. 107: 242–251.

Goicoechea, N., S. Merino and M. Sánchez-Díaz M. 2004. Contribution of arbuscular mycorrhizal fungi (AMF) to the adaptations exhibited by the deciduous shrub *Anthyllis cytisoides* under water deficit. Physiol. Plantarum. 122: 453–464.

Gross, J. 1991. Pigments in vegetables: chlorophylls and carotenoids. AVI Publishing Co., Inc., Westport, CT, U.S.A.

Harlan, J. 1986. Lettuce and the sycamore: sex and romance in ancient Egypt. Econ. Bot. 40: 4–15.

Hart, D.J. and K.J. Scott. 1995. Development and evaluation of an HPLC method for the analysis of carotenoids in foods, and the measurement of the carotenoid content in vegetables and fruits commed in the UK. Food Chem. 54: 101–111.

IPCC. 2007a. Climate change and its impacts in the near and long term under different scenarios. *In:* Core Writing Team; R.K. Pachauri and A. Reisinger [eds.]. Climate Change 2007: Synthesis Report. Contribution of Working Groups I, II and III to the Fourth Assessment Report of the Intergovernmental Panel on Climate Change, IPCC, Geneva, Switzerland. pp. 44–54.

IPCC. 2007b. Observed changes in climate and their effects. *In:* Core Writing Team; R.K. Pachauri and A. Reisinger [eds.]. Climate Change 2007: Synthesis Report. Contribution of Working Groups I, II and III to the Fourth Assessment Report of the Intergovernmental Panel on Climate Change, IPCC, Geneva, Switzerland. pp. 30–33.

Jaganath, I.B. and A. Crozier. 2010. Dietary flavonoids and phenolic compounds. *In:* C.G. Fraga [ed.]. Plant Phenolics and Human Health. Biochemistry, Nutrition, and Pharmacology. Wiley & Sons, Inc. Hoboken, New Jersey, U.S.A. pp. 1–50.

Jifon, J.L., J.H. Graham, D.L. Drouillard and J.P. Syvertsen. 2002. Growth depression of mycorrhizal *Citrus* seedlings grown at high phosphorus supply is mitigated by elevated CO_2. New Phytol. 153: 133–142.

Kendall, C.W.C., A. Esfahani and D.J.A. Jenkins. 2010. The link between dietary fibre and human health. Food Hydrocolloid. 24: 42–48.

Kohler, J., J.A. Hernández, F. Caravaca and A. Roldán. 2008. Plant-growth promoting rhizobacteria and arbuscular mycorrhizal fungi modify alleviation biochemical mechanisms in water-stressed plants. Funct. Plant Biol. 35: 141–151.

Li, Q. and C. Kubota. 2009. Effects of supplemental light quality on growth and phytochemicals of baby leaf lettuce. Environ. Exp. Bot. 67: 59–64.

Lizarazo, K., B. Fernández-Marín, J.M. Becerril and J.I. García-Plazaola. 2010. Ageing and irradiance enhance vitamin E content in green edible tissues from crop plants. J. Sci. Food Agr. 90: 1994–1999.

Llorach, R., A. Martínez-Sánchez, F.A. Tomás-Barberán, M.I. Gil and F. Ferreres. 2008. Characterisation of polyphenols and antioxidant properties of five lettuce varieties and escarole. Food Chem. 108: 1028–1038.

Loladze, I. 2002. Rising atmospheric CO_2 and human nutrition: toward globally imbalanced plant stoichiometry? TRENDS Ecol. Evol. 17: 457–461.

Maiani G., M.J. Periago Castón, G. Catasta, E. Toti, I. Goñi Cambrodón, A. Bysted, F. Granado-Lorencio, B. Olmedilla-Alonso, P. Knuthsen, M. Valoti, V. Böhm, E. Mayer-Miebach, D. Behsnilian and U. Schlemmer. 2009. Carotenoids: actual knowledge of food sources, intakes, stability and bioavailability and their protective role in humans. Mol. Nutr. Food Res. 53: S194–S218.

Mangels, A.R., J.M. Holden, G.R. Beecher, M.R. Forman and E. Lanza. 1993. Carotenoid contents of fruits and vegetables: an evaluation of analytical data. J. Am. Diet Assoc. 93: 284–296.

Marulanda, A., R. Azcón and J.M. Ruiz-Lozano. 2003. Contribution of six arbuscular mycorrhizal isolates to water uptake by *Lactuca sativa* plants under drought stress. Physiol. Plant. 119: 523–533.

Mou, B. 2005. Genetic variation of beta-carotene and lutein contents in lettuce. J. Am. Soc. Hort. Sci. 130: 870–876.

Mou, B. 2009. Nutrient content of lettuce and its improvement. Curr. Nutr. Food Sci. 5: 242–248.

Mulabagal, V., M. Ngouajio, A. Nair, Y. Zhang, A.L. Gottumukkala and M.G. Nair. 2010. *In vitro* evaluation of red and green lettuce (*Lactuca sativa*) for functional food properties. Food Chem. 118: 300–306.

Munné-Bosch, S. and L. Alegre. 2004. Die and let live: leaf senescence contributes to plant survival under drought stress. Funct. Plant Biol. 31: 203–216.

Naidu, K.A. 2003. Vitamin C in human health and disease is still a mystery? An overview. Nutr. J. 2: 7–17.

Nicolle, C., A. Carnat, D. Fraisse, J.-L. Lamaison, E. Rock, H. Michel, P. Amouroux and C. Remesy. 2004a. Characterisation and variation of antioxidant micronutrients in lettuce (*Lactuca sativa* folium). J. Sci. Food Agr. 84: 2061–2069.

Nicolle, C., N. Cardinault, E. Gueux, L. Jaffrelo, E. Rock, A. Mazur, P. Amouroux and C. Rémésy. 2004b. Health effect of vegetable-based diet: lettuce consumption improves cholesterol metabolism and antioxidant status in the rat. Clin. Nutr. 23: 605–614.

Oh, M.-M., H.N. Trick and C.B. Rajashekar. 2009. Secondary metabolism and antioxidants are involved in environmental adaptation and stress tolerance in lettuce. J. Plant Physiol. 166: 180–191.

Oliveira, V.F., L.B.P. Zaindan, M.R. Braga, M.P.M. Aidar and M.A.M. Carvalho. 2010. Elevated CO_2 atmosphere promotes plants growth and inulin production in the cerrado species *Vernonia herbacea*. Funct. Plant Biol. 37: 223–231.

Pantin, F., T. Simonneau and B. Muller. 2012. Coming of leaf age: control of growth by hydraulics and metabolics during leaf ontogeny. New Phytol. 196: 349–366.

Rao, A.V. and L.G. Rao. 2007. Carotenoids and human health. Pharmacol. Res. 55: 207–216.

Romani, A., P. Pinelli, C. Galardi, G. Sani, A. Cimato and D. Heimler. 2002. Polyphenols in greenhouse and open-air-grown lettuce. Food Chem. 79: 337–342.

Ross, J.A. and C.M. Kasum. 2002. Dietary flavonoids: bioavailability, metabolic effects, and safety. Annu. Rev. Nutr. 22: 19–34.

Ruiz-Lozano, J.M. and R. Azcón. 1995. Hyphal contribution to water uptake in mycorrhizal plants as affected by the fungal species and water status. Physiol. Plant. 95: 472–478.

Said, S.A., H.A.E. Kashef, M.M.E.L. Mazar and O. Salama. 1996. Phytochemical and pharmacological studies on *Lactuca sativa* seed oil. Fitoterapia. 67: 215–219.

Seeram, N.P. 2008. Berry fruits: compositional elements, biochemical activities, and the impact of their intake on human health, performance, and disease. J. Agric. Food Chem. 56: 627–629.

Smith, S.E. and D.J. Read. 1997. Mycorrhizal symbiosis. 2nd edn. Academic Press. London.

Stintzing, F.C. and R. Carle. 2004. Functional properties of anthocyanins and betalains in plants, food, and in human nutrition. Trends Food Sci. Tech. 15:19–38.

Strack, D. and T. Fester. 2006. Isoprenoid metabolism and plastid reorganization in arbuscular mycorrhizal roots. New Phytol. 172: 22–34.

Szeto, Y.T., T.C. Kwok and I.F. Benzie. 2004. Effects of a long-term vegetarian diet on biomarkers of antioxidant status and cardiovascular disease risk. Nutrition. 20: 863–866.

Thomas, R.B. and B.R. Strain. 1991. Root restriction as a factor in photosynthetic acclimation of cotton seedlings grown in elevated carbon dioxide. Plant Physiol. 96: 627–634.

Toussaint, J.P., F.A. Smith and S.E. Smith. 2007. Arbuscular mycorrhizal fungi can induce the production of photochemicals in sweet basil irrespective of phosphorus nutrition. Mycorrhiza. 17: 291–297.

Treseder, K.K. and M.F. Allen. 2000. Mycorrhizal fungi have a potential role in soil carbon storage under elevated CO_2 and nitrogen deposition. New Phytol. 147: 189–200.

Tsabedze, M.W. and P.K. Wahome. 2010. Influence of different irrigation regimes on production of lettuce (*Lactuca sativa* L.). American-Eurasian J. Agric. & Environ. Sci. 8: 233–238.

Tunesi, S., V. Poggi and C. Gessa. 1999. Phosphate adsorption and precipitation in calcareous soils: the role of calcium ions in solution and carbonate minerals. Nutr. Cycl. Agroecosyst. 553: 219–227.

Ulrichs, C., G. Fischer, C. Büttner and I. Mewis. 2008. Comparison of lycopene, β-carotene and phenolic contents of tomato using conventional and ecological horticultural practices, and arbuscular mycorrhizal fungi (AMF). Agron. Colomb. 26: 40–46.

Watson, R.T., H. Rodhe, H. Oeschger and U. Siegenthaler. 1990. Greenhouse gases and aerosols. *In:* J.T. Houghton, G.J. Jenkins and J.J. Ephraums [eds.]. Climate Change: the IPCC Scientific Assessment. Cambridge University Press. Cambridge. pp. 1–40.

West, C.E. and E.J. Poortvliet. 1993. The carotenoid content of foods with special reference to developing countries. Department of Human Nutrition, Wageningen Agricultural University. The Netherlands.

You, Q., B. Wang, F. Chen, Z. Huang, X. Wang and P.G. Luo. 2011. Comparison of anthocyanins and phenolics in organically and conventionally grown blueberries in selected cultivars. Food Chem. 125: 201–208.

Zuccarini, P. 2007. Mycorrhizal infection ameliorates chlorophyll content and nutrient uptake of lettuce exposed to saline irrigation. Plant Soil Environ. 53: 283–289.

17

Ectomycorrhizal Plants: Methods, Applications and Certification

P. Alvarado,[1] *M. Honrubia*[2] and *J.L. Manjón*[1,*]

Introduction

Mycorrhizal fungi establish symbiotic, mutualistic relationships with the root systems of up to 90 percent plant species in natural conditions (see also Chapter 15 in this volume). Ectomycorrhizal fungi are especially important for most plants used in forestry, exchanging sugar, water or mineral metabolites with their hosts, and also making them more resistant to different types of biotic and abiotic stress, such as biological diseases, drought or mineral toxicity.

Methods for Production of Ectomycorrhizal Plants

The controlled production of mycorrhized plants is achieved through a group of techniques designed to obtain plants with their root systems enriched in specific fungi with commercial, forestry or agricultural

[1]Department of Plant Biology, University of Alcalá, Madrid, Spain.
[2]Department of Plant Biology, University of Murcia, Murcia, Spain.
*Corresponding author: josel.manjon@uah.es

interests. In the previous decade, a number of methods of inoculation and controlled mycorrhization were developed for different symbiotic ecto- and endo-mycorrhizal systems (Rai 2001). For research purposes, seeds or clonal plants can be used, while the fungal inoculum can be obtained from spores or *in vitro* grown mycelium from solid or liquid cultures (Morte and Honrubia 2009). However, at the greenhouse scale, mycelium cultures of the most interesting but usually slow-growing fungal species do not yield enough quantity for inoculation. A number of commercial products based on these techniques are currently available for customers, going from *ad hoc* mycorrhizal inocula for agriculture, to greenhouse inoculated ectomycorrhizal plants for forestry.

The history of truffle culture can be divided into four periods (Ceruti 1990). The first one, direct culture or "sowing", was used by ancient Romans up to the early 19th century. It is explicitly mentioned by Ciccarello, a doctor from Bevagna (Italy), in his 1564 treatise, where he suggested a simple sowing of truffles in adequate areas in order to grow them. The second period (extending for the rest of the 19th century), can be described as that of indirect culture. Host plants were established in the adequate environments of truffles, in the hope that these were spontaneously colonized by these fungi. Around 1810, Joseph Talon was successful in obtaining truffles in *Fagaceae* plantations, and therefore he is considered the first truffle-cultivator. In the 19th century, black truffles were far more abundant than they are today, allowing this method to succeed in the proper areas. Ecosystem alterations of a different nature during the 20th century probably led to a general recession in natural truffle habitats (Grente 1972), making spontaneous mycorrhization, even in the best soils, harder or impossible.

The third period can be considered the reaction of the 20th century to this situation. The new methods seek to inoculate young plantlets under semi-sterile conditions with fruiting bodies and/or mycelium cultures. Francolini (1931) was maybe the first to consider the possibility of producing mycorrhizal greenhouse plants by sowing truffle spores in young plants that are then transferred to the field. Between 1960 and 1970, Italian researchers such as Fassi, Palenzona, Fontana, Ceruti and Torini described the mycorrhizae of *Tuber* and managed to obtain the first mycorrhizal synthesis in aseptic plants, paving the way for the large-scale production of mycorrhizal plants. In France, the first plants obtained under this system were produced around 1980 (Chevalier and Grente 1979). In 1982, Bencivenga published a critical review of known mycorrhization methods, and the first commercial greenhouses selling mycorrhized plants appeared.

The fourth period (from the late 20th century to date), continues using the third-period methodology, but specifically focuses on culture conditions, trying to maintain mycorrhizal infection in soil, and develop the most favourable conditions for the formation of fruiting bodies. It can be said that it emerges as an answer to culture trial failures such as that of the introduction of *T. melanosporum* in Burgundy, where this truffle was displaced by *T. aestivum* or *T. mesentericum*, typical of more closed plant covers with other host species, or the problems found in New Zealand plantations designed around calcium carbonate amendments of suboptimal soils. In 1988, the Second International Congress on Truffles took place in Spoleto (Italy), confirming the general interest on a scientific approach of truffles and truffle-culture in general, and culture conditions of truffles in particular.

Nowadays, and probably in the future fifth period, greater efforts will be made on the different methods of irrigation, labouring and amendment of the soil, but also on establishing plantations with selected or even clonal plant hosts and fungi, molecular monitoring of the development of fungi in soil and its relation with fruiting or molecular traceability of strains.

Applications of Ectomycorrhyzal plants

Ectomycorrhizal plants are currently used for producing edible fungi (Karwa et al. 2011), afforestation (Pagano and Cabello 2011), crop amelioration (Gianinazzi et al. 2010), ecological restoration (Bücking 2011) or even biofuel production (Rooney et al. 2009), among other uses. This growing industry demands technical support for evaluating methods and products (Vosátka 2009).

One of the products most in demand are plants mycorrhized with truffles of the genus *Tuber*, especially the highly priced species *T. melanosporum*, *T. aestivum*, *T. magnatum* and *T. borchii*. Truffle culture is mainly developed in the Mediterranean basin, especially in Italy, Spain (Fig. 17.1) and France. In Italy, up to 120,000 mycorrhized seedlings are established every year in an area of 300 ha. The planted species are above all *T. melanosporum* (80 percent) and *Tuber aestivum* (15 percent). There are very few (5 percent) plantations of other species such as *T. magnatum*, *T. brumale*, *T. borchii* or *T. macrosporum*. After the 1968 Congress at Spoleto, the spreading of mycorrhizal techniques gave rise to several patents designed to control the production of mycorrhizal seedlings with black truffle in the greenhouse sector. Today, most Italian tree nurseries have agreements with public institutions for controlling and certifying mycorrhizal plants. The acquired experience and the research conducted led to the publication of an objective method for the analysis and certification of mycorrhizal plants by Bencivenga et al. (1987).

In Spain, black truffle has been seen in the provinces of Álava, Albacete, Barcelona, Burgos, Castellón, Ciudad Real, Cuenca, Gerona,

Figure 17.1 Black truffle plantation in Soria (Spain).

Granada, Guadalajara, Huesca, Jaén, La Rioja, Lérida, Murcia, Navarra, Segovia, Soria, Tarragona, Teruel, Valencia and Zaragoza. It can be found along the Catalan Coastal mountain ranges as well as the Pre-Pyrenean, Iberian, and Betic mountain ranges (Reyna 2007), apparently restricted to the eastern part of the Iberian Peninsula, with basic soils. The first large-scale plantations of *Tuber melanosporum* took place at Castellón and Teruel in 1967, and Soria in 1968. The most important was that of Aroth-Catesa, Abejar (Soria) with up to 630 ha. Production can vary greatly, commonly ranging between 20–30 and 100–140 kg/ha (only 0.5–3 kg/ha reported in French plantations!). However, global yields can only reach 7 tons in a bad year, but can also reach 100 tons in the best ones. The prices paid to the collectors range from euros 700 to 200, respectively, with average wages of around 350–400 euros/ kg. However, prices paid by final consumers can be even higher. No plantations of *T. magnatum* or *T. borchii* have been established in Spain to date, despite the fact that its edapho-climatic conditions seem to be adequate. On the other hand, a successful method to cultivate desert truffles (*Terfezia* spp., *Tirmania* spp. and *Picoa* spp.) from sporal or mycelial inocula and micropropagated or seed sowing seedlings has been developed in Spain (Morte et al. 2008, 2009, Honrubia et al. 2012), and protected under a patent. This technology includes spore and seed germination, micropropagation, control of mycorrhization in greenhouses, field plantations and management of orchards. Mycorrhizal plants start to produce one or two years after plantation, reaching a high productivity as high as 600 kg/ha of truffles.

In France, black truffle production seemed to reach its maximum at the end of 19th century, with large stretches of land under cultivation. As mentioned above, this was followed by a conspicuous decrease of production in the 20th century, with a minimum yield of 27 tons in 1980. In the 60s, the community of truffle growers and scientists became aware of the problem, and started to encourage the active cultivation of truffles. The surface of natural truffles land was reduced from around 80.000 ha in 1970 to only 35.000 ha in 2000. The surface covered by older plantations (more than 30 yr) decreased from 10,000 ha to 1,000 ha in the period 1970–2005, while that of medium age plantations (between 10 and 30 yr) reached 10.000 ha, and that of young plantations (10 yr or less) increased to 7,000 ha. Hence, it could be said that French plantations are currently young or very young, due to a big planting effort of around 900 ha/yr. Prices seem a bit higher than those in Spain, i.e., around 1,000 euros/kg, so that the total profit amounts to around euros 70 million to some 15,000 farmers. Only 2,500 of these farmers attend the official markets regularly, where 80 percent of the production is sold.

Certification of Ectomycorrhizal Plants

There are few procedures for evaluating mycorrhized plants (Bencivenga et al. 1987, Abourouh 1994, Fischer and Colinas 1996, Miko and Gažo 2007), which are not followed on a regular basis by experts, who commonly rely on their own adaptations of these methods (Bach et al. 2010), or even rely on other subjective criteria. The development of an objective, reproducible and cheap evaluation protocol is one of the main purposes on this field of research. Such a method is especially necessary for plants mychorrhized with edible fungi with high added value, such as the species in the genus *Tuber*, where the existence of very similar species (e.g., *T. melanosporum* and *T. indicum*), could lead to the involuntary introduction of putatively dangerous allochthonous species which could compete and displace the local ones (Murat et al. 2008, Bonito et al. 2011, Moreno et al. 2012). On the other hand, the mycorrhizal status of these plants introduced in the field could be directly related to the success of a plantation. Even though there are few experimental works on this topic, it is commonly believed that the amount of fungal structures such as mycelium or mycorrhizae (trophic roots) could be related to the fungal growth rate and production onset (Garbaye and Wilhelm 1984). This rate could also condition the probability of the plant being infected by other fungi from the local mycobiota of the planting site, which could eventually displace the inoculated fungus. Additionally, some forestry quality measurements of the plants such as aerial-radical biomass ratio, their respective architecture, as well as their nutritional and phytosanitary status, could all influence the final success of the whole plantation, affecting the survival chance of individual plants, their growth rates or the production onset delay. Alvarado (2011) performed a comparative study of the mycorrhizal status of *Quercus ilex* plants mycorrhized with *Tuber melanosporum* produced in different tree nurseries of Spain (Fig. 17.2). All seedlings inoculated, as well as most non-inoculated, presented mycorrhizal tips of some kind, although their abundance seemed strongly dependent on substrate type, mycorrhization procedures and plant age (Fig. 17.3). This is in accordance with the hypothesis that mycorrhization occurs spontaneously in tree nurseries only to a very low extent, at least if the substrate is previously sterilized, mycorrhization techniques are critical for the final quality of the product (Pruett et al. 2009). Mycorhization measures in also revealed strong differences among distinct suppliers Alvarado (2011). He suggested that the quality standards for a 2

Figure 17.2 Black truffle seedlings production by Spanish nurseries on different substrates and containers.

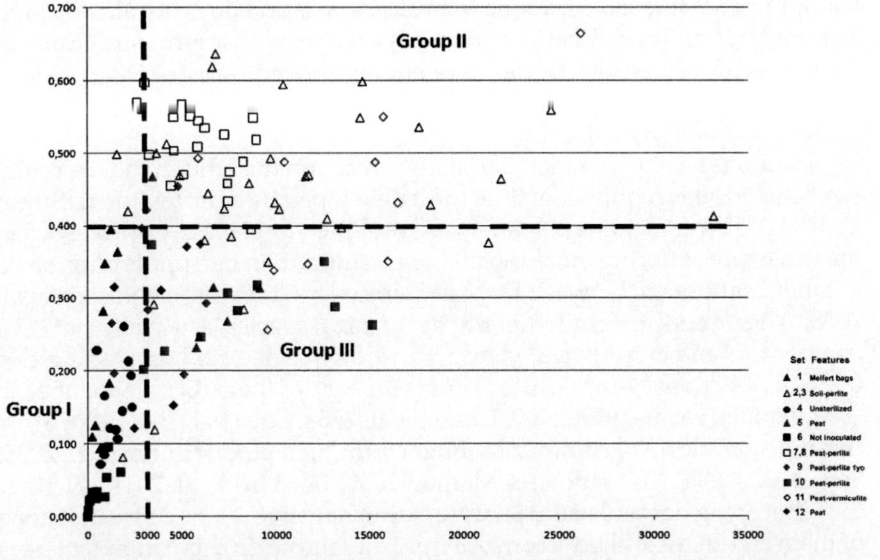

Figure 17.3 Mycorrhization measurements of individual plants of *Quercus ilex* inoculated with *Tuber melanosporum* (Alvarado 2011). Group I represents "suboptimal" plants, Group II is made up of "optimal" plants, while Group III is made up of intermediate "acceptable" plants.

yr-old plant of *Q. ilex* subsp. *ballota* inoculated with *T. melanosporum* were established above 30–40 percent mycorrhization percentage, and 3,000–4,000 mycorrhizal tips per plant. It is hard to say how much these variables will really influence plantation success or production onset delay, since this would require long experimental times (6–10 yr until production onset). It is highly probable that environmental conditions (irrigation, slope, soil type, planting season, climate, and disturbance, among others) have an influence or interaction with these parameters. It would be more suitable to say that mycorrhization measurements are necessary but not sufficient variables for the plantation success.

A complete procedure for evaluating the status of mycorrhization should be able to answer three main questions. Are the absolute and relative quantities of mycorrhizae adequate? Are there any other putatively dangerous species present in the roots? And finally, do plants meet the phytosanitary standards? The existing procedures for mycorrrhiza evaluation commonly include three main steps. First, some kind of cleaning of the root system is performed to remove the substrate adhered to it. Next, mycorrhizal samples are taken through different procedures. Finally, these samples are identified macro-microscopically, with the aid of a stereomicroscope and an optical microscope. For example, the Fischer-Colinas method suggests identifying 250 root tips of up to 12 plants in a plot. Unfortunately, morphotyping mycorrhizal tips is a time-consuming task and needs intensive training of professionals and does not allow a high standardization level. Even for most experts, morphological structures in some species (especially those more closely related) are not easy or even impossible to be distinguished (Salomoni 1994, Giomaro et al. 2000, Rinaldi et al. 2008, Manjón et al. 2009).

Methods based on molecular analysis can, on the other hand, be easily reproduced, and require less time to be developed by a professional (Suz et al. 2008a). Through molecular analysis, multiple samples can be processed at the same time, offering more reliable predictions than morphotyping, since a stable relationship between DNA sequences and taxa exists (Nilsson et al. 2008). The presence of contaminant mycorrhizal species, especially putative competitors and closely related taxa, can either be detected with a single or few PCR reactions with specific primers for each of them (Gandeboeuf et al. 1997, Paolocci et al. 1997, 2000, Douet et al. 2004, Suz et al. 2006, 2008b), or through some kind of community fingerprinting method (Burke et al. 2005, Suz et al. 2008a, Alvarado and Manjón 2009, Cordero et al. 2011). Neither morphotyping nor molecular analyses are exhaustive, in the sense that none of them can analyze all mycorrhizal tips in a sample. In this context, there is a possibility that a contaminant may not be detected under both methods. Additionally, the molecular analysis relies on DNA mixtures of a different quality extracted with a limited efficiency from environmental samples.

The most popular molecular methods used for detecting *Tuber melanosporum* in roots include amplification with more or less specific primers in simple or multiplex reactions (Séjalon-Delmas et al. 2000) which can in turn be cleaved with restriction enzymes (Cordero et al. 2011), on the other hand, community analysis methods, such as the analysis of cleaved or uncleaved fragments produced with universal primers (Alvarado 2011). Comparing both approaches, the specific primers could be preferred if only one or few well-known species are to be detected, while community analysis could be more useful if several or many species are to be detected at once, if specific primers are not available, or if the fungi to be detected are not known *a priori*. However, unknown peaks detected can only be identified by cloning-sequencing or massive sequencing. In addition, real ARISA-TRFLP fragment sizes observed in root samples or fruiting bodies are usually different from those expected after *in silico* predictions, presenting a small average error of 2–3 bp (Dickie and FitzJohn 2007, Riccioni et al. 2008, Alvarado 2011). ARISA fragment size (ribotype) can be shared between two or more different species, especially if these are phylogenetically related. In turn, TRFLP profiles (obtained after digesting ARISA fragments with endonucleases) allow discriminating plants inoculated with different fungi presenting a very similar ITS size (Alvarado 2011), but differing in as little as 1,937 percent of their bases (11 bp in 568). In some lineages, this lies below the species-level threshold, so TRFLPs could readily discriminate between cryptic clades or even strains of the same species.

The geographical origin of the fungal strain inoculated in commercial plants has been said to be important in order to ensure the success of plantations (Zambonelli et al. 2010). Some Spanish public organisms, such as the Junta de Andalucía (BOJA 123 de 27/06/2005, Anexo 7) require that truffle strains inoculated in mycorrhized plants to come from the same geographical areas where they are going to be planted. Black truffles seem to present little intraspecific variability of ITS, microsatellites or simple sequence repeats (SSRs), sequence characterized amplified regions (SCARs) or randomly amplified polymorphic DNA (RAPDs) (Dupré 1997, Gandeboeuf 1997, Bertault et al. 1998, Callot et al. 1999, Bertault et al. 2001, Murat et al. 2003). However, Murat et al. (2004) found up to 10 distinct ITS haplotypes in France, and the two more frequent ones were present in 60 and 28 percent of all samples. The remaining ones were restricted to specific localities. This was interpreted as a result of a putative genetic bottleneck suffered by black truffle populations in Europe, a hypothesis already proposed by Bertault et al. (1998). According to these authors, a few alleles became the most frequent after the habitat contraction resulting from glaciations, heading towards the Italian Peninsula as a putative refuge. The new data found by Murat et al. (2004) were interpreted as post-glacial variations occurred during the re-colonization of France and the Iberian Peninsula. Allelic differences led to the recognition of two distinct bio-geographical areas for black truffle: the Atlantic and the Rhone valley populations. Recently, Riccioni et al. (2008) extended the ITS haplotype analysis with new populations from France, Spain and Italy, used a new set of SSR microsatellites, and the highly variable amplified fragment length polymorphisms (AFLPs). Their results showed that southernmost populations present infrequent microsatellite loci, suggesting that these are not the result of a post-glacial re-colonization, but the residues of the former genetic diversity. AFLP diversity supported these results, presenting Shannon indices close or equal to 1 (maximum diversity) in the southernmost populations (Abruzzi, Sardinia, Navarra). This is the first work to suggest that the Iberian Peninsula could act as a putative glacial refuge during the last glaciation period. Alvarado (2011) conducted a new study on samples coming from different localities from central and southern Spain, and found that central populations did not show a higher degree of genetic diversity than those throughout Europe. Deviant alleles were present, but at a very low frequency compared to other Spanish regions such as Andalucía or Navarra. Intermediate populations were also found between these areas (e.g., Albacete). It was suggested that some haplotypes could be linked with specific geographical areas, thus allowing to trace back the origin of some truffles. This author also applied for the first time this type of analysis on mycorrhizal samples from inoculated trees with great success. SSRs were used to determine if a plant was inoculated with a strain coming from Andalucia or central Spain.

Finally, another important feature to be certified in mycorrhized plants could be mating types (*MAT*). It has been shown (Martin et al. 2010, Rubini et al. 2011a) that *T. melanosporum* presents different *MAT* types that govern the outcross between individuals, although both types could compete with each other, since recent studies on their dynamics show that usually one of them becomes dominant in both field and tree nursery stages (Rubini et al. 2011b). These authors suggest that strain displacement could influence the fungus viability or growth rate, highlighting the importance of producing plots inoculated with either *MAT*(+) or *MAT*(–) strains. In the near future, a certification protocol could include this kind of molecular testing.

Conclusions

A system for evaluating plants of *Q. ilex* mycorrhized with *T. melanosporum* based both on visual inspection (morpho-anatomy) and molecular analysis of the samples is needed. Evaluation should be based on three aspects:

1) Plant quality: the plants should be visually inspected, both their aerial part (leaves, neck), and their radical system (root ball, root development). A plant should meet basic standards, such as a size proportional to its age, a non aberrant aerial and radical architecture, as well as an adequate ratio between them, an adequate phytosanitary status (no pathogens or plagues), not etiolated, and an adequate hardening to resist planting. A unidirectional relation between an acceptable number of mycorrhizae and an adequate plant quality exists, as well as a bidirectional relation between a reduced number of mycorrhizae and a poor quality of the plant (reduced aerial part, small neck diameter, weak root system). Some other physiological quality standards such as the measurement of water stress through chlorophyll fluorescence have been used to assess not only plants quality standard, but also their mycorrhization status. We strongly discourage the use of these measures, since they only provide temporary views of the status of the plant in the day or days previous to measurement.
2) Mycorrhization measures: root tips should be extracted through any fast and reliable method, and the absolute and relative quantity of mycorrhizal tips in 10–12 plants per 1,000 should be estimated (Fischer and Colinas 1996). Optimal values for 2 yr-old plants of *Quercus ilex* plants mycorrhizal with *Tuber melanosporum* are suggested to be above 4,000 mycorrhizae and 40 percent mycorrhization, whereas the acceptability threshold is established above 3,000 mycorrhizae and 30 percent mycorrhization. However, these values could be very different in other mycorrhizal associations, demanding specific research focused on establishing quality standards for them.

3) Molecular test: the detection of inoculated fungi and of any other putative contaminants in the root system can be achieved through different methods, such as the amplification with specific primers, and also by community fingerprinting methods such as fragment analysis.

Acknowledgments

This work was financed by the Spanish Ministry of Education and Culture FPU grant AP2006-00890, and the Spanish "Plan Nacional" Research Project AGL2009-12884-C03-03. The authors thank Dr. M. Bencivenga, Dr. G. Moreno, Dr. B. Moreno-Arroyo, Dr. X. Parladé and A. Rodríguez for providing some mycorrhizal plants and truffle samples. They also thank Mr. Juan M. Barbero for proofreading the manuscript in English.

References

Abourouh, M. 1994. Les évaluations quantitatives des mycorhizes en pépinière et sur le terrain. Cahiers Options Mediterranéennes. 20: 51–61.

Alvarado, P. 2011. Caracterización molecular de *Tuber melanosporum* y otros ascomycetes hipogeos mediante análisis de fragmentos ARISA-TRFLP. PhD Thesis. Alcalá de Henares, Spain.

Alvarado, P. and J.L. Manjón. 2009. Selection of enzymes for terminal restriction fragment length polymorphism analysis of fungal internally transcribed spacer sequences. Appl. Environ. Microbiol. 75: 4747–4752.

Bach, I., S. Bordács and S. Szlávik. 2010. Development of an official method to control the quality of mycorrhized forestry material in Hungary. Öst. Zeitsch. f. Pilzk. 19: 227–229.

Bencivenga, M., A.M. Ferrara, A. Fontana, B. Granetti, G. Gregori, G. Lo Bue, M. Palenzona, E. Rebaudengo, A. Tocci and A. Zambonelli. 1987. Valutazione dello stato di micorrizazione di piante tartufigene. Proposta di un metodo. Ministero dell´Agricultura e Foreste, Roma.

Bertault, G., M. Raymond, A. Berthomieu, G. Callot and D. Fernandez. 1998. Trifling variation in truffles. Nature. 394: 734.

Bertault, G., F. Rousset, D. Fernandez, A. Berthomieu, M.E. Hochberg, G. Callot and M. Raymond. 2001. Population genetics and dynamics of the black truffle in a man-made truffle field. Heredity. 86: 451–458.

Bonito, G., J.M. Trappe, S. Donovan and R. Vilgalys. 2011. The Asian black truffle *Tuber indicum* can form ectomycorrhizas with North American host plants and complete its life cycle in non-native soils. Fung. Ecol. 4: 83–93.

Bücking, H. 2011. Ectomycoremediation: an eco-friendly technique for the remediation of polluted sites. *In:* M. Rai and A. Varma [eds.]. Diversity and Biotechnology of Ectomycorrhizae, Soil Biology 25, Springer-Verlag. Berlin.

Burke, D.J., K.J. Martin, P.T. Rygiewicz and M.A. Topa. 2005. Ectomycorrhizal fungi identification in single and pooled root samples: terminal restriction fragment length polymorphism (TRFLP) and morphotyping compared. Soil Biol. Biochem. 37: 1683–1694.

Callot, G., P. Byé, M. Raymond, J.C. Pargney, A. Parguey-Leduc, M.C. Janex-Favre, R. Moussa and L. Pagès. 1999. La Truffe, la Terre, la Vie. INRA Press. Paris.

Ceruti, A. 1990. Evoluzione delle conoscenze biologiche sul genere Tuber. Atti del II Congresso Internazionale sul Tartufo. Spoleto, 24–27 Novembre 1988. Comunità montana dei Monti Martani e del Serano, Spoleto.

Chevalier, G. and J. Grente. 1979. Application pratique de la symbiose ectomycorhizienne: production à grande échelle de plants mycorhizés par la truffe. Mushroom Sc. 10: 483–505.

Cordero, C., P. Cáceres, G. González, K. Quiroz, C. Bravo, R. Ramírez, P.D.S. Caligari, B. Carrasco and R. García-González. 2011. Molecular tools for rapid and accurate detection of black truffle (*Tuber melanosporum* Vitt.) in inoculated nursery plants and comercial plantations in Chile. Chilean J. Agr. Res. 71: 488–494.

Dickie, I.A. and R.G. FitzJohn. 2007. Using terminal restriction fragment length polymorphism (T-RFLP) to identify mycorrhizal fungi: a methods review. Mycorrhiza. 17: 259–270.

Douet, J.P., M. Castroviejo, D. Mabru, G. Chevalier, C. Dupré, F. Bergougnoux, J.M. Ricard and J. Médina. 2004. Rapid molecular typing of *Tuber melanosporum*, *T. brumale* and *T. indicum* from tree seedlings and canned truffles. Anal. Bioanal. Chem. 379: 668–673.

Dupré, C. 1997. De nouveaux outils au service de la trufficulture, Mémoire d'Ingénieur diplômé par l'Etat, ENESAD. Dijon.

Fischer, C. and C. Colinas. 1996. Methodology for certification of *Quercus ilex* seedlings inoculated with *Tuber melanosporum* for commercial application. 1st International Conference on Mycorrhizae. Berkeley, California, U.S.A.

Francolini, F. 1931. Tartuficoltura e rimboschimenti. Federazione italiana dei consorzi agrari, Piacenza.

Gandeboeuf, D. 1997. Caractérisation et identification moléculaire de différentes espèces du genre *Tuber*, Thèse de Doctorat de l'Université Blaise Pascal. Clermont-Ferrand.

Gandeboeuf, D., C. Dupré, P. Roeckel-Drevet, P. Nicolas and G. Chevalier. 1997. Typing *Tuber* ectomycorrhizae by polymerase chain amplification of the internal transcribed spacer of rDNA and the sequence characterized amplified region markers. Can. J. Microbiol. 43: 723–728.

Garbaye, J. and M.E. Wilhelm. 1984. Influence de la mycorhization acquise en pépinière sur la mycorhization de jeunes plantations de chêne. Acta Oecol. 5: 151–161.

Gianinazzi, S., A. Gollotte, M.-N. Binet, D. van Tuinen, D. Redecker and D. Wipf. 2010. Agroecology: the key role of arbuscular mycorrhizas in ecosystem services. Mycorrhiza. 20: 519–530.

Giomaro, G., A. Zambonelli, D. Sisti, M. Cecchini, V. Evangelista and V. Stocchi. 2000. Anatomical and morphological characterization of mycorrhizas of five strains of *Tuber borchii* Vittad. Mycorrhiza. 10: 107–114.

Grente, J. 1972. Perspectives por une trufficulture moderne. INRA, Clermont-Ferrand.

Honrubia, M., A. Morte and A. Gutiérrez. 2012. Las Terfezias. Un cultivo para el desarrollo rural en regiones áridas y semi-áridas. *In:* S. Reyna (coord.). Truficultura, Fundamentos y Técnicas. Cap. 11. 2nd edn. Mundi-Prensa, Madrid.

Karwa, A., A. Varma and M. Rai. 2011. Edible ectomycorrhizal fungi: cultivation, conservation and challenges. *In:* M. Rai and A. Varma [eds.]. Diversity and Biotechnology of Ectomycorrhizae, Soil Biology 25. Springer-Verlag. Berlin.

Manjón, J.L., L.G. García-Montero, P. Alvarado, G. Moreno and G. Di Massimo. 2009. *Tuber pseudoexcavatum* versus *T. pseudohimalayense*—new data on the molecular taxonomy and mycorrhizae of Chinese truffles. Mycotaxon. 110: 399–412.

Martin, F., A. Kohler, C. Murat, R. Balestrini, P.M. Coutinho, O. Jaillon, B. Montanini, E. Morin, B. Noel, R. Percudani et al. 2010. Périgord black truffle genome uncovers evolutionary origins and mechanisms of symbiosis. Nature. 464: 1033–1038.

Miko, M. and J. Gažo. 2007. Methodical assessment of host tree seedlings inoculated with *Tuber aestivum* Vitt. for application in agroforestry. Acta Fytotech. Zootech. 1: 12–16.

Moreno, G., P. Alvarado and J.L. Manjón. 2012. Phylogenetic affiliation of *Choiromyces magnusii* and *C. venosus* Tuberaceae (Ascomycota) from Spain. Mycol. Prog. 11: 463–471.

Morte, A. and M. Honrubia. 2009. Biotechnology for the industrial production of ectomycorrhizal inoculum and mycorrhizal plants. *In:* A. Varma and V. Neeraj [eds.]. Text Book on Molecular Biotechnology, I.K. International Publishing House, New Delhi, Mumbai, Bangalore.

Morte, A., M. Honrubia and A. Gutiérrez. 2008. Biotechnology and Cultivation of Desert Truffles. *In:* A. Varma [ed.]. Mycorrhiza. Biology, Genetics, Novel Endophytes and Biotechnology 38, Springer-Verlag, New York.

Morte, A., M. Zamora, A. Gutiérrez and M. Honrubia. 2009. Desert truffle cultivation in semiarid Mediterranean areas. *In:* V. Gianinazzi-Pearson and C. Azcón [eds.]. Mycorrhizas: functional processes and ecological impact. Springer-Verlag. Heidelberg.

Murat, C., P. Luis, J. Diez, C. Delaruelle, P. Gillot, C. Dupré, G. Chevalier and F. Martin. 2003. Diversité génétique des populations de *Tuber melanosporum* Vittad. Les Actes du BRG. 4: 467–478.

Murat, C., J. Díez, P. Luis, C. Delaruelle, C. Dupré, G. Chevalier, P. Bonfante and F. Martin. 2004. Polymorphism at the ribosomal DNA ITS and its relation to postglacial re-colonization routes of the Perigord truffle *Tuber melanosporum*. New Phytol. 164: 401–411.

Murat, C., E. Zampieri, A. Vizzini and P. Bonfante. 2008. Is the Perigord black truffle threatened by an invasive species? We dreaded it and it has happened! New Phytol. 178: 699–702.

Nilsson, R.H., E. Kristiansson, M. Ryberg, N. Hallenberg and K.-H. Larsson. 2008. Intraspecific ITS variability in the kingdom fungi as expressed in the international sequence databases and its implications for molecular species identification. Evol. Bioinf. 4: 193–201.

Pagano, M.C. and M.N. Cabello. 2011. Mycorrhizal interactions for reforestation: constraints to dryland agroforest in Brazil. ISRN Ecology Article ID 890850.

Paolocci, F., A. Rubini, B. Granetti and S. Arcioni. 1999. Rapid molecular approach for a reliable identification of *Tuber* spp. ectomycorrhizae. FEMS Microbiol. Ecol. 28: 23–30.

Pruett, G.E., J.N. Bruhn and J.D. Mihail. 2009. Greenhouse production of Burgundy truffle mycorrhizae on oak roots. New Forests. 37: 43–52.

Rai, M.K. 2001. Current advances in mycorrhization in micropropagation. *In Vitro* Cellular and Developmental Biology. Plant. 37: 158–167.

Reyna, S. 2007. Truficultura. Fundamentos y técnicas. Mundi-Prensa. Madrid.

Riccioni, C., B. Belfiori, A. Rubini, V. Passeri, S. Arcioni and F. Paolocci. 2008. *Tuber melanosporum* outcrosses: analysis of the genetic diversity within and among its natural populations under this new scenario. New Phytol. 180: 466–478.

Rinaldi, A.C., O. Comandini and T.W. Kuyper. 2008. Ectomycorrhizal fungal diversity: separating the wheat from the chaff. Fungal Div. 33: 1–45.

Rooney, D.C., K. Killham, G.D. Bending, E. Baggs, M. Weih and A. Hodge. 2009. Mycorrhizas and biomass crops: opportunities for future sustainable development. Trends Plant Sci. 14: 542–549.

Rubini, A., B. Belfiori, C. Riccioni, E. Tisserant, S. Arcioni, F. Martin and F. Paolocci. 2011a. Isolation and characterization of MAT genes in the symbiotic ascomycete *Tuber melanosporum*. New Phytol. 189: 710–722.

Rubini, A., B. Belfiori, C. Riccioni, S. Arcioni, F. Martin and F. Paolocci. 2011b. *Tuber melanosporum*: mating type distribution in a natural plantation and dynamics of strains of different mating types on the roots of nursery-inoculated host plants. New Phytol. 189: 723–735.

Salomoni, S. 1994. La certificazione delle piante in tartuficoltura. Bosco e Ambiente. 2: 15–19.

Séjalon-Delmas, N., C. Roux, M. Martins, M. Kulifaj, G. Bécard and R. Dargent. 2000. Molecular tools for the identification of *Tuber melanosporum* in agroindustry. J. Agric. Food Chem. 48: 2608–2613.

Suz, L.M., M.P. Martín and C. Colinas. 2006. Detection of *Tuber melanosporum* DNA in soil. FEMS Microbiol. Lett. 254: 251–257.

Suz, L.M., A.M. Azul, M.H. Morris, C.S. Bledsoe and M.P. Martín. 2008a. Molecular and morphotyping methods to characterize ectomycorrhizal roots and hyphae in soil. *In:* C.S. Nautiyal and P. Dion [eds.]. Mycorrhizal Manual. Springer-Verlag, Berlin Heidelberg.

Suz, L.M., M.P. Martín, D. Oliach, C. Fischer and C. Colinas. 2008b. Mycelial abundance and other factors related to truffle productivity in *Tuber melanosporum-Quercus ilex* orchards. FEMS Microbiol. Lett. 285: 71–78.

Vosátka, M. 2009. Basic aspects of mycorrhizal biotechnology: fungal inocula production and application. *In:* A.K. Chauhan and A. Varma [eds.]. A textbook of molecular biotechnology. IK International, New Delhi.

Zambonelli, A., F. Piattoni and M. Iotti. 2010. What makes a good truffle infected tree? Öst. Zeitschr. f. Pilzk. 19: 201–207.

18

Metagenomics of Plant-Microorganism Interaction: Source of Novel Recombinant Genes for Biotechnological Applications

A. García-Villaraco, A. Bonilla, D.G. Seco de Herrera* and *E. Algar*

Introduction

The term metagenomics was coined in 1998 by Jo Handelsman to describe a new field which analyzed the genetic material collected directly from environmental samples. Currently, metagenomics broadly comprises any research involving the application of modern genomic techniques to the direct study of biological communities in their natural environment avoiding isolation and culturing of individual organisms. Although metagenomics currently focuses on the genetic material of microorganisms, metagenomic technologies may be applied to other groups of organisms, including extinct and fossilized animals.

San Pablo CEU University. Department of Biology, Rhizosphere biotechnology Group, Ctra. Boadilla del Monte Km 5,3, 28668, Boadilla del Monte, Madrid, Spain.
*Corresponding author: anabec.fcex@ceu.es

Rhizosphere

In 1904 Hiltner defined the rhizosphere as an effect; from there on, this effect has grown to be one of the most complex ecosystems on Earth. The study of the rhizosphere requires a multidisciplinary approach with a wide array of methodologies from all areas of knowledge and research. From the most basic knowledge, study of the interactions from a biophysical and biogeochemical perspective, to molecular levels, as the study of communication systems, metabolic adaptation and multitrophic interactions, until complex physiological levels like the development of symbiotic systems (mycorrhizal and nitrogen fixation), growth promotion and plant health.

The study of biophysical and biogeochemical processes have identified changes in the abiotic components of the system, mainly due to the ability of the organisms in the rhizosphere to carry out metabolic processes unimaginable outside the rhizosphere without the genetic contribution of plants and microorganisms that inhabit this system. This is the case of nutrient mobilization, such as iron and phosphorus, or metabolic activities related to the immobilization of heavy metals or biodegradation of xenobiotic pollutants such as pesticides, polyaromatic hydrocarbons and other compounds highly stable and dangerous for human health. Therefore, the rhizosphere constitutes a system especially suitable for obtaining culturable microorganisms or genes with a great biotechnological application oriented to nutrient mobilization and bioremediation or phytorhizoremediation.

The study of plant microorganisms communication processes has progressed largely in parallel with the study of sophisticated signaling mechanisms that occur in the formation and development of symbiotic systems. The rhizosphere probably holds the most specialized signaling exchanges between microorganisms and plants. In turn, knowledge of molecular communication systems allows understanding of multitrophic relationships established in the rhizosphere and the possibility to extend and relate them to higher levels of organization, such as physiological processes at the plant level.

Today, it is a widely accepted fact that certain strains of rhizosphere bacteria, referred to as plant growth promoting rhizobacteria (PGPR), stimulate plant growth and fitness (Kloepper 1980). Thanks to the knowledge of communication signals between rhizosphere organisms, we are able to understand, at least partially, the PGPR mechanisms of action. In the 90s, the interaction of bacteria with plants was simply thought of as being an effect, but today it is recognized as a process with a high level of complexity in which at least two genomes share information without sharing the same spaces from a cellular perspective. The degree of influence among the organisms involved in the interaction is so important that the biotechnological management of the microorganism has become an essential target for improving the functions of the plant.

The above mentioned multidisciplinary approach to the rhizosphere has unraveled a number of highly complex interactions that develop under high pressure, and therefore, the specialization process selects finely-tuned communication strategies involved in a wide array of different processes ranging from biogeochemical cycles to the most complex secondary metabolic pathways in plants.

It is already a widely known fact that PGPR, broadly speaking, can improve plant fitness. Beyond their effect on nutrition or their biocontrol capacity, some PGPR strains are able to effectively protect plants from pathogens, triggering a response in the plant that makes it resistant to further pathogen attack (Van Loon et al. 1998, see also Chapter 12 in this volume). It spreads throughout the plant although the bacteria interact only in a small area; this protective mechanism is called induced systemic response (ISR), to differentiate it from the systemic acquired response (SAR) (Lawton et al. 1995, Ryals et al. 1996). In the SAR response a pathogenic microorganism virulently interacts with the plant; if the plant is capable of defending itself, the pathogen cannot produce lethal damage, inducing systemic resistance also known as systemic acquired resistance, and is also effective against subsequent pathogen attacks. Therefore, plant protection is triggered by a pathogen (SAR) or non-pathogenic agent (ISR).

The discovery of systemic induction is of great importance since an immune response is proposed. In the same way that a vaccine may cause some mild side effects when administered, the plant may also suffer some similar symptoms, which confirm that contact has been established between beneficial bacteria and the plant. When these bacteria are effective, a metabolic response in the plant is developed as a warning signal and occasionally it may result in reduced plant growth. This alert state is called "priming" (Conrath et al. 2002), and it will not be observed until the plant is challenged with the pathogen or other stress factors, when a much more intense and effective response is detected on primed plants (van Hulten et al. 2006). It has been reported that the ability to induce systemic resistance of a given system depends on the plant-beneficial bacteria-pathogen, and is a highly specific response. However, the protective effect of these bacteria may be broader and confer protection against other biotic and abiotic stress factors. For example, some strains can be considered "broad spectrum" beneficial strains since they can protect against pathogens in different plant species. For example *Chryseobacterium balustinum* (Aur9, CECT 5399) protects *A. thaliana* against *P. syringae* DC3000 under controlled conditions (Ramos Solano et al. 2008a) and in *Oryza sativa* it provides a great protection against *Pyricularia grisea* in rice fields of the Guadalquivir marshes, further increasing productivity (Lucas et al. 2009). Strain Aur9 also protects *Solanum lycopersicum* and *Capsicum annuum* plants against *Fusarium oxysporum* and *Rhizoctonia solani* (Domenech et al. 2006).

Induction of systemic resistance was described as a defensive mechanism in plants and is well known that the defensive metabolism is part of the secondary metabolism. Interestingly, a great part of secondary metabolites are pharmacologically active and currently, 80 percent of drugs that are used come from plants. Therefore, the application of the concept of systemic induction to plants with medicinal interest would involve an improvement in the quality of raw material for the pharmaceutical industry. In this regard, improvement in phytopharmaceutical contents by applying bacteria has been demonstrated in certain plant species, such as *Glycine max* for isoflavones (Ramos Solano et al. 2010), *Digitalis lanata* for cardenolides (Gutierrez Mañero et al. 2003) or *Hypericum perforatum* for antidepressants (Gutierrez Mañero et al. 2012). Moreover, there is another large group of plants that play an essential role for health through diet, by its high content of secondary metabolites with beneficial activity for human health, termed as functional foods. Within this group are species such as soybeans, berries, tomatoes, etc., which constitute an important part of our diet and contribute decisively to prevent disease rather than healing. In addition, an important field of nutritional supplements is being developed with all these plants, so specific stimulation of secondary metabolism in each crop is a biotechnological challenge. As a final consideration to systemic induction in plants with bacteria is that secondary metabolism is also a defensive metabolism, hence an effective stimulation will result in reduction of agrochemicals, contributing to sustainable agriculture, and simultaneously, a high value food product, with a higher bioactive content will be obtained. Thus, the application of beneficial bacteria in the field is presented as a biotechnological tool of great interest to the food and pharmaceutical industry.

As mentioned above, the ability of a given bacterium to induce systemic resistance depends on the plant/beneficial bacteria/pathogen system and is a highly specific response. Therefore, it is of great interest to identify the bacterial determinants (named elicitors) able to trigger this effect on the plants. Identification of bacterial elicitors as the key to induction of secondary metabolism opens several areas for action with clear biotechnological applications. First, use of elicitors in agriculture would overcome the lack of results attributable to poor bacterial performance (root colonization, interaction with other environmental variables, etc.). This would have added value when developing products, since the loss of viability of the inoculant with time or by poor maintenance would not be a limiting factor in the economic viability of the product. Moreover, the use of beneficial microorganisms is limited to plants, while elicitors may be extended to other systems, such as plant cell culture. Plant cell culture allows preventing seasonal variability due to environmental fluctuations, so it is of great interest for industry (Gutierrez Mañero et al. 2012). However, one of the limitations of secondary metabolites production in cell culture systems is the lack of environmental stimuli that trigger secondary metabolism in field conditions. Consistenly, bacterial elicitors could be used in order to improve the yield of compounds of interest (Gutierrez Mañero et al. 2012).

In view of the above, application of beneficial bacteria or derived elicitors as inducers of plant secondary metabolism appears as a biotechnological tool of great interest and great potential for the food and pharmaceutical industry. Consequently, on the basis of the current knowledge regarding mechanisms of interaction that involve plant physiology, microbial primary and secondary metabolism, and plant-microorganism interaction systems, it is possible to apply biotechnological processes that provide a great benefit in areas of interest today such as:

- Bioremediation: biological systems for pollutants cleanup and ecosystems recovery.
- Industry: production of new or improved enzymes to optimize industrial processes.
- Pharmacy: obtaining new or better products with biological activity.
- Agriculture: including crop production from a classical view, i.e., assessing yield, and from the modern point of view which considers issues such as environmental sustainability of production by reducing the agrochemical inputs and environmental impact.

As previously mentioned, PGPR are beneficial rhizobacteria that improve plant fitness. Bacterial elicitors can reproduce these beneficial effects. The advantages of using elicitors better than microorganisms have been observed and one step further may be taken towards development of recombinant products derived from novel genes from culturable bacteria but also from the great source of genes provided by non-culturable bacteria. It could be affirmed that the rhizosphere is no longer just as a source of bacteria of interest but also a source of genes of interest, through metagenomic approaches.

Metagenomics

Metagenomics works with total DNA and therefore, it enables the study of non-culturable microorganisms, essential information to get a true picture of wild biological communities, since more than 99 percent of microorganisms from environmental samples cannot be grown in the laboratory (Torsvik et al. 1990, Whitman et al. 1998). This limitation is due to several causes like the absence of symbionts, nutrients, or surfaces; excess inhibitory compounds;

incorrect combinations of temperature, pressure or atmospheric gas composition; accumulation of toxic waste products of its own metabolism; inherently low growth rate. On the other hand the culturable fraction (1%) is usually grown under artificial conditions different from their natural ecological niches, and as a result, the expression of genes, proteins and metabolic profiles observed in the laboratory are probably different than those expressed in their original environments.

Therefore, metagenomics will enable us to study the unknown ecological functions of microbial communities, trying to unravel the biological functions that are present in an environment or which organisms are present. Additionally, by comparing data from similar or different environments, the fundamental rules of microbial ecology or the mechanisms by which microorganisms adapt to environmental conditions or how does horizontal transfer occur may be studied. Moreover, search for new genes and proteins (new families of proteins) will be possible and analysis of diversity within known families may be carried out as basic knowledge studies. Identification of new genes and proteins has a really relevant degree for biotechnological application.

The initial metagenomics investigations focused on analysis of individual genes, such as the gene of the small subunit ribosomal RNA. In 1985, Pace et al. first proposed to clone DNA sequences directly from environmental samples and in 1991 the first cloning of this type with the 16S rRNA gene of marine picoplankton was achieved. However, the analysis of microbial diversity by 16S rRNA gene sequences only allows to establish phylogenetic relationships, but physiology, abundance or ecological significance, are not studied.

As previously mentioned, the plant-microorganisms interaction in the rhizosphere is crucial for plant health and soil fertility. Most of what we know about these complex interactions has been derived from the study of microorganisms in pure culture. Nowadays, it is possible to explore the interactions between plants and microorganisms through culture-independent study of the microbial communities. While high-throughput analysis of DNA sequences is an important tool for these studies, the wide richness and diversity of these communities requires the use of functional metagenomics strategies involving a wide variety of screening methodologies to discover and study the key biological processes, which are still to be known.

It is essential to have a thorough understanding of the PGPR mechanisms of action, to manipulate the rhizosphere in order to maximize the processes that influence plant productivity. However, to date, metagenomics have not been widely applied to the study of plant-microbe interactions. Therefore, this is a very interesting topic to study.

Metagenomic analysis of the rhizosphere may on one hand, allow the genetic characterization of its microbial communities and to get a deeper understanding of the activities and chemical signaling involved in plant-microbe interactions, and on the other hand, allow the discovery of new genes of industrial, pharmaceutical and agricultural interest. The plant has carried out a great selection among soil microorganisms based on the best skills in such a competitive environment. Therefore, since the rhizosphere has been demonstrated as an excellent source of specialized microorganism, it seems likely to be an excellent place for new genes of interest.

A successful metagenomics approach of the plant-microbe interaction that targets function needs an appropriate design. Screenings should be based on current understanding of this interaction, allowing the expression of bacterial traits involved in the interaction with the plant, like ACC deaminase, phosphate solubilization, nitrogen fixation, quorum sensing, siderophores, chitinases, auxins, gibberellins or antibiotics production. Detection of novel genes coding for known activities related to the interaction with plants is clearly a worthwhile effort, but real progress will be made towards the discovery of currently unknown mechanisms of the plant-microorganism interaction.

The great microbial diversity provides an unlimited source of new genes of interest for a broad spectrum of biotechnological applications. Thus, using functional metagenomic libraries of non-culturable microorganisms will result in compounds with targeted biological activity, like the antibiotics indirubin and turbomycins (Gillespie et al. 2002, Lim et al. 2005). Uncultured marine microorganisms are also a rich source of secondary metabolites different from the terrestrial ones. Thus, some pharmaceutical drugs have been isolated from marine microorganisms such as, cephalosporins, cytarabine and vidarabine (Rayl 1999). Metagenomics have also provided new genetic information of enzymes with industrial interest, such as esterases/lipases, proteases, amylases, nitrilases, cellulases and others. These enzymes and knowledge of their structures and functions have enormous value to pharmaceutical and industrial applications. Similarly, the discovery of novel genes with PGPR abilities may be applied agronomically by synthesizing the enzyme of interest in bioreactors for application to crops, reducing chemicals used.

In this chapter, the focus will be on metagenomics of terrestrial environmental samples (soil), especially in the rhizosphere.

Metagenomics Strategies

A metagenomic study recalls information from all microbial communities within an ecosystem. A careful experimental design should be laid out, since the tools used in the study may condition its outcome.

Target ecosystems have high biodiversity, irrespective of being marine or terrestrial, the total number of microorganisms on Earth has been estimated in 10^{30} (Turnbaugh and Gordon 2008, Singh et al. 2009). Prokaryotic microorganism account for 10^6–10^9 species (Sleator et al. 2008), representing a great reserve of genomic diversity still to be studied, it probably holds a number of new enzymes or metabolic capabilities. Soil from forests in a continental area contains 4. 10^7 microorganism per gram, while in agronomic soils or weeds, there are 2. 10^9 microorganism per gram, with a number of different bacterial species that ranges from 2,000 to 18,000. Difference between these figures confirms the high variability and richness in the different ecosystems. Microorganism number and variability are determined by plant species and soil physical-chemical properties, especially in the rhizosphere. Metagenomics appear as a challenging alternative to overcome the limited capacity to culture microorganisms, allowing isolation and study of non-cultivable microorganisms.

There is a wide array of methodologies available today for isolation and purification of DNA, vectors for metagenomic libraries, functional screening for identification of metabolites or enzymatic activities or genes of interest, and finally sequencing and further analyses of sequences. Keeping all these factors in mind, those parameters that should be addressed in a metagenomic study and considerations depending on the purpose of the study will be reviewed here.

Sampling

The first step and probably one of the most important, is the experimental design of the sampling. Different parameters such as sampling time, sampling site, mass of sample, and number of replicates must be considered.

Temporal Heterogeneity

Both for diversity analysis of microbial communities and for the discovery of new functional features, it is essential to consider the moment of sampling. Since weather conditions will not be the same throughout the different seasons (summer-winter, wet-dry), environmental conditions can dramatically affect the presence of microbial communities. A parameter directly related to environmental conditions is the aqueous fraction, an element which in turn governs the values of other physical-chemical properties of the substrate. It has also been observed that plants show different physiological activities throughout its life cycle, varying the composition of root exudates, and therefore disrupting the microbial populations present in the rhizosphere. Because of this situation, sampling should always be associated with a study of the physicochemical properties of the substrate for subsequent evaluation with results obtained from a metagenomic study.

Spatial Heterogeneity

Besides the temporal parameter, the spatial variable also has a vital role during any sampling performed. Several studies have demonstrated the importance of sampling at various points in the study plot, since the physicochemical properties, or the presence of different plants, can vary the sampling outcome. Small differences may play an important role, by creating micro-habitats, altering the presence of microbial populations that in turn would alter results in the metagenomic study of microbial communities structure, or in the functional approach. Therefore, to overcome this problem, randomly-taken samples should be a minimum number of five, and pooled to obtain a single sample, so variability in physicochemical properties of the substrate can be reduced. This process should be repeated three times (replicates), to minimize variability among samples (Baker et al. 2009).

Sample Size

The minimum amount of substrate required per sample will depend on the type of metagenomic study to be performed. Different analyses have been conducted with the mass of the sample ranging from 0.01 g to 10 g. resulting in differences in the structure of bacterial communities in the samples of lower mass (below 0.25 g) (Kang and Mills 2006). Consistently, several studies related to the *Archaea* communities structure revealed a high heterogeneity in samples with a mass of substrate sampled between 0.1 g and 1.0 g (Nicol et al. 2003). Therefore, the minimum sample size should be over 1 g of substrate. Different studies have shown that the rhizosphere is formed by a very small region of substrate attached to the roots of plants. It has been estimated that for each centimeter of a typical root approximately 20 mg of substrate is attached (Jacobsen 2004) and therefore, an amount of 250 cm of roots to obtain a mass of 5 g of sample would be required. In plants in which this magnitude of roots can be a problem, pooling adjacent plant rhizospheres is required to solve this problem.

Isolation and Purification of Metagenomic DNA

Nowadays, there are a large number of commercial kits and protocols that enable the isolation and purification of microbial DNA. In turn, it is possible to choose between two strategies for the isolation of DNA from the soil sampled. Direct Extraction: proceed to lysis of the bacteria located in the substrate, thereafter isolating and purifying DNA. A second approach is the Indirect Extraction: bacteria are isolated of the substrate prior to proceeding to lysis and purification of DNA.

Either of these two isolation techniques must confirm some quality parameters such as DNA quantity and purity, integrity of the molecule, size of DNA fragments obtained, and representation of the DNA obtained from samples. The displayed variables are of vital importance for proper implementation of metagenomic libraries or sequencing of the samples obtained. The direct extraction technique is the most widely used, although depending on the purpose of the metagenomic study one technique or the other should be applied. The choice of one of these methodologies could vary quality parameters required for the study to be developed.

Direct Extraction

The different microorganisms are lysed with the rest of the matrix that makes up the substrate by physical, chemical and enzymatic processes. The use of vigorous bead-beating has a better outcome compared with freezing and thawing or heating of the sample with microwaves, as it has been found for the lysis of Gram-positive bacteria and spores (Kang and Mills 2006). The use of detergents such as SDS, which disintegrates the plasma membrane of cells, facilitates lysis. Other products of enzymatic nature that are usually used in this methodology, are lysozyme or other nonspecific proteases.

After cell lysis, the negatively-charged DNA molecules can interact with divalent cations present in the substrate (Ca^{2+}, Mg^{2+}) which in turn, can bind to negatively-charged clay particles. Thus, to increase yield, washes with monovalent cations to displace divalent cations should be performed. Besides the different particles in the substrate, the presence of organic matter in the form of humus has to be assessed. Humus is a phase of decaying organic matter due to microorganisms and different abiotic processes. Humus is composed mainly of humic acids, although different proportions of fulvic acids and humins can be assessed. They are very stable compounds with a half-life of 2,000 yr, but considerably improve different physicochemical properties of soil (Lombard et al. 2011). The problem is they react similar to nucleic acids, and therefore may be co-extracted with DNA. Such organic compounds are strong inhibitors of enzymatic reactions, especially when performing a metagenomic library or sequencing metagenomic DNA (Daniel 2005). Therefore, humus should be removed through the steps of precipitation, electrophoresis, and chromatography, but all leads to a significant loss of performance in the DNA obtained. The amount of extracted DNA may vary between 1 μg and 500 μg per gram of soil sampled, being between 10 and 100 times the amount in comparison with the indirect method (Gabor et al. 2003). The average yield of an extraction after the purification of organic matter (humic acids) is around 45 percent. In turn, if this second purification of DNA is not performed, purity by this method is generally lower compared to the indirect method, due to the presence of humic acids in the extraction process. DNA integrity also decreases as compared with indirect extraction. Moreover, after lysing by physical processes (bead-beating) the resulting DNA molecules present a smaller size (from 0.5 to 20 Kb), especially from Gram negative bacteria (Van Elsas et al. 2008).

Indirect Extraction

To avoid the negative aspects of direct DNA extraction (coextraction with humic acids and the limited size of molecules), it is preferable to use the indirect method in a number of cases. This technique requires the separation of the microorganisms of the aggregates that comprise the substrate for subsequent cell lysis. The elements present in the surfaces of cells confer them different physicochemical properties, particularly characterized by positive-negative charges and hydrophobic and hydrophilic characteristics. These properties have a very active role in the aggregation of cells to the soil matrix. Increasing pH, as well as the use of monovalent cations, associated with smooth pulses of homogenizers, promotes the disintegration of the bacteria from the substrate. Then, bacteria can be isolated by differential centrifugation followed by density gradient centrifugation. Once isolated, cells are lysed by the processes discussed above. Using this methodology it is possible to obtain DNA fragments with sizes between 10 Kb and 1000 Kb and a degree of purity far superior to that of the direct technique, which allows performing large-insert metagenomic libraries. However, due to the application of mild disintegration methodologies, the persistence of cells attached to the aggregates of the substrate after treatment is high. Therefore, the yield of cells obtained by this method decreases by 20–50 percent (Robe et al. 2003).

The indirect DNA extraction method is preferred for large-insert metagenomic library construction. Cloning of large metagenomic DNA fragments (upper than 40 kb) favors the recovery of complete biosynthetic pathways and decreases the required size of the library to cover the same amount of DNA. While direct DNA extraction is commonly used for PCR based analysis of soil or for direct sequencing, owing to its speed and the apparently more exhaustive DNA extraction achieved.

Metagenomic Analysis of DNA

Metagenomic Library Construction

The construction of different metagenomic libraries involves cloning of DNA fragments obtained in vectors. This will allow applying different methodologies to obtain a representation of the microbial community present in a sample or to obtain new metabolites or features of interest.

The type of vector used in the construction of the library will depend on two factors: the size of the fragment to be cloned, and the type of metagenomic study to be performed. The size of cloned fragments will govern the use of one or another type of vector. Thus, libraries can be classified into two groups: small-insert libraries which use plasmids (under 7 Kb) and bacteriophages (up to 15 Kb), and large-insert libraries which can be performed with cosmids (up to 40 Kb) and bacterial artificial chromosomes (BACs, from 80 Kb).

Libraries using small size DNA fragments present some advantages, such as enhanced stability of the inserts, easy handling and packaging of the fragments, higher probability in the expression of the fragments, isolation of single genes or sequences forming small operons, as well as an advantage in sequencing by the shot-gun (Sanger) method. On the contrary, formation of such libraries increases the likelihood of obtaining contamination or false negatives, and the need for the study of large number of clones to obtain a positive clone (Daniel 2005).

Using libraries with fragments larger than 15 Kb increases the probability to discover gene clusters or entire metabolic pathways as well as complex operons. But the use of such libraries is highly complex in their handling, for several reasons like the requirement of large DNA fragments of high purity, the packaging process, limitation in the expression levels of the cloned fragments and other technical difficulties (Daniel 2005, Leveau 2007).

As described above, the methodology used in the extraction and purification of DNA may determine the size of DNA molecules. By direct extraction, smaller fragments (7 Kb) are obtained as compared to the indirect extraction, so the use of a method or another has to be defined depending on the type of library to be used in the study.

Next is host selection, a process of great importance. To date, *Escherichia coli* has been widely used as a host because it is easily handled and the vast knowledge about it. However, it has a limited capacity to express heterologous genes, such as those in the *Actinobacteria* class (7 percent)

(Gabor et al. 2004) and also, is negatively affected by expression of metabolites or enzyme activities. Therefore, the use of different hosts such as *Streptomyces*, *Pseudomonas* or *Rhizobium leguminosarum*, could reveal genes or metabolites activities located in libraries which have not been expressed in *E. coli* (Lorenz and Eck 2005).

Various studies have determined the number of clones required to have a representation of the total prokaryotic genomes present in a gram of substrate, being 10^6 the number of BAC type clones with an average fragment size of 100 Kb. This estimation would be true as long as the different microorganisms were in a similar proportion. If this is not the case, the necessary library size would increase to 10^{11} BAC clones or would require the formation of a library with 10,000 Gbp (Daniel 2005). Another study determined that, to identify one gene encoding for an antibiotic in a 5-Mb bacterial genome within 100 bacterial species, around 57,500 inserts with an average size of 40 Kbp would be required (Leveau 2007).

Metagenomic Screening

Two different experimental procedures have been used for discovery and identification of new genes, Expression-independent (Sequence-Screening) and Expression-dependent (Functional-Screening) (Fig. 18.1).

Expression-Independent Screening. Expression-independent screening is based on existing information stored in DNA sequence databases. The isolated DNA sample is compared with the sequences available in various databases for identification of sequences of the gene or locus of interest. To this end, screening can be done with two approaches i) metagenomic library development and further study (cloning dependent) or ii) direct analysis of isolated DNA fragments (cloning independent). Searching and identification of these sequences can be performed with polymerase chain reaction (PCR) with specific primers derived from known sequences, or degenerated primers or with microarrays. With different target probes a wide variety of commercial microarrays are available to analyze structure, functional activity and dynamics of microbial communities, but also to related microbial communities with ecosystem processes and functions (He et al. 2011).

Another way to identify sequences of interest is through massive parallel sequencing through new platforms like pyrosequencing, or the classical Sanger system sequencing. This strategy can be applied to clones of different sizes obtained from the metagenomic library. However, it can also be performed directly without cloning, with metagenomic DNA molecules obtained after isolation and purification from the substrates sampled. It should be noted that the larger the fragments of DNA, the

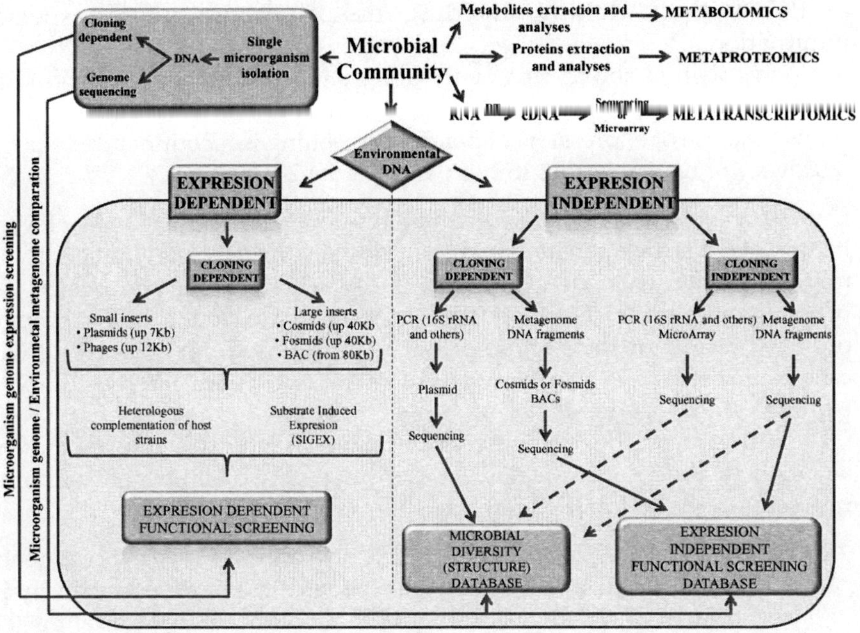

Figure 18.1 Metagenomic approaches.

greater the chance of recovering clusters of complete genes that may encode complex metabolic pathways. The advantage of using metagenomic libraries through the different vectors is the ability to select clones of interest and then perform an expression-dependent screening (Fig. 18.1). In this way, both methods of screening could be combined to confirm the activity of the coding sequence.

An example of the independent-expression screening is its use in studies of structural diversity of microbial populations in an ecosystem. Amplification by PCR of the ribosomal small subunit (16S rRNA gene) and subsequent sequencing may allow identification of different species present in an ecological niche (Woese et al. 1990). Also, how the rhizosphere governs the structure of bacterial communities can be studied (Sanguien et al. 2006). Other genes have been used for the development of phylogenetic trees, such as the *recA*, *recB* genes, encoding proteins involved in homologous recombination, genes encoding chaperones, Heat Shock Protein 70, or Tu or G elongation factors (EF-Tu, EF-G) related to the translation process on ribosomes (Venter et al. 2004).

Currently, the greatest advantage of this type of study is the possibility of sequencing purified DNA fragments (metagenomic DNA) or amplifying by PCR without cloning, avoiding loss of unrepresentative species information.

As explained above, this methodology is based on comparison with known sequences, so chances to obtain new sequences encoding functions is low. The use of expression-dependent screening is recommended for the discovery of new activities (Simon and Daniel 2011).

Expression-Dependent Screening. Expression-dependent screening is relevant in industry. It is the main method used for searching new enzymatic activity for the development of new products. This type of screening requires the use of metagenomic libraries, therefore it is necessary to clone DNA fragments obtained in one of the vectors previously proposed. Fragment size will determine the type of vector, and in turn, this will define the possibility to express cloned fragments.

Expression of cloned fragments that show an activity of interest may be achieved by different strategies. A description of two of the most common processes described in the literature follows.

1. *Host Heterologous Complementation.* This process aims to express genes of interest (up to 15 Kb) located in the vector, and detects activity in host phenotypic changes or in growth medium. A key element is the host capacity for gene expression, and product viability afterwards. Bacteriophage expression system/*E. coli* is usually used due to its high expression capability, together with easy construction and subsequent expression in the host. The system is based on lambda vector T7 bacteriophage RNA polymerase (T7 RNApol), which directs the selective transcription of genes cloned downstream of the major T7 late promoter. The T7RNApol is characterized by very high activity elongating messenger RNA (mRNA) chains about five times faster than the *E. coli* RNA polymerase (Short et al. 1988, Lussier et al. 2011).
2. *Substrate-Induced Gene Expression Screening (SIGEX).* Some gene encoding enzymes for catabolic reactions are induced by the same enzyme that catabolizes the substrates. In this situation, the substrate activates transcription regulators or promoters that are located in the proximity of the regulated genes. Thus, activation of these genes is induced by the substrate. This method is based on the use of an expression vector that includes the *gfp* gene encoding the green fluorescent protein (GFP). If the fragment cloned in this vector provides the substrate-inducible promoter and the gene encoding substrate catabolism, the system will express GFP. Thus, protein expression will indicate an expression activity induced by a substrate (substrate-induced gene-expression activity). This methodology has been effective for identifying diverse operons and genes involved in degradation of organic compounds (Uchiyama and Watanabe 2007).

Sequencing Technologies

Nowadays, the classical and new sequencing technologies enable the acquisition of gigabases of sequence information in just a few hours. In a metagenomic study, different sequencing methods can be used, from conventional capillary-based (Sanger method) to the next-generation sequencing techniques, like 454 GS FLX pyrosequencing (Roche), SOLID system (Applied Biosystems) and the Genome Analyzer Solexa (Illumina). Each of the technologies strikes a different balance between costs, read length, data volume and data generation rate (Voelkerding et al. 2009, Mardis 2008, Shendure and Ji 2008, Metzker 2010)

The first sequencing method used in a metagenomic study was shotgun by Sanger method. With the rapid improvement of sequencing technologies and concomitant decreasing cost new sequencing platforms were generated, and even direct sequencing of metagenomic DNA has become feasible. Nowadays, pyrosequencing (Roche 454 GS FLX Titanium) allows 1–1.5 10^6 reads ~400 bp in length in 10 hr. Additional advantages of pyrosequencing are the low DNA amount required per sample; being a cloning-independent approach sequencing is done after DNA purification, and small DNA fragments (300-800 bp) are adequate (Margulies et al. 2005, Bentley 2006, MacLean et al. 2009).

Sequence Analysis Methods

Powerful sequencing yields a high amount of data that requires large computers for being analyzed. This need has boosted bioinformatics development. The first step is to obtain an individual sequence assembling the thousand sequences, irrespective of the target being a gene, a full genome or many genomes from microbial populations. The next step is comparing the data with a database for further analysis. Nowadays, there are several databases available like Greengenes, SILVA or Ribosomal Database Project II (RDPII). These databases have a lot of information from cultivable microorganisms based on the 16S rRNA gene, and are useful for studies on phylogeny and structural diversity of microbial communities.

Sequence analysis can be approached from two points of view, either by similarities or based on composition. The first type is used to compare sequences and provides similarities or homologies between sequences, and also studies sequence elements. This is associated to informatics software like Basic Local Alignement Search Tool (BLAST), MEGAN, CARMA, Sort-ITEMS, GeneDB, Artemis, Glimmer, FGenesB; all these programs identify coding sequences, promoters, ending sequences, operons, tRNA and rRNA (Simon and Daniel 2011). The limit of this approach is that it relies on the information available in the database, since most of it belongs either to cultivable microorganisms or to unidentified sequences.

The second approach for sequence analysis is based on sequence composition, frequency and distribution. In this approach, G/C content, type of codon used, and frequency of specific oligonucleotides in the sequence (Bentley et al. 2006) are compared to known data associated to taxonomic groups. The informatics tools suitable for this approach are PhyloPythia, TETRA, and TACOA (Teeling et al. 2004, McHardy et al. 2007 and Diaz et al. 2009).

The genetic information hidden in the rhizosphere is enormous, but there is still a limited ability to analyze all genomes within a sample. As presented through the text, metagenomics appears as a strong tool to find new molecules or enzymatic activities relevant for industry, pharmacy, agriculture or just to study interactions of microbial communities in their habitats, and such a broad scope of interest shows in the increased research on this topic, with encouraging applications relevant for society.

Metagenomic Applications

The great potential of metagenomic supported by the development of new sequencing techniques, has resulted in an exponential increase in the number of new products discovered with a practical application in a wide range of areas, from industry to agriculture.

Investigations with practical aims will be reviewed, focusing on those using expression-dependent techniques. A special interest will be given to rhizosphere studies, since the plant-microorganism relationship is very close and needs a special fine-tuning. Applications will be divided into four major groups, aimed towards bioremediation, industry, pharmacy and agriculture.

Metagenomics Applied to Bioremediation

One of the most dramatic consequences of industrial development in the last 200 yr is the degradation of the environment, especially due to uncontrolled dumping of toxic compounds, either natural compounds (fossil fuels or heavy metals discharged at high concentration by human activities), or chemically synthesized compounds (the so-called "xenobiotics", like halogenated, chlorinated or nitroaromatic compounds) (George et al. 2010). Fortunately, nature has provided microorganisms that can naturally degrade practically any compound, but efficiency and speed to cut down these environmental pollutant compounds is still a limitation.

Bioremediation (and its variations, like biostimulation, bioaugmentation, etc.) is one of the most attractive and most promising techniques to remove pollutants from the environment. More specifically, rhizosphere metagenomics (especially from contaminated rhizospheres) appears as a promising area to obtain new enzymes and systems for maximum efficient degradation of dangerous substances to the environment.

On the banks of the Tinto River (Spain), an area with high metal concentration, metagenomic libraries from the rhizosphere of the endemic heather, *Erica andevalensis,* were constructed and genes encoding for nickel resistance were found (Mirete et al. 2007). Rhizosphere bacterial communities from metal-impacted sites display different metal resistances (Idris et al. 2004).

Rhizosphere bacteria can also produce siderophores and chelators, which may reduce the toxic effects on the plants caused by excess metal concentrations (Wenzel 2009, Burd 2000). As previously stated in this chapter, in the expression-independent methods, clones are identified based on the presence of conserved regions in phylogenetic marker genes or in functional genes coding for key processes (George et al. 2010). Screening for the existence of genes encoding specific biodegradative properties in environmental communities has been carried out using gene-specific PCR, like phenol hydroxylase genes and BTEX mono-oxygenase genes (Futamata et al. 2001, Hendrickx et al. 2006). With the use of microarrays, it is now possible to develop massive parallel screening of metagenomic libraries for biodegradative genes (He et al. 2007).

One example could be the polychlorinated biphenyls, a class of synthetic aromatic compounds, which are widely used in the electronics industry and as components of adhesives and plastic materials. These compounds are thermally and chemically stable and found as pollutants throughout the ecosystem and food chain due to their lipophilic behavior, and constitute a serious health problem to mammals. From the polychlorinated-biphenyl contaminated soil, bacteria and functional genes were identified which were associated with biphenyl degradation in the root zone of an Austrian pine (*Pinus nigra* L.) by using Stable Isotope Probing (SIP). SIP is a method that provides ^{13}C-labeled substrate to a microbial community separating ^{13}C-labeled DNA from active organisms, that used the substrate from ^{12}C-DNA of the inactive majority (Uhlíka et al. 2009); this is combined with microarrays technology, especifically Geochip (Leigh et al. 2007, He et al. 2007), aiming to provide direct links between biogeochemical processes and functional activities of microbial communities in a variety of environments.

Furthermore, successful investigations have been developed with non-rhizospheric soil, like the discovery of nitrilases for the elimination of natural and xenobiotic nitrile (Robertson et al. 2004). Due to the high complexity and richness of rhizosphere ecosystem, it is to be expected that research in this regard, both by expression-dependent and independent approaches, will result in important discoveries.

Metagenomics Applied to Industry

The economic potential that represents the possibility of obtaining new enzymes to improve industrial processes making them more efficient, optimal, versatile (with greater flexibility in temperature, pH, etc.) and more environmentally sustainable, makes this branch of biology in of outstanding relevance. In addition to better enzymes, it is also possible to discover new ones in a lot of industrial areas, like new esterases (Peng et al. 2011), amylases, proteases or cellulases (Berlemont et al. 2011). As mentioned before, two possibilities exist depending on whether the objective is to seek completely new compounds or to improve existing ones. In the first case, an expression-dependent method should be used, but in the second one, an expression-independent method should be selected. Furthermore, according to the size of the insert, vector, or the host used, it is possible to observe large differences in the results obtained from the same sample of soil (Gabor et al. 2004), which gives an idea of metagenomics great potential.

Lipolytic Enzymes

Lipolytic enzymes have a lot of applications not only in soap and cleaning compounds manufacturing, but also in agri-food industry and for biodiesel production. Although it is possible to search for a lot of lipolytic activities, esterases are among the most researched due to their great interest and industrial versatility.

Lee et al. (2010) identified a new esterase in red pepper and strawberry rhizospheres in Korea, from recombinant plasmids libraries. A total of 14 recombinant fosmids exhibited lipolytic activity, from which the recombinant plasmid with the maximum lipolytic activity was further analyzed and purified exhibiting a maximum enzymatic activity towards *p*-nitrophenyl butyrate, indicating its esterase nature that could be a member of a novel family of lipolytic enzymes.

Another example of success are the two esterases isolated after an expression-dependent method from Colombian high Andean forestal soil. These esterases show a preference for short-length fatty acids, with specific activity against *p*-nitrophenyl-butyrate and are active at low temperature (Jiménez et al. 2012).

Lignolytic Enzymes

One way of getting clean energy is by biomass production. In this field, metagenomics have also seen major improvements. DeAngelis et al. (2010) found novel lignocellulose-degrading enzymes, important for the deconstruction of lignocellulosic biomass and fuel; the originality of this study was to perform soil enrichment increasing the success rates of finding new enzymes. Enrichment was done by adding to Puerto Rican soil dried ground switchgrass, or lignin, or green waste compost switchgrass (*Panicum virgatum*) and simulated thermophilic composting in a bioreactor to select organisms able to degrade lignin (Allgaier 2010).

Pectate Lyases

Pectate lyase is an enzyme present in some microorganisms, mainly phytopathogenic. It breaks pectin molecules that constitute plant cell walls, resulting in the production of galacturonic acid monomers, rhamnose and dextrin groups. These enzymes are very important in the agri-food industry, for example, in juice production or cotton fabric.

In 2004, Solbak et al. isolated from decomposing soil organic matter a new pectate lyase (which is used to remove pectin from crude cotton fabric during its boil off). Only some of the enzymes discovered were optimal for cotton processing due to the pH range, temperature, etc., but they could be improved by single-site mutations and thus developing enzymes that can improve cotton production.

Industrial Dyes

Lim et al. (2005) sampled different forests soil where pine, oak and beech trees were dominant and found two ORFs (Open Reading Frame) responsible for indirubin and its isomer, indigo blue. These two important dyes, that also have antimicrobial activity, were found through expression-based screening methods, detecting phenotypic changes in colonies color (purple or brown) in several clones.

Metagenomics Applied to Pharmacy

Due to the strong selective pressure that exists in the rhizosphere, secondary metabolites and antibiotics can be used by microorganisms, among other strategies, to successfully colonize it (Mazzola et al. 1992). These compounds have a great potential for the development of new drugs, or to improve existing ones. Metagenomics are particularly appealing to natural product researchers because the genetic information needed to encode for the production of bacterial secondary metabolites is typically clustered on bacterial chromosomes. It is therefore possible to envision capturing complete small molecule biosynthetic gene clusters on individual or a small number of overlapping enviromental DNA clones (Handelsman 1998). A good example is the study of Chung et al. (2008), who constructed two fosmids libraries from pine rhizosphere and found an antifungal gene cluster using *Saccharomyces cerevisiae* as a target fungus.

Although the rhizosphere is expected to be an ecosystem with a great diversity in secondary metabolites, there are not many studies on this topic. In non-rizhospheric soil, Gillespie et al. (2002) found two antibiotics, turbomycin A and turbomycin B, through a library in *Escherichia coli* with DNA extracted directly from soil. In this area, other studies that have searched for a family of long-chain acylamino acid antibiotics from a metagenomic library of DNA extracted directly from soil samples can be mentioned (Brady and Clardy 2000, Brady et al. 2004).

Another way to improve the processes is developing new enzymes that produce the same compounds, but through more efficient routes. One example is the study by Eschenfeldt et al. (2001), who discovered the new enzymes 2,5-diketo-D-gluconic acid reductases (DKGRs) in forestal soil libraries to develop a new and better vitamin C (ascorbic acid) synthesis pathway.

Knowledge of new genes influencing antibiotic resistance could be very useful in the pharmaceutical field (Allen 2010). To illustrate this field is the study carried out with Alaska soil, where 14 clones with antibiotic resistance in *E. coli* where found (Allen et al. 2009). These clones codified for genes of different β-lactamases classes, which hydrolyze β-lactam antibiotics, like penicillin o cephalosporin.

Metagenomics Applied to Agriculture

The interaction between the plants and microorganisms in the rhizosphere makes this ecosystem a source of genes with a great potential, especially for agricultural application in order to improve crop yields, protect from pests and improve the quality of agricultural products. This could be performed by searching for genes responsible for PGPR activities, among others, and then applying the recombinant products to the soil.

A competent chitinase could be used as effective fungicide thanks to the fungal cell-wall composition. A good example is the study of Ikeda et al. (2007), who used T-RFLP (Terminal-restriction fragment length polymorphism) and clone library analysis to identify chitinase genes in the maize rhizosphere. They found novel groups of bacterial chitinase genes and large differences in chitinase gene diversity between the bulk and rhizosphere soil. Another good example could be that of Hjort et al. (2010), who also searched for chitinases in a phytopathogen-suppressive soil.

An activity that could improve plant "fitness" and, as a result, improve crop production is ACC deaminase activity. In fact, deamination of the ethylene precursor 1-aminocyclopropane-1-carboxylic acid (ACC) is a relevant beneficial trait found in PGPR, since lowering ethylene levels in plants results in increased plant root and shoot length, increasing its biomass, and protecting plants from inhibitory effects of ethylene levels (Glick et al. 2007). Nikolic et al. (2011) analyzed ACC deaminase genes (acdS) of bacterial endophytes colonizing field grown potato plants. Through PCR analysis, they discovered the presence of two types of *acdS* genes, the dominant one showing high homology to an *acdS* gene derived from *Pseudomonas fluorescens*.

The search for siderophores is particularly interesting due to its triple function, nutritional, systemic resistance inductor (ISR) and biocontrol (Bakker et al. 2007, Ramos Solano et al. 2008b, see also Chapter 12 in this volume). Despite its relevance for agriculture, the only example to date that has been carried out in a marine environment, through expression-dependent methods, and resulted in a cluster composed by 5 ORF that synthesized vibrioferrina, a siderophore discovered in *Vibrio parahaemolyticus* (Fujita et al. 2011).

In a similar way, bioavailability of phosphorous could be a key process for plant nutrition improvement. In fact, 80 percent of organic P is in myo-inositol hexakisphosphate (phytate) form (Turner et al. 2002), not available for plants. However, certain microorganisms can make it available for plants by releasing phytases, implying a key process for recycling phosphorus in the biosphere. In fact, using a bacteria with high phytase activity, could improve phytate phosphorus absorption by plants (Richardson 2001). Despite the relevance of this metagenomic application, to date there is only one investigation in this field (Lim et al. 2007). An expression-independent method was used to determine the distribution of the four known classes of phytases in the microbial world by bioinformatic studies (NCBI and the CAMERA databases).

In addition to the above mentioned possibilities, there are more for PGPR activities from culturable and non-culturable rhizosphere microorganism in agriculture. For example, compounds able to modify the plant's hormonal balance such as auxins, gibberellins or their precursors. Another target could be molecules that break *quorum sensing* in pathogenic microorganisms or activate ISR (as exopolysaccharides or lipopolysaccharides), both for field pests biocontrol.

To conclude this chapter, it is important to note that transference of knowledge from basic research to industry is one of the great challenges of metagenomics, that in recent years is beginning to show its first results funded by companies that are marketing new enzymes as phytases, cellulases, pectate lyases, or xylanases. These enzymes are making substantial improvement to industrial processes, placed on the global market for industrial enzymes that is forecast to reach US$ 3.74 billion by the year 2015 (Global Industry Analysts 2011).

Acknowledgements

AGL2009-08324 and CM S0505.

References

Allen, H.K., L.A. Moe, J. Rodbumrer, A. Gaarder and J. Handelsman. 2009. Functional metagenomics reveals diverse β-lactamases in a remote Alaskan soil. ISME J. 3: 243–251.

Allen, H.K., J. Donato, H.H. Wang, K.A. Cloud-Hansen, J. Davies and J. Handelsman. 2010. Call of the wild: antibiotic resistance genes in natural environments. Nature Reviews Microbiology. 8: 251–259.

Allgaier, M., R.A. Park, N. Ivanova and P. D'haeseleer. 2010. Targeted discovery of glycoside hydrolases from a switchgrass-adapted compost community. PLoS ONE. 5: 1–9.

Baker, K., S. Langenheder, G. Nicol, D. Ricketts, K. Killham, C. Campbel and J. Prosser. 2009. Environmental and spatial characterisation of bacterial community composition in soil to inform sampling strategies. Soil Biol. Biochem. 41: 2292–2298.

Bakker, P.A.H., M. Corné, J. Pieterse and L.C. van Loon. 2007. Induced systemic resistance by fluorescent *Pseudomonas* spp. Phytopathology. 97: 239–243.

Bentley, S.D. and J. Parkhill. 2004. Comparative genomic structure of prokaryotes. Annu. Rev. Genet. 38: 771–792.

Bentley, S.D. 2006. Whole-genome re-sequencing. Curr. Opin. Genet. Dev. 16: 545–552.

Berlemont, R., D. Pipers, M. Delsaute, F. Angiono, G. Feller, M. Galleni and P. Power. 2011. Exploring the Antarctic soil metagenome as a source of novel cold-adapted enzymes and genetic mobile elements. Rev. Argent. Microbiol. 43: 94–103.

Brady, S.F. and J. Clardy. 2000. Long-chain n-acyl amino acid antibiotics isolated from heterologously expressed environmental DNA. J. Am. Chem. Soc. 122: 12903–12904.

Brady, S.F., C.J. Chao and J. Clardy. 2004. Long-chain n-acyltyrosine synthases from environmental DNA. Appl. Environ. Microbiol. 70: 6865–6870.

Burd, G.I., D.G. Dixon and B.R. Glick. 2000. Plant growth-promoting bacteria that decrease heavy metal toxicity in plants. Can. J. Microbiol. 46: 237–245.

Chung, E.J., H.K. Lim, J. Kim, G.J. Choi, E.J. Park, M.H. Lee, Y.R. Chung and S. Lee. 2008. Forest soil metagenome gene cluster involved in antifungal activity expression in *Escherichia coli*. Appl. Environ. Microbiol. 74: 723–730.

Conrath, U., C.M.J. Pieterse and B. Mauch-Mani. 2002. Priming in plant-pathogen interactions. Trends Plant Sci. 7: 210–216.

Daniel, R. 2005. The metagenomic of soil. Nature 5: 470–478.

DeAngelis, K.M., J.M. Gladden, M. Allgaier, P. D'haeseleer, J.L. Fortney, A. Reddy, P. Hugenholtz, S.W. Singer, J.S. Gheynst, W.L. Silver, B.A. Simmons and T.C. Hazen. 2010. Strategies for enhancing the effectiveness of metagenomic-based enzyme discovery in lignocellulolytic microbial communities. BioEnergy research. 3: 146–161.

Diaz, N., L. Krause, A. Goesmann, K. Niehaus and T. Nattkemper. 2009. TACOA: taxonomic classification of environmental genomic fragments using a kernelized nearest neighbor approach. BMC Bioinformatics. 10: 56–72.

Domenech, J., M.S. Reddy, J.W. Kloepper, B. Ramos Solano and J. Gutiérrez Mañero. 2006. Combined application of the biological product LS213 with *Bacillus*, *Pseudomonas* or *Chryseobacterium* for growth promotion and biological control of soil-borne diseases in pepper and tomato. Biocontrol. 51: 245–258.

Eschenfeldt, W.H., L. Stols, H. Rosenbaum, Z.S. Khambatta, E. Quaite-Randall, S.W., D. Kilgore, J. Trent and M. Donnelly. 2001. DNA from uncultured organisms as a source of 2,5-diketo-dgluconic acid reductases. Appl. Environ. Microbiol. 67: 4206–4214.

Fujita, J.M., N. Kimura, A. Sakai, Y. Ichikawa, T. Hanyu and M. Otsuka. 2011. Cloning and heterologous expression of the *Vibrioferrin* biosynthetic gene cluster from a marine metagenomic library. Biosci. Biotechnol. Biochem. 75: 2283–2287.

Futamata, H., S. Harayama and K. Watanabe. 2001. Group-specific monitoring of phenol hydroxylase genes for a functional assessment of phenol stimulated trichloroethylene bioremediation. App. Environ. Microbiol. 67: 4671–4677.

Gabor, M., E. Vries and D. Janssen. 2003. Efficient recovery of environmental DNA for expression cloning by indirect methods. FEMS Microbiol. Ecol. 44: 153–163.

Gabor, E.M., W.B.L. Alkema and D.B. Janssen. 2004. Quantifying the accessibility of the metagenome by random expression cloning techniques. Environ. Microbiol. 6: 879–886.

Galleni, P. and Power. 2011. Exploring the Antarctic soil metagenome as a source of novel cold-adapted enzymes and genetic mobile elements. Rev. Argent. Microbiol. 43: 94–103.

George, I., B. Stenuit and S. Agathos. 2010. Application of Metagenomics to Bioremediation. *In*: D. Marco [ed.]. Application of metagenomics to bioremediation. Metagenomics. Theory, methods and application. Caister Academic Academic Press. Norfolk, UK. pp. 107–140.

Gillespie, D.E., S.F. Brady, A.D. Bettermann, N.P. Cianciotto, M.R. Liles, M.R. Rondon, J.C. Robert, M. Goodman and J. Handelsman. 2002. Isolation of antibiotics turbomycin a and b from a metagenomic library of soil microbial DNA. App. Environ. Microbiol. 68: 4301–4306.

Glick, B.R., B. Todorovic, J. Czarny, Z. Cheng, J. Duan and B. McConkey. 2007. Promotion of plant growth by bacterial ACC Deaminase. CRC Crit. Rev. Plant. Sci. 26: 227–242.

Global Industry Analysts, Inc. 2011. A global strategic business report. San José, California, USA. http://www.strategyr.com.

Gutierrez Mañero, F.J., B. Ramos, J.A. Lucas García, A. Probanza and M.L. Barrientos Casero. 2003. Systemic induction of the biosynthesis of terpenic compounds in *Digitalis lanata*. J. Plant Physiol. 160: 105–113.

Gutiérrez Mañero, F., E. Algar, M.S. Martín Gómez, M.D. Saco Sierra and B. Ramos Solano. 2012. Elicitation of secondary metabolism in *Hypericum perforatum* by rhizosphere bacteria and derived elicitors in seedlings and shoot cultures. Pharm. Biol. (in press).

Handelsman, J., M.R. Rondon, S.F. Brady, J. Clardy and R.M. Goodman. 1998. Molecular biological access to the chemistry of unknown soil microbes: a new frontier for natural products. Chem. Biol. 5: 245–R249.

He, Z., T.J. Gentry, C.W. Schadt, L. Wu, J. Liebich, S.C. Chong, Z. Huang, W. Wu, B. Gu, P. Jardine, C. Criddle and J. Zhou. 2007. GeoChip: a comprehensive microarray for investigating biogeochemical, ecological and environmental processes. ISME J. 1: 67–77.

He, Z., J. Van Nostrand and J. Zhou. 2011. Applications of functional gene microarrays for profiling microbial communities. Curr. Opin. Biotech. 23: 1–7.

Hendrickx, B., W. Dejonghe, F. Faber, W. Boënne, L. Bastiaens, W. Verstraete, E.M. Top and D. Springael. 2006. PCR-DGGE method to assess the diversity of BTEX mono-oxygenase genes at contaminated sites. FEMS Microb. Ecol. 55: 262–273.

Hiltner, L. 1904. Über neuere Erfahrungen und Probleme auf dem Gebiete der Bodenbakteriologie unter besonderer Berücksichtigung der Gründüngung und Brache. Arbeiten der Deutschen Landwirtschaftlichen Gesellschaft. 98: 59–78.

Hjort, K., M. Bergström, M.F. Adesina, J.K. Jansson, K. Smalla and S. Sjöling. 2010. Chitinase genes revealed and compared in bacterial isolates, DNA extracts and a metagenomic library from a phytopathogen-suppressive soil. FEMS Microbiol. Ecol. 71: 197–207.

Idris, R., R. Trifonova, M. Puschenreiter, W.W. Wenzel and A. Sessitsch. 2004. Bacterial communities associated with flowering plants of the Ni hyperaccumulator *Thlaspi goesingense*. Appl. Environ. Microbiol. 70: 2667– 2677.

Ikeda, S., N. Ytow, H. Ezura, K. Minamisawa, K. Miyashita and T. Fujimura. 2007. Analysis of molecular diversity of bacterial chitinase genes in the maize rhizosphere using culture independent methods. Microbes Environ. 22: 71–77.

Jacobsen, C. 2004. Purification of microbial genes from soil and rhizosphere by magnetic capture hybridization and subsequent amplification of target genes by PCR. *In:* G.A. Kowalchuk, F. De Bruijn, I. Head, A. Akkermans and J. Van Elsas [eds.]. Molecular microbial ecology manual. Springer, Dordrecht, The Netherlands. pp. 181–188.

Jiménez, D.J., J.S. Montaña, D. Álvarez and S. Baena. 2012. A novel cold active esterase derived from Colombian high Andean forest soil metagenome. World. J. Microbiol. Biotechnol. 28: 361–370.

Kang, S. and A. Mills. 2006. The effect of sample size in studies of soil microbial community structure. J. Microbiol. Meth. 66: 242–250.

Kloepper, J.W., J. Leong, M. Teintze and M.N. Schroth. 1980. Enhanced plant growth by siderophores produced by plant growth-promoting rhizobacteria. Nature. 286: 885–886.

Lawton, K., K. Weymann, L. Friedrich, B. Vernooij, S. Uknes and J. Ryals. 1995. Systemic acquired resistance in Arabidopsis requires salicylic acid but not ethylene. Mol. Plant. Microbe. In. 8: 863–870.

Lee, M.H., K.S. Hong, S. Malhotra, J.P.E.C. Hwang, H.K. Choi, Y.S.K.W. Tao and S. Lee. 2010. A new esterase EstD2 isolated from plant rhizosphere soil metagenome. Appl. Microbiol Biotechnol. 88: 1125–1134.

Leigh, M.B., V.H. Pellizari, O. Uhlık, R. Sutka, J. Rodrigues, N.E. Ostrom, J. Zhou and J.M. Tiedje. 2007. Biphenyl-utilizing bacteria and their functional genes in a pine root zone contaminated with polychlorinated biphenyls (PCBs). ISME J. 1: 134–148.

Leveau, J. 2007. The magic and menace of metagenomics: prospects for the study of plant growth-promoting rhizobacteria. Eur. J. Plant Pathol. 119: 279–300.

Lim, B.L., P. Yeung, C. Cheng and J.E. Hill. 2007. Distribution and diversity of phytate-mineralizing bacteria. ISME J. 1: 321–330.

Lim, H.K., E.J. Chung, J. Kim, G.J. Choi, K.S. Jang, Y.R. Chung, K.Y. Cho and S. Lee. 2005. Characterization of a forest soil metagenome clone that confers indirubin and indigo production on *Escherichia coli*. Appl. Environ. Microbiol. 71: 7768–7777.

Lombard, N., E. Prestat, J. Van Elsas and P. Simonet. 2011. Soil-specific limitations for access and analysis of soil microbial communities by metagenomics. FEMS Microb. Ecol. 78: 31–49.

Lorenz, P. and J. Eck. 2005. Metagenomics and industrial applicayions. Nat. Rev. Microbiol. 3: 510–516.

Lucas, J.A., B. Ramos Solano, F. Montes, J. Ojeda, M. Megias and F.J. Gutierrez Mañero. 2009. Use of two PGPR strains in the integrated management of blast disease in rice (*Oryza sativa*) in Southern Spain. Field Crop. Res. 114: 404–410.

Lussier, F., O. Chambenoit, A. Cote, J. Hupe, F. Denis, P. Juteau, R. Beaudet and F. Shareck. 2011. Construction and functional screening of a metagenomic library using a T7 RNA polymerase-based expression cosmid vector. J. Ind. Microbiol. Biotechnol. 38: 1321–1328.

MacLean, D., J. Jones and D. Studholme. 2009. Application of 'next-generation'sequencing technologies to microbial genetics. Nat. Rev. Microbiol. 7: 287–296.

Mardis, E. 2008. Next-generation DNA sequencing methods. Annu. Rev. Genomics Hum. Genet. 9: 387–402.

Margulies, M., M. Egholm, W. Altman, S. Attiya, J. Bader, L. Bemben, J. Berka, M. Braverman, Y. Chen, Z. Chen, S. Dewell, L. Du, J. Fierro, X. Gomes, B. Goodwin, W. He, S. Helgesen, C. Ho, G. Irzyk, S. Jando, M. Alenquer, T. Jarvie, K. Jirage, J. Kim, J. Knight, J. Lanza, J. Leamon, S. Lefkowitz, M. Lei, J. Li, K. Lohman, H. Lu, V. Makhijani, K. McDade, M. McKenna, E. Myers, E. Nickerson, J. Nobile, R. Plant, B. Puc, M. Ronan, G. Roth, G. Sarkis, J. Simons, J. Simpson, M. Srinivasan, K. Tartaro, A. Tomasz, K. Vogt, G. Volkmer, S. Wang, Y. Wang, M. Weiner, P. Yu, R. Begley and J. Rothberg. 2005. Genome sequencing in microfabricated highdensity picolitre reactors. Nature. 437: 376–380.

McHardy, A., H. Martín, A. Tsirigos, P. Hugenholtz and I. Rigoutsos. 2007. Accurate phylogenetic classification of variable-length DNA fragments. Nat. Methods. 4: 63–72.

Mazzola, M., R. James Cook, L. Thomashow, D.M. Weller and I.S. Pierson. 1992. Contribution of phenazine antibiotic biosynthesis to the ecological competence of *Fluorescent Pseudomonads* in soil habitats. Appl. Environ. Microbiol. 58: 2616–2624.

Metzker, M. 2010. Sequencing technologies—the next generation. Nat. Rev. Genet. 11: 31–46.

Mirete, S., C.G. de Figueras and J.E. González-Pastor. 2007. Novel nickel resistance genes from the rhizosphere metagenome of plants adapted to acid mine drainage. Appl. Environ. Microbiol. 73: 6001–6011.

Nicol, G., L. Glover and J. Prosser. 2003. Spatial analysis of archaeal community structure in grassland soil. Appl. Environ. Microb. 69: 7420–7429.

Nikolic, B., H. Schwab and A. Sessitsch. 2011. Metagenomic analysis of the 1-aminocyclopropane-1-carboxylate deaminase gene (acdS) operon of an uncultured bacterial endophyte colonizing *Solanum tuberosum.* L. Arch. Microbiol. 193: 665–676.

Pace, N.R., D.A. Stahl, D.J. Lane and G.J. Olsen. 1985. Analyzing natural microbial populations by rRNA sequences. ASM News. 51: 4–12.

Peng, Q., X. Zhang, M. Shang, X. Wang, G. Wang, B. Li, G. Guan, Y. Li and Y. Wang. 2011. A novel esterase gene cloned from a metagenomics library from neritic sediments of the South China Sea. Microb. Cell. Fact. 10: 95–106.

Ramos Solano, B., J. Barriuso Maicas, M.T. Pereyra De La Iglesia, J. Domenech and F.J. Gutierrez Mañero. 2008a. Systemic disease protection elicited by plant growth promoting rhizobacteria strains: relationship between metabolic responses, systemic disease protection and biotic elicitors. Phytopathology. 98: 451–457.

Ramos Solano, B., B. Maicas and J. Gutiérrez Mañero. 2008b. Physiological and molecular mechanisms of PGPRs. *In:* I. Ahamd, J. Pichtel and S. Hayat [eds.]. En: Plant-Bacteria interactions. Strategies and techniques to promote plant growth. Wiley. New York. pp. 41–54.

Ramos Solano, B., E. Algar, A. García-Villaraco, J. García Cristóbal, J.A. Lucas García and F.J. Gutiérrez Mañero. 2010. Biotic elicitation of isoflavone metabolism with plant growth promoting rhizobacteria in early stages of development in *Glycine max* var. Osumi. J. Agr. Food Chem. 58: 1484–1492.

Rayl, A.J.S. 1999. Oceans: medicine chests of the future? The Scientist. 13: 1.

Richardson, A.E. 2001. Prospects for using soil microorganisms to improve the acquisition of phosphorus by plants. J. Plant Physiol. 28: 897–906.

Robe, P., R. Nalin, C. Capellano, T. Vogel and P. Simonet. 2003. Extraction of DNA from soil. Eur. J. Soil Biol. 39: 183–190.

Robertson, D.E., J.A. Chaplin, G. DeSantis, M. Podar, M. Madden, E.C. Richardson, A. Milan, M. Miller, D.P. Weiner, K. Wong, J. McQuaid, B. Farwell, L.A. Preston, X. Tan, M.A. Snead, M. Keller, E. Mathur, P.L. Kretz, M.J. Burk and J.M. Short. 2004. Exploring nitrilase sequence space for enantioselective catalysis. Appl. Environ. Microbiol. 70: 2429–2436.

Ryals, J.A., U.H. Neuenschwander, M.G. Willits, A. Molina, H.Y. Steiner and M.D. Hunt. 1996. Systemic acquired resistance. Plant Cell. 8: 1809–1819.

Sanguin, H., B. Remenant, A. Dechesne, J. Thioulouse, T. Vogel, X. Nesme, Y. Moënne-Loccoz and G. Grundmann. 2006. Potential of a 16S rRNA-based taxonomic microarray for analyzing the rhizosphere effects of maize on Agrobacterium spp. and bacterial communities. Appl. Environ. Microbiol. 72: 4302–4312.

Shendure, J. and H. Ji . 2008. Next-generation DNA sequencing. Nat. Biotechnol. 26: 1135–45.

Short, J., J. Fernandez, M., J. Sorge and W. Huse. 1988. λ ZAP: a bacteriophage λ expression vector with in vivo excision properties. Nucleic Acids Res. 16: 7583–600.

Simon, C. and R. Daniel. 2011. Metagenomic analyses: Past and future trends. Appl. Environ. Microb. 77: 1153–1161.

Singh, B., C. Campbell, S. Sorenson and J. Zhou. 2009. Soil Genomics. 7: 756–757.

Singh, J., A. Behal, N. Singla, A. Joshi, N. Birbian, S. Singh, V. Bali and N. Batra. 2009. Metagenomics: Concept, methodology, ecological inference and recent advances. Biotechnol. J. 4: 480–494.

Sleator, R., C. Shortall and C. Hill. 2008. Metagenomics. Lett. Appl. Microbiol. 47: 361–366.

Solbak A.I., T.H. Richardson, R.T. McCann, K.A. Kline, F. Bartnek, G. Tomlinson, X. Tan, L. Parra-Gessert, G.J. Frey, M. Podar, P. Luginbu, K.A. Gray, E.J. Mathur, D.E. Robertson, M.J. Burk, G.P. Hazlewood, J.M. Short and J. Kerovuo. 2005. Discovery of pectin-degrading enzymes and directed evolution of a novel pectate lyase for processing cotton fabric. J. Biol. Chem. 10: 9431–9438.

Teeling, H., A. Meyerdierks, M. Bauer, R. Amann and F.O. Glöckner. 2004. Application of tetranucleotide frequencies for the assignment of genomic fragments. Environ. Microbiol. 6: 938–947.

Torsvik, V., J. Goksoyr and F.L. Daae. 1990. High diversity in DNA of soil bacteria. Appl. Environ. Microbiol. 56: 782–787.

Turnbaugh, P. and J.I. Gordon. 2008. An invitation to the marriage of metagenomics and metabolomics. Cell. 134: 708–713.

Turner, B.L., M.J. Paphazy, P.M. Haygarth and I.D. McKelvie. 2002. Inositol phosphates in the environment. Philos. Trans. R. Soc. Lond. B. Biol. Sci. 357: 449–469.

Uchiyama, T. and K. Watanabe. 2007. Substrate-induced gene expression (SIGEX) screening of metagenome libraries. Nature Protocols. 3: 1202–1212.

Uhlíka, O., K. Jecnáa, M.B. Leighc, M. Mackováa and T. Maceka. 2009. DNA-based stable isotope probing: a link between community structure and function. Sci. Total Environ. 407: 3611–3619.

van Elsas, J., R. Costa, J. Jansson, S. Sjoling, M. Bailey, R. Nalin, T. Vogel TM and L. Van Overbeek. 2008. The metagenomics of disease-suppressive soils experiences from the METACONTROL project. Trends Biotechnol. 26: 591–601.

van Hulten, M., M. Pelser, L.C. van Loon, M.J.P. Corné and J. Ton. 2006. Cost and benefits of priming for defense in Arabidopsis. Proc. Natl. Acad. Sci. U.S.A. 103: 5602–5607.

van Loon, L.C., P.A.H.M. Bakker and C.M.J. Pieterse. 1998. Systemic resistance induced by rhizosphere bacteria. Ann. Rev. Phytopathol. 36: 453–483.

Venter, J., K. Remington, J.F. Heidelberg, A.L. Halpern, D. Rusch, J.A. Eisen, D. Wu, I. Paulsen, K.E. Nelson, W. Nelson, D.E. Fouts, S. Levy, A.H. Knap, M.W. Lomas, K. Nealson, O. White, J. Peterson, J. Hoffman, R. Parsons, H. Baden-Tillson, C. Pfannkoch, Y. Rogers and H.O. Smith. 2004. Environmental genome shotgun sequencing of the Sargasso Sea. Science. 304: 66–74.

Voelkerding, K., S. Dames, J. Durtschi. 2009. Next-generation sequencing: from basic research to diagnostics. Clin. Chem. 55: 641–58.

Wenzel, W.W. 2009. Rhizosphere processes and management in plant-assisted bioremediation (phytoremediation) of soils. Plant Soil. 321: 385–408.

Whitman, W.B., D.C. Coleman and W.J. Wiebe. 1998. Prokaryotes: the unseen majority. Proc. Natl. Acad. Sci. U.S.A. 95: 6578–6583.

Woese, C., O. Kandler and M. Wheelis. 1990. Towards a natural system of organisms: proposal for the domains Archea, Bacteria, and Eucarya. Proc. Natl. Acad. Sci. U.S.A. 87: 4576–4579.

Color Plate Section

Chapter 2

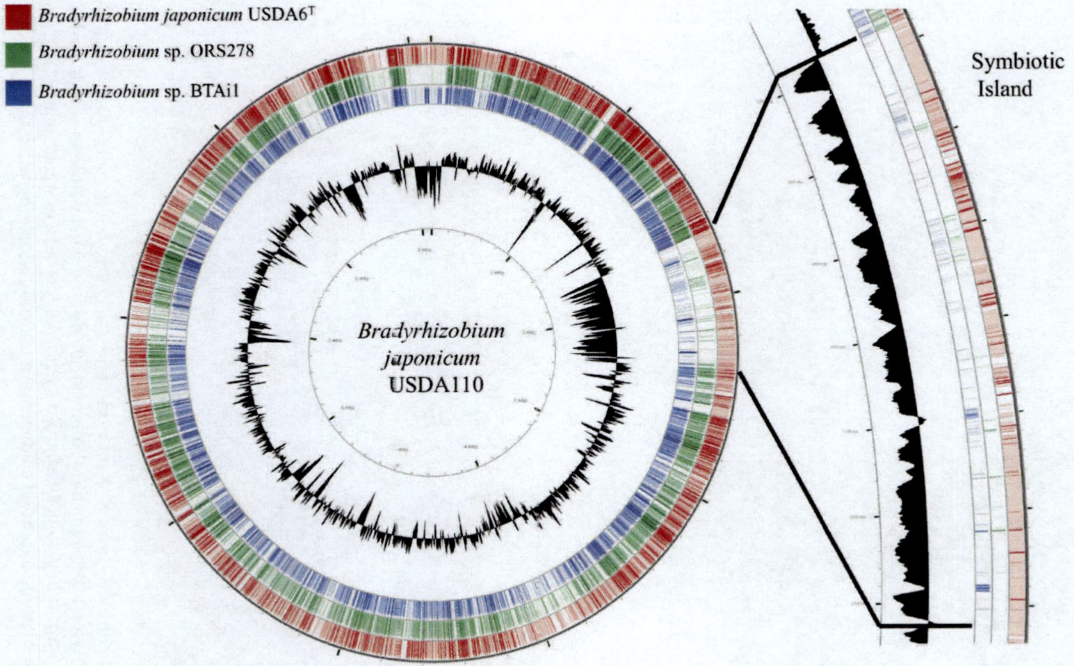

Figure 2.4 Comparative Genome Atlas of *Bradyrhizobium japonicum* USDA110. The comparative genome atlas was constructed using CGview software (Grant and Stothard 2008) to map pairwise blastN alignments between *Bradyrhizobium* strains. The chromosome of *B. japonicum* USDA110 (genome positions in Mb represented in the innermost circle) was compared to *B. japonicum* USDA6^T (red), *Bradyrhizobium* sp. ORS278 (green) and *Bradyrhizobium* sp. BTAi1 (blue). The average G+C content is shown in the second circle. The region harboring the symbiotic island was zoomed 6x.

Chapter 6

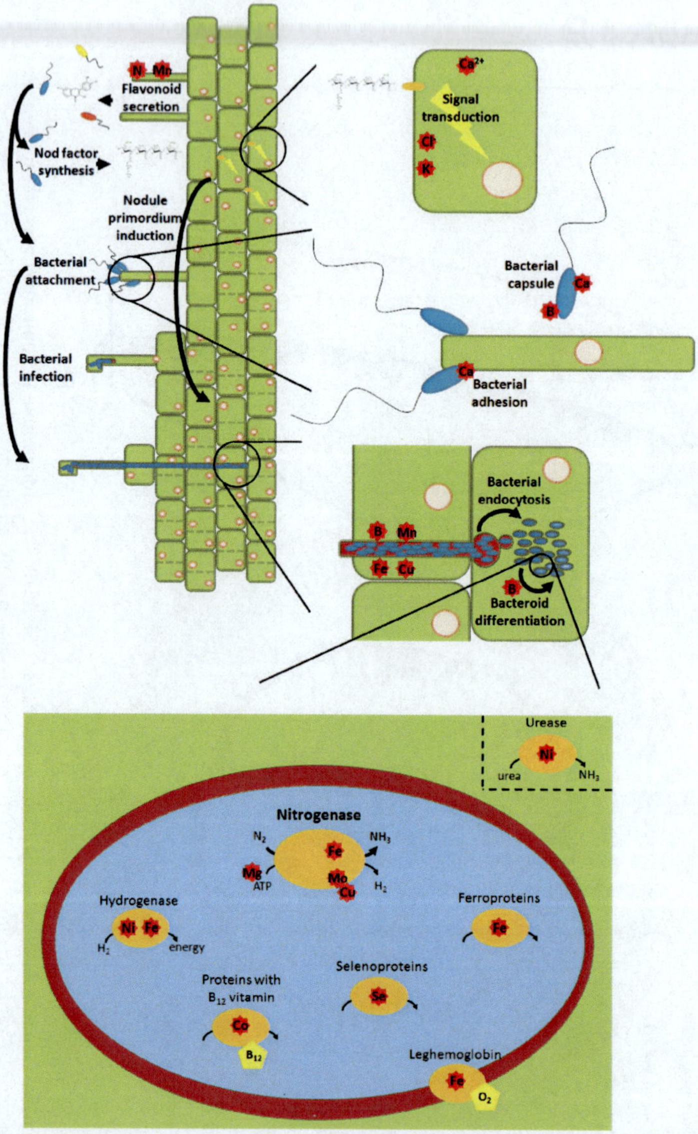

Figure 6.1 Effects of different elements on nodulation and enzymatic activities inside the symbiosome. The drawing represents the symbiosis development in an indeterminate nodule, where the elements with a key role during the symbiosis are schematized with a red star, although the functions can be extended to determinate nodule-forming plants.

Chapter 13

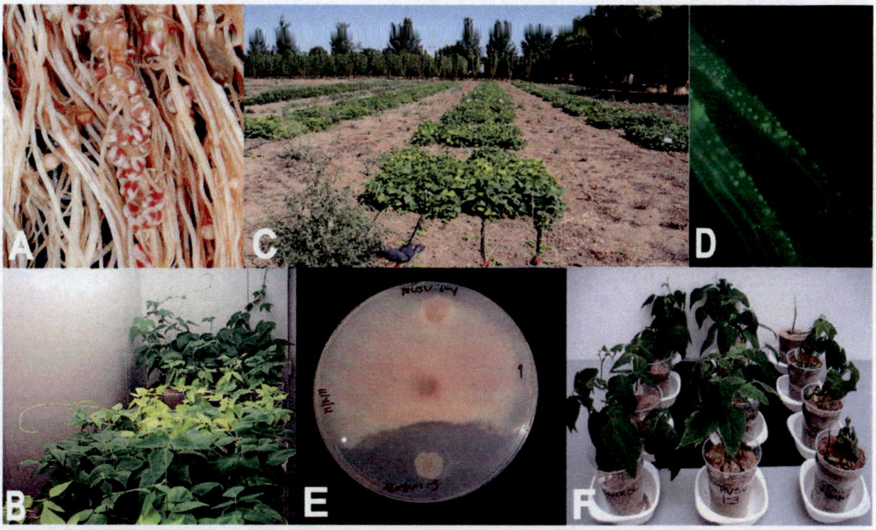

Figure 13.2 A. Nodules formed in the common bean's (*Phaseolus vulgaris* L.) roots, after inoculation under axenic conditions of seedlings with *Rhizobium leguminosarum* strains isolated from commercial common bean fields in León–Spain. B. *Phaseolus vulgaris* L. plants inoculated under axenic conditions and in the absence of mineral nitrogen, with different strains isolated from commercial common beans fields in León–Spain. The effectiveness of each strain is evidenced by the colour of the plant and its growth. C. Field pilot trials (small scale) to select, out of several strains pre-selected in axenic conditions, those with better adaptation to the environment. D. *Rhizobium leguminosarum* colonizing the surface of the rizosphere of barley (*Hordeum vulgare* L.). Bacteria have been transformed with a fluorescent GFP. E. Biocontrol assay of *Fusarium oxysporum* f. sp. phaseoli *in vitro* (in the centre of the Petri dish) with two different bacterial strains isolated from surface sterilized roots of the common bean and pepper (*Capsicun annuum* L.). F. *In vivo* assay of biocontrol of *Fusarium oxysporum* f. sp. phaseoli in the common bean. The four plants on the left side are the non-inoculated control, the four on the right side were inoculated with the pathogen, and the four in the middle were inoculated with the pathogen and a strain of *Pseudomonas brassicacearum* subsp. *neoaurantiaca* isolated from common bean.

Chapter 14

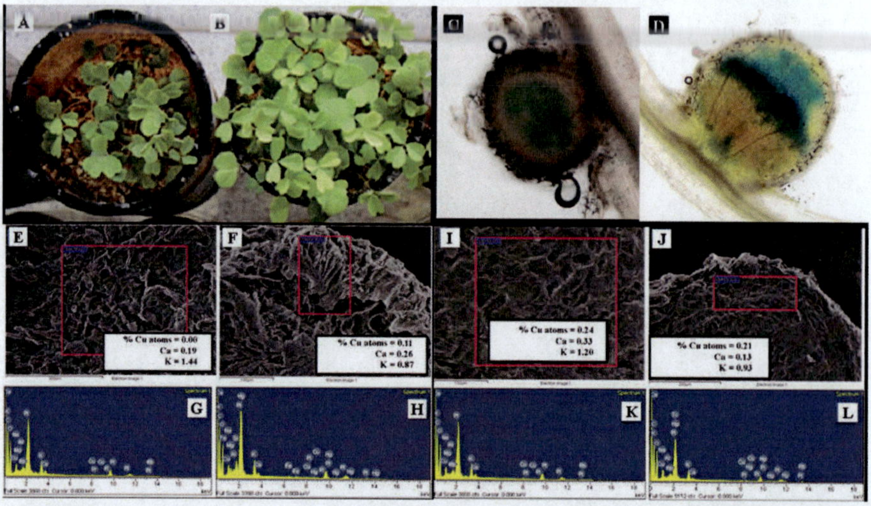

Figure 14.1 Effect of genetically modified *Sinorhizobium medicae* resistant to copper on *Medicago truncatula* growth and nodulation under Cu stress. Plants were grown in copper 200 μM and inoculated with wild type *Sinorhizobium medicae* MA11 or genetically modified *S. medicae* (*copAB*) containing the copper resistance genes *copAB* from *Pseudomonas fluorescens* under the control of the *pnifH* promoter. (A, B) Aspect of plants grown in 200 μCu, (A) non-inoculated, (B) inoculated with *S. medicae* (*copAB*). (C, D) Aspects of nodules stained with X-gal, (C) nodules formed by wild type *S. medicae* MA11, (D) nodules formed by *S. medicae* (*copAB*), (note the dark cortex in C). (E, F) Accumulation of Cu determined by SEM-EDX in nodules formed by the wild type strain, (E) central tissue, (F) cortex. (G, H) SEM-EDX spectra of the tissues shown in E, F, respectively. (I, J) Accumulation of Cu determined by SEM-EDX in nodules formed by the modified strain, (I) central tissue, (J) cortex. (K, L) SEM-EDX spectra of the tissues shown in I, J, respectively. (Note the increase in Cu accumulation from not detected to 0.24 units in nodules formed by the genetically modified strain).